MicroRNA in Development and in the Progression of Cancer

Shree Ram Singh · Pranela Rameshwar
Editors

MicroRNA in Development and in the Progression of Cancer

Springer

Editors
Shree Ram Singh
National Cancer Institute
Frederick
Maryland
USA

Pranela Rameshwar
New Jersey Medical School, Rutgers
Newark
New Jersey
USA

ISBN 978-1-4899-8064-9 ISBN 978-1-4899-8065-6 (eBook)
DOI 10.1007/978-1-4899-8065-6
Springer New York Heidelberg Dordrecht London

Library of Congress Control Number: 2014934840

Printed on acid free paper

Springer is part of Springer Science+Business Media (www.springer.com)

Preface

In recent years, microRNAs (miRNAs) have been emerged as important players in physiological as well as malignant processes. Early research in developmental timing in *Caenorhabditis elegans* (*C. elegans*) led to the discovery of *lin-4,* the first miRNA gene. Since then, studies have shown shown that miRNAs can participate in several developmental and disease processes, including embryogenesis, organ development, cellular proliferation, differentiation, developmental timing, cell cycle regulation, stem cell fate determination, aging, host–pathogen interactions, various human diseases, including heart diseases, muscular disorders and neurodegenerative diseases, diabetes, hypertension, renal dysfunction, chronic hepatitis, acquired immune deficiency syndrome (AIDS), autoimmune disorders, cancer, obesity, and apoptosis.

miRNAs are a class of endogenous, small, non-protein-coding RNA molecules (~22 nucleotides), which are novel posttranscriptional regulators of gene expression. Using molecular cloning and bioinformatics prediction strategies, hundreds of miRNAs have been identified in worms, flies, fish, frogs, mammals, and flowering plants. The human genome may encode over 1000 miRNAs, which may target about 60 % of mammalian genes and are abundant in many human cell types. Since we have hundreds of miRNAs, the major challenge is now to understand their specific biological function. In fact, the experimental evidence suggests that signaling pathways could be ideal candidates for miRNA-mediated regulation. It is interesting to note that several studies suggests that miRNAs affect the responsiveness of cells to signaling molecules such as WNT, Notch, TGF-β, and EGFR.

Altered expression of particular miRNAs has been implicated in the onset and development of cancer and could be used as potential biomarkers for disease diagnosis. Recently, many studies have found that miRNAs have crucial regulatory roles in cancer stem cells (CSCs), a kind of tumor initiating cells (TICs) and dormancy. Findings also suggest that DNA methylation may be important in regulating the expression of many miRNAs in several cancer-initiating cells. Several miRNAs are known to be either upregulated or downregulated in CSCs when compared to non-cancerous cells from the same tissues. CSCs are a small subpopulation of cells identified in a variety of tumors and involved in self-renewal, multilineage differentiation, chemoresistance, and tumorigenesis. Therefore, these fields have been highlighted as important advancements due to their potential to further elucidate diseases and therapeutic perspectives.

miRNAs have the capacity to function as oncogenes or tumor suppressors. In addition, due to their small size and molecular properties, we can manipulate them as targets and therapeutics in several diseases, including cancer treatment. *MicroRNA in Development and in the Progression of Cancer* is divided in three parts. It provides a more complete understanding of miRNA function, summarizes the recent progress, provides insights by which miRNAs regulate normal development and diseases (including cancers) and the fate of stem cells. It also presents the prospect of the great potential of miRNAs in CSCs and therapeutic advances for cancer treatment.

We would like to thank Fiona Sarne, acquisitions editor, Cancer Research at Springer, for editorial guidance and assistance throughout preparation of the book for publication. We would also like to express our sincere appreciation and gratitude to the contributors for sharing their precious expertise with the microRNA and cancer research community. The editors and the authors hope that the collection of chapters in this work will provide a comprehensive overview of this important field.

Shree Ram Singh
Pranela Rameshwar

Contents

Contributors

Aamir Ahmad Department of Pathology, Barbara Ann Karmanos Cancer Institute, Wayne State University School of Medicine, 740 Hudson Webber Cancer Research Center, Detroit, MI, USA

Omozusi Andrews Department of Biological Sciences, Vanderbilt University, Nashville, TN, USA

Sumit Arora Department of Urology, Veterans Affairs Medical Center, University of California, San Francisco, CA, USA

Bin Bao Department of Pathology, Barbara Ann Karmanos Cancer Institute, Wayne State University School of Medicine, 740 Hudson Webber Cancer Research Center, Detroit, MI, USA

Bernadette Bibber Graduate School of Biomedical Sciences and New Jersey Medical School, Rutgers, Newark, NJ, USA

Wendy B Bollag Charlie Norwood VA Medical Center, Augusta, GA, USA

Department of Physiology, Medical College of Georgia, Georgia Regents University, Augusta, GA, USA

Departments of Cell Biology and Anatomy, Medicine, Oral Biology and Orthopaedic Surgery, Medical College of Georgia, Georgia Regents University, Augusta, GA, USA

Brian Chan Mouse Cancer Genetics Program, and Basic Research Laboratory, Center for Cancer Research, National Cancer Institute, Frederick, MD, USA

Krishna Chauhan Department of Biochemistry, University of Mississippi Medical Center, Jackson, MS, USA

Cancer Institute, University of Mississippi Medical Center, Jackson, MS, USA

Daohong Chen Departments of Surgery, Indiana University School of Medicine, Indianapolis, IN, USA

Vivek Choudhary Charlie Norwood VA Medical Center, Augusta, GA, USA

Department of Physiology, Medical College of Georgia, Georgia Regents University, Augusta, GA, USA

Stephen S. Chung Human Oncology and Pathogenesis Program, New York, NY, USA

Department of Medicine and Leukemia Service, Memorial Sloan-Kettering Cancer Center, New York, NY, USA

Neetu Dahiya Department of Obstetrics and Gynecology, University of Virginia, Charlottesville, VA, USA

Rajvir Dahiya Department of Urology, Veterans Affairs Medical Center, University of California, San Francisco, CA, USA

Heba Degheidy OSEL, CDRH, FDA, White Oak, MD, USA

Joseph S. Fernandez-Moure Department of Surgery and Surgical Advanced Technologies Lab, The Methodist Hospital Research Institute, Houston, TX, USA

Mauro Ferrari Department of Nanomedicine, Houston Methodist Hospital and Houston Methodist Research Institute, Houston, TX, USA

Lyric Forney Mouse Cancer Genetics Program, and Basic Research Laboratory, Center for Cancer Research, National Cancer Institute, Frederick, MD, USA

Allen C. Gao Department of Urology and UC Davis Comprehensive Cancer Center, University of California Davis Medical Center, Sacramento, CA, USA

Steven J. Greco Department of Medicine, Hematology/Oncology, New Jersey Medical School, Rutgers, Newark, NJ, USA

Meg Gullotto Department of Medical Illustration, Georgia Regents University, Augusta, GA, USA

Hiroshi Hirata Department of Urology, Veterans Affairs Medical Center, University of California, San Francisco, CA, USA

Hailiang Hu Department of Pathology and Laboratory Medicine, Jonsson Comprehensive Cancer Center, David Geffen School of Medicine at UCLA, Los Angeles, CA, USA

Ashok Jain Department of Natural Sciences, Albany State University, Albany, GA, USA

Siddha Kasar Pathology and Laboratory Medicine, New Jersey Medical School, Newark, NJ, USA

Dejuan Kong Department of Pathology, Barbara Ann Karmanos Cancer Institute, Wayne State University School of Medicine, 740 Hudson Webber Cancer Research Center, Detroit, MI, USA

Yiwei Li Department of Pathology, Barbara Ann Karmanos Cancer Institute, Wayne State University School of Medicine, 740 Hudson Webber Cancer Research Center, Detroit, MI, USA

Marilyn Lightfoote OSEL, CDRH, FDA, White Oak, MD, USA

Shahana Majid Department of Urology, Veterans Affairs Medical Center, University of California, San Francisco, CA, USA

Gerald Marti OSEL, CDRH, FDA, White Oak, MD, USA

Reo Maruyama Department of Molecular Biology, Sapporo Medical University, Sapporo, Japan

Smruti Mehta Pathology and Laboratory Medicine, New Jersey Medical School, Newark, NJ, USA

Ryan D. Mohan Stowers Institute for Medical Research, Kansas City, MO, USA

Jessian L. Munoz Department of Medicine, Hematology/Oncology, New Jersey Medical School, Rutgers, Newark, NJ, USA

Nagalakshmi Nadiminty Department of Urology and UC Davis Comprehensive Cancer Center, University of California Davis Medical Center, Sacramento, CA, USA

George R. Nahas Graduate School of Biomedical Sciences and New Jersey Medical School, Rutgers, Newark, NJ, USA

Harikrishna Nakshatri Departments of Surgery, Indiana University School of Medicine, Indianapolis, IN, USA

Biochemistry & Molecular Biology, Indiana University School of Medicine, Indianapolis, IN, USA

Abigail F. Olena Department of Biological Sciences, Vanderbilt University, Nashville, TN, USA

J. Griffin Orr Cancer Institute, University of Mississippi Medical Center, Jackson, MS, USA

Christopher Y. Park Human Oncology and Pathogenesis Program, New York, NY, USA

Departments of Pathology and Laboratory Medicine, New York, NY, USA

Shyam A. Patel Graduate School of Biomedical Sciences and New Jersey Medical School, Rutgers, Newark, NJ, USA

James G. Patton Department of Biological Sciences, Vanderbilt University, Nashville, TN, USA

Patrice Penfornis Cancer Institute, University of Mississippi Medical Center, Jackson, MS, USA

Radhika Pochampally Department of Biochemistry, University of Mississippi Medical Center, Jackson, MS, USA

Cancer Institute, University of Mississippi Medical Center, Jackson, MS, USA

Stephanie Rager Mouse Cancer Genetics Program, and Basic Research Laboratory, Center for Cancer Research, National Cancer Institute, Frederick, MD, USA

Pranela Rameshwar Graduate School of Biomedical Sciences and New Jersey Medical School, Rutgers, Newark, NJ, USA

Elizabeth Raveche Pathology and Laboratory Medicine, New Jersey Medical School, Newark, NJ, USA

Sharanjot Saini Department of Urology, Veterans Affairs Medical Center, University of California, San Francisco, CA, USA

Monika Sangwan University of Toronto, Genetics and Development Division, Toronto, Ontario, Canada

Fazlul H. Sarkar Department of Pathology, Barbara Ann Karmanos Cancer Institute, Wayne State University School of Medicine, Detroit, MI, USA

Department of Oncology, Barbara Ann Karmanos Cancer Institute, Wayne State University School of Medicine, Detroit, MI, USA

Lisa Sato Department of Medical Illustration, Georgia Regents University, Augusta, GA, USA

Shree Ram Singh Mouse Cancer Genetics Program, and Basic Research Laboratory, Center for Cancer Research, National Cancer Institute, Frederick, MD, USA

Garima Sinha Graduate School of Biomedical Sciences and New Jersey Medical School, Rutgers, Newark, NJ, USA

Hiromu Suzuki Department of Molecular Biology, Sapporo Medical University, Sapporo, Japan

Ennio Tasciotti Department of Nanomedicine, Houston Methodist Hospital and Houston Methodist Research Institute, Houston, TX, USA

Muthusawamy Thangraju Cancer Research Center, Georgia Health Science University, Augusta, GA, USA

Ramakumar Tummala Department of Urology and UC Davis Comprehensive Cancer Center, University of California Davis Medical Center, Sacramento, CA, USA

Chingiz Underbayev Pathology and Laboratory Medicine, New Jersey Medical School, Newark, NJ, USA

Krishna C Vallabhaneni Cancer Institute, University of Mississippi Medical Center, Jackson, MS, USA

Jeffrey Van Eps Department of Nanomedicine, Houston Methodist Hospital and Houston Methodist Research Institute, Houston, TX, USA

Department of Surgery and Surgical Advanced Technologies Lab, The Methodist Hospital Research Institute, Houston, TX, USA

Bradley K. Weiner Department of Nanomedicine, Houston Methodist Hospital and Houston Methodist Research Institute, Houston, TX, USA

Department of Surgery and Surgical Advanced Technologies Lab, The Methodist Hospital Research Institute, Houston, TX, USA

Safak Yalcin Human Oncology and Pathogenesis Program, New York, NY, USA

Eiichiro Yamamoto Department of Molecular Biology, Sapporo Medical University, Sapporo, Japan

Yao Yuan Pathology and Laboratory Medicine, New Jersey Medical School, Newark, NJ, USA

Part I
The Functions and Mechanisms of MicroRNAs

Chapter 1
miRNA Biogenesis and Function

Abigail F. Olena and James G. Patton

Abstract miRNAs are small noncoding RNAs that bind the 3′ untranslated regions (UTRs) of mRNA targets and, acting with associated proteins, facilitate translation repression and degradation of target mRNAs. Since their discovery in *Caenorhabditis elegans,* miRNAs and their accessory proteins have been shown to be conserved throughout phylogeny. miRNAs exert their regulatory functions in myriad biological settings, from development and growth to disease. In exploring the mechanism of miRNA biogenesis and function, both canonical and noncanonical, it is possible to gain a broader understanding of how miRNAs work in different biological states, including cancer. Here, we provide an overview of miRNA discovery, biogenesis, and function.

Keywords MicroRNA · Biogenesis · Drosha · Dicer · RNA-induced silencing complex (RISC) · Argonautes

1 The Discovery of miRNAs

The first evidence that small RNAs play a regulatory role was published in 1993. Rosalind C. Lee and Rhonda L. Feinbaum, working in the laboratory of Victor Ambros, and Bruce Wightman and Ilho Ha, working in Gary Ruvkun's laboratory, showed in concurrent publications that *lin-4,* known to be necessary for regulation of the heterochronic gene, *lin-14,* in *Caenorhabditis elegans* was not a protein coding gene, but a small RNA with complementarity to the *lin-14* 3′ UTR (untranslated region) [1, 2].

lin-14 is expressed in a temporal gradient and regulates the normal sequence of cell lineage during *C. elegans* development [3, 4]. Before the seminal work from the Ambros and Ruvkun laboratories, it was known that *lin-4* mutants affect cell lineage decisions with reiteration of larval cell fates during later stages (as do *lin-14* gain of function mutants) and that *lin-4* is a negative regulator of *lin-14* [5, 6]. The

J. G. Patton (✉) · A. F. Olena
Department of Biological Sciences, Vanderbilt University, Nashville, TN, USA
e-mail: james.g.patton@vanderbilt.edu

S. R. Singh, P. Rameshwar (eds.), *MicroRNA in Development and in the Progression of Cancer,* DOI 10.1007/978-1-4899-8065-6_1,

Ambros group cloned the *lin-4* locus and created mutants to disrupt potential open reading frames (ORFs). However, even when the putative ORFs were disrupted, the constructs were still able to rescue *lin-4* mutants. Coupled with the fact that *lin-4* sequences are conserved in other *Caenorhabditis* species, it was concluded that the *lin-4* product does not function as a protein. The *lin-4* product was then shown by northern blot and RNAase protection assays to encode two small RNAs of ~61 and ~22 nucleotides (nt) in length. Finally, the Ambros group showed that sequences in the 3′ UTR of *lin-14,* known from previous work to be necessary for its negative temporal regulation [7], are complementary to sequences in *lin-4* [1].

Ruvkun's group showed that the 3′ UTR of *lin-14* is sufficient to confer temporal regulation to a *lacZ* reporter and that regulation of the reporter was not recapitulated in *lin-4* mutants. They also identified nucleotides in the 3′ end of *lin-4* complementary to seven sites in the *lin-14* 3′ UTR, which were found to be conserved in *C. briggsae.* Based on these findings, the group hypothesized that the seven binding sites in the *lin-14* 3′ UTR are bound cooperatively by *lin-4* to control downregulation of *lin-14* posttranscriptionally at precise developmental time points [2].

The discovery of *lin-4* presented a fascinating new possibility for posttranscriptional regulation of gene expression but because *lin-4* is not conserved in higher eukaryotes, it was unclear whether such regulation was restricted to worms. In 2000, the Ruvkun group published evidence for the existence of another small RNA, *let-7,* which also plays a role in developmental timing by regulating the expression of several heterochronic genes, including *lin-14* and *lin-41* [8]. In contrast to *lin-4, let-7* is conserved from arthropods to vertebrates and the developmental timing of *let-7* expression is also conserved in *Drosophila,* zebrafish, and mollusks [9]. These RNAs were dubbed small temporal RNAs (stRNAs), based upon the expectation that discovery of other small RNAs would play a similar role in developmental timing [8–10]. Subsequent work has since revealed multiple small RNAs in mammalian genomes, as well as in the genomes of fish, flies, worms, and plants. These small RNAs do not always facilitate developmental timing events, so the name stRNAs gave way to the more general term microRNAs (miRNAs). It is now apparent that miRNAs function in multiple biological processes, from growth and maintenance to apoptosis [11–21].

2 Understanding miRNA Biogenesis

2.1 Canonical miRNA Biogenesis: Dicer

With the discovery that miRNAs are conserved in higher eukaryotes, work quickly began to understand how miRNAs are transcribed and processed. The discovery of RNA interference (RNAi) mediated by short, double-stranded RNAs (dsRNAs) by Fire et al. in 1998 provided unexpected insight into miRNA biogenesis [22]. Small

RNAs with regulatory functions were studied in plants [23], and the mechanism of mRNA degradation by small dsRNAs was further explored in other systems, namely *Drosophila,* as well [24–27]. In work published in January 2001, Bernstein et al. identified an enzyme that they termed Dicer, an RNase III superfamily member containing a PAZ domain (later shown to be an RNA-binding domain), two RNase III domains, and an amino-terminal helicase domain [28]. They showed that Dicer is capable of producing 22-nt-long small interfering RNA (siRNA) fragments from long dsRNA (the initial step of RNAi) and that transfecting cells with long dsRNAs complementary to Dicer mRNA in S2 cells abrogates the ability to silence a number of genes. They suggested, therefore, that Dicer is the enzyme responsible for the cleavage of long dsRNAs into 22-nt-long "guide strands" necessary for RNAi.

Because the initial discoveries of miRNAs detected RNAs of ~70 and ~22 nt on northern blots [1, 8, 9], several groups hypothesized that the 70 nt form is a precursor to the 22 nt RNA. The discovery of Dicer as the enzyme responsible for cleavage of long dsRNA into 22 nt siRNAs provided an excellent candidate for the enzyme that creates 22 nt mature miRNAs from the longer 70 nt precursors. Grishok et al. showed that dsRNAs targeting the *C. elegans* Dicer homolog *dcr-1* caused a loss of the 22 nt RNAs and accumulation of the 70 nt forms for both *lin-4* and *let-7,* suggesting that the 70 nt RNA is indeed a precursor to the 22 nt form and that Dicer is necessary for its processing [29]. Hutvágner et al. demonstrated in HeLa cells, *Drosophila* pupae, and *Drosophila* extracts that Dicer is necessary for *let-7* maturation from a precursor stem–loop structure into the 22 nt mature RNA [30]. Ketting et al. then showed in *C. elegans dcr-1* mutants that *let-7* 22 nt RNA levels are reduced, while the 70 nt form of *let-7* accumulates [31]. Recombinant human Dicer was later shown to generate both ~21–23 nt products from long dsRNA and mature *let-7* from *pre-let-7* transcripts [32]. All of this evidence taken together indicated that Dicer is the conserved RNase III family member responsible for generating the 22 nt mature miRNA duplexes from ~70 nt precursor miRNAs (pre-miRNA).

Insight into Dicer's structure revealed how it generates mature miRNAs. Zhang et al. generated mutations in putative catalytic residues in human Dicer and concluded that Dicer has a single dsRNA processing center containing two RNase III sites for cleavage of phosphodiester bonds on opposite RNA strands, thereby yielding mature dsRNAs with 2 nt 3′ overhangs [33]. By solving the crystal structure of *Giardia intestinalis* Dicer, MacRae et al. showed that the dsRNA-binding PAZ domain of Dicer is ~65 Å from the processing center, which is the distance needed to accommodate ~25 base pairs (bp) of dsRNA, the length of *Giardia* small RNAs [34]. They therefore concluded that Dicer acts as a molecular ruler that measures a specified distance from the end of the dsRNA contained in the PAZ domain to position the RNase III domains for cleavage and generation of ~25 nt small RNAs. Their data suggested that the vertebrate Dicer processing site is positioned so as to allow cleavage of the precursor stem loop to generate mature miRNA duplexes of approximately 22 bp. This hypothesis was confirmed by V. Narry Kim's laboratory using immunopurified human Dicer and radiolabeled synthetic pre-miRNAs [35]. Kim's group further showed that the 5′ and 3′ ends of pre-miRNAs are anchored by the PAZ domain and regions surrounding it, and that Dicer measures ~22 nt from the 5′ end in order to determine the cleavage site. Gu et al. later suggested that Dicer recognizes the ends of the miRNA as well as the loop, facilitating precise precursor cleavage and avoidance of off-target effects [36].

Dicer cleavage of pre-miRNAs has since been shown to require the action of cofactors. In two studies using human cell lines, the human immunodeficiency virus transactivating response dsRNA-binding protein (TRBP or TARBP2) was shown to associate with Dicer and to be required for miRNA biogenesis and posttranscriptional gene silencing [37, 38]. Three groups independently showed that the protein Loquacious (Loqs), the *Drosophila* homolog of TRBP, is found in complex with Dicer-1 (the *Drosophila* Dicer that is responsible for pre-miRNA cleavage). Loqs deficiency in S2 cells leads to the accumulation of pre-miRNAs, while loss of Loqs in the germ line of male and female flies causes miRNA processing defects as well as sterility [39–41].

2.2 Canonical miRNA Biogenesis: Drosha

Early cloning and genomic studies provided additional insight into the genomic organization of miRNAs and thus their biogenesis (also reviewed in [42]). Lau et al. (in *C. elegans*), Lagos-Quintana et al. (in *Drosophila* and human cells), and Mourelatos et al. (in human cells) showed that some miRNAs are found in clusters in the genome [11, 12, 43]. The *Drosophila miR-3/miR-6* cluster, which contains *miR-3, miR-4, miR-5,* and three copies of *miR-6*, is not only encoded in close proximity in the genome, but expression of these miRNAs is also temporally coordinated [12]. In *C. elegans*, miRNAs in four clusters, including those encoded by the *miR-35–miR-41* cluster, are also expressed in a temporally coordinated fashion during embryo and young adult stages [11]. This suggested that miRNAs might be transcribed as polycistronic primary transcripts. Genomic data also showed that miRNAs are sometimes found antisense to protein coding genes and in intergenic regions, which further indicated that they must (at least in some cases) be individual transcriptional units [11–13, 43].

In 2002, V. Narry Kim's group used HeLa cell total RNA to perform reverse transcription polymerase chain reaction (RT-PCR) with primers outside the precursor sequences for two miRNA clusters (*miR-23-27-24-2* and *miR-17-18-19a-20-19b-1*) and one individual miRNA (*miR-30*) and observed bands larger than the single precursors [44]. This evidence suggested that both clustered and individual miRNAs are transcribed as longer units containing multiple stem–loop secondary structures, which Kim's group termed primary miRNAs (pri-miRNAs). Kim's group developed an in vitro processing assay, which showed that pri-miRNAs up to several kilobases long are the forerunners of miRNAs and are processed in at least two sequential steps: first to create ~70 nt precursors, a step which they demonstrated most likely happens in the nucleus, and then again in the cytoplasm by Dicer to create ~22 nt miRNAs. Just 1 year later, Kim's group identified another RNase III family member that is responsible for nuclear cleavage of pri-miRNAs and termed this as enzyme Drosha [45]. They showed that immunopurified Drosha cleaves pri-miRNA into pre-miRNA in vitro, resulting in hairpin dsRNAs with 2 nt overhangs at the 3′ ends, characteristic of RNase III enzymes. They also showed that RNAi against Drosha in HeLa cells abrogates pri-miRNA processing, resulting in increased detection of pri-miRNA transcripts and a corresponding decrease in pre-miRNAs and mature miRNAs.

Interestingly, similar to Dicer, Drosha requires a dsRNA-binding protein cofactor. This factor, called DiGeorge syndrome critical region gene 8 (DGCR8) in humans and Pasha in *Drosophila* and *C. elegans,* was shown in human cells, S2 cells, and *C. elegans* to be in complex with Drosha. Depletion or mutation of DGCR8/Pasha in vitro and in vivo leads to an accumulation of pri-miRNA transcripts and a loss of mature miRNAs. The Drosha–DGCR8/Pasha complex is called the Microprocessor [46–49]. The processing of pri-miRNAs by the Microprocessor occurs cotranscriptionally [50–52].

In order to understand how the Microprocessor recognizes its substrate, Bryan Cullen's group used a cell-free system to show that RNA hairpins with loops $\geq$ 10 nt and stems of ~ 30 bp are preferentially processed [53]. Cullen's group demonstrated in vitro that the Microprocessor requires its pri-miRNA substrates to have a hairpin of at least 80 nt and flanking single-stranded RNA (ssRNA) regions at least 10 nt, though in cells, the requirement for the length of flanking ssRNA is at least 40 nt. They also showed that the Drosha dsRNA-binding domain (dsRBD) has a very low affinity for RNA [54]. To clearly ascertain the molecular basis for Microprocessor recognition of its pri-miRNA substrate, V. Narry Kim's laboratory examined the predicted primary structures of known miRNAs to determine a generalized structure for pri-miRNAs: hairpins containing a loop, approximately three helical turns, and 5′ and 3′ ssRNA elements. They then used an in vitro processing assay to show that the pri-miRNA loop can be replaced by two ssRNA segments, implying that it is the instability of the loop and not the loop itself that is essential for processing. Kim's group next confirmed that the ssRNA elements at both the 5′ and 3′ ends of the hairpin are essential, and showed that deletion of base pairs from the helix nearest the ssRNA elements changed the site of cleavage. These data suggested that the ssRNA elements are used as a landmark in processing. Finally, they showed that it is DGCR8 that preferentially binds to pri-miRNAs to facilitate cleavage by Drosha within the Microprocessor [55].

2.3 Canonical miRNA Biogenesis: Nuclear Export

Since, miRNA processing had been demonstrated to have both nuclear and cytoplasmic localization, the question of how pre-miRNAs leave the nucleus was the next to be answered. Yi et al. showed using RNAi in 293T cells against the nucleocytoplasmic transport factor Exportin5 (Exp5), which is Ran-guanosine triphosphate (GTP) dependent, that pre-miRNA, and mature miRNA presence and function in the cytoplasm is decreased as compared to mock-transfected cells [56]. They also showed in vitro that Exp5 binds the *miR-30* precursor in the presence of Ran-GTP. Bohnsack et al. showed independently that pre-miRNA export is sensitive to deficits of Ran-GTP in the nucleus and used affinity columns with bound pre-miRNAs to recover Exp5 [57]. Additionally, they demonstrated that they could block pre-miRNA export from *Xenopus* oocyte nuclei using antibodies raised against *Xenopus* Exp5. Lund et al. added to the weight of evidence for the role of Exp5 in pre-miRNA export by showing a decrease in cytoplasmic *let-7a* upon treatment of HeLa cells with siRNAs against Exp5 [58]. They also showed in HeLa cell extract that the binding of pre-miRNAs to Exp5 is highly specific and resistant to competition by small RNAs with similar secondary structures to pre-miRNAs. All these data taken together confirmed that Exp5 is the nucleocytoplasmic transport factor responsible for pre-miRNA export from the nucleus, and that its role in the process is Ran-GTP dependent.

2.4 Canonical miRNA Biogenesis: Transcription

With the knowledge of compartmentalized, stepwise processing of miRNAs in hand, the field next turned to the question of how miRNAs are transcribed. Both the Kim and Cullen laboratories used the efficient m7G cap-binding properties of eIF4E to isolate capped RNAs and analyzed the content of the resulting RNA pool using RT-PCR with pri-miRNA specific primers [59, 60]. Both groups showed that pri-miRNAs were present in capped RNA fractions. Each group followed a similar principle to show the presence of pri-miRNAs in a HeLa cell complementary DNA (cDNA) library generated using an oligo-dT primer [60] and in a group of RNAs enriched with oligo-dT beads [59]. These experiments provided strong evidence that pri-miRNAs are polyadenylated. As 5′ capping and 3′ polyadenylation are hallmarks of RNA polymerase II transcription, the Kim laboratory used the Pol II-specific inhibitor α-amanatin on HeLa cells and compared the transcription levels of several pri-miRNAs to their transcription levels in untreated cells. In all of the pri-miRNAs they tested, a decrease in pri-miRNA transcription was observed upon α-amanatin treatment. The final piece of evidence that Pol II is responsible for pri-miRNA transcription came with the analysis of the promoter of the *miR-23a-27a-24-2* cluster, which was shown to be bound by Pol II in vivo [59]. Overall, miRNAs are transcribed as long primary transcripts that are then processed in two steps, one in the nucleus and the second in the cytoplasm (Fig. 1.1).

3 Noncanonical miRNA Biogenesis

The large majority of miRNAs follow the biogenesis pathways as described, but two noncanonical pathways have also been discovered, a Microprocessor-independent pathway and a Dicer-independent pathway. For an in-depth review of these topics,

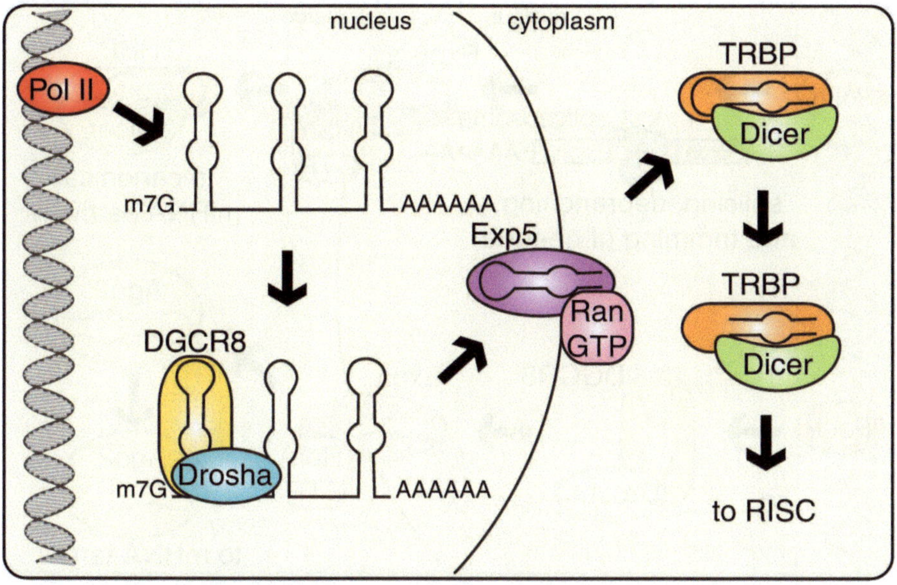

Fig. 1.1 Canonical miRNA biogenesis. miRNAs are transcribed as primary transcripts by RNA polymerase II. Pri-miRNAs contain one or more stem loops, which are recognized and cleaved by Drosha and its cofactor DGCR8. The resulting pre-miRNA is exported from the nucleus by Exportin 5, a process dependent upon Ran-GTP. In the cytoplasm, the pre-miRNA is processed by Dicer and its cofactor TRBP, resulting in a mature miRNA duplex containing two 21–22-nt-long strands. This miRNA duplex is then incorporated into the RNA-induced silencing complex (RISC)

see the excellent review by Jr-Shiuan Yang and Eric C. Lai [61]. For an overview of noncanonical miRNA biogenesis, see Fig. 1.2.

3.1 Microprocessor-Independent miRNA Biogenesis

Genomically, miRNAs have been found in exons, introns, between genes (intergenic), and as part of either mono- or polycistronic RNAs [62, 63]. In the special case of mirtrons, discovered by Okamura et al. and Ruby et al. in *Drosophila* and *C. elegans,* mature miRNAs are derived from short intronic hairpins with splice sites on each end so that the sequence of the pre-miRNA corresponds exactly to the sequence of the intron [64, 65]. Instead of being dependent upon Drosha processing, these short intronic hairpins are excised by the spliceosome. After processing by the lariat-debranching enzyme, they fold into hairpin secondary structures resembling pre-miRNAs that are recognized by Exp5 for nuclear export, thus rejoining the canonical miRNA biogenesis pathway.

Since their initial discovery, mirtrons have also been identified in mammals by computational prediction and by high-throughput sequencing in RNA from human and rhesus macaque brains [66], from DGCR8-null mouse embryonic stem

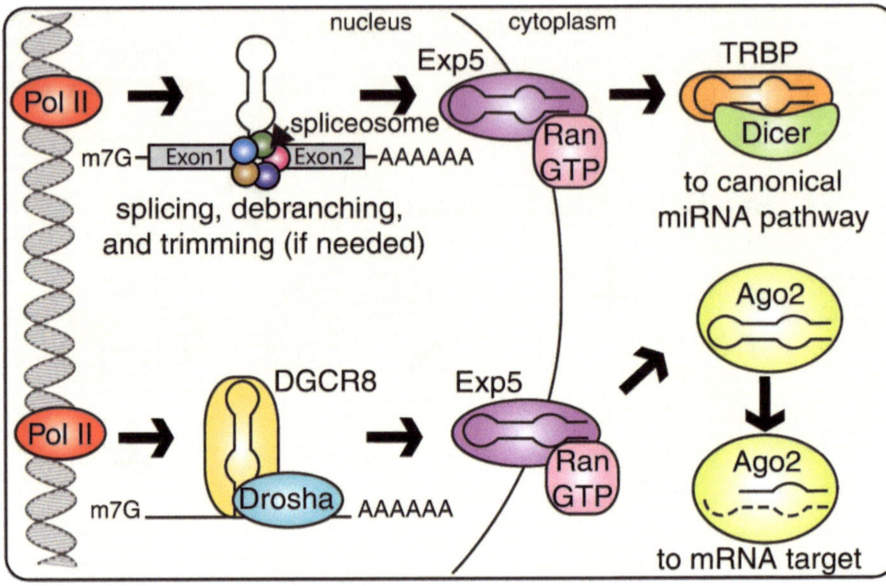

Fig. 1.2 Noncanonical miRNA biogenesis. At the top of the figure, mirtron biogenesis is depicted. Mirtrons are transcribed as introns of protein coding genes by RNA Pol II. The spliceosome splices these mirtrons, then they are debranched and undergo exosomal trimming (if necessary). After these initial processing steps, mirtrons are treated as canonical miRNAs and transported to the cytoplasm by Exp5 to be processed by Dicer and TRBP and incorporated into the RISC. At the bottom of the figure, Dicer-independent miRNA biogenesis is depicted. *miR-451,* a Dicer-independent miRNA, is transcribed by RNA Pol II, processed by Drosha/DGCR8, and exported from the nucleus by Exp5. Once in the cytoplasm, *miR-451* is bound by Argonaute 2 (Ago 2) and is cleaved to yield the functional, mature, single-stranded miRNA

cells [67], from Drosha-null murine T cells [68], and in the chick [69]. Chong et al. showed that the majority of Drosha-independent miRNAs are not canonical mirtrons and appear to come from long introns or independent transcriptional units, suggesting that splicing is not solely responsible for the generation of this class of Drosha-independent miRNAs [68]. Ruby et al., Babiarz et al., and Glazov et al., also found evidence for tailed mirtrons, in which one of the two ends of the putative pre-miRNA does not correspond to a splice site, confirming that another processing event must be necessary for pre-miRNA generation [65, 67, 69].

One such tailed mirtron, *miR-1017,* was discovered by Ruby et al. in *C. elegans* [65] and more closely examined by Flynt et al. in *Drosophila* [70]. High-throughput sequencing data from male *Drosophila* heads [71] contained approximately 6,000 reads for *miR-1017,* but also 14 reads for *miR-1017** (for an explanation of this nomenclature, see Sect. 3.2), whose sequence aligns exactly with the 5′ splice donor sequence, consistent with splicing being involved in *miR-1017/1017** biogenesis. In examining various *Drosophila* genomes, extensive variability was detected in the tail sequence (the sequence preceding the 3′ splice acceptor) of the *miR-1017* mirtron suggesting that the tail sequence is not important when generating a functional *miR-1017*. Flynt and colleagues then used RNAi in S2 cells to show that the RNA exosome, a 3′ to 5′ exonuclease complex with multiple subunits, is responsible for the removal of this mirtron's tail. The group also found genomic evidence for five more 3′-tailed mirtrons and suggested that these are processed by the RNA exosome as well [70].

3.2 Dicer-Independent miRNA Biogenesis

The year 2010 brought the discovery of a Drosha-dependent, Dicer-independent miRNA, *miR-451* [72–74]. Cleavage of the loop of pre-miRNAs by vertebrate Dicer generally yields mature miRNA duplexes with two paired RNA strands of ~22 nt, corresponding to either side of the pre-miRNA stem. In the case of *miR-451,* the miRNA comes from a portion of the 5′ stem and the loop of the putative hairpin. *miR-451* is conserved across phylogeny, and was specifically examined in mouse [72], zebrafish [73] and *Drosophila* [74]. These three groups showed independently that Argonaute 2 (Ago2), initially shown to be involved in miRNA function (for more on this, see Sect. 3.1), is responsible for cleaving the hairpin and releasing a version of *miR-451,* which must then be trimmed by an as yet unknown exonuclease to generate the functional, mature miRNA [72–74].

4 Regulation of miRNA Biogenesis

miRNA biogenesis has the potential for significant regulation, based on the multiple steps and players involved. Regulation at the transcriptional level is similar to other RNA polymerase II-transcribed genes. Using Pol II ChIP, some miRNA genes have been shown to have unique promoters, but the majority share general features with promoters of protein coding genes, modulated by elements that can be several kilobases away [75]. By combining nucleosome mapping with chromatin signatures, Ozsolak et al. identified proximal promoters for 175 human miRNAs and confirmed the regulation of nine miRNAs by the microphthalmia-associated transcription factor (MITF) in melanoma cell lines [76]. They also showed that some intronic miRNAs, which were generally assumed to be transcribed with their host genes, have their own promoters. Based on bioinformatic and biochemical examination of miRNAs regulated by transcription factors, it was shown that a transcription factor that regulates an miRNA often regulates that miRNA's targets as well [77].

Regulation of miRNA production at the level of posttranscriptional processing is also widespread. In one important example of a regulatory loop, the interaction of the splicing factor SF2/ASF with *pri-miR-7* was shown to promote Drosha cleavage of the primary miRNA, which when fully processed targets the SF2/ASF 3′ UTR [78]. Lin28, a highly conserved RNA-binding protein implicated in oncogenesis, has been shown to block the maturation of *let-7* at both the primary and precursor levels [79–85] (reviewed in: [86]). Forman et al. demonstrated experimentally that *let-7* targets Dicer within its coding sequence, with obvious implications for the regulation of miRNA biogenesis by miRNAs [87]. miRNA sequence editing by adenosine deaminases that act on RNA (ADARs) can alter the base pairs of miRNAs at the primary and precursor levels, which can affect processing [88, 89]. Examples of the regulation of miRNA biogenesis are much more extensive than covered here and are reviewed in depth by Davis et al. and Krol et al. [90, 91].

5 Understanding miRNA Function

The discovery of the regulation of *lin-14* by *lin-4* coupled with the dependence of that regulation on elements in the 3′ UTR of *lin-14* that are complementary to *lin-4* provided initial insight into the mechanism of miRNA function [1, 2]. What follows is an overview of the proteins that are involved in this mechanism, as well as information on how miRNA structure informs miRNA strand selection and mRNA target recognition, and how miRNAs and their associated proteins affect posttranscriptional gene silencing. For an overview, see Fig. 1.3.

5.1 Assembly of the RNA-Induced Silencing Complex

The discovery of proteins that allow miRNAs to alter target gene expression was buoyed by work from the siRNA field. Studies using cultured mammalian and *Drosophila* cells and cell/embryo lysates showed that one strand of the siRNA duplex is the effector in RNAi and that it cooperates with a (then unknown) nuclease(s), which cleaves mRNAs at sites complementary to the siRNA. The complex in which the siRNA and nuclease act was termed the RNA-induced silencing complex (RISC) [24, 27, 92–94]. By purifying RISC from *Drosophila* cells, Hammond et al. showed that one of the components of the RISC that copurifies with siRNA is Ago2, a member of the Argonaute family [95].

Argonaute family members have a PIWI and PAZ (Piwi/Argonaute/Zwille) domain [96], whose functions were elucidated with structural studies. Three groups solved the structures of the *Drosophila* Ago1 and Ago2 PAZ domains and demonstrated that the PAZ domains interact with the 3′ overhangs of dsRNA and prefer to interact with RNA rather than DNA. The structural data also suggested that the human Ago and Dicer PAZ domains are structurally similar and form a cleft for binding RNA [97–99]. When structures of the PAZ domain bound to RNA and DNA were solved, it was shown that PAZ binding stabilizes dsRNA duplexes by securing the 2 nt 3′ overhangs in a conserved binding cleft and by binding the phosphodiester backbone of the overhang-containing strand. Also, the data indicated that the PAZ domain contributes to specific recognition of dsRNA duplexes within the RISC [100, 101]. The first structure of full-length Ago2 (from *Pyrococcus furiosus*)

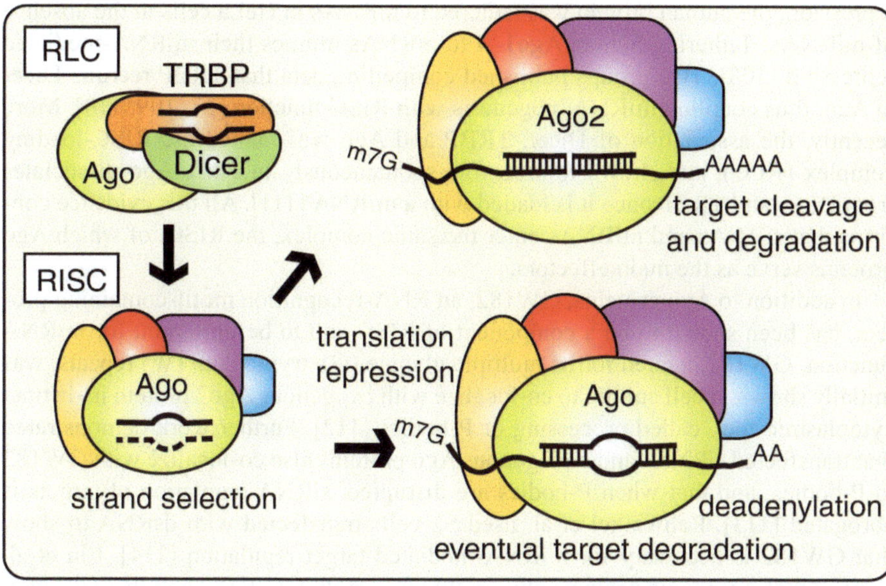

Fig. 1.3 miRNA function. Dicer, TRBP, and Argonaute form the RISC loading complex (RLC), which facilitates assembly of the RISC and incorporation of the mature miRNA duplex. Once incorporated into the RISC, the nonfunctional strand of the miRNA duplex is degraded and the functional strand is guided to its target, generally in the 3′ UTR of a messenger RNA. If the miRNA pairs perfectly with its target (pictured *top right*), the miRNA–mRNA duplex is cleaved by Ago2 between nucleotides 10 and 11 of the miRNA strand, which leads to the target's rapid degradation. If the miRNA pairs imperfectly with its target (depicted *bottom right*), the interaction can lead to translation repression, potentially occurring at the initiation step by interference with the mRNA cap, but still poorly understood, and to deadenylation; both outcomes may lead to eventual target degradation

indicated that Ago2 is composed of four domains, previously unknown N-terminal (N) and MIDdle (MID) domains, as well as the PIWI and PAZ domains[102]. The structure showed that the N, PIWI, and MID domains form a crescent over which the PAZ domain is held by a stalk-like region. Most importantly, the PIWI domain structure is similar to RNAse H and is therefore responsible for the cleavage of mRNAs complementary to siRNAs, the catalytic component of RISC. Further analysis of the four mammalian Argonautes (Ago1–4) demonstrated that only Ago2 is able to cleave mRNA substrates [103].

While the structural work was done using siRNA–Ago pairs, miRNAs were also found to assemble in complexes containing Ago2 along with a helicase, Gemin3, and Gemin4 [43]. Though these complexes are highly similar to the RISC, they were initially termed microribonucleoproteins (miRNPs). In human cell extracts, *let-7* was detected in RISCs but imperfect miRNA:mRNA pairing leads to translation repression rather than mRNA cleavage [104]. siRNAs that do not pair perfectly with mRNA targets were also shown to translationally repress their targets, as opposed to cleaving them [105–107]. To assess whether Ago facilitates translation repression, the human protein was tethered to mRNAs in HeLa cells in the absence of miRNAs. Tethering human Ago1–4 to mRNAs mimics their miRNA-mediated repression [108]. Two groups published compelling data that TRBP recruits Dicer to Ago, thus coupling miRNA biogenesis with RISC function [37, 109, 110]. More recently, the association of Dicer, TRBP and Ago was named the RISC-loading complex (RLC), and shown to assemble spontaneously in vitro. Ago dissociates from Dicer and TRBP once it is loaded with a miRNA [111]. All this evidence confirms that siRNAs and miRNAs enter the same complex, the RISC, of which Ago proteins serve as the main effectors.

In addition to Ago proteins, GW182, an RNA-recognition motif-containing protein, has been shown to be a component of RISC and to be important for miRNA function. GW182, named for its multiple glycine (G)–tryptophan (W) repeats, was initially shown in cell culture to co-localize with exogenous Ago2 protein in distinct cytoplasmic foci, called processing or P-bodies [112]. Further work demonstrated that transfected siRNAs and endogenous Ago proteins also co-localize with GW182 in P-bodies, and that when P-bodies are disrupted, siRNA regulation of targets is abrogated [113]. Rehwinkel et al. used S2 cells transfected with dsRNA to show that GW182 is necessary for miRNA-mediated target regulation [114]. Liu et al. showed that Argonaute proteins physically interact with GW182 and that silencing GW182 perturbs miRNA function [115]. Consistent with this, Ding et al. showed that a *C. elegans* protein-sharing homology with GW182 is responsible for targeting the worm Ago homolog to P-bodies [116]. The GW182–Ago interaction has since then been established to be crucial for miRNA target repression [117]. It is likely that Ago proteins interact in the RISC with other partners. Indeed, in a proteomics screen, an RNA helicase, MOV10, and another RNA-recognition motif-containing protein, TNCR6B, were found to interact with Ago proteins [118].

5.2 Selection of the Functional miRNA Strand

Martinez et al. showed that only one of the two strands of the siRNA duplex is retained in the RISC and works with the RISC components to effect sequence-specific silencing [119]. This raised the question as to how one strand is chosen over the other. Direct cloning of miRNAs readily detected the presence of mature miRNA strands but in some cases, the other strand (denoted miRNA*) was also detected albeit at low levels [20]. The presence of the star strand was a good first indication that there is a preference for choosing which side of the pre-miRNA is incorporated into RISC. By designing a wide variety of synthetic siRNA duplexes, Schwarz et al. showed that for the majority of cases, the strand with the less stably paired 5′ end is the one incorporated into the RISC [120]. To extend this paradigm to miRNAs, putative precursors of known miRNAs were analyzed to determine whether predictions based on 5′ stability agree with strand selection in vivo. The data are consistent with the model that the strand with lower internal thermodynamic stability at the 5′ terminus is generally the functional strand [121].

5.3 Canonical miRNA Target Recognition

The majority of early work on miRNA–target interactions showed that miRNAs do not bind perfectly with their targets [1, 2, 8, 122]. Furthermore, miRNAs can bind combinatorially, so more than one miRNA usually regulates a target, and miRNAs may target several mRNAs [123]. In order to facilitate target prediction, several groups undertook to find guidelines for miRNA–target pairing. Lewis et al. found pairing between nucleotides 2–8 (from the 5′ end) of the miRNA and the mRNA 3′ UTR to be the most reliable predictor of a verifiable miRNA–mRNA interaction, and referred to this region as the "seed" of the miRNA, calling a Watson–Crick base-paired interaction between the miRNA seed and an mRNA a "seed match" [124]. The seed rule was extended by analyzing miRNA seed sequences and finding an overrepresentation of adenosines on either side of seed sequences [125].

Using reporter assays, Doench et al. confirmed that miRNAs cooperatively bind to targets and that the 5′ region of the miRNA is extremely important for miRNA–target pairing [126]. They also showed that miRNA 3′ end pairing with the 3′ UTR of the mRNA is only important if the interaction at the 5′ end is less stable. Kiriakidou et al. used mutagenesis of luciferase reporter constructs containing single MREs to show the requirement for a bulge in the center of miRNAs bound to targets (whose maximum length is context dependent) [127]. Brennecke et al. then showed in vivo that a single 8mer at the 5′ end of the miRNA pairing perfectly with the mRNA 3′ UTR is able to confer target repression, with a 7mer conferring slightly less repression [128]. When two 8- or 7mers were present, target regulation increased. They also distinguished between canonical miRNA recognition sites (MREs), which have strong base pairing at both the 5′ and 3′ ends of the miRNA; seed sites, which have strong base pairing at the 5′ end with minimal 3′ end pairing, and 3′; and compensatory sites, which have mismatches or G:U wobbles in the seed but strong base pairing at the 3′ end. They also found that the 3′ end of the miRNA is responsible for differences in mRNA target recognition by members of miRNA family members, which share identical seed sequences, but with differences at their 3′ ends.

To further refine target prediction, the Bartel group concluded that miRNA repression could be context dependent with several factors increasing the likelihood of target regulation [129]. MREs are positioned near AU-rich sequences or close to MREs for other miRNAs increase the efficiency of miRNA–mRNA pairing. In addition, positioning within the 3′ UTR can also affect pairing: Stronger sites are usually noncentrally located MREs in long UTRs and/or those located at least 15 nt from the end of the coding sequence. They also corroborated other data that showed a role for the 3′ end of the miRNA, narrowing it down to pairing between nt 13 and 16. Cooperation between MREs was also experimentally demonstrated in other systems [130]. In addition to confirming that MREs are found close to the beginning or end of long human 3′ UTRs, Gaidatzis et al. observed that MRE location is evolutionarily conserved within 3′ UTRs [131]. They suggested that this conservation of location could provide greater MRE accessibility for RISC and therefore have implications for miRNA function. Nielsen et al. further refined the importance of the context of the MRE within the UTR by showing that an adenosine is generally opposite miRNA nt 1 in the 3′ UTR and an adenosine or uridine is generally opposite miRNA nt 9 [132]. Moreover, they showed that increased sequence conservation approximately 50 nt up and downstream of the seed match in the 3′ UTR increases target repression.

6 Noncanonical miRNA Target Recognition

Perhaps not surprisingly, as previous work showed the ability of noncanonical MREs to regulate targets [128], several examples of functional noncanonical MREs have emerged. Li et al. demonstrated that *dispatched homolog 2* is regulated by *miR-214* during zebrafish development through the action of three noncanonical MREs in its 3′ UTR that act combinatorially to confer repression [133]. The Bartel group identified a new class of miRNA–target binding, in which the binding interaction does not feature strong seed pairing, but instead features at least 11 contiguous Watson–Crick base pairs starting at nt 4 of the miRNA [134]. In vitro under certain buffer conditions, these sites can lead to mRNA cleavage, but in vivo they mostly work to repress translation. In the same work, more miRNAs were identified that facilitate target cleavage (as siRNAs do) than was originally thought, though the number of miRNAs acting in this fashion is still low compared to how the vast majority of miRNAs function. Chi et al. used data from RNA libraries to show that *miR-124,* a common neuronal miRNA, binds its targets with a G-bulge in the 3′ UTR, between nt 5 and 6 of the miRNA [135]. They further showed that nucleation bulges are evolutionarily conserved, functional in the murine brain, and present in *C. elegans* as well as in mice. Finally, Loeb et al. used a whole transcriptome approach to identify all the targets of *miR-155* in murine T cells [136]. They found that 40 % of miRNA–target interactions are noncanonical, with the majority of these interactions having one mismatch in the seed region. Interestingly, *miR-155* is still able to regulate gene expression using a noncanonical site. The discovery of noncanonical sites is necessary for facilitating understanding of miRNA function. Perhaps more importantly, the discovery of noncanonical sites highlights the difficulty in predicting miRNA targets and the necessity for experimental validation of predicted targets. See Fig. 1.4 for a summary of the ways miRNAs recognize their targets.

7 Mechanisms of miRNA-Mediated Gene Silencing

While direct miRNA-mediated cleavage of mRNA targets is rare in vertebrates [137], examples abound of miRNA regulation of gene expression through both translation repression and mRNA deadenylation leads to degradation. Early evidence suggested that the levels of mRNA targets are unaffected by miRNA repression [138], consistent with translation repression models. Initial work used polysome

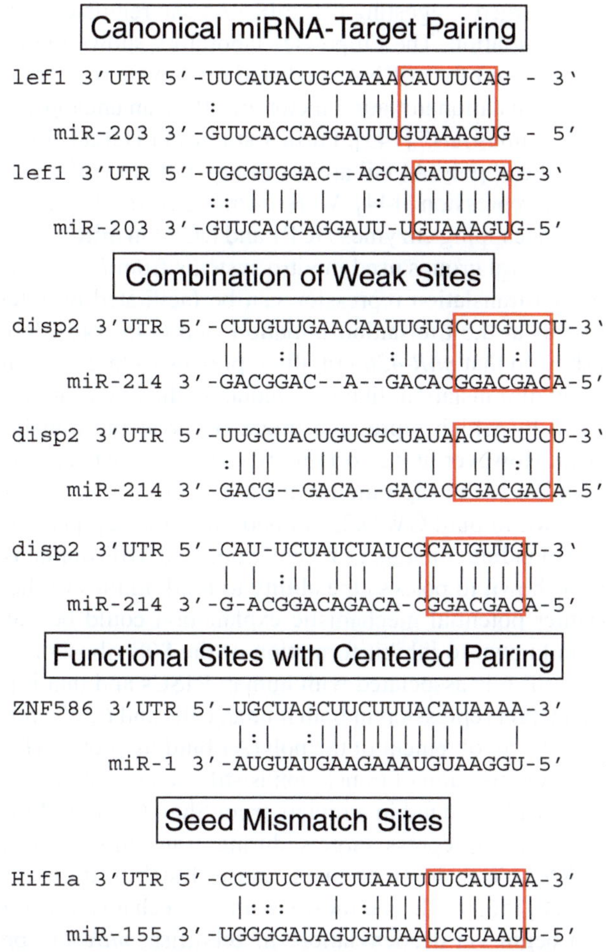

Fig. 1.4 Examples of miRNA target recognition. The pairing of *miR-203* and its target during zebrafish fin regeneration, lef1, is shown at *top*. Straight lines represent Watson–Crick base pairs, while colons represent G:U wobble pairing. The 3′UTR of lef1 contains two miRNA target sites, both of which are good examples of canonical miRNA–target pairing, with perfect matches in the seed region (nt 2–8 of the miRNA, boxed in *red*, adapted from [174]). Second from *top*, the disp2 3′ UTR contains three *miR-214* target sites, all of which would be considered weak based on lack of perfect pairing in the seed region (boxed in *red*), but the three weak sites have been shown to act collaboratively to facilitate *miR-214* regulation of disp2 [133]. Second from *bottom* is an example of centered pairing between *miR-1* and its target ZNF586 [134]. At *bottom*, one example of a *miR-155* target (Hif1a) with mismatches in the seed region (boxed in *red*); Hif1a also contains a canonical pairing site for *miR-155* (not pictured) [136]

profiling in *C. elegans* to show that mRNAs and miRNAs are often associated with active ribosomes [138, 139]. This work was corroborated in human cell lines and suggested that translation repression can occur after initiation [140–142].

Evidence also emerged suggesting that miRNAs can regulate their targets at the level of translation initiation. The Filipowicz laboratory showed in HeLa cells that a 5′ 7meG cap is required for miRNA-mediated mRNA repression, and that artificially tethering the translation initiation factor eIF-4E to an uncapped mRNA allows repression by endogenous *let-7* [143]. Further studies in HeLa cells confirmed that the 7meG cap and the 3′ poly(A) tail are necessary, but not sufficient, for miRNA-mediated translation repression [144]. Work from the Izarraulde laboratory showed involvement of the decapping enzymes DCP1 and DCP2 in miRNA-mediated target repression [114]. Several groups used in vitro systems from *Drosophila* and human cells to confirm that translation repression can be facilitated by interference with cap-binding proteins at the translation initiation step [145–148]. Recent evidence from zebrafish showed that *miR-430* initially represses its targets in the developing embryo at the level of translation, likely by reducing the rate of initiation [149].

The mechanism by which repression might occur at the translation initiation step is still unclear. However, it was originally suggested that Ago2's MID domain is significantly similar to the cap-binding domain of eIF-4E [150], but this domain has since been shown to bind GW182, as opposed to the cap [117–151]. The data are somewhat confounding, though, as recently purified MID domains from several species have been shown to possess the ability to bind mimics of the 7meG cap in vitro [152]. Another potential mechanistic explanation could be the involvement of eIF-6, a protein known to inhibit formation of the 80S ribosome. Chendrimada et al. showed that eIF-6 is associated with human RISCs and that it plays a role in miRNA-mediated target repression in both human cells and *C. elegans* [153]. Other evidence points to the involvement of the poly(A)-binding protein (PAB) [154], but the mode of miRNA repression of translation is still vague. A more in-depth discussion of these issues can be found in a review from Shuo Gu and Mark Kay [155].

While some miRNAs repress targets during translation, evidence has also emerged that miRNAs can, in other cases, cause destabilization and degradation of targeted mRNAs [156, 157]. The most common mechanism of mRNA destabilization seems to begin via deadenylation. In zebrafish, *miR-430* promotes rapid deadenylation and clearance of maternal mRNAs during the transition from maternal to zygotic transcription [158]. Consistent with this, miRNA-mediated mRNA degradation requires the CCR4:NOT deadenylase [159–162]. In mammalian cells, miRNAs are able to direct rapid deadenylation of targets [163], and in cell-free systems miRNAs can deadenylate mRNAs [148]. In *Drosophila* cells, 60 % of targets of Ago are regulated by CAF1 or NOT1 deadenylases, indicating that deadenylation of miRNA targets is widespread [164].

It seems likely that the use of these two modes of target repression is context dependent, and in some cases, both mechanisms are used in concert. Recent data support the idea that deadenylation and translational repression may not be mutually exclusive [165, 166]. In zebrafish, for example, *miR-430* first translationally represses and then destabilizes target mRNAs [149]. A deeper discussion of these possibilities and a more thorough look at the evidence for all modes of miRNA-mediated target repression can be found in excellent reviews from Fabian et al. and Huntzinger et al. [167, 168].

8 Regulation of miRNA Function

Just as miRNA biogenesis is highly regulated, so too is miRNA function. The obvious place to begin with regulation of function is with the proteins involved in assembly and function of the RISC. Proteomic analysis of Ago-associated proteins in human cells demonstrated that many proteins associate with Ago proteins, raising the possibility for regulation at the level of RISC assembly and function [169]. Regulation of the Ago proteins themselves also presents opportunities for regulation. For instance, Adams et al. showed that Ago2 expression is responsive to the epidermal growth factor receptor (EGFR)/mitogen-activated protein kinase (MAPK) signaling pathway in cell culture, indicating one possible regulatory mechanism for Ago2 expression [170]. miRNA function can also be affected at the level of the miRNA sequence. RNA editing via adenosine deaminases acting on RNA (ADARs) not only can affect processing, as discussed in Sect. 2.3, but can also affect miRNA target recognition. For example, Kawahara et al. showed in mice null for ADARs that *miR-376,* which has a highly editable site in its seed region, targets a different set of genes than in mice that have ADARs and presumably edited *miR-376* [171]. Furthermore, even though no prominent examples of RNA editing affecting strand selection exist in the literature, one could imagine that editing may affect stability at the 5′ ends of miRNAs and miRNA*s, which could impact which strand is incorporated into the RISC with subsequent downstream effects on mRNA targeting. A more comprehensive discussion of miRNA function regulation can be found in reviews by Davis et al. and Krol et al. [90, 91].

9 Conclusions

Following their discovery in *C. elegans,* miRNAs have come to assume prominent roles in the regulation of gene expression in many biological contexts. Their unique discovery as functional, nonprotein coding RNAs has spurred research into other classes of noncoding RNAs that are likely to have equally diverse and widespread functions [172, 173]. Research is ongoing in the areas of miRNA biogenesis and function, especially how these processes are regulated. As new work emerges, it is obvious there is much to learn. Moving forward, it is important to recognize that the rapid progress that has been made needs to be tempered by the recognition that what seems like dogma based on initial studies might turn out to be much more complex as the complete story becomes known. Nevertheless, it seems clear that a complete understanding of miRNA biogenesis and function will provide further insight into the role of miRNAs during development and disease.

References

1. Lee RC, Feinbaum RL, Ambros V. The C. elegans heterochronic gene lin-4 encodes small RNAs with antisense complementarity to lin-14. Cell. 1993;75(5):843–54.
2. Wightman B, Ha I, Ruvkun G. Posttranscriptional regulation of the heterochronic gene lin-14 by lin-4 mediates temporal pattern formation in C. elegans. Cell. 1993;75(5):855–62.
3. Ruvkun G, Giusto J. The Caenorhabditis elegans heterochronic gene lin-14 encodes a nuclear protein that forms a temporal developmental switch. Nature. 1989;338(6213):313–9. doi:10.1038/338313a0.
4. Ambros V. A hierarchy of regulatory genes controls a larva-to-adult developmental switch in C. elegans. Cell. 1989;57(1):49–57.
5. Ambros V, Horvitz HR. The lin-14 locus of Caenorhabditis elegans controls the time of expression of specific postembryonic developmental events. Genes Dev. 1987;1(4):398–414.
6. Arasu P, Wightman B, Ruvkun G. Temporal regulation of lin-14 by the antagonistic action of two other heterochronic genes, lin-4 and lin-28. Genes Dev. 1991;5(10):1825–33.
7. Wightman B, Burglin TR, Gatto J, Arasu P, Ruvkun G. Negative regulatory sequences in the lin-14 3'-untranslated region are necessary to generate a temporal switch during Caenorhabditis elegans development. Genes Dev. 1991;5(10):1813–24.
8. Reinhart BJ, Slack FJ, Basson M, Pasquinelli AE, Bettinger JC, Rougvie AE, Horvitz HR, Ruvkun G. The 21-nucleotide let-7 RNA regulates developmental timing in Caenorhabditis elegans. Nature. 2000;403(6772):901–6.
9. Pasquinelli AE, Reinhart BJ, Slack FJ, Martindale MQ, Kuroda MI, Maller B, Hayward DC, Ball EE, Degnan B, Müller P, Spring J, Srinivasan A, Fishman M, Finnerty J, Corbo J, Levine M, Leahy P, Davidson E, Ruvkun G. Conservation of the sequence and temporal expression of let-7 heterochronic regulatory RNA. Nature. 2000;408(6808):86–89. doi:10.1038/35040556.
10. Slack FJ, Basson M, Liu Z, Ambros V, Horvitz HR, Ruvkun G. The lin-41 RBCC gene acts in the C. elegans heterochronic pathway between the let-7 regulatory RNA and the LIN-29 transcription factor. Mol Cell. 2000;5(4):659–69.
11. Lau NC, Lim LP, Weinstein EG, Bartel DP. An abundant class of tiny RNAs with probable regulatory roles in Caenorhabditis elegans. Science. 2001;294(5543):858–62.
12. Lagos-Quintana M, Rauhut R, Lendeckel W, Tuschl T. Identification of novel genes coding for small expressed RNAs. Science. 2001;294(5543):853–8.
13. Lee RC, Ambros V. An extensive class of small RNAs in Caenorhabditis elegans. Science. 2001;294(5543):862–4. doi:10.1126/science.1065329.
14. Lagos-Quintana M, Rauhut R, Yalcin A, Meyer J, Lendeckel W, Tuschl T. Identification of tissue-specific microRNAs from mouse. Curr Biol. 2002;12(9):735–9.
15. Lagos-Quintana M, Rauhut R, Meyer J, Borkhardt A, Tuschl T. New microRNAs from mouse and human. RNA. 2003;9(2):175–9.
16. Ambros V. MicroRNA pathways in flies and worms: growth, death, fat, stress, and timing. Cell. 2003;113(6):673–6.
17. Aravin AA, Lagos-Quintana M, Yalcin A, Zavolan M, Marks D, Snyder B, Gaasterland T, Meyer J, Tuschl T. The small RNA profile during drosophila melanogaster development. Dev Cell. 2003;5(2):337.
18. Houbaviy HB, Murray MF, Sharp PA. Embryonic stem cell-specific MicroRNAs. Dev Cell. 2003;5(2):351–8.
19. Lim LP, Glasner ME, Yekta S, Burge CB, Bartel D. Vertebrate microRNA genes. Science. 2003;299(5612):1540. doi:10.1126/science.1080372.
20. Lim LP, Lau NC, Weinstein EG, Abdelhakim A, Yekta S, Rhoades MW, Burge CB, Bartel DP. The microRNAs of Caenorhabditis elegans. Genes Dev. 2003;17(8):991–1008.
21. Reinhart BJ, Weinstein EG, Rhoades MW, Bartel B, Bartel DP. MicroRNAs in plants. Genes Dev. 2002;16(13):1616–26.

22. Fire A, Xu S, Montgomery MK, Kostas SA, Driver SE, Mello CC. Potent and specific genetic interference by double-stranded RNA in Caenorhabditis elegans. Nature. 1998;391(6669):806–11.
23. Hamilton AJ, Baulcombe DC. A species of small antisense RNA in posttranscriptional gene silencing in plants. Science. 1999;286(5441):950–2.
24. Hammond SM, Bernstein E, Beach D, Hannon GJ. An RNA-directed nuclease mediates post-transcriptional gene silencing in drosophila cells. Nature. 2000;404(6775):293–6. doi:10.1038/35005107.
25. Parrish S, Fleenor J, Xu S, Mello C, Fire A. Functional anatomy of a dsRNA trigger: differential requirement for the two trigger strands in RNA interference. Mol Cell. 2000;6(5):1077–87.
26. Yang D, Lu H, Erickson JW. Evidence that processed small dsRNAs may mediate sequence-specific mRNA degradation during RNAi in Drosophila embryos. Curr Biol. 2000;10 (19):1191–200.
27. Zamore PD, Tuschl T, Sharp PA, Bartel DP. RNAi: double-stranded RNA directs the ATP-dependent cleavage of mRNA at 21–23 nucleotide intervals. Cell. 2000;101(1):25–33.
28. Bernstein E, Caudy AA, Hammond SM, Hannon GJ. Role for a bidentate ribonuclease in the initiation step of RNA interference. Nature. 2001;409(6818):363–6.
29. Grishok A, Pasquinelli AE, Conte D, Li N, Parrish S, Ha I, Baillie DL, Fire A, Ruvkun G, Mello CC. Genes and mechanisms related to RNA interference regulate expression of the small temporal RNAs that control C. elegans developmental timing. Cell. 2001;106 (1):23–34.
30. Hutvágner G, McLachlan J, Pasquinelli AE, Bálint E, Tuschl T, Zamore PD. A cellular function for the RNA-interference enzyme Dicer in the maturation of the let-7 small temporal RNA. Science. 2001;293(5531):834–8. doi:10.1126/science.1062961.
31. Ketting RF, Fischer SE, Bernstein E, Sijen T, Hannon GJ, Plasterk RHA. Dicer functions in RNA interference and in synthesis of small RNA involved in developmental timing in C. elegans. Genes Dev. 2001;15(20):2654–9. doi:10.1101/gad.927801.
32. Provost P, Dishart D, Doucet J, Frendewey D, Samuelsson B, Radmark O. Ribonuclease activity and RNA binding of recombinant human Dicer. EMBO J. 2002;21(21):5864–74.
33. Zhang H, Kolb FA, Jaskiewicz L, Westhof E, Filipowicz W. Single processing center models for human Dicer and bacterial RNase III. Cell. 2004;118(1):57.
34. MacRae IJ, Zhou K, Li F, Repic A, Brooks AN, Cande WZ, Adams PD, Doudna JA. Structural basis for double-stranded RNA processing by Dicer. Science. 2006;311(5758):195–8.
35. Park JE, Heo I, Tian Y, Simanshu DK, Chang H, Jee D, Patel DJ, Kim VN. Dicer recognizes the 5′ end of RNA for efficient and accurate processing. Nature. 2011;475(7355):201–5. doi:10.1038/nature10198.
36. Gu S, Jin L, Zhang Y, Huang Y, Zhang F, Valdmanis Paul N, Kay Mark A. The loop position of shRNAs and pre-miRNAs is critical for the accuracy of dicer processing invivo. Cell. 2012;151(4):900–11. doi:10.1016/j.cell.2012.09.042.
37. Chendrimada T, Gregory R, Kumaraswamy E, Norman J, Cooch N, Nishikura K, Shiekhattar R. TRBP recruits the Dicer complex to Ago2 for microRNA processing and gene silencing. Nature. 2005;436(7051):740–4. doi:10.1038/nature03868.
38. Haase AD, Jaskiewicz L, Zhang H, Lainé S, Sack R, Gatignol A, Filipowicz W. TRBP, a regulator of cellular PKR and HIV-1 virus expression, interacts with dicer and functions in RNA silencing. EMBO Rep. 2005;6(10):961–7. doi:10.1038/sj.embor.7400509.
39. Förstemann K, Tomari Y, Du T, Vagin VV, Denli AM, Bratu DP, Klattenhoff C, Theurkauf WE, Zamore PD. Normal microRNA maturation and germ-line stem cell maintenance requires Loquacious, a double-stranded RNA-binding domain protein. PLoS Biol. 2005;3(7):e236. doi:10.1371/journal.pbio.0030236.
40. Saito K, Ishizuka A, Siomi H, Siomi MC. Processing of pre-microRNAs by the Dicer-1-Loquacious complex in Drosophila cells. PLoS Biol. 2005;3(7):e235.
41. Jiang F, Ye X, Liu X, Fincher L, McKearin D, Liu Q. Dicer-1 and R3D1-L catalyze microRNA maturation in Drosophila. Genes Dev. 2005;19(14):1674–9.
42. Olena AF, Patton JG. Genomic organization of microRNAs. J Cell Physiol. 2009;222(3):540–5. doi:10.1002/jcp.21993.

43. Mourelatos Z, Dostie J, Paushkin S, Sharma A, Charroux B, Abel L, Rappsilber J, Mann M, Dreyfuss G. miRNPs: a novel class of ribonucleoproteins containing numerous microRNAs. Genes Dev. 2002;16(6):720–8.
44. Lee Y, Jeon K, Lee J-T, Kim S, Kim VN. MicroRNA maturation: stepwise processing and subcellular localization. EMBO J. 2002;21(17):4663–70.
45. Lee Y, Ahn C, Han J, Choi H, Kim J, Yim J, Lee J, Provost P, Rådmark O, Kim S, Kim VN. The nuclear RNase III Drosha initiates microRNA processing. Nature. 2003;425(6956):415–9. doi:10.1038/nature01957.
46. Han J, Lee Y, Yeom KH, Kim YK, Jin H, Kim VN. The Drosha-DGCR8 complex in primary microRNA processing. Genes Dev. 2004;18(24):3016–27.
47. Denli AM, Tops BBJ, Plasterk RHA, Ketting RF, Hannon GJ. Processing of primary microRNAs by the microprocessor complex. Nature. 2004;432(7014):231–5. doi:10.1038/nature03049.
48. Gregory RI, Yan K-P, Amuthan G, Chendrimada T, Doratotaj B, Cooch N, Shiekhattar R. The Microprocessor complex mediates the genesis of microRNAs. Nature. 2004;432(7014):235–40. doi:10.1038/nature03120.
49. Landthaler M, Yalcin A, Tuschl T. The human DiGeorge syndrome critical region gene 8 and Its D. melanogaster homolog are required for miRNA biogenesis. Curr Biol. 2004;14(23):2162–7.
50. Kim Y-K, Kim VN. Processing of intronic microRNAs. EMBO J. 2007;26(3):775–83. doi:10.1038/sj.emboj.7601512.
51. Morlando M, Ballarino M, Gromak N, Pagano F, Bozzoni I, Proudfoot NJ. Primary microRNA transcripts are processed co-transcriptionally. Nat Struct Mol Biol. 2008;15(9):902–9.
52. Pawlicki JM, Steitz JA. Primary microRNA transcript retention at sites of transcription leads to enhanced microRNA production. J Cell Biol. 2008;182(1):61–76. doi:10.1083/jcb.200803111.
53. Zeng Y, Yi R, Cullen BR. Recognition and cleavage of primary microRNA precursors by the nuclear processing enzyme Drosha. EMBO J. 2005;24(1):138–48. doi:10.1038/sj.emboj.7600491.
54. Zeng Y, Cullen BR. Efficient processing of primary microRNA hairpins by Drosha requires flanking nonstructured RNA sequences. J Biol Chem. 2005;280(30):27595–60.
55. Han J, Lee Y, Yeom K-H, Nam J-W, Heo I, Rhee J-K, Sohn SY, Cho Y, Zhang B-T, Kim VN. Molecular basis for the recognition of primary microRNAs by the Drosha-DGCR8 complex. Cell. 2006;125(5):887–901. doi:10.1016/j.cell.2006.03.043.
56. Yi R, Qin Y, Macara IG, Cullen BR. Exportin-5 mediates the nuclear export of pre-microRNAs and short hairpin RNAs. Genes Dev. 2003;17(24):3011–6.
57. Bohnsack MT, Czaplinski K, Gorlich D. Exportin 5 is a RanGTP-dependent dsRNA-binding protein that mediates nuclear export of pre-miRNAs. RNA. 2004;10(2):185–91.
58. Lund E, Güttinger S, Calado A, Dahlberg JE, Kutay U. Nuclear export of microRNA precursors. Science. 2004;303(5654):95–8. doi:10.1126/science.1090599.
59. Lee Y, Kim M, Han J, Yeom K-H, Lee S, Baek SH, Kim VN. MicroRNA genes are transcribed by RNA polymerase II. EMBO J. 2004;23(20):4051–60. doi:10.1038/sj.emboj.7600385.
60. Cai X, Hagedorn CH, Cullen BR. Human microRNAs are processed from capped, polyadenylated transcripts that can also function as mRNAs. Rna. 2004;10(12):1957–66.
61. Yang J-S, Lai EC. Alternative miRNA biogenesis pathways and the interpretation of core miRNA pathway mutants. Mol Cell. 2011;43(6):892–903. doi:10.1016/j.molcel.2011.07.024.
62. Thatcher EJ, Bond J, Paydar I, Patton JG. Genomic organization of zebrafish microRNAs. BMC Genomics. 2008;9:253. doi:10.1186/1471–2164-9–253.
63. Griffiths-Jones S, Saini HK, van Dongen S, Enright AJ. miRBase: tools for microRNA genomics. Nucleic Acids Res. 2008;36:D154–8 (Database issue).
64. Okamura K, Hagen JW, Duan H, Tyler DM, Lai EC. The mirtron pathway generates microRNA-class regulatory RNAs in Drosophila. Cell. 2007;130(1):89–100. doi:10.1016/j.cell.2007.06.028.

65. Ruby JG, Jan CH, Bartel DP. Intronic microRNA precursors that bypass Drosha processing. Nature. 2007;448(7149):83–6. doi:10.1038/nature05983.
66. Berezikov E, Chung WJ, Willis J, Cuppen E, Lai EC. Mammalian mirtron genes. Mol Cell. 2007;28(2):328–36. doi:10.1016/j.molcel.2007.09.028.
67. Babiarz JE, Ruby JG, Wang Y, Bartel DP, Blelloch R. Mouse ES cells express endogenous shRNAs, siRNAs, and other microprocessor-independent, Dicer-dependent small RNAs. Genes Dev. 2008;22(20):2773–85. doi:10.1101/gad.1705308.
68. Chong MM, Zhang G, Cheloufi S, Neubert TA, Hannon GJ, Littman DR. Canonical and alternate functions of the microRNA biogenesis machinery. Genes Dev. 2010;24(17):1951–60. doi:10.1101/gad.1953310.
69. Glazov EA, Cottee PA, Barris WC, Moore RJ, Dalrymple BP, Tizard ML. A microRNA catalog of the developing chicken embryo identified by a deep sequencing approach. Genome Res. 2008;18(6):957–64.
70. Flynt AS, Greimann J, Chung W, Lima C, Lai EC. MicroRNA biogenesis via splicing and exosome-mediated trimming in Drosophila. Mol Cell. 2010;38(6):900–7.
71. Chung WJ, Okamura K, Martin R, Lai EC. Endogenous RNA interference provides a somatic defense against Drosophila transposons. Curr Biol. 2008;18(11):795–802. doi:10.1016/j.cub.2008.05.006.
72. Cheloufi S, Dos Santos C, Chong M, Hannon G. A dicer-independent miRNA biogenesis pathway that requires Ago catalysis. Nature. 2010;465(7298):584–9.
73. Cifuentes D, Xue H, Taylor DW, Patnode H, Mishima Y, Cheloufi S, Ma E, Mane S, Hannon GJ, Lawson ND, Wolfe SA, Giraldez AJ. A novel miRNA processing pathway independent of Dicer requires Argonaute2 catalytic activity. Science. 2010;328(5986):1694–8. doi:10.1126/science.1190809.
74. Yang JS, Maurin T, Robine N, Rasmussen KD, Jeffrey KL, Chandwani R, Papapetrou EP, Sadelain M, O'Carroll D, Lai EC. Conserved vertebrate mir-451 provides a platform for Dicer-independent, Ago2-mediated microRNA biogenesis. Proc Natl Acad Sci U S A. 2010;107(34):15163–8. doi:10.1073/pnas.1006432107.
75. Corcoran DL, Pandit KV, Gordon B, Bhattacharjee A, Kaminski N, Benos PV. Features of mammalian microRNA promoters emerge from polymerase II chromatin immunoprecipitation data. PLoS One. 2009;4(4):e5279.
76. Ozsolak F, Poling LL, Wang Z, Liu H, Liu XS, Roeder RG, Zhang X, Song JS, Fisher DE. Chromatin structure analyses identify miRNA promoters. Genes Dev. 2008;22(22):3172–83. doi:10.1101/gad.1706508.
77. Wang Y, Li X, Hu H. Transcriptional regulation of co-expressed microRNA target genes. Genomics. 2011;98(6):445–52. doi:10.1016/j.ygeno.2011.09.004.
78. Wu H, Sun S, Tu K, Gao Y, Xie B, Krainer AR, Zhu J. A splicing-independent function of SF2/ASF in microRNA processing. Mol Cell. 2010;38(1):67–77. doi:10.1016/j.molcel.2010.02.021.
79. Viswanathan SR, Daley GQ, Gregory RI. Selective blockade of microRNA processing by Lin28. Science. 2008;320(5872):97–100. doi:10.1126/science.1154040.
80. Viswanathan SR, Powers JT, Einhorn W, Hoshida Y, Ng TL, Toffanin S, O'Sullivan M, Lu J, Phillips LA, Lockhart VL, Shah SP, Tanwar PS, Mermel CH, Beroukhim R, Azam M, Teixeira J, Meyerson M, Hughes TP, Llovet JM, Radich J, Mullighan CG, Golub TR, Sorensen PH, Daley GQ. Lin28 promotes transformation and is associated with advanced human malignancies. Nat Genet. 2009;41(7):843–8. doi:10.1038/ng.392.
81. Heo I, Joo C, Cho J, Ha M, Han J, Kim VN. Lin28 mediates the terminal uridylation of let-7 precursor MicroRNA. Mol Cell. 2008;32(2):276–84. doi:10.1016/j.molcel.2008.09.014.
82. Hagan JP, Piskounova E, Gregory RI. Lin28 recruits the TUTase Zcchc11 to inhibit let-7 maturation in mouse embryonic stem cells. Nat Struct Mol Biol. 2009;16(10):1021–5. doi:10.1038/nsmb.1676.
83. Jones MR, Quinton LJ, Blahna MT, Neilson JR, Fu S, Ivanov AR, Wolf DA, Mizgerd JP. Zcchc11-dependent uridylation of microRNA directs cytokine expression. Nat Cell Biol. 2009;11(9):1157–63. doi:10.1038/ncb1931.

84. Heo I, Joo C, Kim Y-K, Ha M, Yoon M-J, Cho J, Yeom K-H, Han J, Kim VN. TUT4 in concert with Lin28 suppresses microRNA biogenesis through pre-microRNA uridylation. Cell. 2009;138(4):696–708. doi:10.1016/j.cell.2009.08.002.
85. Newman M, Thomson J, Hammond S. Lin-28 interaction with the Let-7 precursor loop mediates regulated microRNA processing. RNA. 2008;14(8):1539–49. doi:10.1261/rna.1155108.
86. Viswanathan SR, Daley GQ. Lin28: a microRNA regulator with a macro role. Cell. 2010;140(4):445–9. doi:10.1016/j.cell.2010.02.007.
87. Forman JJ, Legesse-Miller A, Coller HA. A search for conserved sequences in coding regions reveals that the let-7 microRNA targets Dicer within its coding sequence. Proc Natl Acad Sci U S A 2008;105(39):14879–84. doi:10.1073/pnas.0803230105.
88. Kawahara Y, Zinshteyn B, Chendrimada TP, Shiekhattar R, Nishikura K. RNA editing of the microRNA-151 precursor blocks cleavage by the Dicer-TRBP complex. EMBO Rep. 2007;8(8):763–9. doi:10.1038/sj.embor.7401011.
89. Kawahara Y, Megraw M, Kreider E, Iizasa H, Valente L, Hatzigeorgiou AG, Nishikura K. Frequency and fate of microRNA editing in human brain. Nucleic Acids Res. 2008;36(16):5270–80. doi:10.1093/nar/gkn479.
90. Davis BN, Hata A. Regulation of MicroRNA biogenesis: a miRiad of mechanisms. Cell Commun Signal. 2009;7:18. doi:10.1186/1478-811X-7-18.
91. Krol J, Loedige I, Filipowicz W. The widespread regulation of microRNA biogenesis, function and decay. Nat Rev Genet. 2010;11(9):597–610.
92. Tuschl T, Zamore PD, Lehmann R, Bartel DP, Sharp PA. Targeted mRNA degradation by double-stranded RNA in vitro. Genes Dev. 1999;13(24):3191–7.
93. Elbashir SM, Lendeckel W, Tuschl T. RNA interference is mediated by 21- and 22-nucleotide RNAs. Genes Dev. 2001;15(2):188–200.
94. Elbashir SM, Harborth J, Lendeckel W, Yalcin A, Weber K, Tuschl T. Duplexes of 21-nucleotide RNAs mediate RNA interference in cultured mammalian cells. Nature. 2001;411(6836):494–8.
95. Hammond S, Boettcher, S., Caudy, A., Kobayashi, R., Hannon, G. Argonaute2, a link between genetic and biochemical analyses of RNAi. Science. 2001;293:1146–50.
96. Cerutti L, Mian N, Bateman A. Domains in gene silencing and cell differentiation proteins: the novel PAZ domain and redefinition of the Piwi domain. Trends Biochem Sci. 2000;25(10):481–2.
97. Yan KS, Yan, S., Farooq A, Han A, Zeng L, Zhou M. Stucture and conserved RNA binding of the PAZ domain. Nature. 2003;426:469–74.
98. Lingel A, Simon B, Izaurralde E, Sattler M. Structure and nucleic-acid binding of the Drosophila Argonaute 2 PAZ domain. Nature. 2003;426(6965):465–9. doi:10.1038/nature02123.
99. Song J, Liu J, Tolia N, Schneiderman J, Smith SK, Martienssen RA, Hannon G, Joshua-Tor L. The crystal structure of the Argonaute2 PAZ domain reveals an RNA binding motif in RNAi effector complexes. Nat Struct Biol. 2003;10(12):1026–32. doi:10.1038/nsb1016.
100. Lingel A, Simon B, Izaurralde E, Sattler M. Nucleic acid 3′-end recognition by the Argonaute2 PAZ domain. Nat Struct Mol Biol. 2004;11(6):576–7. doi:10.1038/nsmb777.
101. Ma JB, Ye K, Patel DJ. Structural basis for overhang-specific small interfering RNA recognition by the PAZ domain. Nature. 2004;429(6989):318–22. doi:10.1038/nature02519.
102. Song J, Smith SK, Hannon G, Joshua-Tor L. Crystal structure of Argonaute and its implications for RISC slicer activity. Science. 2004;305(5689):1434–7. doi:10.1126/science.1102514.
103. Liu J. Argonaute2 is the catalytic engine of mammalian RNAi. Science. 2004;305(5689):1437–41. doi:10.1126/science.1102513.
104. Hutvágner G, Zamore PD. A microRNA in a multiple-turnover RNAi enzyme complex. Science. 2002;297(5589):2056–60. doi:10.1126/science.1073827.
105. Doench JG, Petersen CP, Sharp PA. siRNAs can function as miRNAs. Genes Dev. 2003;17(4):438–42.

106. Saxena S, Jonsson ZO, Dutta A. Small RNAs with imperfect match to endogenous mRNA repress translation. Implications for off-target activity of small inhibitory RNA in mammalian cells. J Biol Chem. 2003;278(45):44312–9. doi:10.1074/jbc.M307089200.
107. Zeng Y, Yi R, Cullen BR. MicroRNAs and small interfering RNAs can inhibit mRNA expression by similar mechanisms. Proc Natl Acad Sci U S A. 2003;100(17):9779–84.
108. Pillai RS, Artus CG, Filipowicz W. Tethering of human Ago proteins to mRNA mimics the miRNA-mediated repression of protein synthesis. RNA. 2004;10(10):1518–25.
109. Gregory RI, Chendrimada TP, Cooch N, Shiekhattar R. Human RISC couples microRNA biogenesis and posttranscriptional gene silencing. Cell. 2005;123(4):631–40.
110. Maniataki E, Mourelatos Z. A human, ATP-independent, RISC assembly machine fueled by pre-miRNA. Genes Dev. 2005;19(24):2979–90. doi:10.1101/gad.1384005.
111. MacRae IJ, Ma E, Zhou M, Robinson CV, Doudna JA. In vitro reconstitution of the human RISC-loading complex. Proc Natl Acad Sci U S A. 2008;105(2):512–7. doi:10.1073/pnas.0710869105.
112. Sen GL, Blau HM. Argonaute 2/RISC resides in sites of mammalian mRNA decay known as cytoplasmic bodies. Nat Cell Biol. 2005;7(6):633.
113. Jakymiw A, Lian S, Eystathioy T, Li S, Satoh M, Hamel JC, Fritzler MJ, Chan EK. Disruption of GW bodies impairs mammalian RNA interference. Nat Cell Biol. 2005;7(12):1267–74.
114. Rehwinkel JAN, Behm-Ansmant I, Gatfield D, Izaurralde E. A crucial role for GW182 and the DCP1:DCP2 decapping complex in miRNA-mediated gene silencing. RNA. 2005;11(11):1640–7.
115. Liu J, Rivas FV, Wohlschlegel J, Yates JR 3rd, Parker R, Hannon GJ. A role for the P-body component GW182 in microRNA function. Nat Cell Biol. 2005;7(12):1261–6.
116. Ding L, Spencer A, Morita K, Han M. The developmental timing regulator AIN-1 interacts with miRISCs and may target the argonaute protein ALG-1 to cytoplasmic P bodies in C. elegans. Mol Cell. 2005;19(4):437–47. doi:10.1016/j.molcel.2005.07.013.
117. Eulalio A, Huntzinger E, Izaurralde E. GW182 interaction with Argonaute is essential for miRNA-mediated translational repression and mRNA decay. Nat Struct Mol Biol. 2008;15(4):346–53. doi:10.1038/nsmb.1405.
118. Meister G, Landthaler M, Peters L, Chen PY, Urlaub H, Luhrmann R, Tuschl T. Identification of novel argonaute-associated proteins. Curr Biol. 2005;15(23):2149–55. doi:10.1016/j.cub.2005.10.048.
119. Martinez J, Patkaniowska A, Urlaub H, Lührmann R, Tuschl T. Single-stranded antisense siRNAs guide target RNA cleavage in RNAi. Cell. 2002;110(5):563–74.
120. Schwarz DS, Hutvágner G, Du T, Xu Z, Aronin N, Zamore PD. Asymmetry in the assembly of the RNAi enzyme complex. Cell. 2003;115(2):199–208.
121. Khvorova A, Reynolds A, Jayasena SD. Functional siRNAs and miRNAs exhibit strand bias. Cell. 2003;115(2):209–16.
122. Vella MC, Choi E-Y, Lin S-Y, Reinert K, Slack FJ. The C. elegans microRNA let-7 binds to imperfect let-7 complementary sites from the lin-41 3′UTR. Genes Dev. 2004;18(2):132–7. doi:10.1101/gad.1165404.
123. Enright AJ, John B, Gaul U, Tuschl T, Sander C, Marks DS. MicroRNA targets in Drosophila. Genome Biol. 2003;5(1):R1.
124. Lewis BP, Shih I-h, Jones-Rhoades MW, Bartel DP, Burge CB. Prediction of mammalian microRNA targets. Cell. 2003;115(7):787–98.
125. Lewis BP, Burge CB, Bartel DP. Conserved seed pairing, often flanked by adenosines, indicates that thousands of human genes are microRNA targets. Cell. 2005;120(1):15–20.
126. Doench JG, Sharp PA. Specificity of microRNA target selection in translational repression. Genes Dev. 2004;18(5):504–11.
127. Kiriakidou M, Nelson PT, Kouranov A, Fitziev P, Bouyioukos C, Mourelatos Z, Hatzigeorgiou A. A combined computational-experimental approach predicts human microRNA targets. Genes Dev. 2004;18(10):1165–78.

128. Brennecke J, Stark A, Russell RB, Cohen SM. Principles of microRNA-target recognition. PLoS Biol. 2005;3(3):e85. doi:10.1371/journal.pbio.0030085.
129. Grimson A, Farh KK-H, Johnston WK, Garrett-Engele P, Lim LP, Bartel DP. MicroRNA targeting specificity in mammals: determinants beyond seed pairing. Mol Cell. 2007;27(1):91–105. doi:10.1016/j.molcel.2007.06.017.
130. Saetrom P, Heale BS, Snove O Jr., Aagaard L, Alluin J, Rossi JJ. Distance constraints between microRNA target sites dictate efficacy and cooperativity. Nucleic Acids Res. 2007;35(7):2333–42. doi:10.1093/nar/gkm133.
131. Gaidatzis D, van Nimwegen E, Hausser J, Zavolan M. Inference of miRNA targets using evolutionary conservation and pathway analysis. BMC Bioinformatics. 2007;8:69.
132. Nielsen CB, Shomron N, Sandberg R, Hornstein E, Kitzman J, Burge CB. Determinants of targeting by endogenous and exogenous microRNAs and siRNAs. Rna. 2007;13(11):1894–910.
133. Li N, Flynt A, Kim HR, Solnica-Krezel L, Patton J. Dispatched Homolog 2 is targeted by miR-214 through a combination of three weak microRNA recognition sites. Nucleic Acids Res. 2008;36(13):4277–85. doi:10.1093/nar/gkn388.
134. Shin C, Nam J-W, Farh KK-H, Chiang HR, Shkumatava A, Bartel DP. Expanding the microRNA targeting code: functional sites with centered pairing. Mol Cell. 2010;38(6):789–802. doi:10.1016/j.molcel.2010.06.005.
135. Chi SW, Hannon GJ, Darnell RB. An alternative mode of microRNA target recognition. Nat Struct Mol Biol. 2012;19(3):321–7. doi:10.1038/nsmb.2230.
136. Loeb Gabriel B, Khan Aly A, Canner D, Hiatt Joseph B, Shendure J, Darnell Robert B, Leslie Christina S, Rudensky Alexander Y. Transcriptome-wide miR-155 binding map reveals widespread noncanonical microRNA targeting. Mol Cell. 2012;48(5):760–70. doi:10.1016/j.molcel.2012.10.002.
137. Yekta S, Shih I-h, Bartel DP. MicroRNA-directed cleavage of HOXB8 mRNA. Science. 2004;304(5670):594–6. doi:10.1126/science.1097434.
138. Olsen PH, Ambros V. The lin-4 regulatory RNA controls developmental timing in Caenorhabditis elegans by blocking LIN-14 protein synthesis after the initiation of translation. Dev Biol. 1999;216(2):671–80. doi:10.1006/dbio.1999.9523.
139. Seggerson K, Tang L, Moss EG. Two genetic circuits repress the Caenorhabditis elegans heterochronic gene lin-28 after translation initiation. Dev Biol. 2002;243(2):215–25.
140. Petersen CP, Bordeleau M-E, Pelletier J, Sharp PA. Short RNAs repress translation after initiation in mammalian cells. Mol Cell. 2006;21(4):533–42. doi:10.1016/j.molcel.2006.01.031.
141. Maroney PA, Yu Y, Fisher J, Nilsen TW. Evidence that microRNAs are associated with translating messenger RNAs in human cells. Nat Struct Mol Biol. 2006;13(12):1102–7.
142. Nottrott S, Simard MJ, Richter JD. Human let-7a miRNA blocks protein production on actively translating polyribosomes. Nat Struct Mol Biol. 2006;13(12):1108–14.
143. Pillai RS, Bhattacharyya SN, Artus CG, Zoller T, Cougot N, Basyuk E, Bertrand E, Filipowicz W. Inhibition of translational initiation by Let-7 MicroRNA in human cells. Science. 2005;309(5740):1573–6.
144. Humphreys DT, Westman BJ, Martin DIK, Preiss T. MicroRNAs control translation initiation by inhibiting eukaryotic initiation factor 4E/cap and poly(A) tail function. Proc Natl Acad Sci U S A. 2005;102(47):16961–6. doi:10.1073/pnas.0506482102.
145. Wang B, Love TM, Call ME, Doench JG, Novina CD. Recapitulation of short RNA-directed translational gene silencing in vitro. Mol Cell. 2006;22(4):553–60.
146. Thermann R, Hentze MW. Drosophila miR2 induces pseudo-polysomes and inhibits translation initiation. Nature. 2007;447(7146):875–8. doi:10.1038/nature05878.
147. Mathonnet G, Fabian MR, Svitkin YV, Parsyan A, Huck L, Murata T, Biffo S, Merrick WC, Darzynkiewicz E, Pillai RS, Filipowicz W, Duchaine TF, Sonenberg N. MicroRNA inhibition of translation initiation in vitro by targeting the cap-binding complex eIF4F. Science. 2007;317(5845):1764–7. doi:10.1126/science.1146067.

148. Wakiyama M, Takimoto K, Ohara O, Yokoyama S. Let-7 microRNA-mediated mRNA deadenylation and translational repression in a mammalian cell-free system. Genes Dev. 2007;21(15):1857–62. doi:10.1101/gad.1566707.
149. Bazzini AA, Lee MT, Giraldez AJ. Ribosome profiling shows that miR-430 reduces translation before causing mRNA decay in zebrafish. Science. 2012;336(6078):233–7. doi:10.1126/science.1215704.
150. Kiriakidou M, Tan GS, Lamprinaki S, De Planell-Saguer M, Nelson PT, Mourelatos Z. An mRNA m7G cap binding-like motif within human Ago2 represses translation. Cell. 2007;129(6):1141–51.
151. Kinch LN, Grishin NV. The human Ago2 MC region does not contain an eIF4E-like mRNA cap binding motif. Biol Direct. 2009;4:2. doi:10.1186/1745–6150-4–2.
152. Djuranovic S, Zinchenko MK, Hur JK, Nahvi A, Brunelle JL, Rogers EJ, Green R. Allosteric regulation of Argonaute proteins by miRNAs. Nat Struct Mol Biol. 2010;17(2):144–50. doi:10.1038/nsmb.1736.
153. Chendrimada TP, Finn KJ, Ji X, Baillat D, Gregory RI, Liebhaber SA, Pasquinelli AE, Shiekhattar R. MicroRNA silencing through RISC recruitment of eIF6. Nature. 2007;447(7146):823–8. doi:10.1038/nature05841.
154. Walters RW, Bradrick SS, Gromeier M. Poly(A)-binding protein modulates mRNA susceptibility to cap-dependent miRNA-mediated repression. RNA. 2010;16(1):239–50. doi:10.1261/rna.1795410.
155. Gu S, Kay MA. How do miRNAs mediate translational repression? Silence. 2010;1(1):11. doi:10.1186/1758–907X-1–11.
156. Lim LP, Lau NC, Garrett-Engele P, Grimson A, Schelter JM, Castle J, Bartel DP, Linsley PS, Johnson JM. Microarray analysis shows that some microRNAs downregulate large numbers of target mRNAs. Nature. 2005;433(7027):769–73.
157. Bagga S, Bracht J, Hunter S, Massirer K, Holtz J, Eachus R, Pasquinelli AE. Regulation by let-7 and lin-4 miRNAs results in target mRNA degradation. Cell. 2005;122(4):553–63.
158. Giraldez AJ, Mishima Y, Rihel J, Grocock RJ, van Dongen S, Inoue K, Enright AJ, Schier AF. Zebrafish MiR-430 promotes deadenylation and clearance of maternal mRNAs. Science. 2006;312(5770):75–9. doi:10.1126/science.1122689.
159. Behm-Ansmant I, Rehwinkel J, Doerks T, Stark A, Bork P, Izaurralde E. mRNA degradation by miRNAs and GW182 requires both CCR4:NOT deadenylase and DCP1:DCP2 decapping complexes. Genes Dev. 2006;20(14):1885–98. doi:10.1101/gad.1424106.
160. Fabian MR, Cieplak MK, Frank F, Morita M, Green J, Srikumar T, Nagar B, Yamamoto T, Raught B, Duchaine TF, Sonenberg N. miRNA-mediated deadenylation is orchestrated by GW182 through two conserved motifs that interact with CCR4-NOT. Nat Struct Mol Biol. 2011;18(11):1211–7. doi:10.1038/nsmb.2149.
161. Braun JE, Huntzinger E, Fauser M, Izaurralde E. GW182 proteins directly recruit cytoplasmic deadenylase complexes to miRNA targets. Mol Cell. 2011;44(1):120–33. doi:10.1016/j.molcel.2011.09.007.
162. Chekulaeva M, Mathys H, Zipprich JT, Attig J, Colic M, Parker R, Filipowicz W. miRNA repression involves GW182-mediated recruitment of CCR4-NOT through conserved W-containing motifs. Nat Struct Mol Biol. 2011;18(11):1218–26. doi:10.1038/nsmb.2166.
163. Wu L, Fan J, Belasco JG. MicroRNAs direct rapid deadenylation of mRNA. Proc Natl Acad Sci U S A. 2006;103(11):4034–9. doi:10.1073/pnas.0510928103.
164. Eulalio A, Huntzinger E, Nishihara T, Rehwinkel J, Fauser M, Izaurralde E. Deadenylation is a widespread effect of miRNA regulation. RNA. 2009;15(1):21–32. doi:10.1261/rna.1399509.
165. Baek D, Villén J, Shin C, Camargo F, Gygi S, Bartel D. The impact of microRNAs on protein output. Nature. 2008;455(7209):64–71. doi:10.1038/nature07242.
166. Guo H, Ingolia NT, Weissman JS, Bartel DP. Mammalian microRNAs predominantly act to decrease target mRNA levels. Nature. 2010;466(7308):835–40. doi:10.1038/nature09267.

167. Fabian MR, Sonenberg N, Filipowicz W. Regulation of mRNA translation and stability by microRNAs. Annu Rev Biochem. 2010;79:351–79. doi:10.1146/annurev-biochem-060308–103103.
168. Huntzinger E, Izaurralde E. Gene silencing by microRNAs: contributions of translational repression and mRNA decay. Nat Reviews Genet. 2011;12(2):99–110. doi:10.1038/nrg2936.
169. Hock J, Weinmann L, Ender C, Rudel S, Kremmer E, Raabe M, Urlaub H, Meister G. Proteomic and functional analysis of Argonaute-containing mRNA-protein complexes in human cells. EMBO Rep. 2007;8(11):1052–60. doi:10.1038/sj.embor.7401088.
170. Adams BD, Claffey KP, White BA. Argonaute-2 expression is regulated by epidermal growth factor receptor and mitogen-activated protein kinase signaling and correlates with a transformed phenotype in breast cancer cells. Endocrinology. 2009;150(1):14–23. doi:10.1210/en.2008–0984.
171. Kawahara Y, Zinshteyn B, Sethupathy P, Iizasa H, Hatzigeorgiou AG, Nishikura K. Redirection of silencing targets by adenosine-to-inosine editing of miRNAs. Science. 2007;315(5815):1137–40. doi:10.1126/science.1138050.
172. Thomson T, Lin H. The biogenesis and function of PIWI proteins and piRNAs: progress and prospect. Annu Rev Cell Dev Biol. 2009;25:355–76. doi:10.1146/annurev.cellbio.24.110707.175327.
173. Mercer TR, Dinger ME, Mattick JS. Long non-coding RNAs: insights into functions. Nat Rev Genet. 2009;10(3):155–9. doi:10.1038/nrg2521.
174. Thatcher EJ, Paydar I, Anderson KK, Patton JG. Regulation of zebrafish fin regeneration by microRNAs. Proc Natl Acad Sci U S A. 2008;105(47):18384–9. doi:10.1073/pnas.0803713105.

Chapter 2
MicroRNAs in Cancer Progression

Omozusi Andrews and James G. Patton

Abstract Tumor initiation and progression have been widely investigated and ongoing work implicates microRNAs (miRNAs) as central players. miRNAs can control proliferation and differentiation as well as apoptosis, consistent with miRNAs functioning as oncogenes or tumor suppressors. Specific miRNAs have been shown to play roles in the overlapping fields of cancer stem cells and chemoresistance. In this review, we summarize existing data to elucidate the role of miRNAs in tumorigenesis and potential strategies using miRNAs for cancer therapy or as biomarkers.

Keywords MicroRNA · Cancer · Anchorage independence · Angiogenesis · Antagomir · Biomarker · Chemotherapy · epithelial-to-mesenchymal transition (EMT) · Lymphoma · Metastasis · Oncogenes · Tumor suppressors

1 Cancer Overview

Cancer consists of numerous diseases characterized by misregulated and/or altered cell division. Most commonly, changes in cell cycle control arise from accumulated mutations leading to chromosomal instability, proliferation, and aggressive metastatic behavior. While some cancers harbor hereditary mutations such as colorectal cancers (Lynch syndrome with mutations in mismatch repair complex proteins) or breast cancers (mutation in BRCA1 and BRCA2 genes), the majority are not hereditary and instead arise from random somatic mutations (http://www.cancer.gov/cancertopics/understandingcancer/genetesting/page27). These mutations typically activate downstream pathways and share specific hallmarks as identified by Hanahan and Weinberg [1]. These hallmarks are defined as acquired functional capabilities that allow cancer cells to survive, proliferate, and disseminate. These shared characteristics include self-sufficiency in growth signals, insensitivity to antigrowth signals, the ability to evade apoptosis, limitless replicative potential, sustained angiogenesis, tissue invasion, and metastasis. Broadly, there are two defining features of cancers: upregulation of oncogenes and downregulation of tumor suppressor

J. G. Patton (✉) · O. Andrews
Department of Biological Sciences, Vanderbilt University, Nashville, TN, USA
e-mail: James.G.Patton@Vanderbilt.edu

S. R. Singh, P. Rameshwar (eds.), *MicroRNA in Development and in the Progression of Cancer*, DOI 10.1007/978-1-4899-8065-6_2,

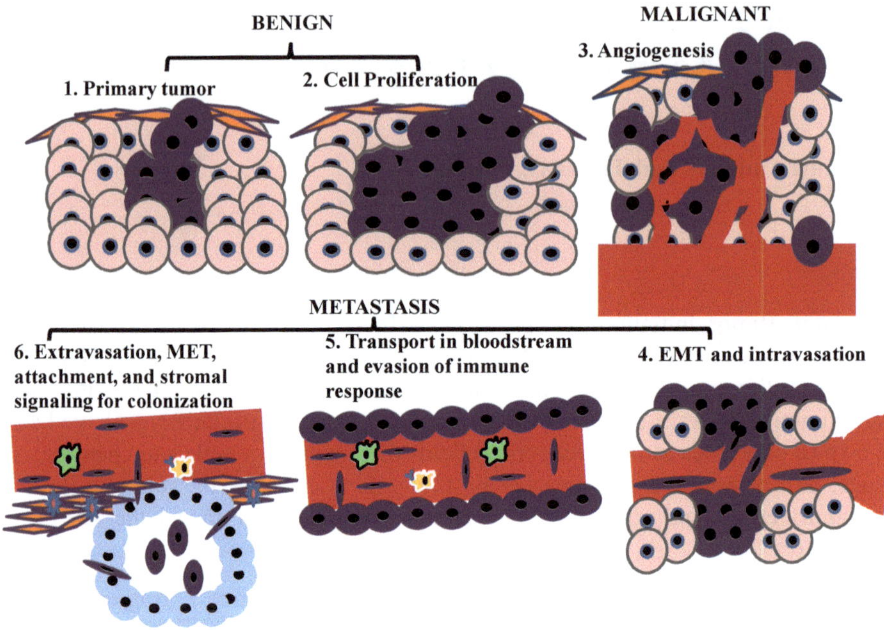

Fig. 2.1 Cancer progression. During the benign stage, cells acquire multiple mutations and form the primary tumor. With increasing tumor growth, angiogenesis is initiated and cells locally invade during the malignant phase. Aggressive metastasis results when tumor cells undergo epithelial-to-mesenchymal transition (EMT), enter blood vessels, and colonize secondary sites through cross talk in a tumor-permissive microenvironment

genes. However, ongoing research has revealed that induction of stromal changes and evasion of host defense mechanisms are critical for tumor progression. Tumor cells aggressively outcompete their neighbors, activate angiogenesis, detach from their primary location, intravasate into the bloodstream, survive immune clearance, extravasate out of blood vessels, and metastasize to distant sites (Fig. 2.1). Currently, additional hallmarks have emerged including reprogramming of energy metabolism and the concept of the tumor microenvironment [2].

Due to the multiple genetic events that occur in succession and accumulate over time, development of cancer typically manifests itself in older adults. According to the World Health Organization, cancer accounted for 7.6 million deaths in 2008 and 70 % of all cancer deaths occurred in low-income and middle-income countries (http://www.who.int/mediacentre/factsheets/fs297/en/index.html). Worldwide studies have attributed 30 % of cancer deaths to five behavioral and dietary risks including high body mass index, low fruit and vegetable intake, lack of physical activity, tobacco use, and alcohol use (http://www.who.int/mediacentre/factsheets/fs297/en/index.html). According to the National Cancer Institute, it is estimated that 1,638,910 men and women will be diagnosed with cancer in 2012 in the USA (American Cancer Society: Cancer Facts and Figures 2012). The median age at death from cancers of all types was 72 years old from 2005 to 2009 (American Cancer Society: Cancer Facts and Figures 2012) with most fatalities due to secondary metastases [3]. However, cancer is an equal opportunity disease, also affecting young children. While the majority of tumors seen in adults arise from epithelial cells, that is, solid tumors, children more commonly develop lymphomas and leukemias. Diagnosis is dependent on the tissue of origin, analysis of cell morphology, and the extent of tumor spread, broadly known as staging (http://www.cancer.gov/cancertopics/factsheet/detection/staging). Therapies are dependent on staging but for a more personalized treatment, tumors can be screened and targeted via detection of specific mutations associated with particular tumors. For example, lung cancers with gene amplification or tyrosine kinase mutations, the epidermal growth factor receptor (EGFR) can be treated with erlotinib, a selective inhibitor that blocks kinase activity [4]. While cancer survival has improved due to radiation administration, chemotherapy, and/or specific inhibitors, many cancers unfortunately become drug resistant, leading to recurrence.

A plausible explanation for the adverse response to chemotherapy is the heterogeneity of cancer cells and the adaptability of specific stem cell-like populations within the primary tumor [5]. Growing research on the underlying mechanism of chemotherapeutic resistance has led to a popular model hypothesizing the existence of cancer stem cells (CSCs) [6, 7]. These rare cells are hypothesized to behave similar to stem cells and are capable of self-renewal, asymmetric cell division, and multipotent differentiation. However, the origin of CSCs remains elusive and it has been postulated that CSCs arise from transforming mutations occurring in either multipotent stem cells, tissue-specific stem cells, progenitor cells, mature cells, or cancer cells [7]. Recent evidence indicates that some CSCs are distinct from stem cells and arise from progenitor cells as evidenced by induction of neurofibroma formation by differentiated glial progeny [8]. Since CSCs are rare with yet uncharacterized molecular signatures, further investigation is necessary to understand their role. If cancer initiation involves CSCs exhibiting stem cell-like behavior, then regulators of stem cells and associated pathways may be linked to cancer development including sonic hedgehog signaling in basal cell carcinoma and glioma, Notch signaling in T-cell acute lymphoblastic leukemia/lymphoma, epidermal growth factor (EGF) pathways in human squamous cell lung cancer, and the canonical Wnt pathway in colorectal cancer [9–12]. These same pathways control embryonic development, so it is conceivable that tumor cells reprogram normal cells to turn on genes required during the early proliferative phases of embryonic growth. In this way, tumor cells take advantage of genetic pathways required for normal development but perturb the balance between normal and aberrant signaling. Recently, stem cell regulators and stem cells themselves have been shown to be posttranscriptionally regulated by noncoding RNA molecules during normal animal development [13]. The focus of this chapter is on microRNAs (miRNAs) and their role in tumor progression and prevention.

2 miRNAs

Found in all metazoans, miRNAs are small RNA molecules that negatively regulate gene expression at the posttranscriptional level [14–16]. After processing of primary and precursor transcripts by the RNAse III enzymes Drosha and Dicer,

Table 2.1 Oncogenic miRNAs

Oncogenic miRNAs	Gene loci	Validated mRNA targets	Cancer	Cellular processes
miR-17–92 cluster	Chromosome 13q31–32	p21, Bim, Pten, CTGF, and Tsp1[42]	B Cell lymphomas, lung cancer	Proliferation, apoptosis, angiogenesis
miR-155	Chromosome 21q21	TP53INP1	Burkitt, Hodgkin's, primary mediastinal and diffuse large-B cell lymphoma, breast cancer	Cell cycle, proliferation, apoptosis
miR-372/373	Chromosome 19	LATS2	Testicular cancer	Proliferation
miR-221/222	Chromosome X	p27 Kip1	Glioblastoma multiforme, prostate cancer, thyroid cancer, HCC	Proliferation

respectively, mature miRNAs interact with Argonaute proteins and other components of the RNA-induced silencing complex (RISC) in the cytoplasm. The RISC facilitates binding of miRNAs to their messenger RNA (mRNA) targets. MiRNA–mRNA binding occurs through imperfect base pairing with the 3′untranslated region (3′UTR) of mRNA, leading to translational repression and/or mRNA degradation. Computational predictions suggest that about one-third of all protein-coding genes are regulated by miRNAs in humans [17]. Initially discovered as regulators of animal development, miRNAs have been shown to control multiple steps in tumorigenesis, including proliferation and differentiation [18, 19]. The following sections will elucidate the lessons learned so far about miRNAs that behave as protumorigenic and antitumorigenic contributors to cancer.

3 miRNAs as Oncogenes

The first report of an oncogenic miRNA (Table 2.1) showed that the *miR-17–92* polycistron cluster is highly expressed in human B cell lymphomas [20]. The *miR-17–92* cluster is located within a genomic locus on the chromosome 13q31 that was previously established as a frequently amplified region in cases of diffuse large B cell lymphoma, follicular lymphoma, mantle cell lymphoma, primary cutaneous B cell lymphoma, and other cancer types [21]. The cluster encodes *miR-17–5p, 17–3p, 18, 19a, 20, 19b-1,* and *92–1*. One of the defining characteristics of an oncogene is chromosomal amplification and as a proof of principle, serial transplant assays are typically used to examine the tumor-forming potential of a gene of interest. Using a mouse model of human B cell lymphoma, where the c-Myc oncogene is overexpressed by the immunoglobulin heavy chain enhancer (Eμ). He and colleagues [20] demonstrated that the *miR-17–19b* cluster accelerated tumorigenesis . Transplantation of reconstituted hematopoietic stem cells (derived from Eμ–Myc mice) with *miR-17–19b* expression into irradiated recipient mice led to lymphoma development, invasion of tumor cells into other organs, and decreased survival.

Tumors continued to form after two rounds of serial transplantation, and analysis of the tumor cell population for markers of pre-B and mature B cells (CD19 and immunoglobulin M (IgM), respectively) suggested that overexpression of the cluster favored transformation of B cell progenitors [20]. Another study implicating the *miR-17–92* cluster in lung tumorigenesis used northern blotting and detected increased *miR-17–92* expression, increased copy number, and functionally enhanced lung cancer cell growth through cell proliferation assays [22]. Thus, the *miR-17–92* cluster can accurately be classified as oncogenic.

Further investigation of oncogenic miRNAs, or "oncomiRs" as they are designated, led to the detection of elevated *miR-155* expression in lymphomas derived from B cells of different developmental stages [23]. Expression of *miR-155* (found on chromosome 21q21) was derived from sequences present in the *bic* RNA, which was previously discovered as a target for insertional mutagenesis in avian B cell lymphomas [24]. *bic* cooperates with c-Myc in enhancing the growth and transformation potential of cultured chicken embryo fibroblasts. In pancreatic ductal adenocarcinoma, *miR-155* expression is upregulated and functional assays identified tumor protein 53-induced nuclear protein 1 (TP53INP1) as its target [25]. Decreased TP53INP1 occurred through translational inhibition, providing a link between *miR-155* and a regulator of cell cycle progression and apoptosis. Further evidence for the oncogenic nature of *miR-155* was observed in breast cancer [26]. Overexpression of *miR-155* triggered constitutive activation of growth pathways including the signal transducers and activators of transcription 3 (STAT3) and Janus kinase (JAK) pathways, and its effects were mediated through targeting the tumor suppressor gene, Socs1 [26]. These experiments provided support for the contribution of miRNAs in tumorigenesis by downregulation of genes controlling cell growth and division, a mandatory prerequisite for primary tumor formation.

While initial findings of pro-tumorigenic miRNAs were mostly derived from DNA copy number analyses, few have been identified through targeted approaches, i.e., genetic screens. Voorhoeve and colleagues (2006) designed a library of vectors expressing the majority of cloned human miRNAs and used microarrays to examine expression [27]. With an oncogenic stress model in which human fibroblasts express a constitutively active Ras, they discovered that vectors encoding *miR-372* and *miR-373* conferred a selective growth advantage to cells that would otherwise undergo a stress response, known as oncogene-induced senescence. Their results indicated that these miRNAs act in cooperation with Ras to promote tumorigenesis and also provided evidence implicating both miRNAs in tumors that retained a wild-type (WT) copy of p53 but were nevertheless sensitive to DNA-damaging agents. Analysis of chemosensitive testicular germ cell tumors harboring WT p53 indicated high expression of *miR-372* and *miR-373* in tumors classified as embryonal carcinoma [27]. Also, germ cell lines failed to undergo growth arrest in the presence of a cell cycle inhibitor and *miR-372* and *miR-373*. Using prediction algorithms to identify targets whose 3′UTRs contained putative binding sites for both miRNAs, the serine threonine kinase large tumor suppressor homolog 2 (LATS2) was identified and validated using a luciferase reporter assay. Negative regulation of LATS2 by *miR-372* and *miR-373* occurs through a combination of RNA ablation and inhibition of protein synthesis. Since LATS2 is a tumor suppressor, its loss in mouse embryonic fibroblasts provides a growth advantage [27]. Together, *miR-372* and *miR-373* fit the oncogenic criteria because downregulation of their mRNA targets prevents exit from the cell cycle, resulting in uncontrolled cell growth.

The classification of miRNAs as oncogenes was further substantiated when specific miRNAs were found to be overexpressed within diverse tumor types [19, 28, 29]. One such miRNA that appears to be ubiquitously required for aggressive metastatic potential is *miR-21*. Encoded on chromosome 17q, *miR-21* was originally identified as an oncogene in human glioblastoma cells [30]. Through analysis of RNA isolated from neoplastic and nonneoplastic glioma samples, *miR-21* was found to be upregulated in gliomas. Inhibition of *miR-21* expression in cultured glioblastoma cells caused a marked increase in apoptosis through activation of the caspase machinery [30]. Additional evidence for pro-tumorigenic *miR-21* was illustrated through microarray analyses that showed significant upregulation of *miR-21* in all tumors (breast, colon, lung, pancreas, prostate, and stomach) irrespective of the disease status [31]. Consistent with this, the anti-apoptotic protein Bcl-2 was found to be a direct target of *miR-21* in breast cancer cells and in a xenograft model of breast cancer [32]. Other tumor suppressor targets of *miR-21* have been identified including phosphatase and tensin homolog (PTEN), which inhibits the oncogene PI3K in AKT-mediated cell proliferation [33]. *miR-21* modulates PTEN levels in lung cancer cells resulting in increased cell growth, migration, and invasion [33].

Several reports have referred to the phenomenon of dependence on a single oncogene as "oncogenic addiction" [34–36]. A landmark in miRNA cancer research was the in vivo demonstration of oncogenic addiction in mice conditionally expressing *miR-21* [37]. Using Cre recombinase and the Tet-off system, overexpression of *miR-21* in hematopoietic tissues accelerated pre-B malignant lymphoid-like tumor formation, whereas loss of *miR-21* resulted in regression of tumors. This clearly demonstrates that *miR-21* is a bona fide oncomiR. Overexpression of *miR-21* in different cancer types (Fig. 2.2) illustrates the dysregulation and dependence on common miRNA pathways and common mRNA targets in the acceleration of tumorigenesis [38–41].

Similar to *miR-21,* the *miR-221/222* family encoded on the X chromosome has been implicated in several cancer types after initial detection in glioblastoma multiforme (GBM) [42]. Global miRNA expression profiles (miRNome) showed increased expression of *miR-221/222* in patient tissue samples and GBM cell lines. Similar microarray analyses uncovered an increase in *miR-221/222* expression in papillary thyroid carcinoma, consistent with decreased Kit receptor expression, a gene with *miR-221/222* binding sites [43]. Also, prostate cancer cells showed increased *miR-221/222* expression, particularly in highly aggressive PC3 cells (cells derived from a distal metastasis) compared to LNCaP cells (slow-growing cells derived from local lymph node metastasis) [44]. Overexpression of both miRNAs in the slowly growing cell line increased the number of cells entering S-phase. For these experiments, targeted inhibition of both miRNAs caused increased expression of $p27^{Kip1}$. Similarly, treatment of tumor-bearing mice with intravenous injection of

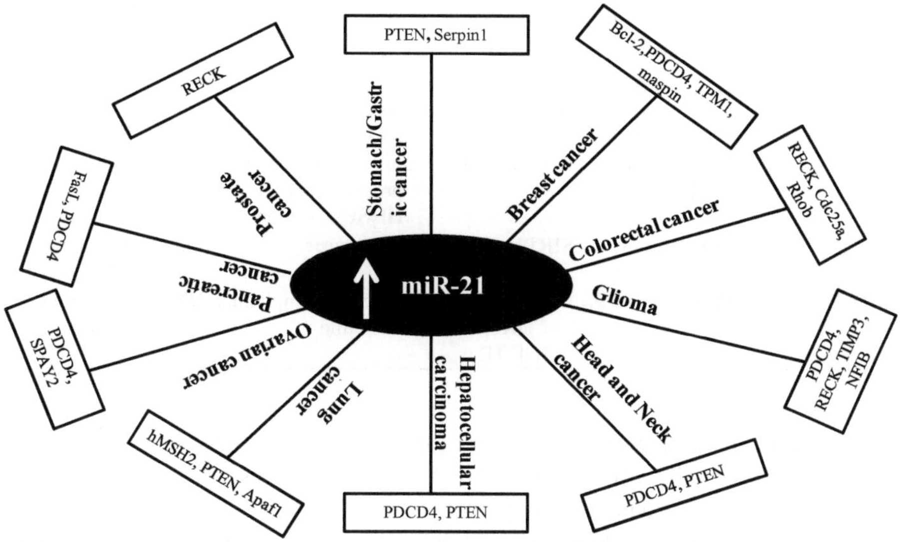

Fig. 2.2 miR-21 expression in cancer. Diverse and redundant mRNA targets of *miR-21* have been identified in multiple tissues. Some targets like Bcl-2 and Apaf1 regulate apoptosis, while others control cell growth and invasion (Cdc25a, PDCD4, PTEN, RECK, Rhob, and TIMP3)

antisense oligonucleotides against *miR-221* resulted in increased survival in mice injected with human hepatocellular carcinoma (HCC) [45].

4 miRNAs as Tumor Suppressors

Located in a cluster on chromosome 13q14.3, the first-described tumor suppressor miRNA was the *miR-15/16* family [46] (Table 2.2). Deletions and translocations in this region were found in ~65 % of B cell chronic lymphocytic leukemia (CLL). An initial inverse correlation was recognized between *miR-15/16* and the anti-apoptotic regulator protein Bcl2, which is overexpressed and a hallmark of CLL [47]. Other *miR-15/16* family members harbor the same 9-base-pair Bcl2-complementary sequence, providing evidence for a putative interaction consistent with experiments that showed that *miR-15/16* posttranscriptionally regulates Bcl2. Re-introduction of *miR-15/16* in a leukemia-derived cell line lacking both miRNAs resulted in strong reduction in Bcl2 levels. Also, cells transfected with plasmids expressing *miR-15/16* demonstrated increased apoptosis as measured by DNA fragmentation, activation of caspases, and terminal deoxynucleotidyl transferase dUTP nick end labeling (TUNEL) staining of individual cells [47].

Tumor growth relies on genetic alterations coupled with external communication to and from the environment. A critical process for tumor expansion is the establishment of stromal progression and the metastatic niche [48]. Tumor cells

Table 2.2 *Tumor suppressor miRNAs*

Tumor suppressor miRNAs	Gene loci	Validated mRNA targets	Cancer	Cellular processes
miR-15/16	Chromosome 13q14.3	Bcl2, FGF-2, FGFR1	Chronic lymphocytic leukemia	Apoptosis, proliferation
let-7	Chromosome 21q21	Ras, Hmga2	Lung, breast, ovarian, prostate cancers, HNSCC, sarcoma[61]	Proliferation
miR-34	Chromosome 1p36	SIRT1	Colon cancer	Epigenetic modification of histones
miR-29	Chromosome X	DNMT3A/3B CDK6, IGF-1R	Mantle cell lymphoma, leukemia	DNA methylation, proliferation

interact with the extracellular matrix (stroma) and upon metastasis, distant stromal–tumor interactions form a niche that facilitates tumor growth. Distinct cell types configure the stromal architecture including fibroblasts that often receive signals (cytokine and chemokine) from tumor cells that prime the stroma for tumor growth. Investigation of prostate cancer–stromal interaction showed downregulation of *miR-15/16* in comparison to stroma surrounding noncancerous tissues [49]. Candidate mRNA targets were screened using bioinformatic prediction algorithms and luciferase reporter assays and found that the 3′UTR of fibroblast growth factor 2 (FGF-2) and fibroblast growth factor receptor 1 (FGFR1) can be silenced by *miR-15/16,* consistent with increased FGF-2 and FGFR1 in prostate cancer [49]. Reconstitution of carcinoma-associated fibroblasts with *miR-15/16* also showed a reduction of FGF-2 and FGFR1. Last, subcutaneous co-injection of prostate cancer cells with fibroblasts transduced with *miR-15/16* expression vectors dramatically reduced tumor growth in immunocompromised mice [49]. Histological analyses showed decreased parenchymal invasion, impaired angiogenic behavior, and reduction of FGF-2 expression in the stroma. These experiments show how miRNAs can regulate multiple aspects of cancer development including primary tumor growth and metastatic spreading.

let-7, the first conserved miRNA discovered in *Caenorhabditis elegans,* plays a pivotal role in regulating animal development and has been implicated in tumorigenesis [50, 51]. In humans, there are 13 *let-7* family members encoded on nine different chromosomes [52]. Microarray analysis of *let-7* levels in lung cancer patients with squamous cell carcinoma showed a significant decrease in let-7 expression and follow-up experiments support the idea that *let-7* negatively regulates Ras, consistent with increased levels of oncogenic Ras in lung cancer [51]. Indeed, *let-7 g* reduced luciferase activity when the K-Ras 3′UTR was fused to luciferase in mouse lung adenocarcinoma cells. *let-7* also repressed the levels of c-Myc using the same reporter system.

Typically, *let-7* loss is due to chromosomal deletions but mutations in the 3′UTR of a *let-7* target can also block *let-7* function [53]. Chromosomal translocations associated with human tumors disrupted repression of the oncogenic chromatin-associated *let-7* target high-mobility group protein A2 (Hmga2) by deletion of 3′UTR regions targeted by *let-7* [53]. *let-7* regulates Hmga2 by interacting with seven conserved 3′UTR elements. Expression of truncated Hmga2 lacking the *let-7* sites induced higher colony formation on soft agar and anchorage-independent behavior. Consistent with this, subcutaneous injection of mouse embryonic fibroblast cells with full length, WT Hmga2 produced no tumors, whereas fibroblasts harboring mutated *let-7* sites or 3′UTR truncations in Hmga2 generated tumors at the sites of injections [53]. These data provide in vivo support for a tumor suppressor function for *let-7* through Hmga2 regulation.

One of the most widely studied anti-tumorigenic proteins is p53, the most commonly mutated gene in numerous cancers [54]. Originally identified as a transcription factor, p53 also acts as a DNA damage response protein and apoptotic regulator. He and colleagues (2007) examined global changes in miRNA levels in p53-deficient mouse embryonic fibroblasts and found that *miR-34* expression correlates with p53 status [55]. Conditional activation of p53 or induction of p53 increased *miR-34* levels. Analysis of p53 binding showed that p53 occupies specific *miR-34* promoter regions. Ectopic expression of *miR-34* inhibited growth of human primary fibroblast cells, in addition to inducing cell cycle arrest in G1 after addition of *miR-34* in immortalized mouse cells and human tumor cells. Upregulation of *miR-34* also produced changes in mRNA expression patterns among a large number of genes implicated in cell proliferation including cyclin-E and CDK-4 [55]. These experiments suggest a novel mechanism of p53-mediated regulation of cell proliferation through activation of *miR-34*. Therefore, *miR-34* can be considered a tumor suppressor and, indeed, the chromosomal region 1p36 encoding *miR-34* is frequently deleted in various cancers including colon cancer [56]. For targets of *miR-34,* increased silent mating type information regulator 1 (SIRT1) expression was observed after *miR-34a* inhibition [57]. SIRT1 is known to regulate apoptosis in response to oxidative and genotoxic stress through nicotinamide adenine dinucleotide (NAD)-dependent deacetylation (reviewed in [58]). Cell survival decreased when *miR-34a* was introduced into colon cancer cells but SIRT1 re-expression partially blocked *miR-34a*-mediated cell death. Interestingly, SIRT1 can deacetylate histones associated with the p53 gene, suggesting a positive feedback loop between p53, *miR-34a,* and SIRT1 [55].

Besides regulating histone acetylation, miRNAs can also regulate DNA methylation. DNA methylation patterns are frequently altered in cancer [59]. The human DNA methyltransferases DNMT3A and DNMT3B are targets of *miR-29* [60]. The *miR-29* family includes *miR-29a, miR-19b-1, miR-29b-II,* and *miR-29,* located on chromosome 7q32, a region frequently deleted in various leukemias. Lung cancer cells transfected with *miR-29* showed a global reduction of DNA methylation and reduced promoter methylation and therefore expression of the tumor suppressor genes FHIT and WWOX (normally hypermethylated and silenced in lung cancer) [60]. In a mantle cell lymphoma model, *miR-29* levels were found to be controlled by c-Myc [61]. Promoter analysis showed that Myc, HDAC3, and EZH2 form a repressive complex to epigenetically repress *miR-29* transcription in Myc-expressing lymphoma cells. Loss of *miR-29* results in upregulation of CDK6 and insulin-like growth factor 1 receptor (IGF-1R), pro-growth genes that promote Myc-associated lymphomagenesis [61]. *miR-29* loss therefore confers a growth advantage, supporting its designation as a tumor suppressor miRNA.

5 miRNA Function in Hallmarks of Cancer

As mentioned above, cancer cells have shared characteristics known as the hallmarks of cancer. Although the majority of the pro-tumorigenic and anti-tumorigenic miRNAs mentioned so far regulate early tumor events including cell growth, apoptosis, and invasion, several findings have also implicated miRNAs in later stages of tumor progression. These include angiogenesis, epithelial-to-mesenchymal transition (EMT), intravasation into the bloodstream, evading the immune response in transit, extravasation, formation of micrometastases, and, eventually, colonization at a secondary site. Angiogenesis is a key step of the malignant tumor phase and *miR-296* is associated with this process via targeting hepatocyte growth factor-regulated tyrosine kinase substrate (HGS) mRNA, resulting in HGS-mediated degradation of the growth factor receptors vascular endothelial growth factor receptor 2 (VEGFR2) and platelet-derived growth factor receptor-beta (PDGFRβ) [62]. Administration of antagomirs targeting *miR-296* resulted in reduced tumor volume. *miR-9* is elevated in human breast cancer cell lines and is also associated with angiogenesis, possibly through the targeting of E-cadherin [63]. Inhibition of E-cadherin by *miR-9* resulted in increased β-catenin and VEGF levels in breast cancer cells. Transplantation of breast cancer cells expressing *miR-9* showed increased MECA-32 (endothelial cell antigen) staining and a greater than tenfold increase in the density of intratumoral microvessels [63]. Mammary fat pad injections of the highly metastatic 4T1 breast cancer cells transfected with an *miR-9* sponge showed a ~50% decrease in lung metastases. These results support a pro-metastatic role for *miR-9* in later stages of tumor progression.

Tumor cells undergo an EMT resulting in the loss of cell adhesion and detachment [64]. As a result of EMT, cancer cells at the invasive front gain enhanced migratory capacity that facilitates intravasation into the bloodstream. Importantly, these cells lose E-cadherin expression while activating mesenchymal gene products like vimentin and N-cadherin. *miR-200* and *miR-205* inhibit the expression of two zinc-finger containing transcription factors, ZEB1 AND ZEB2 that target and repress E-cadherin [65]. Human epithelial kidney cells transfected with inhibitors against *miR-200* displayed minimal E-cadherin expression, increased ZEB1 and ZEB2 expression, elongated cell morphology, and increased invasion. Transfection of *miR-200a/b* and *miR-205* into MDCK-Pez-transformed cells resulted in re-expression of E-cadherin and a rounded epithelial-like morphology, in addition to loss of ZEB1 mRNA and protein levels. In breast cancer cells, a reciprocal pattern was observed between *miR-200* and ZEB1 and ZEB2 with a loss of *miR-200* in mesenchymal-like cell lines compared to epithelial-like cells [65]. Last, patient samples of less invasive ductal tumors were found to express *miR-200* with coincidently increased

E-cadherin levels compared to sarcomatoid metaplastic tumors. This suggests a selection against *miR-200* expression during the EMT step of the metastatic cascade.

Using quantitative polymerase chain reaction (PCR), a number of miRNAs were identified that are selectively deleted in rare metastatic foci [66]. Mice inoculated with aggressive malignant cells from a patient with metastatic breast cancer developed lung and bone metastases, and analysis of these metastatic cell derivatives showed that *miR-335, miR-126, and miR-206* were downregulated. Clinical analysis also revealed a significant association between the absence of these miRNAs and survival. A follow-up study on *miR-126* in breast tumor progression indicated that inhibition of *miR-126* led to enhanced lung and systemic metastases [67]. The effects of *miR-126* are due in part to targeting of insulin-like growth factor binding protein 2 (IGFBP2) and c-Mer tyrosine kinase (MERTK), which have *miR-126* binding sites in their 3′UTRs. Knockdown of IGFBP2 or MERTK in cells lacking *miR-126* significantly suppressed endothelial recruitment and metastatic colonization. Enzyme-linked immunosorbent assay (ELISA) analysis of conditioned media from metastatic breast cancer cells confirmed a twofold increase in secreted IGFBP2 compared to the less metastatic parental cells. Finally, *miR-126* was shown to regulate MERTK, as secreted MERTK from metastatic cells acted as a decoy receptor for the GAS6 ligand with inhibition of *miR-126* [67]. These data provide evidence for the opposing role miRNAs play in the aggressive stages of tumor progression.

Recent observations have renewed interest in altered metabolic profiles in cancer. New results have resurrected Warburg's observation in 1924 that cancer cells metabolize glucose distinct from normal cells, mainly through aerobic glycolysis, referred to as the Warburg effect [68]. Nevertheless, mitochondrial adenosine triphosphate (ATP) production and glutamine metabolism also play a role in tumor cell behavior [69]. A search of Myc-responsive genes altered in the mitochondrial proteome led to the identification of mitochondrial glutaminase (GLS) which catalyzes the conversion of glutamine to glutamate necessary for tricarboxylic acid (TCA) cycle-mediated production of ATP. Reverse experiments in human B cell lines showed that loss of Myc decreased GLS expression while restoration of Myc increased GLS levels. Small interfering RNAs (siRNAs) against GLS attenuated Myc-induced cell proliferation but the regulation was indirect, suggesting other modes of regulation. Analysis of the GLS 3′UTR revealed binding motifs for *miR-23-a/b*. Both miRNAs were previously characterized as direct downstream targets of Myc using chromatin immunoprecipitation assays and luciferase reporter assays [69]. Thus, Myc suppression of *miR-23a/b* enhances glutamine catabolism through increased GLS expression.

6 miRNAs and Chemoresistance

In addition to serving as regulators of the cell cycle and proliferation, *miR-221/222* was shown to confer resistance to specific inhibitors in a model of breast cancer [70]. Tumors of the breast are classified according to the presence or absence of hormone receptors, particularly the estrogen receptor (ER). The ER cascade becomes stimulated as the downstream target genes of the ER are activated following addition and binding of estradiol. Perturbation of the ER pathway is commonly employed as ER-positive tumors (belonging to the luminal category) respond more favorably to antiestrogen therapy, whereas highly aggressive ER-negative tumors (including triple-negative and basal-like breast cancer) are nonresponsive to antiestrogen therapy [71]. Unfortunately, emergence of antiestrogen resistance is observed in some patients with ER-positive tumors [72]. Several inhibitors of ER signaling have been used for cancer therapy including tamoxifen and fulvestrant [73]. *miR-221/222* expression is increased in fulvestrant-resistant breast cancer cell lines, and overexpression of both miRNAs can induce estrogen-independent growth [70]. Apparently, *miR-221/222* controls the transforming growth factor beta (TGF-β) signaling pathway to generate a growth inhibitory loop during development of fulvestrant resistance.

A population of perpetual CSCs has been attributed to the formation of chemoresistance in some cancer patients undergoing chemotherapy [5, 71]. Serial passage of cells in engrafted mice treated with chemotherapy led to the enrichment of breast tumor-initiating cells [74]. *let-7* levels were shown to be reduced in these $CD44^+$ and CD24– expressing cells, indicative of stem cells. The effect of *let-7* on self-renewal was illustrated using the mammosphere formation assay [75]. Briefly, stem cells grown in suspension form spherical colonies in a manner similar to the three-dimensional (3D) environment of the tumor. Lentiviral delivery of *let-7* resulted in a reduction in mammosphere formation [74]. Reduced *let-7* expression favored an increase in cell proliferation while inhibiting differentiation. Inhibition of known *let-7* targets showed that a *let-7*–Ras interaction mediated self-renewal, whereas differentiation was regulated via repression of HMGA2 expression. Transplantation of mammospheric cells expressing *let-7* led to a reduction in tumor formation in mice and slower growth rates. Thus, *let-7* may play an important role regulating stem cell maintenance and pathways mediating chemoresistance.

7 Therapeutic and Clinical Implications of miRNAs

Diverse studies have investigated miRNAs as diagnostic, prognostic, and therapeutic tools in cancer [15, 76–78]. Recent evidence of circulating miRNAs in cancer patient serum prompted work to determine whether circulating miRNAs can be utilized as markers for cancer detection and/or disease progression. Tumor-associated miRNAs were first detected in the serum of patients with large B-cell lymphomas [79]. Increased levels of *miR-10b* were also detected in serum from patients with metastatic breast cancer and inhibition of *miR-10b* expression prevented invasion and migration but not cell viability [80]. Similarly, *miR-21* serum levels were found to be elevated in patients with chemoresistant hormone-refractory prostate cancer (HRPC) but lower in serum of patients with localized prostate cancer [81]. High *miR-21* expression was also associated with high prostate-specific antigen (PSA) levels. Likewise, *miR-24* was found to be increased in the sera from patients with tongue squamous cell carcinoma [82]. Last, patients with colorectal cancer and benign adenomas displayed increased plasma *miR-92a* levels, suggesting that early detection of circulating *miR-92a* could help prevent or delay transformation of benign lesions to malignant tumors [83]. Overall, numerous tumor-associated extracellular miRNAs have been uncovered, and continuing research will aid in understanding their functional role in cancer progression and validation as bona fide biomarkers.

Besides serum biomarkers, miRNA expression profiles might help with disease prognosis. Pancreatic cancer miRNA expression patterns have been developed to try to classify pancreatic adenocarcinoma from benign tissue, normal pancreatic tissue, and chronic pancreatitis [84]. Hierarchical clustering led to the identification of four major groups with distinct miRNA profiles, in addition to finding candidate tumor-related miRNAs. In lung cancer, differential miRNA expression identified six miRNAs (*hsa-miR-205, has-miR-99b, hsa-miR-203, hsa-miR-202, hsa miR-102,* and *hsa-miR-204* precursor) showing the greatest difference in expression between two common types of non-small cell lung cancer [85]. Also, a worse prognostic impact on lung adenocarcinoma patients was observed with *miR-155* expression while reduced *let-7a2* expression correlated with poor survival [85]. For prostate cancer, analysis of 51 miRNAs identified upregulated and downregulated miRNAs allowing tumor classification based on androgen dependence, either hormone naïve or hormone refractory [86]. Last, HCC tumors had decreased *miR-26* expression, and lower *miR-26* expression correlated with shorter survival [87]. Together, these data demonstrate differential miRNA expression in various tumors and support an emerging prognostic role for miRNAs in cancer.

As a therapeutic strategy for combating tumorigenesis, both delivery and inhibition of miRNA activity have shown promising potential. Using a murine model of human lung cancer driven by an activated K-Ras mutation, lung delivery of a *let-7* mimetic through inhalation resulted in a 66 % reduction in tumor burden [88]. *let-7* was also shown to induce cell death in an apoptotic-independent manner [88]. In a murine model of HCC driven by Myc activation, delivery of *miR-26a* significantly reduced or completely ablated tumors [89]. In breast cancer, *miR-10b* was identified as a suppressor of breast cancer metastasis [90]. Systemic treatment of breast tumor-bearing mice with an *miR-10b* antagomir suppressed formation of lung metastases, mediated via targeting Hoxd10. These experiments suggest that development of miRNA-based therapies might provide powerful approaches to reducing tumor burden and dissemination.

8 Conclusions and Future Perspectives

The role that miRNAs play in the initiation and progression of tumorigenesis has opened the door to understanding the regulation of gene expression during tumor formation and created the possibility for new therapeutic approaches to cancer treatment (Fig. 2.3). miRNAs can act as both oncogenes and tumor suppressors to regu-

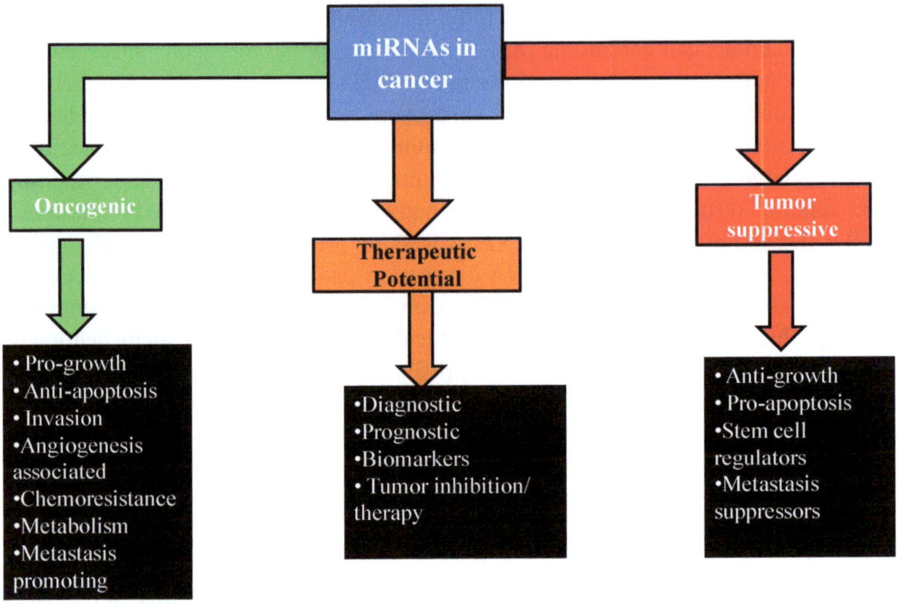

Fig. 2.3 miRNA therapeutic strategies in cancer treatment

late hallmarks of cancer. Because miRNAs can regulate multiple genes, the key to future work is to identify specific targets that directly regulate decisions controlling proliferation and/or apoptosis. However, as targets are being identified, expression profiles of miRNAs within distinct stages of tumors or in serum might allow for the development of novel biomarkers to assist in diagnostic and prognostic outcomes. Personalized medicine based on miRNA expression and the use of miRNA therapeutic strategies has the potential to provide powerful new tools in cancer treatment.

References

1. Hanahan D, Weinberg RA. The hallmarks of cancer. Cell. 2000;100(1):57–70.
2. Hanahan D, Weinberg RA. Hallmarks of cancer: the next generation. Cell. 2011;144(5):646–74.
3. Pantel K, Brakenhoff RH. Dissecting the metastatic cascade. Nat Rev Cancer. 2004;4(6):448–56.
4. Perez-Soler R, Chachoua A, Hammond LA, Rowinsky EK, Huberman M, Karp D, Rigas J, Clark GM, Santabarbara P, Bonomi P. Determinants of tumor response and survival with erlotinib in patients with non–small-cell lung cancer. J Clin Oncol. 2004;22(16):3238–47.
5. Reya T, Morrison SJ, Clarke MF, Weissman IL. Stem cells, cancer, and cancer stem cells. Nature. 2001;414(6859):105–11.
6. Visvader JE. Cells of origin in cancer. Nature. 2011;469(7330):314–22.
7. Wu XZ. Origin of cancer stem cells: the role of self-renewal and differentiation. Ann Surg Oncol. 2008;15(2):407–14.
8. Joseph NM, Mosher JT, Buchstaller J, Snider P, McKeever PE, Lim M, Conway SJ, Parada LF, Zhu Y, Morrison SJ. The loss of Nf1 transiently promotes self-renewal but not tumorigenesis by neural crest stem cells. Cancer Cell. 2008;13(2):129–40.

9. Fan H, Oro AE, Scott MP, Khavari PA. Induction of basal cell carcinoma features in transgenic human skin expressing Sonic Hedgehog. Nat Med. 1997;3(7):788–92.
10. Aster JC, Blacklow SC, Pear WS. Notch signalling in T-cell lymphoblastic leukaemia/lymphoma and other haematological malignancies. J Pathol. 2011;223(2):262–73.
11. Hendler FJ, Ozanne BW. Human squamous cell lung cancers express increased epidermal growth factor receptors. J Clin Invest. 1984;74(2):647–51.
12. Segditsas S, Tomlinson I. Colorectal cancer and genetic alterations in the Wnt pathway. Oncogene. 2006;25(57):7531–7.
13. Leal JA, Lleonart ME. MicroRNAs and cancer stem cells: Therapeutic approaches and future perspectives. Cancer Lett. 2013;338(1):174–83.
14. He L, Hannon GJ. MicroRNAs: small RNAs with a big role in gene regulation. Nat Rev Genet. 2004;5(7):522–31.
15. Bartel D. MicroRNAs: genomics, biogenesis, mechanism, and function. Cell. 2004;116:281–97.
16. Bushati N, Cohen SM. microRNA functions. Annu Rev Cell Dev Biol. 2007;23:175–205.
17. Lewis BP, Burge CB, Bartel DP. Conserved seed pairing, often flanked by adenosines, indicates that thousands of human genes are microRNA targets. Cell. 2005;120(1):15–20.
18. Lee YS, Dutta A. MicroRNAs in cancer. Annu RevPathol. 2009;4:199–227.
19. Lujambio A, Calin GA, Villanueva A, Ropero S, Sanchez-Cespedes M, Blanco D, Montuenga LM, Rossi S, Nicoloso MS, Faller WJ, Gallagher WM, Eccles SA, Croce CM, Esteller M. A microRNA DNA methylation signature for human cancer metastasis. Proc Natl Acad Sci U S A. 2008;105(36):13556–61.
20. He L, Thomson JM, Hemann MT, Hernando-Monge E, Mu D, Goodson S, Powers S, Cordon-Cardo C, Lowe SW, Hannon GJ, Hammond SM. A microRNA polycistron as a potential human oncogene. Nature. 2005;435(7043):828–33.
21. Ota A, Tagawa H, Karnan S, Tsuzuki S, Karpas A, Kira S, Yoshida Y, Seto M. Identification and characterization of a novel gene, C13orf25, as a target for 13q31-q32 amplification in malignant lymphoma. Cancer Res. 2004;64(9):3087–95.
22. Hayashita Y, Osada H, Tatematsu Y, Yamada H, Yanagisawa K, Tomida S, Yatabe Y, Kawahara K, Sekido Y, Takahashi T. A polycistronic microRNA cluster, miR-17-92, is overexpressed in human lung cancers and enhances cell proliferation. Cancer Res. 2005;65(21):9628–32.
23. Eis PS, Tam W, Sun L, Chadburn A, Li Z, Gomez MF, Lund E, Dahlberg JE. Accumulation of miR-155 and BIC RNA in human B cell lymphomas. Proc Natl Acad Sci U S A. 2005;102(10):3627–32.
24. Tam W, Ben-Yehuda D, Hayward WS. bic, a novel gene activated by proviral insertions in avian leukosis virus-induced lymphomas, is likely to function through its noncoding RNA. Mol Cell Biol. 1997;17(3):1490–1502.
25. Gironella M, Seux M, Xie MJ, Cano C, Tomasini R, Gommeaux J, Garcia S, Nowak J, Yeung ML, Jeang KT, Chaix A, Fazli L, Motoo Y, Wang Q, Rocchi P, Russo A, Gleave M, Dagorn JC, Iovanna JL, Carrier A, Pebusque MJ, Dusetti NJ. Tumor protein 53-induced nuclear protein 1 expression is repressed by miR-155, and its restoration inhibits pancreatic tumor development. Proc Natl Acad Sci U S A. 2007;104(41):16170–5.
26. Jiang S, Zhang HW, Lu MH, He XH, Li Y, Gu H, Liu MF, Wang ED. MicroRNA-155 functions as an OncomiR in breast cancer by targeting the suppressor of cytokine signaling 1 gene. Cancer Res. 2010;70(8):3119–27.
27. Voorhoeve PM, le Sage C, Schrier M, Gillis AJ, Stoop H, Nagel R, Liu YP, van Duijse J, Drost J, Griekspoor A, Zlotorynski E, Yabuta N, De Vita G, Nojima H, Looijenga LH, Agami R. A genetic screen implicates miRNA-372 and miRNA-373 as oncogenes in testicular germ cell tumors. Cell. 2006;124(6):1169–81.
28. Esquela-Kerscher A, Slack FJ. The age of high-throughput microRNA profiling. Nat Methods. 2004;1(2):106–7.
29. Dalmay T, Edwards DR. MicroRNAs and the hallmarks of cancer. Oncogene. 2006;25(46):6170–5.
30. Chan JA, Krichevsky AM, Kosik KS. MicroRNA-21 is an antiapoptotic factor in human glioblastoma cells. Cancer Res. 2005;65(14):6029–33.

31. Volinia S, Calin GA, Liu CG, Ambs S, Cimmino A, Petrocca F, Visone R, Iorio M, Roldo C, Ferracin M, Prueitt RL, Yanaihara N, Lanza G, Scarpa A, Vecchione A, Negrini M, Harris CC, Croce CM. A microRNA expression signature of human solid tumors defines cancer gene targets. Proc Natl Acad Sci U S A. 2006;103(7):2257–61.
32. Si ML, Zhu S, Wu H, Lu Z, Wu F, Mo YY. miR-21-mediated tumor growth. Oncogene. 2007;26(19):2799–803.
33. Liu ZL, Wang H, Liu J, Wang ZX. MicroRNA-21 (miR-21) expression promotes growth, metastasis, and chemo- or radioresistance in non-small cell lung cancer cells by targeting PTEN. Mol Cell Biochem. 2013;372(1–2):35–45.
34. Weinstein IB. Cancer. Addiction to oncogenes–the Achilles heal of cancer. Science. 2002;297(5578):63–4.
35. Jain M, Arvanitis C, Chu K, Dewey W, Leonhardt E, Trinh M, Sundberg CD, Bishop JM, Felsher DW. Sustained loss of a neoplastic phenotype by brief inactivation of MYC. Science. 2002;297(5578):102–4.
36. Chin L, Tam A, Pomerantz J, Wong M, Holash J, Bardeesy N, Shen Q, O'Hagan R, Pantginis J, Zhou H, Horner JW 2nd, Cordon-Cardo C, Yancopoulos GD, DePinho RA. Essential role for oncogenic Ras in tumour maintenance. Nature. 1999;400(6743):468–72.
37. Medina PP, Nolde M, Slack FJ. OncomiR addiction in an in vivo model of microRNA-21-induced pre-B-cell lymphoma. Nature. 2010;467(7311):86–90.
38. Krichevsky AM, Gabriely G. miR-21: a small multi-faceted RNA. J Cell Mol Med. 2009;13(1):39–53.
39. Wang P, Zou F, Zhang X, Li H, Dulak A, Tomko RJ Jr., Lazo JS, Wang Z, Zhang L, Yu J. microRNA-21 negatively regulates Cdc25A and cell cycle progression in colon cancer cells. Cancer Res. 2009;69(20):8157–65.
40. Zhong Z, Dong Z, Yang L, Gong Z. miR-21 induces cell cycle at S phase and modulates cell proliferation by down-regulating hMSH2 in lung cancer. J Cancer Res Clin Oncol. 2012;138(10):1781–88.
41. Reis ST, Pontes-Junior J, Antunes AA, Dall'Oglio MF, Dip N, Passerotti CC, Rossini GA, Morais DR, Nesrallah AJ, Piantino C, Srougi M, Leite KR. miR-21 may acts as an oncomir by targeting RECK, a matrix metalloproteinase regulator, in prostate cancer. BMC Urology. 2012;12:14.
42. Ciafre SA, Galardi S, Mangiola A, Ferracin M, Liu CG, Sabatino G, Negrini M, Maira G, Croce CM, Farace MG. Extensive modulation of a set of microRNAs in primary glioblastoma. Biochem Biophys Res Commun. 2005;334(4):1351–8.
43. He H, Jazdzewski K, Li W, Liyanarachchi S, Nagy R, Volinia S, Calin GA, Liu CG, Franssila K, Suster S, Kloos RT, Croce CM, de la Chapelle A. The role of microRNA genes in papillary thyroid carcinoma. Proc Natl Acad Sci U S A. 2005;102(52):19075–80.
44. Galardi S, Mercatelli N, Giorda E, Massalini S, Frajese GV, Ciafre SA, Farace MG. miR-221 and miR-222 expression affects the proliferation potential of human prostate carcinoma cell lines by targeting p27Kip1. J Biol Chem. 2007;282(32):23716–24.
45. Park JK, Kogure T, Nuovo GJ, Jiang J, He L, Kim JH, Phelps MA, Papenfuss TL, Croce CM, Patel T, Schmittgen TD. miR-221 silencing blocks hepatocellular carcinoma and promotes survival. Cancer Res. 2011;71(24):7608–16.
46. Calin GA, Dumitru CD, Shimizu M, Bichi R, Zupo S, Noch E, Aldler H, Rattan S, Keating M, Rai K, Rassenti L, Kipps T, Negrini M, Bullrich F, Croce CM. Frequent deletions and down-regulation of micro-RNA genes miR15 and miR16 at 13q14 in chronic lymphocytic leukemia. Proc Natl Acad Sci U S A. 2002;99(24):15524–9.
47. Cimmino A, Calin GA, Fabbri M, Iorio MV, Ferracin M, Shimizu M, Wojcik SE, Aqeilan RI, Zupo S, Dono M, Rassenti L, Alder H, Volinia S, Liu CG, Kipps TJ, Negrini M, Croce CM. miR-15 and miR-16 induce apoptosis by targeting BCL2. Proc Natl Acad Sci U S A. 2005;102(39):13944–9.
48. Whiteside TL. The tumor microenvironment and its role in promoting tumor growth. Oncogene. 2008;27(45):5904–12.
49. Musumeci M, Coppola V, Addario A, Patrizii M, Maugeri-Sacca M, Memeo L, Colarossi C, Francescangeli F, Biffoni M, Collura D, Giacobbe A, D'Urso L, Falchi M, Venneri MA, Muto

G, De Maria R, Bonci D. Control of tumor and microenvironment cross-talk by miR-15a and miR-16 in prostate cancer. Oncogene. 2011;30(41):4231–42.
50. Boyerinas B, Park SM, Hau A, Murmann AE, Peter ME. The role of let-7 in cell differentiation and cancer. Endocr Relat Cancer. 2010;17(1):F19–36.
51. Johnson SM, Grosshans H, Shingara J, Byrom M, Jarvis R, Cheng A, Labourier E, Reinert KL, Brown D, Slack FJ. RAS is regulated by the let-7 microRNA family. Cell. 2005;120(5):635–47.
52. Roush S, Slack FJ. The let-7 family of microRNAs. Trends Cell Biol. 2008;18(10):505–16.
53. Mayr C, Hemann MT, Bartel DP. Disrupting the pairing between let-7 and Hmga2 enhances oncogenic transformation. Science. 2007;315(5818):1576–9.
54. Finlay CA, Hinds PW, Levine AJ. The p53 proto-oncogene can act as a suppressor of transformation. Cell. 1989;57(7):1083–93.
55. He L, He X, Lim LP, de Stanchina E, Xuan Z, Liang Y, Xue W, Zender L, Magnus J, Ridzon D, Jackson AL, Linsley PS, Chen C, Lowe SW, Cleary MA, Hannon GJ. A microRNA component of the p53 tumour suppressor network. Nature. 2007;447(7148):1130–4.
56. Di Vinci A, Infusini E, Peveri C, Risio M, Rossini FP, Giaretti W. Deletions at chromosome 1p by fluorescence in situ hybridization are an early event in human colorectal tumorigenesis. Gastroenterology. 1996;111(1):102–7.
57. Yamakuchi M, Ferlito M, Lowenstein CJ. miR-34a repression of SIRT1 regulates apoptosis. Proc Natl Acad Sci U S A. 2008;105(36):13421–6.
58. Bosch-Presegue L, Vaquero A. The dual role of sirtuins in cancer. Genes Cancer. 2011;2(6):648–62.
59. El-Osta A. DNMT cooperativity–the developing links between methylation, chromatin structure and cancer. Bioessays. 2003;25(11):1071–84.
60. Fabbri M, Garzon R, Cimmino A, Liu Z, Zanesi N, Callegari E, Liu S, Alder H, Costinean S, Fernandez-Cymering C, Volinia S, Guler G, Morrison CD, Chan KK, Marcucci G, Calin GA, Huebner K, Croce CM. MicroRNA-29 family reverts aberrant methylation in lung cancer by targeting DNA methyltransferases 3A and 3B. Proc Natl Acad Sci U S A. 2007;104(40):15805–10.
61. Zhang X, Zhao X, Fiskus W, Lin J, Lwin T, Rao R, Zhang Y, Chan JC, Fu K, Marquez VE, Chen-Kiang S, Moscinski LC, Seto E, Dalton WS, Wright KL, Sotomayor E, Bhalla K, Tao J. Coordinated silencing of MYC-mediated miR-29 by HDAC3 and EZH2 as a therapeutic target of histone modification in aggressive B-Cell lymphomas. Cancer Cell. 2012;22(4):506–23.
62. Wurdinger T, Tannous BA, Saydam O, Skog J, Grau S, Soutschek J, Weissleder R, Breakefield XO, Krichevsky AM. miR-296 regulates growth factor receptor overexpression in angiogenic endothelial cells. Cancer Cell. 2008;14(5):382–93.
63. Ma L, Young J, Prabhala H, Pan E, Mestdagh P, Muth D, Teruya-Feldstein J, Reinhardt F, Onder TT, Valastyan S, Westermann F, Speleman F, Vandesompele J, Weinberg RA. miR-9, a MYC/MYCN-activated microRNA, regulates E-cadherin and cancer metastasis. Nat Cell Biol. 2010;12(3):247–56.
64. Kalluri R, Weinberg RA. The basics of epithelial-mesenchymal transition. J Clin Invest. 2009;119(6):1420–8.
65. Gregory PA, Bert AG, Paterson EL, Barry SC, Tsykin A, Farshid G, Vadas MA, Khew-Goodall Y, Goodall GJ. The miR-200 family and miR-205 regulate epithelial to mesenchymal transition by targeting ZEB1 and SIP1. Nat Cell Biol. 2008;10(5):593–601.
66. Tavazoie SF, Alarcon C, Oskarsson T, Padua D, Wang Q, Bos PD, Gerald WL, Massague J. Endogenous human microRNAs that suppress breast cancer metastasis. Nature. 2008;451(7175):147–52.
67. Png KJ, Halberg N, Yoshida M, Tavazoie SF. A microRNA regulon that mediates endothelial recruitment and metastasis by cancer cells. Nature. 2012;481(7380):190–4.
68. Vander Heiden MG, Cantley LC, Thompson CB. Understanding the Warburg effect: the metabolic requirements of cell proliferation. Science. 2009;324(5930):1029–33.
69. Gao P, Tchernyshyov I, Chang TC, Lee YS, Kita K, Ochi T, Zeller KI, De Marzo AM, Van Eyk JE, Mendell JT, Dang CV. c-Myc suppression of miR-23a/b enhances mitochondrial glutaminase expression and glutamine metabolism. Nature. 2009;458(7239):762–5.

70. Rao X, Di Leva G, Li M, Fang F, Devlin C, Hartman-Frey C, Burow ME, Ivan M, Croce CM, Nephew KP. MicroRNA-221/222 confers breast cancer fulvestrant resistance by regulating multiple signaling pathways. Oncogene. 2011;30(9):1082–97.
71. Polyak K. Heterogeneity in breast cancer. J Clin Invest. 2011;121(10):3786–8.
72. Miller TW, Balko JM, Arteaga CL. Phosphatidylinositol 3-kinase and antiestrogen resistance in breast cancer. J Clin Oncol. 2011;29(33):4452–61.
73. Higgins MJ, Baselga J. Targeted therapies for breast cancer. J Clin Invest. 2011;121(10):3797–803.
74. Yu F, Yao H, Zhu P, Zhang X, Pan Q, Gong C, Huang Y, Hu X, Su F, Lieberman J, Song E. let-7 regulates self renewal and tumorigenicity of breast cancer cells. Cell. 2007;131(6):1109–23.
75. Dontu G, Al-Hajj M, Abdallah WM, Clarke MF, Wicha MS. Stem cells in normal breast development and breast cancer. Cell Prolif. 2003;36(Suppl 1):59–72.
76. Calin GA, Croce CM. MicroRNA signatures in human cancers. Nat Rev Cancer. 2006;6(11):857–66.
77. Bader AG, Brown D, Stoudemire J, Lammers P. Developing therapeutic microRNAs for cancer. Gene Ther. 2011;18(12):1121–6.
78. Mo MH, Chen L, Fu Y, Wang W, Fu SW. Cell-free circulating miRNA biomarkers in cancer. J Cancer. 2012;3:432–48.
79. Lawrie CH, Gal S, Dunlop HM, Pushkaran B, Liggins AP, Pulford K, Banham AH, Pezzella F, Boultwood J, Wainscoat JS, Hatton CS, Harris AL. Detection of elevated levels of tumour-associated microRNAs in serum of patients with diffuse large B-cell lymphoma. Brit J Haematol. 2008;141(5):672–5.
80. Heneghan HM, Miller N, Lowery AJ, Sweeney KJ, Newell J, Kerin MJ. Circulating microRNAs as novel minimally invasive biomarkers for breast cancer. Ann Surg. 2010;251(3):499–505.
81. Zhang HL, Yang LF, Zhu Y, Yao XD, Zhang SL, Dai B, Zhu YP, Shen YJ, Shi GH, Ye DW. Serum miRNA-21: elevated levels in patients with metastatic hormone-refractory prostate cancer and potential predictive factor for the efficacy of docetaxel-based chemotherapy. Prostate. 2011;71(3):326–31.
82. Lin SC, Liu CJ, Lin JA, Chiang WF, Hung PS, Chang KW. miR-24 up-regulation in oral carcinoma: positive association from clinical and in vitro analysis. Oral Oncol. 2010;46(3):204–8.
83. Huang Z, Huang D, Ni S, Peng Z, Sheng W, Du X. Plasma microRNAs are promising novel biomarkers for early detection of colorectal cancer. Int J Cancer. 2010;127(1):118–26.
84. Lee EJ, Gusev Y, Jiang J, Nuovo GJ, Lerner MR, Frankel WL, Morgan DL, Postier RG, Brackett DJ, Schmittgen TD. Expression profiling identifies microRNA signature in pancreatic cancer. Int J Cancer. 2007;120(5):1046–54.
85. Yanaihara N, Caplen N, Bowman E, Seike M, Kumamoto K, Yi M, Stephens RM, Okamoto A, Yokota J, Tanaka T, Calin GA, Liu CG, Croce CM, Harris CC. Unique microRNA molecular profiles in lung cancer diagnosis and prognosis. Cancer Cell. 2006;9(3):189–98.
86. Porkka KP, Pfeiffer MJ, Waltering KK, Vessella RL, Tammela TL, Visakorpi T. MicroRNA expression profiling in prostate cancer. Cancer Res. 2007;67(13):6130–5.
87. Ji J, Shi J, Budhu A, Yu Z, Forgues M, Roessler S, Ambs S, Chen Y, Meltzer PS, Croce CM, Qin LX, Man K, Lo CM, Lee J, Ng IO, Fan J, Tang ZY, Sun HC, Wang XW. MicroRNA expression, survival, and response to interferon in liver cancer. New Engl J Med. 2009;361(15):1437–47.
88. Trang P, Medina PP, Wiggins JF, Ruffino L, Kelnar K, Omotola M, Homer R, Brown D, Bader AG, Weidhaas JB, Slack FJ. Regression of murine lung tumors by the let-7 microRNA. Oncogene. 2010;29(11):1580–7.
89. Kota J, Chivukula RR, O'Donnell KA, Wentzel EA, Montgomery CL, Hwang HW, Chang TC, Vivekanandan P, Torbenson M, Clark KR, Mendell JR, Mendell JT. Therapeutic microRNA delivery suppresses tumorigenesis in a murine liver cancer model. Cell. 2009;137(6):1005–17.
90. Ma L, Reinhardt F, Pan E, Soutschek J, Bhat B, Marcusson EG, Teruya-Feldstein J, Bell GW, Weinberg RA. Therapeutic silencing of miR-10b inhibits metastasis in a mouse mammary tumor model. Nat Biotechnol. 2010;28(4):341–7.

Chapter 3
MicroRNA, DNA Repair, and Cancer

Hailiang Hu

Abstract Genomic instability is a hallmark of cancer cells and one of the underlying mechanisms is probably caused by the failure to repair DNA damages that have been passed on to the progeny cells. Cells have evolved many types of DNA repair mechanisms to counteract the DNA damages induced by exogenous insults, such as ionizing radiation, ultraviolet radiation, and chemical reagents, or endogenous stimuli-like reactive oxygen species (ROS). These repair mechanisms constitute an elaborate genome maintenance system to protect genomic integrity and therefore defend tumorigenesis. Most recently, microRNAs (miRNAs) have been reported to be a new class of regulators that modulate the DNA damage response pathways by targeting the protein components of response machinery. Here, we summarize and highlight the miRNAs that have been shown to regulate the different DNA repair pathways and discuss their roles in carcinogenesis and implications in cancer therapy.

Keywords MicroRNA · DNA damage response · DNA repair · Double-strand breaks

1 Introduction

Genomic instability is a hallmark of cancer cells and is thought to be an underlying factor responsible for the other six acquired hallmarks of cancer [1, 2]. Cells have evolved an elaborate genome maintenance system to protect genomic integrity and resolve the defects in DNA [3, 4]. DNA repair mechanisms lie in the core of this genome maintenance system. Different DNA repair mechanisms have been developed by cells to counteract various types of DNA lesions. For example, base excision repair (BER) fixes the small chemical alterations of DNA bases; mismatch repair (MMR) replaces the mispaired DNA, while nucleotide excision repair (NER) removes an oligonucleotide containing the damaged bases [5, 6]. Single-strand breaks (SSBs) are repaired by single-strand break repair (SSBR), whereas double-strand breaks (DSBs) are processed either by nonhomologous end joining (NHEJ) or homologous recombination repair (HRR) [7, 8]. These repair mechanisms constitute a critical

H. Hu (✉)
Department of Pathology and Laboratory Medicine, Jonsson Comprehensive Cancer Center, David Geffen School of Medicine at UCLA, Los Angeles, CA 90095, USA
e-mail: hhu@mednet.ucla.edu

S. R. Singh, P. Rameshwar (eds.), *MicroRNA in Development and in the Progression of Cancer*, DOI 10.1007/978-1-4899-8065-6_3,

defense mechanism against genomic instability and subsequent tumorigenesis [9, 10]. The mutations in the genes encoding the protein components of the DNA repair machinery leading to cancer have been well documented. For instance, BRCA1 and BRCA2, two important DSB HRR proteins, are involved in hereditary breast cancers [11]. Most of these genes behave like tumor suppressors during tumorigenesis.

MicroRNAs (miRNAs) are single-stranded small RNAs of 19–25 nucleotides in length that have been known to be involved in many normal physiological or abnormal pathological processes, including cancers. miRNAs function either as oncogenes or as tumor suppressors during tumor development [12, 13]. More than 700 human miRNAs have been identified, but the function for most of them still needs to be characterized [14]. Around 20–30% of human genes are predicted to be regulated by miRNAs [14]. Therefore, one miRNA can regulate multiple genes and one gene can be controlled by multiple miRNAs. miRNAs are found within or near genomic fragile sites and more than 50% human miRNAs are found at or near the cancer-associated genomic regions [15, 16], implicating the involvement of miRNAs in cancer development. Recently, miRNAs are reported to be involved in DNA damage response (DDR) and modulate the response of cancer cells to cytotoxic treatments, including radio/chemotherapy [17–19]. Here, we summarized the DNA repair-associated miRNAs and discussed their potential role in maintaining the genomic integrity and cancer development.

2 MicroRNA Biogenesis, Regulation and Cancer

2.1 MiRNA Biogenesis: Core Components

MiRNAs are noncoding small RNAs that silence gene expression by either cleaving target messenger RNAs (mRNAs) or repressing translation [20]. The biogenesis of miRNAs comprises three steps: transcription, processing/maturation, and degradation (Fig. 3.1). First, miRNAs are transcribed by RNA polymerase II into pri-miRNAs with the aid of transcription factors [21–24]. Second, pri-miRNAs are 5′ capped and 3′ polyadenylated and further cleaved into pre-miRNA by the Drosha/DGCR8 microprocessor complex [25–27]. Pre-miRNA is then exported from nucleus to cytoplasm by exportin-5 and Ran–guanine triphosphate (GTP) [28, 29]. In the cytoplasm, pre-miRNA is cleaved by Dicer/transactivation response (TAR) RNA-binding protein (TRBP) to an imperfect miRNA/miRNA* duplex of around 20–25 nt in size [30]. Only one strand of the duplex is incorporated into RNA-induced silencing complex (RISC/Argonaute (AGO) 1–4) to bind to 3′-untranslated region (UTR) of target gene and suppress its expression while the other strand is normally degraded. The RISC-loaded mature miRNA is protected from degradation by AGO proteins [31–33]. Finally, after finishing its task, the mature single-strand miRNA will be degraded by the 5′-3′ exoribonuclease XRN2 [34] or the 3′-5′ exoribonucleases, such as human polynucleotide phosphorylase (hPNPase) [35] and nuclear exosome [36].

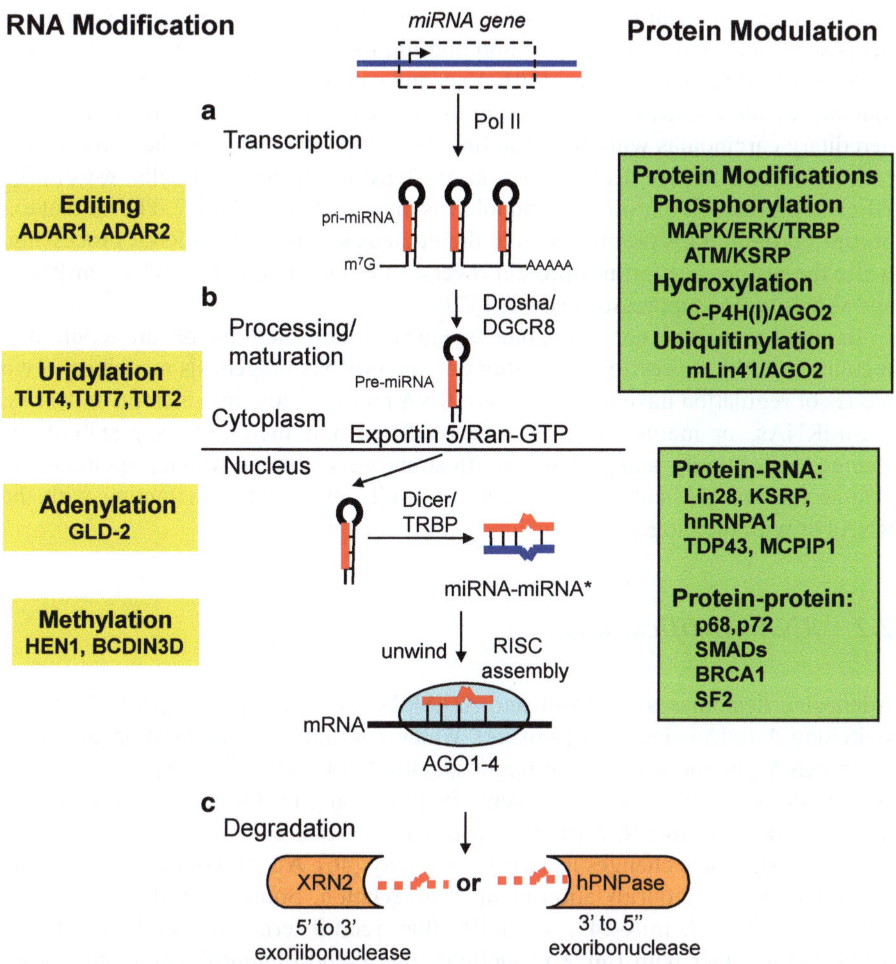

Fig. 3.1 miRNA biogenesis and regulation. The biogenesis of miRNA comprises three steps: transcription (**a**), processing and maturation (**b**), and degradation (**c**). The proteins involved in these steps constitute the core machinery of miRNA biogenesis [17]. RNA modification, including A-to-I editing, uridylation, adenylation, and methylation on pri-mRNAs, pre-miRNAs, or mature miRNAs, changes the miRNA stability and/or target specificity. Protein modulation, such as protein modification (phosphorylation, hydroxylation, ubiquitinylation), protein–RNA interaction, or protein–protein interaction, regulates the core miRNA biogenesis and alters the miRNA expression levels

The core components of miRNA biogenesis machinery include Drosha/DGCR8, exportin5/Ran5–GTP, Dicer/TRBP, AGO1–4, and XRN2/hPNPase [37]. Loss of expression or gain of function caused by mutations in these core genes results in dysregulation of many miRNAs, therefore contributing to the development of tumors. For example, loss of Dicer and Drosha expression is found in breast cancer samples [38], implying that Dicer and Drosha function as suppressors of breast cancer progression. Furthermore, loss of Dicer is a predictor of better response for breast cancers to chemotherapy and to endocrine therapy [39]. Ovarian cancer patients with both high Dicer expression and high Drosha expression are associated with increased median survival [40]. A second example is that reduced TRBP protein expression caused by mutations in the *TARBP2* gene is found in sporadic and hereditary carcinomas with microsatellite instability and results in the defect of the processing of miRNAs. Reintroduction of TRBP in the deficient cells restores the efficient production of miRNAs and inhibits tumor cell growth [41]. The mutant exportin-5 protein traps pre-miRNAs in the nucleus and reduces miRNA processing, while the restored exportin-5 protein reverses the impaired export of pre-miRNAs and shows tumor suppressor features [42].

In addition to these core components, many other protein factors are reported to regulate miRNA expression by modulating the miRNA biogenesis machinery. Two modes of regulation have been reported: RNA modification, in which pri-miRNAs, pre-miRNAs, or mature miRNAs are modified and therefore their stability is changed accordingly, and protein modification/interaction, in which protein factors change the miRNA stability or target specificity by directly interacting with the RNAs or with the biogenesis machinery protein components.

2.2 RNA Modifications

Adenosine deaminases (ADAR1 and ADAR2) edit some human pri-miRNAs by switching A (adenosine) to I (inosine), which can block the maturation of pri- or pre-miRNAs and/or change their target specificity [43–45] (Fig. 3.1). For example, A-to-I editing of pri-miR-142 prevents its processing by Drosha but promotes its degradation by Tudo-SN, a ribonuclease, while A-to-I editing of pre-miR-376 in the "seed" sequence changes its target specificity [46]. A 3′ end of pre-miRNAs can be modified by oligouridylation or monouridylation. Some pre-miRNAs, including let-7, miR-107, miR-143, and miR-200c, recruit terminal uridylyltransferase 4 (TUT4) together with Lin28 to facilitate the 3′-end oligouridylation and subsequent degradation in stem cells [47]. Prototypic pre-miRNAs usually have a 2-nt 3′ overhang that can be recognized by Dicer for processing, while some pre-miRNAs (let-7 and miR-105) only acquire a 1-nt 3′ overhang from Drosha processing and therefore require a 3′-end monouridylation for Dicer processing. TUT7, TUT4, and TUT2, as the terminal uridylyl transferases, are responsible for pre-miRNA monouridylation. Monouridylation occurs in somatic cells to promote let-7 biogenesis, while oligouridylation inhibits let-7 in embryonic stem cells [48]. In addition to uridylation, the 3′-end of pre-miRNA can also undergo adenylation. A liver-specific miRNA, miR-122, is added to an adenosine at the 3′-end by GLD2, a regulatory cytoplasmic poly(A) polymerase, which is required for the selective stabilization of miR-122 in the liver [49, 50]. Interestingly, miR-122 is found to downregulate cytoplasmic polyadenylation element-binding protein (CPEB), which promotes polyadenylation/translation on the 3′-UTR of p53 mRNA by recruiting Gld4, a second noncanonical poly(A) polymerase. This newly identified pathway Gld2/miR-122/CPEB/Gld4/p53 results in cellular senescence of primary human diploid fibroblasts

[51]. A 2′-O-methylation on the 3′-terminal ribose is another major mechanism that increases the stability of small RNAs. HUA ENHANCER1 (HEN1), an RNA methyltransferase, has been reported to methylate miRNAs and small interfering RNAs (siRNAs) in plants and *Drosophila*, Piwi-interacting RNAs (piRNAs) in animals [52–55]. Another RNA methyltransferase, BCDIN3D, can O-methylate 5′-monophosphate of pre-miRNAs and negatively regulates miRNA maturation. Specifically, BCDIN3D phospho-dimethylates pre-miR-145 and inhibits the Dicer processing for pre-miR-145 [56].

2.3 Protein Modifications

The core protein component TRBP can be phosphorylated by the mitogen-activated protein kinase (MAPK) Erk, and this phosphorylation enhances miRNA production by increasing stability of the Dicer/TRBP complex [57] (Fig. 3.1). KH-type splicing regulatory protein (KSRP) has been shown to be serine phosphorylated by ataxia telangiectasia mutated (ATM) protein in response to DNA damage, and this phosphorylation facilitates the miRNA processing for a subset of miRNAs [58]. These studies suggest a general principle wherein signaling pathways can achieve their biological outcome through regulating the miRNA machinery. AGO proteins are essential components of the RISCs. It is reported that AGO2 can be hydroxylated by the type I collagen prolyl-4-hydroxylase (C-P4H(I)) and this hydroxylation is important for AGO2 stability and efficient RNA interference [59]. AGO2 can also be ubiquitinylated by an E3 ubiquitin ligase Mouse Lin41 (mLin41) and destined for degradation. Therefore, mLin41 acts as an inhibitor of the miRNA pathway by targeting AGO2 for ubiquitinylation. But mLin41 also cooperates with Lin28 in suppressing let-7 activity independent of its E3 ligase activity, revealing a dual control mechanism regulating let-7 in stem cells [60].

2.4 Protein–RNA Interactions

Lin28 Lin28 is a pluripotency factor with two isoforms Lin28A and Lin28B. Both Lin28A and Lin28B can downregulate let-7 miRNA expression but with different mechanisms (Fig. 3.1). Lin28A recognizes a tetranucleotide sequence motif (GGAG) in the terminal loop and recruits TUT4 to add an oligouridine tail to the pre-let-7, which blocks Dicer processing [47]. Lin28A also uses TUT7 as an alternative TUTase that redundantly controls let-7 biogenesis in embryonic stem cells with TUT4 [61]. Lin28B represses let-7 processing through a TUT4-independent mechanism. Lin28B functions in the nucleus by sequestering pri-let-7 transcripts and inhibiting their processing by Drosha/TRBP complex. Furthermore, Lin28A and Lin28B are exclusively expressed in human breast tumors: Lin28A is overexpressed in HER2-overexpressing breast tumors, whereas Lin28B is overexpressed in triple-negative breast tumors, suggesting that the different mechanisms that Lin28A and Lin28B employed may contribute to the different types of breast tumors [62].

KSRP KSRP is a key mediator of mRNA decay and can regulate the biogenesis of a subset of miRNAs by forming complexes with Drosha or Dicer [63]. KSRP binds to the terminal loop of pre-miRNAs and promotes their maturation. The target mRNAs by KSRP-induced miRNAs have been shown to be involved in specific biological programs, including proliferation, apoptosis, and differentiation [63]. Interestingly, KSRP is also a key player that transduces DNA damage signaling to miRNA biogenesis. The ATM kinase directly binds to and phosphorylates KSRP, leading to enhanced interaction between KSRP and pri-miRNAs and increased KSRP activity in miRNA processing, suggesting a novel mechanism by which DNA damage signaling is linked to miRNA biogenesis [58].

hnRNP A1 Heteronuclear ribonucleoprotein A1 (hnRNP A1) can negatively regulate let-7a by binding to the terminal loop of pri-let-7a and inhibiting its processing by Drosha. The binding of hnRNP A1 to let-7a interferes with the binding of KSRP, thereby having an antagonistic role in the regulation of let-7a expression [64, 65]. Additionally, hnRNP A1 not only binds to the pri-miR-18a and reduces its processing by Drosha but also involves in the mature miR-18a-mediated repression of target genes, suggesting the new role for general RNA-binding proteins as auxiliary factors to facilitate the processing of specific miRNAs [66].

TDP-43 TAR DNA-binding protein-43 (TDP-43), a homolog to hnRNPs, is known to be involved in RNA processing and its mutation and abnormal cellular distribution is a key feature of amyotrophic lateral sclerosis (ALS) and frontotemporal lobar degeneration (FTLD), two neurodegenerative diseases [67]. TDP-43 facilitates the production of a subset of pre-miRNAs by interacting with the Drosha complex and binding directly to the relevant pri-miRNAs. Furthermore, cytoplasmic TDP-43, which interacts with the Dicer complex, promotes the processing of some pre-miRNAs via binding to their terminal loops. The involvement of TDP-43 in miRNA biogenesis is indispensable for neuronal outgrowth [68]. The *Drosophila* TDP-43 (dTDP-43) has also been reported as controlling the precision of sensory organ precursor (SOP) specification through acting on miR-9a, suggesting a novel role for endogenous TDP-43 in neurodegeneration diseases via miRNAs [69].

MCPIP1 Monocyte chemotactic protein-induced protein 1 (MCPIP1) is a ribonuclease that acts as a broad suppressor of miRNA activity and biogenesis. MCPIP1 counteracts with Dicer in miRNA processing and suppresses miRNA biosynthesis via cleavage of the terminal loops of pre-miRNAs. The balance between Dicer-mediated processing and MCPIP1-mediated destroying modulates miRNA biogenesis and potentially affects the normal and pathological miRNA regulation [70].

2.5 Protein–Protein Interactions

p68 (DDX5) and p72 (DDX17) Both p68 and p72 are DEAD-box RNA helicase subunits that are required for efficient RNA splicing and miRNA processing. They are found in the Drosha complex and are required for the recognition of a subset of pri-miRNAs in Drosha-mediated processing [71]. In particular, in response to DNA damage, p53 interacts with the Drosha-processing complex through the association with p68 and facilitates the processing of pri-miRNAs to pre-miRNAs, including miR-16–1, miR-143, and miR-145. Inactive p53 mutants interfere with the p68/Drosha interaction, leading to attenuation of miRNA-processing activity [72]. miR-143 and miR-145 belong to a subset of miRNAs whose expression is controlled by p53 and p68/p72. The combination of miR-143 and miR-145 inhibits the expression of c-Myc, whereas miR-145 downregulates p72 expression, forming a feedback loop to prevent overproduction of a subset of tumor suppressive miRNAs by repressing their own modulators p68/p72 [73]. The overexpression of p68/p72 has been reported in three major human cancers (colon, breast, prostate), strongly suggesting their proto-oncoprotein properties [74].

SMADs Smads, the signal transducers of transforming growth factor (TGF)-β/bone morphogenetic protein (BMP), promote the expression of a subset of miRNAs by facilitating the cleavage by Drosha. A majority of TGF-β/BMP-regulated miRNAs (T/B-miRs) contain a consensus sequence Smad binding element (R-SBE) within the stem region of the primary transcripts of T/B-miRs (pri-T/B-miRs), to which Smads directly bind. Mutation of the R-SBE abrogates TGF-β/BMP-induced recruitment of Smads, Drosha, and DGCR8 to pri-T/B-miRs and impairs their processing, whereas introduction of R-SBE to unregulated pri-miRNAs is sufficient to recruit Smads and to allow regulation by TGF-β/BMP [75, 76].

BRCA1 BRCA1 accelerates the processing of pri-miRNAs. BRCA1 increases the expressions of both precursor and mature forms of let-7a-1, miR-16–1, miR-145, and miR-34a. BRCA1 binds directly to Drosha and p68 and also recognizes the RNA secondary structure and directly binds with pri-miRNAs via a DNA-binding domain. BRCA1 regulates miRNA biogenesis via the Drosha/DGCR8 complex, suggesting a novel function of BRCA1 in miRNA biogenesis, which may be linked to its tumor suppressor mechanism and maintenance of genomic stability [77]. BRCA1 can also epigenetically repress miR-155 expression via its association with HDAC2, which deacetylates histones H2A and H3 on the miR-155 promoter. The R1699Q variant of BRCA1, a potentially moderate-risk variant, does not impair DNA damage repair but abrogates the repression of miR-155. This demonstrates a new mode of tumor suppression by BRCA1 and suggests that miR-155 is a potential therapeutic target for BRCA1-deficient tumors [78].

SF2 The splicing factor serine/arginine-rich splicing factor 2 (SF2/ASF) is found to regulate about 40 miRNAs expression by miRNA deep sequencing [79]. SF2/ASF and one of its upregulated miRNAs (miR-7) form a negative feedback loop: SF2/ASF promotes miR-7 maturation, and mature miR-7 in turn targets the 3′-UTR of SF2/ASF to repress its translation. Direct interaction between SF2/ASF and pri-miR-7 facilitates Drosha cleavage and enhances its expression, which is independent of SF2/ASF's splicing function. Other miRNAs, including miR-221 and miR-222, may also be regulated by SF2/ASF through a similar mechanism [79].

Taken together, the miRNA biogenesis is subject to multiple levels of regulation to tightly control its activity (Fig. 3.1). Loss of activity of the core components and other regulatory factors leads to the dysregulation of a subset of miRNAs or a specific miRNA. It will be interesting to see how the core miRNA biogenesis machinery is regulated by specific factors to achieve its specificity of regulation on a subgroup of miRNAs or a specific single miRNA. Of particular interest to the DNA repair is that the miRNA biogenesis regulatory proteins, p68, BRCA1, and ATM/KSRP, are already known as DNA repair factors, suggestive of an intrinsic link of DDR with miRNA biogenesis.

3 MicroRNAs in DNA Repair

It has been known that miRNA expression can be modulated by different types of DNA damage. For example, ionizing radiation (IR)-induced DNA damage (mainly the DSBs) induces miRNA expression. However, no obvious overlap of IR-responsive miRNAs has been found among different cell lines, suggesting that IR-responsive miRNA profiles might be cell type specific [17]. Different DNA-damaging agents, such as ultraviolet (UV) radiation, H_2O_2, and chemical compounds, have been reported to modulate miRNA expression but induce a different miRNA response [80–82]. In addition to being regulated by DNA damage, miRNAs can also regulate DDR by targeting the protein components of DDR pathways. Herein, we will summarize the miRNAs that modulate the DDR after briefly reviewing the different types of DNA repair mechanisms.

3.1 DNA Repair Mechanisms

Cells have developed different DNA repair mechanisms to deal with DNA damages caused by endogenous stimuli and exogenous environmental insults. The major endogenous DNA damage molecules are reactive oxygen species (ROS), which are the by-products of normal cellular metabolism and can cause oxidative stress. Oxidative stress has been demonstrated to play a significant role in the etiology of many diseases, including DNA repair deficiency disorders [83]. An excess of ROS may lead to the formation and accumulation of mutagenic, toxic, and genome-destabilizing DNA lesions. To repair and resolve such lesions, cells have developed several DNA repair pathways, such as NER, BER and MMR.

NER and BER are two pathways responsible for repair of the majority of DNA lesions induced by ROS. NER is a multistep repair pathway that specifically fixes the oxidative-modified DNA bases, such as 8-oxoguanine, thymine glycol (Tg), and cyclodeoxyadenosine. NER involves five steps: recognition, recruiting, incision, synthesis, and ligation. Basically, the DNA damage is recognized by the repair factors xeroderma pigmentosum complementation group A (XPA), replication protein A (RPA) and XPC, which recruit the helicases XPB and XPD to the damaged sites to unwind DNA. Then XPG and XPF–excision repair cross-complementing (ERCC1) nucleases are recruited for the 3′ and 5′ incisions, respectively, to remove

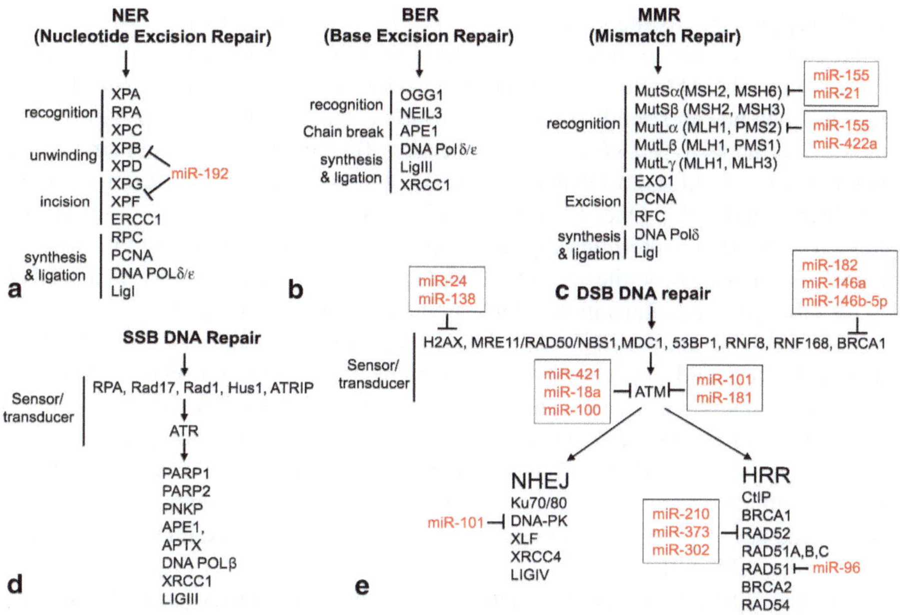

Fig. 3.2 DNA repair machinery and miRNAs. The core protein components of different repair pathways are summarized: **a** nucleotide excision repair, **b** base excision repair, **c** mismatch repair, **d** single-strand break DNA repair, and **e** double-strand break DNA repair. miRNAs that have been shown to regulate the core proteins of these different repair pathways are indicated in *red*

the damaged DNA. Then repair synthesis starts to fill the gap by replication factor C (RPC), proliferating cell nuclear antigen (PCNA) and DNA polymerases POLδ/ε using the other strand as a template and finally the newly synthesized strand is ligated by DNA ligase I (Fig. 3.2a) [84].

BER is a simpler version of NER in the case that the oxidative damage is confined to a base. BER essentially involves three steps: (1) Recognition: the modified base is recognized and cleaved by an appropriate DNA glycosylase, such as oxoguanine glycosylase (OGG1) for 8-oxo-G damage or endonuclease VIII-like 3 (NEIL3) for Tg damage. (2) Chain break: the AP endonuclease (APE)1 is recruited to the 5′ side of the base to break the chain. (3) Fill-up and ligation: DNA POLδ/ε comes to fill up one nucleotide gap and the DNA ligase III/XRCC1 complex arrives to seal the nick (Fig. 3.2b) [85, 86].

Mispaired bases of DNA are recognized by heterodimeric complexes of mutator S (MutS)-related proteins, which then recruit the downstream MutL-related proteins to facilitate MMR. Different heterodimeric complexes of MutS and MutL are reported to be responsible for recognition and repair of different types of mismatched DNA. For example, MutSα, composed of melanocyte-stimulating hormones (MSH)2 and MSH6, initiates the repair of base–base mismatches and small insertion–deletion loops (IDLs). MutSβ, a heterodimer of MSH2 and MSH3, repairs both small loops in addition to large loop mismatches of approximately ten nucleotides. MutLα (composed of MutL homolog 1 (MLH1) and postmeiotic segregation

2 (PMS2) proteins) is the primary complex for mismatch correction while MutLβ heterodimer (composed of MLH1 and PMS1 proteins) plays a minor role. Furthermore, MutLγ (MLH1 and MLH3) acts as a backup for MutLα in the repair of base–base mismatches and small IDLs. MutS/MutL forms a sliding clamp with PCNA and replication factor C (RFC) to allow the identification of the daughter strand, and exonuclease 1 (EXO1), a DNA exonuclease, enters the DNA structure to remove daughter-strand DNA. Once the mismatch is removed, the activity of EXO1 is suppressed by MutL, thus terminating DNA excision. Upon completion of this process, a DNA polymerase δ synthesizes DNA in place of the excised sequence with a DNA ligase I that joins any gaps in the DNA sequence (Fig. 3.2c) [5, 87].

SSB and DSB are usually induced by exogenous stimulators such as UV, IR or chemical compounds. ATM and Ataxia Telangiectasia and Rad3 related (ATR) are serine/threonine kinases that transduce these SSB and DSB DNA damage signals to downstream events. ATM is primarily activated in response to DSBs, whereas ATR is mainly involved in SSB and stalled replication forks. ATM/ATR coordinate downstream events such as cell cycle, DNA repair, and apoptosis by phosphorylating a wide set of protein substrates and impact a variety of cellular physiologies (Fig. 3.2d, e) [88, 89].

To repair SSB, ATR has to be activated. Single-strand DNA (ssDNA) is stabilized by replication protein A (RPA) binding and then recruits Rad17 to load Rad9–Rad1–Hus1 (9–1–1) complex and ATR-interacting protein (ATRIP)–ATR complex onto DNA, during which ATR is activated. In SSBR, poly(ADP-ribosyl)ation of ssDNA by poly(ADP-ribose) polymerase (PARP)1/2 is the first step and thought to aid in recruiting other DNA repair proteins, such as XRCC1 and ligase III to the site and promote SSB repair following DNA end processing by XRCC1-interacting proteins such as DNA Polβ, polynucleotide kinase (PNK), apurinic/apyrimidinic (AP) endonuclease 1 (APE1), Aprataxin (APTX), and Aprataxin and PNKP like factor (APLF) (Fig. 3.2d) [90].

DNA DSBs are highly toxic to cells and can drive genomic instability [91]. Failure to properly execute DSB repair is known to accelerate tumorigenesis and is associated with several genetic disorders [92]. ATM activation is required for DSB repair. The MRE11-RAD50-NBS1 (MRN) complex acts as a sensor of DSB to recruit ATM to DNA damage site. The phosphorylation of histone variant H2A histone family, member X (H2AX) by ATM results in the recruitment of MDC1 as a scaffold to further recruit 53BP1 and a series of ubiquitin ligases ring finger protein 8 (RNF8), RNF168 and BRCA1 to initiate DSB DNA repair (Fig. 3.2e) [93].

NHEJ and HRR represent two major DSB repair pathways in different cell cycle phase. NHEJ occurs in the G0/G1 phase while HRR in the late S/G2 phase. Up to 90 % of DSBs are repaired by NHEJ in G1 phase of the cell cycle by an ATM-independent mechanism. Six core proteins required for NHEJ have been identified to date, Ku70, Ku80, DNA-dependent protein kinase, catalytic subunits (DNA-PKcs), X-ray repair cross-complementing protein 4 (XRCC4), XRCC4-like factor (XLF), and DNA ligase IV (LIGIV), which are assembled as two steps: the Ku heterodimer (Ku70/80) binds to DSB ends and recruits DNA-PKcs and consequently coordinate end processing with rejoining by recruiting XRCC4, XLF, and LIGIV (Fig. 3.2e) [94]. DSBs can also be repaired by homologous recombination (HR)-mediated pathways. Repair is initiated by CtBP-interacting protein (CtIP)–BRCA1 complex-mediated resection of a DSB to provide 3′ssDNA overhangs [95] and followed by strand invasion and strand displacement, which is mediated by RAD52, RAD51 paralogs (-A, -B, -C). DNA resynthesis of the broken portion with the undamaged sister molecules as a template is then mediated by RAD51–BRCA2 complex and RAD54 (Fig. 3.2e) [96].

3.2 Core MMR Proteins-miR-155, miR-422a, miR-21

Defects in MMR can lead to genomic instability and cause hereditary colorectal cancer as well as 10–40 % of sporadic colorectal and gastric cancers [97] (Fig. 3.2c). A human miRNA, miR-155, is reported to negatively regulate MSH2, MSH6, and MLH1, the three core MMR proteins that affect the recognition and repair of mismatch DNA. Overexpression of miR-155 induces the genomic instability, including elevated mutation rates and microsatellite instability (MSI), by targeting these MMR core proteins. The inversed correlation between miR-155 overexpression and downregulation of MLH1 or MSH2 protein expression is found in human colorectal cancer, suggesting that miR-155 modulation of MMR might be a mechanism of colorectal cancer pathogenesis [98]. Another miRNA, miR-422a, is found to downregulate MutLα (MLH1–PMS2 heterodimer) levels by suppressing MLH1 expression through its 3′-UTR. Interestingly, MutLα stimulates the conversion of pri-miR-422a to pre-miR-422a, thereby forming a feedback loop that regulates the level of both molecules [99]. MMR repair-deficient cells display a characteristic of reduced 5-fluorouracil (FU)-induced G2/M damage arrest and apoptosis. Overexpression of miR-21 in cells exhibits this cellular phenotype and miR-21 is found to downregulate the core MMR recognition protein complex, MSH2 and MSH6. A high level of miR-21 is inversely correlated with reduced MSH2 protein expression in a number of human tumors including colorectal cancer. Moreover, xenograft studies demonstrate that miR-21 overexpression dramatically reduces the therapeutic efficacy of 5-FU. These studies suggest that the downregulation of the MMR gene by miR-21 overexpression may be an important clinical indicator of therapeutic efficacy in colorectal cancer [100].

3.3 NER Proteins–miR-192

NER has been found to be inhibited by hepatitis B virus (HBV) [101] (Fig. 3.2a). miRNA expression profiling of HBV-infected hepatocellular cells identified that miR-192 is significantly upregulated in these infected cells. Furthermore, overexpressing miR-192 inhibits cellular NER by downregulating XPB and XPF, two key factors in NER. These results indicate that persistent HBV infection might trigger NER impairment in part through upregulation of miR-192, which suppresses the levels of XPB and XPF. It provides new insights into the effect of chronic HBV infection on NER and genetic instability in cancer [102].

3.4 H2AX-miR-24, miR-138

The phosphorylation of H2AX (γ-H2AX) is the initial step for a cascade of DSB response, functioning to link the damaged DNA to the DNA repair machinery. γ-H2AX foci formation is also an indicator for DSB. Modulation of H2AX expression is, therefore, important for the DSB detection and repair. miR-24 is identified by miRNA array to be upregulated during postmitotic differentiation of hematopoietic cell lines and downregulates the expression of H2AX. miR-24-mediated suppression of H2AX renders hematopoietic cells hypersensitive to γ-irradiation and genotoxic drugs, which might account for the reduced capacity to repair DSB in terminally differentiated hematopoietic cells [103]. By screening a library of human miRNA mimics to inhibit γ-H2AX foci formation, miR-138 directly targets the histone H2AX 3′-UTR to reduce H2AX expression and induces chromosomal instability after DNA damage. Overexpression of miR-138 inhibits HRR and enhances cellular sensitivity to multiple DNA-damaging agents. Reintroduction of H2AX in miR-138 overexpressing cells abrogates miR-138-mediated hypersensitivity. This study suggests that miR-138 is an important regulator of genomic stability and a potential therapeutic agent to improve the efficacy of radiotherapy and chemotherapy with DNA-damaging agents [104].

3.5 BRCA1-miR182, miR-146a

BRCA1 is a strong breast cancer susceptibility gene, and germline mutations in the BRCA1 gene predispose women to breast cancer. The BRCA1 protein plays a critical role in DSB DDR by detecting the DNA damage and promoting HRR [105]. BRCA1 deficient cells show genome instability and are intrinsically sensitive to PARP inhibitors [106]. BRCA1 modulates the miRNA biogenesis by interacting with Drosha and p68 (Fig. 3.1). However, BRCA1 per se is subjected to miRNA modulation. In a pull-down assay, BRCA1 transcripts are reported to be enriched in the AGO/miR-182 complex. Overexpression of miR-182 leads to the hypersensitivity of cells to IR and impairs HRR. These impaired DNA repair phenotypes can be rescued by introducing back BRCA1 protein, suggesting that BRCA1 mediates the effects of miR-182 on DNA repair. On the other hand, inhibition of miR-182 increases BRCA1 protein levels and protects cells from IR-induced cell death. miR-182-overexpressing breast tumor cells are hypersensitive to PARP1 inhibitors, similar to the BRCA1 deficient cell. Conversely, inhibiting miR-182 enhances BRCA1 levels and induces resistance to PARP1 inhibitors. These results suggest that miR-182-mediated downregulation of BRCA1 affects DNA repair and may impact breast cancer therapy [107].

Low levels of BRCA1 protein is also found in about one-third of sporadic breast cancers. Two miRNAs, miR-146a and miR-146b-5p, have been shown to bind to 3′-UTR of BRCA1 and downregulate BRCA1 expression. The miR-146a/miR-146b-5p-mediated BRCA1 increases cell proliferation and reduces HRR. Furthermore, the highest levels of miR-146a and/or miR-146b-5p are found in basal-like mammary tumor epithelial cell lines, and in triple negative breast tumors, which are the closest to tumors arising in carriers of BRCA1 mutations. This work provides further evidence for the involvement of miRNAs in sporadic breast cancer through downregulation of BRCA1 [108].

3.6 ATM-miRNA421, miR-18a, miR-100, miR-101, miR-181

ATM is the chief transducer in the DSB signaling and mutations in this gene lead to a typical genomic instability disorder ataxia telangiectasia [109]. This disease also displays hypersensitivity to ionizing radiation, suggesting that modulation of ATM protein could alter cellular radiosensitivity [110, 111]. By using the target prediction program, miR-421 is reported to suppress ATM expression by targeting the 3′-UTR of ATM transcripts. Ectopic expression of miR-421 results in S-phase cell-cycle checkpoint changes and an increased sensitivity to ionizing radiation. This is the first study to show that ATM is subject to miRNA regulation and miR-421–ATM pathway might contribute to the DDR in a variety of cells given the broad expression pattern of ATM [112]. Interestingly, a squamous carcinoma cell line SKX exhibits a pronounced radiosensitivity after IR with enhanced levels of miR-421. Transfection of SKX cells with either anti-miR-421 inhibitor or a miRNA-insensitive ATM vector restores the ATM expression and abrogates the hyperradiosensitivity. This is the first report describing miRNA-mediated downregulation of ATM leading to clinically manifest tumor radiosensitivity [113].

Another miRNA, miR-18a, is found to be overexpressed in breast cancer cell lines and breast cancer patients' tissue samples. The overexpression of miR-18a reduces HRR and sensitizes breast cancer cells to IR treatment in a similar way to ATM siRNA. Ectopically expressing miR-18a downregulates ATM expression by directly targeting the ATM–3′-UTR and abrogates the IR-induced cell-cycle arrest. On the other hand, inhibition of miR-18a leads to augmentation of DNA damage repair, increase of HRR efficiency and reduced cellular radiosensitivity [114]. This work provides a second miRNA that regulates cellular radiosensitivity through modulation of ATM protein level.

A low level of ATM is found in a human malignant glioma cell line with hyperradiosensitivity while miR-100 is highly expressed in this cell line. The 3′-UTR of ATM contains a binding site for miR-100. Knocking down miR-100 promotes ATM expression while overexpressing miR-100 reduces ATM expression and sensitizes these cells to IR. These results indicate that miR-100 could be another miRNA to target ATM and sensitize tumor cells to IR [115].

By combining the program prediction and the experiment validation, miR-101 could efficiently target DNA-PKcs and ATM via binding to the 3′-UTR of DNA-PKcs or ATM mRNA. Upregulating miR-101 efficiently reduced the protein levels of DNA-PKcs and ATM in the tumor cells and, most importantly, sensitized the tumor cells to radiation both in vitro and in vivo [116].

Reduced ATM expression has been found in TGF-β-induced breast cancer mammospheres and this is thought to be mediated by miR-181, which is upregulated by

TGF-β at the posttranscriptional level. Overexpression of miR-181 or depletion of ATM is sufficient to induce sphere formation in breast cancer cells, suggesting that the miR-181–ATM pathway is involved in the TGF-β-induced cancer stem cells [117].

3.7 *Rad51–miR-96*

The DNA repair protein RAD51 lies in the core of HRR by promoting DNA synthesis. MiR-96 is reported to target RAD51 on its coding region instead of 3′-UTR. Overexpression of miR-96 in human cancer cells reduces the levels of RAD51, decreases the efficiency of homologous recombination, and enhances sensitivity to the PARP inhibitor and to cisplatin. This study suggests that miR-96 can regulate chemosensitivity by repressing RAD51 and may serve as a therapeutic candidate to improve chemotherapeutic efficacy by increasing the sensitivity of cancer cells to DNA damage [118].

3.8 *RAD52–miR-210, 373, miR-302*

Two miRNAs, miR-210 and miR-373, are upregulated in a hypoxia-inducible factor-1 alpha (HIF1α)-dependent manner in hypoxic cells. Bioinformatics analyses suggested that these miRs could regulate factors implicated in DNA repair pathways. Overexpression of miR-210 is found to suppress the levels of RAD52, which is a key factor in HRR; the forced expression of miR-373 leads to a reduction in the NER protein, RAD23B, as well as in RAD52. Consistent with these results, both RAD52 and RAD23B are found to be downregulated in hypoxia, but in both cases, the hypoxia-induced downregulation could be partially reversed by antisense inhibition of miR-210 and miR-373. Importantly, luciferase reporter assays indicate that miR-210 is capable of interacting with the 3′-UTR of RAD52 and that miR-373 can act on the 3′-UTR of RAD23B. These results indicate that hypoxia-inducible miR-210 and miR-373 play roles in modulating the expression levels of key proteins involved in the HRR and NER pathways, providing new mechanistic insight into the effect of hypoxia on DNA repair and genetic instability in cancer [119]. miR-302 is downregulated in irradiated breast cancer cells. Additionally, the expression levels of miR-302a are inversely correlated with those of AKT1 and RAD52, two critical regulators of radioresistance. More promisingly, miR-302a sensitizes radioresistant breast cancer cells to radiation therapy and reduces the expression of AKT1 and RAD52. These data suggest that miR-302 is a potential sensitizer to radiotherapy [120].

4 miRNAs in Other DNA Damage Response Events

DNA repair is one of the most important events in DDR. There are some other events upstream or downstream of DNA repair, such as cell-cycle arrest to allow cells to repair damaged DNA or apoptosis if the DNA damage is not able to be fixed. miRNAs are reported to be involved in these events. For example, overexpression of miR-106b promotes cell-cycle progression, whereas loss of function reverses this phenotype. The cyclin-dependent kinase inhibitor p21 is a direct target of miR-106b and the miR-106b-mediated p21 downregulation overrides a doxorubicin-induced DNA damage checkpoint [121]. miR-21 is induced by DNA damage, negatively regulates G1/S transition, and participates in DNA damage-induced G2/M checkpoint. This is achieved by the downregulation of CDC25A, a cell-cycle regulator. miR-21 suppresses CDC25A expression through a defined sequence in the 3′-UTR of CDC25A [122].

p53 plays an important role in the DNA damage-induced apoptosis. Computational predictions suggest that several miRNAs are involved in the posttranscriptional regulation of p53. miR-504 downregulates human p53 through its direct binding to two sites in the p53 3′-UTR. Overexpression of miR-504 decreases p53 protein levels and regulates p53 transcriptional activity, p53-mediated apoptosis, and cell-cycle arrest in response to stress [123]. miR-125b is another negative regulator of p53 by targeting the 3′-UTR of p53 mRNA. Overexpression of miR-125b represses the endogenous level of p53 protein and suppresses apoptosis [124]. miR-138 directly targets the 3′-UTR of p53, significantly decreasing the expression of p53 and its downstream genes. Interestingly, the ectopic expression of miR-138 significantly improves the efficiency of induced pluripotent stem (iPS) cell generation by downregulating p53 expression [125].

There are other reported miRNAs that regulate the expression of core protein components of the DDR pathways, including miR-449a/b and miR-16 both targeting CDC25A [80, 126], miR-195 targeting Wee1 [127], miR-124a targeting CDK2 [128], and miR-100 targeting PLK1 [129].

5 Conclusions and Future Prospective

Here, we reviewed the DNA repair mechanisms that cells have developed to process the different types of DNA damages and summarized the core protein components for these DNA repair machineries. Only a few proteins to date are known to be regulated by miRNAs and no miRNAs have been reported to regulate BER and SSBR (Fig. 3.2). It is estimated that 20–30% of human proteins are regulated by miRNAs. Therefore, we envision that there will be more proteins in the DDR pathways that can be regulated by miRNAs.

Defects in these DNA repairs have been shown to cause different types of cancer with characteristics of genomic instability. Targeting DNA repair has already been used to treat cancers and the miRNA-mediated negative regulation of the DNA repair proteins is becoming a promising strategy to overcome the resistance developed by these cancer treatments, such as radiation therapy or chemotherapy. For instance, overexpression of miR-155 may induce colorectal cancer by modulating MMR protein expression [98] and miR-421 overexpression in squamous carcinoma tumors leads to clinical hyperradiosensitivity by targeting ATM expression [113]. These studies suggest that miRNAs are attractive therapeutic candidates to improve cancer treatment.

However, the challenge for using these miRNAs to treat cancers is how to deliver these miRNA mimics or anti-miRs specifically to tumors. Multiple technologies have been developed to achieve systemic delivery of miRNA mimics or anti-miRs, including the application of chemically modified oligonucleotides, lentiviral-based delivery, or nanoparticle-based delivery. However, each miRNA is known to target many transcripts and each target gene can be regulated by multiple miRNAs, as demonstrated by ATM, which can be regulated by multiple miRNAs (miR-421, miR-18a, miR-100, miR-101, and miR-181), and miR-101 can downregulate DNA-PK in addition to ATM (Fig. 3.2). Therefore, the safety of using miRNA for therapy needs to be extensively scrutinized and evaluated in animal model before clinical trials.

References

1. Hanahan D, Weinberg RA. The hallmarks of cancer. Cell. 2000;100(1):57–70.
2. Hanahan D, Weinberg RA. Hallmarks of cancer: the next generation. Cell. 2011;144(5):646–74.
3. Harper JW, Elledge SJ. The DNA damage response: ten years after. Mol Cell. 2007;28(5):739–45.
4. Misteli T, Soutoglou E. The emerging role of nuclear architecture in DNA repair and genome maintenance. Nat Rev Mol Cell Biol. 2009;10(4):243–54.
5. Jiricny J. The multifaceted mismatch-repair system. Nat Rev Mol Cell Biol. 2006;7(5):335–46.
6. Barnes DE, Lindahl T. Repair and genetic consequences of endogenous DNA base damage in mammalian cells. Annu Rev Genet. 2004;38:445–76.
7. Caldecott KW. Single-strand break repair and genetic disease. Nat Rev Genet. 2008;9(8):619–31.
8. West SC. Molecular views of recombination proteins and their control. Nat Rev Mol Cell Biol. 2003;4(6):435–45.
9. Bartek J, Bartkova J, Lukas J. DNA damage signalling guards against activated oncogenes and tumour progression. Oncogene. 2007;26(56):7773–9.
10. Jackson SP, Bartek J. The DNA-damage response in human biology and disease. Nature. 2009;461(7267):1071–8.
11. Li ML, Greenberg RA. Links between genome integrity and BRCA1 tumor suppression. Trends Biochem Sci. 2012;37(10):418–24.
12. Calin GA, Croce CM. MicroRNA-cancer connection: the beginning of a new tale. Cancer Res. 2006;66(15):7390–4.
13. Garzon R, et al. MicroRNA expression and function in cancer. Trends Mol Med. 2006;12(12):580–7.
14. MicroRNAs BDP. genomics, biogenesis, mechanism, and function. Cell. 2004;116(2):281–97.
15. Calin GA, et al. Human microRNA genes are frequently located at fragile sites and genomic regions involved in cancers. Proc Natl Acad Sci U S A. 2004;101(9):2999–3004.
16. Sevignani C, et al. MicroRNA genes are frequently located near mouse cancer susceptibility loci. Proc Natl Acad Sci U S A. 2007;104(19):8017–22.
17. Hu H, Gatti RA. MicroRNAs: new players in the DNA damage response. J Mol Cell Biol. 2011;3(3):151–8.
18. Wan G, et al. miRNA response to DNA damage. Trends Biochem Sci. 2011;36(9):478–84.
19. Hummel R, Hussey DJ, Haier J. MicroRNAs: predictors and modifiers of chemo- and radiotherapy in different tumour types. Eur J Cancer. 2010;46(2):298–311.
20. Bartel DP. MicroRNAs: target recognition and regulatory functions. Cell. 2009;136(2):215–33.
21. Lau NC, et al. An abundant class of tiny RNAs with probable regulatory roles in Caenorhabditis elegans. Science. 2001;294(5543):858–62.
22. Lagos-Quintana M, et al. Identification of novel genes coding for small expressed RNAs. Science. 2001;294(5543):853–8.

23. Baskerville S, Bartel DP. Microarray profiling of microRNAs reveals frequent coexpression with neighboring miRNAs and host genes. RNA. 2005;11(3):241–7.
24. Rodriguez A, et al. Identification of mammalian microRNA host genes and transcription units. Genome Res. 2004;14(10A):1902–10.
25. Han J, et al. The Drosha-DGCR8 complex in primary microRNA processing. Genes Dev. 2004;18(24):3016–27.
26. Kim YK, Kim VN. Processing of intronic microRNAs. Embo J. 2007;26(3):775–83.
27. Morlando M, et al. Primary microRNA transcripts are processed co-transcriptionally. Nat Struct Mol Biol. 2008;15(9):902–9.
28. Yi R, et al. Exportin-5 mediates the nuclear export of pre-microRNAs and short hairpin RNAs. Genes Dev. 2003;17(24):3011–6.
29. Lund E, et al. Nuclear export of microRNA precursors. Science. 2004;303(5654):95–8.
30. Chendrimada TP, et al. TRBP recruits the Dicer complex to Ago2 for microRNA processing and gene silencing. Nature. 2005;436(7051):740–4.
31. Okamura K, et al. Distinct roles for Argonaute proteins in small RNA-directed RNA cleavage pathways. Genes Dev. 2004;18(14):1655–66.
32. Rand TA, et al. Argonaute2 cleaves the anti-guide strand of siRNA during RISC activation. Cell. 2005;123(4):621–9.
33. Meister G, et al. Human Argonaute2 mediates RNA cleavage targeted by miRNAs and siRNAs. Mol Cell. 2004;15(2):185–97.
34. Chatterjee S, Grosshans H. Active turnover modulates mature microRNA activity in Caenorhabditis elegans. Nature. 2009;461(7263):546–9.
35. Das SK, et al. Human polynucleotide phosphorylase selectively and preferentially degrades microRNA-221 in human melanoma cells. Proc Natl Acad Sci U S A. 2010;107(26):11948–53.
36. Pawlicki JM, Steitz JA. Nuclear networking fashions pre-messenger RNA and primary microRNA transcripts for function. Trends Cell Biol. 2010;20(1):52–61.
37. Murphy D, Dancis B, Brown JR. The evolution of core proteins involved in microRNA biogenesis. BMC Evol Biol. 2008;8:92.
38. Yan M, et al. Dysregulated expression of dicer and drosha in breast cancer. Pathol Oncol Res. 2012;18(2):343–8.
39. Khoshnaw SM, et al. Loss of Dicer expression is associated with breast cancer progression and recurrence. Breast Cancer Res Treat. 2012;135(2):403–13.
40. Merritt WM, et al. Dicer, Drosha, and outcomes in patients with ovarian cancer. N Engl J Med. 2008;359(25):2641–50.
41. Melo SA, et al. A TARBP2 mutation in human cancer impairs microRNA processing and DICER1 function. Nat Genet. 2009;41(3):365–70.
42. Melo SA, et al. A genetic defect in exportin-5 traps precursor microRNAs in the nucleus of cancer cells. Cancer Cell. 2010;18(4):303–15.
43. Yang W, et al. Modulation of microRNA processing and expression through RNA editing by ADAR deaminases. Nat Struct Mol Biol. 2006;13(1):13–21.
44. Wulff BE, Nishikura K. Modulation of microRNA expression and function by ADARs. Curr Top Microbiol Immunol. 2012;353:91–109.
45. Vesely C, et al. Adenosine deaminases that act on RNA induce reproducible changes in abundance and sequence of embryonic miRNAs. Genome Res. 2012;22(8):1468–76.
46. Nishikura K. Functions and regulation of RNA editing by ADAR deaminases. Annu Rev Biochem. 2010;79:321–49.
47. Heo I, et al. TUT4 in concert with Lin28 suppresses microRNA biogenesis through pre-microRNA uridylation. Cell. 2009;138(4):696–708.
48. Heo I, et al. Mono-uridylation of pre-microRNA as a key step in the biogenesis of group II let-7 microRNAs. Cell. 2012;151(3):521–32.
49. D'Ambrogio A, et al. Specific miRNA Stabilization by Gld2-Catalyzed Monoadenylation. Cell Rep. 2012;2(6):1537–45.
50. Katoh T, et al. Selective stabilization of mammalian microRNAs by 3' adenylation mediated by the cytoplasmic poly(A) polymerase GLD-2. Genes Dev. 2009;23(4):433–8.

51. Burns DM, et al. CPEB and two poly(A) polymerases control miR-122 stability and p53 mRNA translation. Nature. 2011;473(7345):105–8.
52. Ji L, Chen X. Regulation of small RNA stability: methylation and beyond. Cell Res. 2012;22(4):624–36.
53. Huang Y, et al. Structural insights into mechanisms of the small RNA methyltransferase HEN1. Nature. 2009;461(7265):823–7.
54. Saito K, et al. Pimet, the Drosophila homolog of HEN1, mediates 2'-O-methylation of Piwi-interacting RNAs at their 3' ends. Genes Dev. 2007;21(13):1603–8.
55. Yang Z, et al. HEN1 recognizes 21-24 nt small RNA duplexes and deposits a methyl group onto the 2' OH of the 3' terminal nucleotide. Nucleic Acids Res. 2006;34(2):667–75.
56. Xhemalce B, Robson SC, Kouzarides T. Human RNA methyltransferase BCDIN3D regulates microRNA processing. Cell. 2012;151(2):278–88.
57. Paroo Z, et al. Phosphorylation of the human microRNA-generating complex mediates MAPK/Erk signaling. Cell. 2009;139(1):112–22.
58. Zhang X, et al. The ATM kinase induces microRNA biogenesis in the DNA damage response. Mol Cell. 2011;41(4):371–83.
59. Qi HH, et al. Prolyl 4-hydroxylation regulates Argonaute 2 stability. Nature. 2008;455(7211):421–4.
60. Rybak A, et al. The let-7 target gene mouse lin-41 is a stem cell specific E3 ubiquitin ligase for the miRNA pathway protein Ago2. Nat Cell Biol. 2009;11(12):1411–20.
61. Thornton JE, et al. Lin28-mediated control of let-7 microRNA expression by alternative TUTases Zcchc11 (TUT4) and Zcchc6 (TUT7). Rna. 2012;18(10):1875–85.
62. Nam Y, et al. Molecular basis for interaction of let-7 microRNAs with Lin28. Cell. 2011;147(5):1080–91.
63. Trabucchi M, et al. The RNA-binding protein KSRP promotes the biogenesis of a subset of microRNAs. Nature. 2009;459(7249):1010–4.
64. Michlewski G, Guil S, Caceres JF. Stimulation of pri-miR-18a Processing by hnRNP A1. Adv Exp Med Biol. 2010;700:28–35.
65. Michlewski G, Caceres JF. Antagonistic role of hnRNP A1 and KSRP in the regulation of let-7a biogenesis. Nat Struct Mol Biol. 2010;17(8):1011–8.
66. Guil S, Caceres JF. The multifunctional RNA-binding protein hnRNP A1 is required for processing of miR-18a. Nat Struct Mol Biol. 2007;14(7):591–6.
67. Buratti E, Baralle FE. Multiple roles of TDP-43 in gene expression, splicing regulation, and human disease. Front Biosci. 2008;13:867–78.
68. Kawahara Y, Mieda-Sato A. TDP-43 promotes microRNA biogenesis as a component of the Drosha and Dicer complexes. Proc Natl Acad Sci U S A. 2012;109(9):3347–52.
69. Li Z, et al. The FTD/ALS-associated RNA-binding protein TDP-43 regulates the robustness of neuronal specification through microRNA-9a in Drosophila. Hum Mol Genet. 2013;22(2):218–25.
70. Suzuki HI, et al. MCPIP1 ribonuclease antagonizes dicer and terminates microRNA biogenesis through precursor microRNA degradation. Mol Cell. 2011;44(3):424–36.
71. Fukuda T, et al. DEAD-box RNA helicase subunits of the Drosha complex are required for processing of rRNA and a subset of microRNAs. Nat Cell Biol. 2007;9(5):604–11.
72. Suzuki HI, et al. Modulation of microRNA processing by p53. Nature. 2009;460(7254):529–33.
73. Takaoka Y, et al. Forced expression of miR-143 represses ERK5/c-Myc and p68/p72 signaling in concert with miR-145 in gut tumors of Apc(Min) mice. PLoS One. 2012;7(8):e42137.
74. Janknecht R. Multi-talented DEAD-box proteins and potential tumor promoters: p68 RNA helicase (DDX5) and its paralog, p72 RNA helicase (DDX17). Am J Transl Res. 2010;2(3):223–34.
75. Davis BN, et al. Smad proteins bind a conserved RNA sequence to promote microRNA maturation by Drosha. Mol Cell. 2010;39(3):373–84.
76. Davis BN, et al. SMAD proteins control DROSHA-mediated microRNA maturation. Nature. 2008;454(7200):56–61.
77. Kawai S, Amano A. BRCA1 regulates microRNA biogenesis via the DROSHA microprocessor complex. J Cell Biol. 2012;197(2):201–8.

78. Chang S, et al. Tumor suppressor BRCA1 epigenetically controls oncogenic microRNA-155. Nat Med. 2011;17(10):1275–82.
79. Wu H, et al. A splicing-independent function of SF2/ASF in microRNA processing. Mol Cell. 2010;38(1):67–77.
80. Pothof J, et al. MicroRNA-mediated gene silencing modulates the UV-induced DNA-damage response. Embo J. 2009;28(14):2090–9.
81. Li G, et al. Alterations in microRNA expression in stress-induced cellular senescence. Mech Ageing Dev. 2009;130(11-12):731–41.
82. Hummel R, et al. Chemotherapy-induced modification of microRNA expression in esophageal cancer. Oncol Rep. 2011;26(4):1011–7.
83. Hegde ML, et al. Oxidative genome damage and its repair: implications in aging and neurodegenerative diseases. Mech Ageing Dev. 2012;133(4):157–68.
84. Nouspikel T. DNA repair in mammalian cells: Nucleotide excision repair: variations on versatility. Cell Mol Life Sci. 2009;66(6):994–1009.
85. Kim YJ, Wilson DM 3rd. Overview of base excision repair biochemistry. Curr Mol Pharmacol. 2012;5(1):3–13.
86. Robertson AB, et al. DNA repair in mammalian cells: Base excision repair: the long and short of it. Cell Mol Life Sci. 2009;66(6):981–93.
87. Pena-Diaz J, Jiricny J. Mammalian mismatch repair: error-free or error-prone? Trends Biochem Sci. 2012;37(5):206–14.
88. Lavin MF. Ataxia-telangiectasia: from a rare disorder to a paradigm for cell signalling and cancer. Nat Rev Mol Cell Biol. 2008;9(10):759–69.
89. Shiloh Y. ATM and related protein kinases: safeguarding genome integrity. Nat Rev Cancer. 2003;3(3):155–68.
90. Ciccia A, Elledge SJ. The DNA damage response: making it safe to play with knives. Mol Cell. 2010;40(2):179–204.
91. Chapman JR, Taylor MR, Boulton SJ. Playing the end game: DNA double-strand break repair pathway choice. Mol Cell. 2012;47(4):497–510.
92. Rass U, Ahel I, West SC. Defective DNA repair and neurodegenerative disease. Cell. 2007;130(6):991–1004.
93. van Attikum H, Gasser SM. Crosstalk between histone modifications during the DNA damage response. Trends Cell Biol. 2009;19(5):207–17.
94. Lieber MR. The mechanism of human nonhomologous DNA end joining. J Biol Chem. 2008;283(1):1–5.
95. Sartori AA, et al. Human CtIP promotes DNA end resection. Nature. 2007;450(7169):509–14.
96. San Filippo J, Sung P, Klein H. Mechanism of eukaryotic homologous recombination. Annu Rev Biochem. 2008;77:229–57.
97. Eshleman JR, Markowitz SD. Mismatch repair defects in human carcinogenesis. Hum Mol Genet. 1996;5 Spec No:1489–94.
98. Valeri N, et al. Modulation of mismatch repair and genomic stability by miR-155. Proc Natl Acad Sci U S A. 2010;107(15):6982–7.
99. Mao G, et al. Modulation of microRNA processing by mismatch repair protein MutLalpha. Cell Res. 2012;22(6):973–85.
100. Valeri N, et al. MicroRNA-21 induces resistance to 5-fluorouracil by down-regulating human DNA MutS homolog 2 (hMSH2). Proc Natl Acad Sci U S A. 2010;107(49):21098–103.
101. Jia L, Wang XW, Harris CC. Hepatitis B virus X protein inhibits nucleotide excision repair. Int J Cancer. 1999;80(6):875–9.
102. Xie QH, et al. MiR-192 inhibits nucleotide excision repair by targeting ERCC3 and ERCC4 in HepG2.2.15 cells. Biochem Biophys Res Commun. 2011;410(3):440–5.
103. Lal A, et al. miR-24-mediated downregulation of H2AX suppresses DNA repair in terminally differentiated blood cells. Nat Struct Mol Biol. 2009;16(5):492–8.
104. Wang Y, et al. MicroRNA-138 Modulates DNA Damage Response by Repressing Histone H2AX Expression. Mol Cancer Res. 2011;9(8):1100–11.

105. Caestecker KW, Van de Walle GR. The role of BRCA1 in DNA double-strand repair: Past and present. Exp Cell Res. 2013;319(5):575–87.
106. Helleday T. Homologous recombination in cancer development, treatment and development of drug resistance. Carcinogenesis. 2010;31(6):955–60.
107. Moskwa P, et al. miR-182-mediated downregulation of BRCA1 impacts DNA repair and sensitivity to PARP inhibitors. Mol Cell. 2011;41(2):210–20.
108. Garcia AI, et al. Down-regulation of BRCA1 expression by miR-146a and miR-146b-5p in triple negative sporadic breast cancers. EMBO Mol Med. 2011;3(5):279–90.
109. Gatti RA, et al. The pathogenesis of ataxia-telangiectasia. Learning from a Rosetta Stone. Clin Rev Allergy Immunol. 2001;20(1):87–108.
110. Gatti RA. The inherited basis of human radiosensitivity. Acta Oncol. 2001;40(6):702–11.
111. Nahas SA, Gatti RA. DNA double strand break repair defects, primary immunodeficiency disorders, and 'radiosensitivity'. Curr Opin Allergy Clin Immunol. 2009;9(6):510–6.
112. Hu H, et al. ATM is down-regulated by N-Myc-regulated microRNA-421. Proc Natl Acad Sci U S A. 2010;107(4):1506–11.
113. Mansour WY, et al. Aberrant overexpression of miR-421 downregulates ATM and leads to a pronounced DSB repair defect and clinical hypersensitivity in SKX squamous cell carcinoma. Radiother Oncol. 2013;106(1):147–54.
114. Song L, et al. miR-18a impairs DNA damage response through downregulation of ataxia telangiectasia mutated (ATM) kinase. PLoS One. 2011;6(9):e25454.
115. Ng WL, et al. Over-expression of miR-100 is responsible for the low-expression of ATM in the human glioma cell line: M059J. DNA Repair (Amst). 2010;9(11):1170–5.
116. Yan D, et al. Targeting DNA-PKcs and ATM with miR-101 sensitizes tumors to radiation. PLoS One. 2010;5(7):e11397.
117. Wang Y, et al. Transforming growth factor-beta regulates the sphere-initiating stem cell-like feature in breast cancer through miRNA-181 and ATM. Oncogene. 2011;30(12):1470–80.
118. Wang Y, et al. MiR-96 downregulates REV1 and RAD51 to promote cellular sensitivity to cisplatin and PARP inhibition. Cancer Res. 2012;72(16):4037–46.
119. Crosby ME, et al. MicroRNA regulation of DNA repair gene expression in hypoxic stress. Cancer Res. 2009;69(3):1221–9.
120. Liang Z, et al. MicroRNA-302 replacement therapy sensitizes breast cancer cells to ionizing radiation. Pharm Res. 2013;30(4):1008–16.
121. Ivanovska I, et al. MicroRNAs in the miR-106b family regulate p21/CDKN1A and promote cell cycle progression. Mol Cell Biol. 2008;28(7):2167–74.
122. Wang P, et al. microRNA-21 negatively regulates Cdc25A and cell cycle progression in colon cancer cells. Cancer Res. 2009;69(20):8157–65.
123. Hu W, et al. Negative regulation of tumor suppressor p53 by microRNA miR-504. Mol Cell. 2010;38(5):689–99.
124. Le MT, et al. MicroRNA-125b is a novel negative regulator of p53. Genes Dev. 2009;23(7):862–76.
125. Ye D, et al. MiR-138 promotes induced pluripotent stem cell generation through the regulation of the p53 signaling. Stem Cells. 2012;30(8):1645–54.
126. Yang X, et al. miR-449a and miR-449b are direct transcriptional targets of E2F1 and negatively regulate pRb-E2F1 activity through a feedback loop by targeting CDK6 and CDC25A. Genes Dev. 2009;23(20):2388–93.
127. Qi J, et al. microRNAs regulate human embryonic stem cell division. Cell Cycle. 2009;8(22):3729–41.
128. Nakamachi Y, et al. MicroRNA-124a is a key regulator of proliferation and monocyte chemoattractant protein 1 secretion in fibroblast-like synoviocytes from patients with rheumatoid arthritis. Arthritis Rheum. 2009;60(5):1294–304.
129. Shi W, et al. Significance of Plk1 regulation by miR-100 in human nasopharyngeal cancer. Int J Cancer. 2010;126(9):2036–48.

Part II
MicroRNAs in Stem Cells and Progression of Cancer

Chapter 4
Role of MicroRNAs in Stem Cell Regulation and Tumorigenesis in *Drosophila*

Stephanie Rager, Brian Chan, Lyric Forney and Shree Ram Singh

Abstract MicroRNAs (miRNAs) are small noncoding RNAs that modulate the expression of target mRNA. They are involved in many biological processes such as developmental timing, differentiation, cell death, immune response, stem cell behavior, and cancer. Growing evidence suggests that miRNAs play vital roles in regulating several aspects of stem cell biology in *Drosophila* including cell division, self-renewal, and differentiation. In recent years, miRNAs have emerged as collaborating factors that promote the activity of oncogenes in tumor development. Here, we present a brief overview on the role of miRNAs in the regulation of stem cell behavior and tumorigenesis in *Drosophila*.

Keywords MicroRNA · Stem cells · Tumorigenesis · *Drosophila*

1 Introduction

MicroRNAs (miRNAs) are small ~22-nucleotide (nt)-long noncoding RNAs, which bind to the 3′ untranslated region (UTR) of target mRNAs to regulate gene expression through translational repression and mRNA degradation [1–4]. miRNA biogenesis is a multistep process [5, 6]. miRNAs are initially transcribed in the nucleus as a primary miRNA transcript (pri-mRNA) by RNA polymerase II [7], which are then processed into precursor miRNAs (pre-mRNAs) by a microprocessor protein complex, the nuclear RNase III Drosha, and a double-stranded RNA-binding domain (dsRBD) protein Pasha [8–13]. The pre-miRNAs are then exported to the cytoplasm by the guanosine triphosphate-bound Ran (RanGTP)-dependent transporter protein Exportin 5 [14, 15], where they are further cleaved by RNase III enzyme Dicer [16–18] and its dsRBD partner Loquacious (Loqs) [19] to generate ~22-nt-long miRNA: miRNA* duplex. Finally, the one strand of this duplex (miRNA) is transferred to the RNA-induced silencing complex (RISC), containing

S. R. Singh (✉) · S. Rager · B. Chan · L. Forney
Mouse Cancer Genetics Program, and Basic Research Laboratory,
Center for Cancer Research, National Cancer Institute, Frederick, MD 21702, USA
e-mail: singhshr@mail.nih.gov

S. R. Singh, P. Rameshwar (eds.), *MicroRNA in Development and in the Progression of Cancer*, DOI 10.1007/978-1-4899-8065-6_4,

Argonaute-1 (Ago-1) for targeting gene expression, and releases the other strand (miRNA*) that undergoes degradation [20, 21].

The first miRNA gene, *lin-4,* and its target *lin-14* were discovered in a screening for genes that control developmental timing in *Caenorhabditis elegans* [22, 23]. Since then a large number of miRNAs conserved from worms to mammals have been identified [24–33]. Experimental studies in the past two decades have demonstrated that miRNAs play a regulatory role in various biological processes including development, tissue homeostasis, cell proliferation, tissue growth, cell death, neurogenesis, metabolism, immunity, cell fate determination, stem cell maintenance, aging, and several diseases including cancer [4, 24, 34–45]. Dysregulation of miRNA pathway results in developmental defects, several human diseases, and cancer. In this chapter, we will mainly focus on the role of miRNAs in regulation of stem cell self-renewal, differentiation, and tumorigenesis in *Drosophila.*

2 miRNAs in Stem Cell Regulation

Stem cells play a critical role in tissue development and homeostasis. There are two major classes of stem cells reported, embryonic stem (ES) cells and adult stem cells (including somatic and germ line). Stem cells are undifferentiated cells and have an enormous capacity for self-renewal and differentiation to form specialized cell types. Stem cells follow both asymmetric and symmetric division. Asymmetric division of stem cells results in the formation of two daughter cells, one retaining the stem cell characteristics and other one differentiating into specialized cell types [46, 47]. Stem cell self-renewal divisions are controlled by intrinsic and extrinsic (niche cells) factors [46, 47]. Failure of stem cell function of tissue maintenance results in degenerative diseases; on the other hand, overproliferation of stem cells results in tumor development and cancer [47]. Stem cells offer a great opportunity to study the growth and differentiation of individual cells into tissues and recent studies suggest that they can be used in the treatment of degenerative diseases and cancer [47].

Studies in recent years demonstrated that miRNAs play an important role in self-renewal and differentiation of stem cells in a variety of animal model systems [4, 41, 48–56]. Here, we focus only on the role of miRNAs in stem cell self-renewal and differentiation of germ-line stem cells (GSCs) and somatic stem cells (SSCs) in *Drosophila.*

3 miRNAs in *Drosophila* GSCs: Self-Renewal and Differentiation

GSCs are a self-renewing population of germ cells that generate haploid gametes. In *Drosophila* ovary and testis, GSCs are anchored around the niche cells (hub cells in testis and cap cells in ovary). Several signaling pathways regulate both male and

Table 4.1 MicroRNA pathway and its function in *Drosophila* stem cells and tumorigenesis

MicroRNA pathway	Function	References
Stem cells		
dicer-1	Reduction in germ-line cyst production and delayed GSC division in ovary	[57]
	Maintenance of GSC and SSC population in ovary	[59]
loqs	GSC maintenance in ovary	[48, 53]
bantam	GSC maintenance and repress PGC differentiation	[60, 64]
	Intestinal stem cell proliferation	[91]
Ago-1	GSC fate, oocyte formation, and GSC division in ovary	[61, 62, 66]
miR-7, miR-278	GSC division and differentiation in ovary and testis	[65, 68]
Mei-P26	Restricts growth and proliferation in the ovarian stem cell lineage	[63, 71]
	Regulates germ cell differentiation in ovary by genetically interacting with vasa	[67]
miR-184	GSC development and differentiation	[69]
miR-275, miR-306	Control stem cell differentiation by regulating Bam in testis	[73]
miR-310/13	Regulation of germ and somatic cell differentiation in testis	[74]
miR-124, let-7, miR-8/miR-200	Neuroblast stem cell division and differentiation	[87–90]
Tumorigenesis		
bantam	Promotes growth by limiting expression of Socs36E	[112]
	Regulates cell proliferation, cell death, and tissue growth	[107, 108, 115, 116]
miR-278	Misexpression in the developing eye causes massive overgrowth because of inhibition of apoptosis	[109]
miR-8/200	Growth inhibition by inducing apoptosis and blocking cell proliferation	[110]
miR-7	Enhances Notch pathway-induced eye overgrowth	[113]

female GSC systems. Recent studies demonstrated that the miRNA pathway plays a crucial role in the GSCs in *Drosophila* reproductive organs [48, 57–75] (Table 4.1).

3.1 miRNA and Female GSC

In the adult *Drosophila* ovary, the anterior tip of each germarium contains two to three GSCs, escort stem cells (ESCs), and follicle stem cells (FSCs). Each germarium contains five to seven nondividing somatic cap cells that physically anchor GSCs. Anterior to the cap cells are eight to ten terminal filament (TF) cells and inner germarium sheath (IGS) cells. GSC through asymmetric division produces a self-renewing GSC, and a differentiating cystoblast (CB) cell, which form an interconnected 16-cell cyst by incomplete cytokinesis. These germ cells become an oocyte and the nurse cells. In addition to GSCs, two to three FSCs reside in the middle of each germarium to proliferate and produce an egg chamber and follicle cells [76].

The role of miRNAs in *Drosophila* stem cells was first demonstrated using ovary GSC systems, where they promote cell division and maintenance of GSCs in their niche [48, 57–62] (Table 4.1). Hatfield et al. [57], using *Drosophila* ovarian GSC systems demonstrated that loss of *dicer*-1, the dsRNaseIII required for miRNA biogenesis, results in marked depletion of developing egg chambers because of the reduction in germ-line cyst production. Further, they found that reduction in cyst production in *dicer*-1 mutant GSCs was not only due to loss of GSCs or a change in their identity but due to a delayed G1-S-phase transition that is dependent on the cyclin-dependent kinase inhibitor Decapo [57]. It has been shown that normal processing of pre-miRNA by Dicer-1 required the dsRBD protein Loqs, which is further demonstrated to be involved in GSC maintenance in *Drosophila* ovary [48]. Further, it has been found that Loqs, Dicer-1, and Ago-1 intrinsically control the self-renewal of GSCs [53, 59]. In addition, Jin and Xie [59] found that Dicer-1 is also required for FSC maintenance in *Drosophila* ovary. Yang et al. [61, 62] found that overexpression of Ago-1 protein leads to GSC overproliferation; however, loss of *Ago-1* results in loss of GSCs, which suggests that *Ago-1* plays an essential and intrinsic role in GSC fate, oocyte formation, and GSC division [66]. Further, they showed that *Ago-1* is not required for *bag of marbles (bam)* silencing and proposed that an Ago-1-dependent miRNA pathway may play a crucial role in repressing GSC/CB [61, 62]. In addition to the role of Dicer-1 in adult GSC maintenance, Shcherbata et al. [60] found that *bantam* miRNA is extrinsically required for GSC maintenance.

Several studies suggest that the miRNA pathway regulates GSC maintenance by repressing *bam* in *Drosophila* [53, 59, 61, 62]. However, the miRNA pathway that controls the balance between self-renewal and differentiation was not clear until Neumuller et al. [63] demonstrated that *mei-p26,* a trim-NHL protein, together with *bam* and by interacting with Ago-1 through the NHL domain inhibits miRNA expression and controls germ cell differentiation [63]. Further, they also demonstrated that *mei-P26* regulates several miRNAs including *bantam*. Further, Liu et al. [67] have demonstrated that vasa promotes germ cell differentiation by genetically interacting with Mei-P26 and activating its translation by binding directly to a (U)-rich motif in its 3′ UTR. Furthermore, Li et al. [71] have shown that Mei-P26 regulates the fates of both GSCs and their differentiating daughters by promoting bone morphogenetic protein (BMP) signaling.

Yu et al. [65] reported that extrinsic signals from the insulin receptor (InR) pathway control Dacapo (Dap) expression through Dicer-1 to regulate GSC division. They found that *dicer-1* can directly regulate Dap levels through the *dap* 3′ UTR in GSCs. Further, in a luciferase assay, they found that *dap* 3′ UTR is targeted by *miR-7, miR-278,* and *miR-309.* Among these miRNAs, they showed that the GSC cell cycle is regulated through *dap* 3′ UTR by *miR-7* and *miR-278.* Furthermore, they showed that *miR-7* and *miR-278* and Dap-based cell cycle regulation in GSCs are controlled by InR signaling [65]. Lovino et al. (69) have demonstrated that *miR-184* controls GSC differentiation by translational repression of decapentaplegic (DPP) receptor Saxophone (Sax) protein levels. Yang et al. [61, 62] have shown that fragile X mental retardation protein (FMRP) interacts with Ago-1 and bantam and is required for GSC maintenance and repressing differentiation, and also needed for repressing primordial germ cell (PGC) differentiation and functions as an extrinsic factor for GSC maintenance in *Drosophila* ovary [64]). Recently, Wang et al. [70] provided the evidence that artificial miRNAs can effectively downregulate endogenous target genes (in this case, *bam, mad, ote,* and *dpp*) in GSCs and somatic cells in *Drosophila* ovary. More recently, Joly et al. [75] identified *mei-P26* mRNA as a direct and major target of Nos/Pum/CCR4-mediated translational repression for *Drosophila* female GSC self-renewal.

3.2 miRNAs and male GSC

The *Drosophila* testis tip harbors two types of stem cells, GSCs and SSCs. Each testis has six to nine GSCs, which are encysted by two SSCs [77, 78]. Both GSCs and SSCs are physically attached to a group of 12 nondividing somatic hub cells [79–82]. Each GSC divides asymmetrically to form two daughter cells, one retaining GSC identity and the other one called gonialblast (GB) initiating differentiation [83, 84] In a similar way, SSCs self-renew and give rise to daughters that differentiate into somatic cyst cells [85]. The GBs undergo four rounds of mitotic division with incomplete cytokinesis to form 16 interconnected spermatogonia; however, the SSCs will grow without further division and form a thin layer around the spermatogonial cyst [86]. Germ cells form spermatocytes and finally undergo meiosis and differentiate into sperm [82].

In addition to their role in GSC self-renewal and differentiation, miRNAs are also known to play a crucial role in GSC and somatic cell differentiation and GSC-niche aging in *Drosophila* testis [68, 72–74] (Table 4.1). Pek and colleagues [68] have shown that Maelstrom (Mael) represses the expression of *miR-7* that targets *bam* through its 3′ UTR. They found that overexpression of *miR-7* in *mael* mutant testes leads to Bam repression, resulting in a differentiation defect. This suggests that Mael ensures proper differentiation of GSC lineage by repressing *miR-7* [68]. Recently, Eun et al. [73] have shown that in the *Drosophila* male GSC lineage, *bam* mRNA, but not Bam, is present in spermatocytes. They found that repression of Bam accumulation is attained by *miR-275* and *miR-306* through the *bam* 3′ UTR. Further, they found that failure to block Bam protein expression in spermatocytes results in spermiogenesis defects and male sterility, which suggests that *miR-275* and *miR-306* downregulate Bam expression to ensure proper spermatid terminal differentiation [73]. Pancratov et al. [74] in a functional screen identified *miR-310/13* cluster (*miR-310* to *miR-313*) as a novel antagonist of the Wingless pathway that directly targets the 3′ UTR of *armadillo* (*arm*) and *pangolin* (*pan*). Interestingly, they found that the *miR-310/13* mutant flies show abnormal germ and somatic cell differentiation in the *Drosophila* testis [74]. In addition to the role of miRNAs in male GSC and somatic cell differentiation, Toledano et al. [72] have demonstrated that the IGF-II messenger RNA-binding protein (Imp) counteracts with Ago-2 and Dicer-2 to regulate *unpaired* (upd) levels and GSC maintenance. Further, they found that Imp expression decreases in the hub cells of aged males because of the targeting of *Imp* by *let-7,* which suggests that proper expression of Imp is essential to protecting *upd* mRNA from degradation [72].

4 miRNAs in *Drosophila* SSCs

In the past few years, miRNAs have emerged as a major player in stem cell regulation in *Drosophila* GSC systems with only very rare reports have described its function in other characterized *Drosophila* stem cell (neuroblast, intestinal and hematopoietic) systems. There are few reports that demonstrated the role of miRNAs in regulation of *Drosophila* neuroblast stem cells; these include *miR-124* [87, 88], *let-7* [89], and *miR-8/miR-200* [90]. Recently, Huang et al. [91] showed that bantam miRNA, which is highly expressed in *Drosophila* intestinal precursor cells (intestinal stem cells (ISCs), enteroblast (EB) cells) and enteroendocrine (ee) cells and weakly expressed in enterocytes (ECs), is essential for *Drosophila* ISC proliferation in response to the Hippo (hpo) signaling pathway. Tokusumi et al. [92] have shown that the germ-line differentiation factor Bam and *miR-7* antagonize the differentiation-promoting function of Yan to maintain the stem-like hematopoietic progenitor state during hematopoiesis in *Drosophila.*

5 miRNAs in Tumorigenesis in *Drosophila*

Emerging evidence suggests that dysfunction of miRNAs is correlated with various human diseases including cancer. It is known that cancer is the result of genetic alternations in oncogenes and tumor suppressors [93, 94]. Recent studies demonstrated that miRNAs are also involved in tumor formation and function as tumor suppressors or oncogenes by modulating the activity of evolutionarily conserved signaling pathways, which are usually dysregulated in human cancers [94–98]. It is also suggested that miRNAs may promote tumorigenesis by regulating the expression of some very important class of genes involved in tumor cell proliferation and apoptosis [99]. Kumar et al. [100] demonstrated that repressing the miRNA maturation by blocking the miRNA biogenesis components, particularly in cancer cells, can promote cell growth, transformation, and tumorigenesis.

Because more than 68 % of the genes involved in human cancer are conserved in *Drosophila* [101, 102], it has become a useful model organism to study cancer research [103–106]. Several key cancer events such as loss of cell polarity, the competition between tumor and normal cells, and metastasis have been demonstrated using *Drosophila* as a model system in the recent years. In the past few years, several miRNA pathways have been identified to regulate the tissue growth, cell proliferation, tumorigenesis, and metastasis in the *Drosophila* tumor model

[107–114]. Several studies demonstrated that bantam miRNA interacts with Hippo, and epidermal growth factor receptor (EGFR) pathways to control tissue growth, cell proliferation, and tumorigenesis [108, 111, 112, 114–116]. Herranz et al. [112] identified growth regulatory miRNA *bantam* and its target, Suppressor of cytokine signaling at 36E (Socs36E), a negative regulator of the Janus kinase/signal transducers and activators of transcription (JAK-STAT) signaling pathway, as cooperating factors in EGFR-driven tumorigenesis and metastasis in a *Drosophila* model for epithelial-to-mensenchymal transformation (EMT). In a misexpression study, it has been found that *Drosophila miR-278/mirvana* in the developing eye causes massive overgrowth, which is partly because of the inhibition of apoptosis [109]. In an overexpression screen, Vallejo et al. [110] identified *Drosophila miR-8* as a potent inhibitor of Notch-induced overgrowth and tumor metastasis. They found that *miR-8* could repress growth by inducing apoptosis and blocking cell proliferation via repressing *serrate* (*Ser*), a notch ligand. In a recent study, Da Ros et al. [113] identified the conserved miRNA *miR-7* that enhances Notch pathway-induced eye overgrowth in *Drosophila.* They found that the *interference hedgehog (ihog)* gene is the functional target of *miR-7* in Notch-mediated tumorigenesis. Further, they found that *miR-7* and Notch pathway cooperatively dampen hedgehog (Hh) signaling through downregulation of its receptors *ihog* and *brother of ihog* (boi). Their study suggests that the genetic cooperation of *miR-7,* Notch, and Hh is probably participating in the development of certain human tumors [113].

6 Conclusion

miRNAs are the key regulatory molecules in several biological processes. miRNAs play crucial roles in the self-renewal and differentiation of stem cells. miRNAs function as oncogenes or tumor suppressors. Abnormal expression of miRNAs results in developmental defects, loss of tissue homeostasis, and tumorigenesis. *Drosophila* provides an ideal model system to study stem cell regulation and tumor formation. Since miRNAs regulate stem cells, tumor-initiating cells, tumor growth, and metastasis, they have an enormous potential to be used as therapeutic targets for human cancers.

Acknowledgements This research was supported by the Intramural Research Program, National Cancer Institute of the National Institutes of Health. Lyric Forney is supported by Werner H. Kirsten Student Intern Program (WHK SIP) of National Cancer Institute at Frederick.

References

1. Plasterk RH. Micro RNAs in animal development. Cell. 2006;124:877–81.
2. Vasudevan S, Tong Y, Steitz JA. Switching from repression to activation: microRNAs can up-regulate translation. Science. 2007;318(5858):1931–4.
3. Flynt AS, Lai EC. Biological principles of microRNA-mediated regulation: shared themes amid diversity. Nat Rev Genet. 2008;9(11):831–42.

4. Gangaraju VK, Lin H. MicroRNAs: key regulators of stem cells. Nat Rev Mol Cell Biol. 2009;10(2):116–25.
5. Lee Y, Jeon K, Lee JT, Kim S, Kim VN. MicroRNA maturation: stepwise processing and subcellular localization. EMBO J. 2002;21(17):4663–70.
6. Thomson T, Lin H. The biogenesis and function of PIWI proteins and piRNAs: progress and prospect. Annu Rev Cell Dev Biol. 2009;25:355–76.
7. Lee Y, Kim M, Han J, Yeom KH, Lee S, Baek SH, Kim VN. MicroRNA genes are transcribed by RNA polymerase II. EMBO J. 2004;23(20):4051–60.
8. Lee Y, Ahn C, Han J, Choi H, Kim J, Yim J, Lee J, Provost P, Rådmark O, Kim S, Kim VN. The nuclear RNase III Drosha initiates microRNA processing. Nature. 2003;425(6956):415–9.
9. Denli AM, Tops BB, Plasterk RH, Ketting RF, Hannon GJ. Processing of primary microRNAs by the Microprocessor complex. Nature. 2004;432(7014):231–5.
10. Landthaler M, Yalcin A, Tuschl T. The human DiGeorge syndrome critical region gene 8 and its *D. melanogaster* homolog are required for miRNA biogenesis. Curr Biol. 2004;14(23):2162–7.
11. Gregory RI, Yan KP, Amuthan G, Chendrimada T, Doratotaj B, Cooch N, Shiekhattar R. The Microprocessor complex mediates the genesis of microRNAs. Nature. 2004;432(7014):235–40.
12. Han J, Lee Y, Yeom KH, Nam JW, Heo I, Rhee JK, Sohn SY, Cho Y, Zhang BT, Kim VN. Molecular basis for the recognition of primary microRNAs by the Drosha-DGCR8 complex. Cell. 2006;125(5):887–901.
13. Martin R, Smibert P, Yalcin A, Tyler DM, Schäfer U, Tuschl T, Lai EC. Drosophila pasha mutant distinguishes the canonical microRNA and mirtron pathways. Mol Cell Biol. 2009;29(3):861–70.
14. Yi R, Qin Y, Macara IG, Cullen BR. Exportin-5 mediates the nuclear export of pre-microRNAs and short hairpin RNAs. Genes Dev. 2003;17(24):3011–6.
15. Bohnsack MT, Czaplinski K, Gorlich D. Exportin 5 is a RanGTP-dependent dsRNA-binding protein that mediates nuclear export of pre-miRNAs. RNA. 2004;10(2):185–91.
16. Hutvágner G, McLachlan J, Pasquinelli AE, Bálint E, Tuschl T, Zamore PD. A cellular function for the RNA-interference enzyme Dicer in the maturation of the let-7 small temporal RNA. Science. 2001;293(5531):834–8.
17. Carmell MA, Hannon GJ. RNase III enzymes and the initiation of gene silencing. Nat Struct Mol Biol. 2004;11(3):214–8.
18. Zhang H, Kolb FA, Jaskiewicz L, Westhof E, Filipowicz W. Single processing center models for human Dicer and bacterial RNase III. Cell. 2004;118(1):57–68.
19. Saito K, Ishizuka A, Siomi H, Siomi MC. Processing of pre-microRNAs by the Dicer-1-Loquacious complex in Drosophila cells. PLoS Biol. 2005;3(7):e235.
20. Okamura K, Ishizuka A, Siomi H, Siomi MC. Distinct roles for Argonaute proteins in small RNA-directed RNA cleavage pathways. Genes Dev. 2004;18(14):1655–66.
21. Chendrimada TP, Gregory RI, Kumaraswamy E, Norman J, Cooch N, Nishikura K, Shiekhattar R. TRBP recruits the Dicer complex to Ago2 for microRNA processing and gene silencing. Nature. 2005;436(7051):740–4.
22. Lee RC, Feinbaum RL, Ambros V. The *C. elegans* heterochronic gene *lin-4* encodes small RNAs with antisense complementarity to *lin-14*. Cell. 1993;75:843–54.
23. Wightman B, Ha I, Ruvkun G. Posttranscriptional regulation of the heterochronic gene *lin-14* by *lin-4* mediates temporal pattern formation in *C. elegans*. Cell. 1993;*75*:855–62.
24. Reinhart BJ, Slack FJ, Basson M, Pasquinelli AE, Bettinger JC, Rougvie AE, Horvitz HR, Ruvkun G. The 21-nucleotide *let-7* RNA regulates developmental timing in *Caenorhabditis elegans*. Nature. 2000;*403:*901–6.
25. Pasquinelli AE, Reinhart BJ, Slack F, Martindale MQ, Kuroda MI, Maller B, Hayward DC, Ball EE, Degnan B, Muller P, et al. Conservation of the sequence and temporal expression of *let-7* heterochronic regulatory RNA. Nature. 2000;408:86–9.
26. Lagos-Quintana M, Rauhut R, Lendeckel W, Tuschl T. Identification of novel genes coding for small expressed RNAs. Science. 2001;294:853–8.

27. Farh KK, Grimson A, Jan C, Lewis BP, Johnston WK, Lim LP, Burge CB, Bartel DP. The widespread impact of mammalian MicroRNAs on mRNA repression and evolution. Science. 2005;310(5755):1817–21.
28. Chen K, Rajewsky N. Natural selection on human microRNA binding sites inferred from SNP data. Nat Genet. 2006;38(12):1452–6.
29. Friedman RC, Farh KK, Burge CB, Bartel DP. Most mammalian mRNAs are conserved targets of microRNAs. Genome Res. 2009;19(1):92–105.
30. Berezikov E. Evolution of microRNA diversity and regulation in animals. Nat Rev Genet. 2011;12(12):846–60.
31. Berezikov E, Robine N, Samsonova A, Westholm JO, Naqvi A, Hung JH, Okamura K, Dai Q, Bortolamiol-Becet D, Martin R, Zhao Y, Zamore PD, Hannon GJ, Marra MA, Weng Z, Perrimon N, Lai EC. Deep annotation of *Drosophila melanogaster* microRNAs yields insights into their processing, modification, and emergence. Genome Res. 2011;21(2):203–15.
32. Xia J, Zhang W. A meta-analysis revealed insights into the sources, conservation and impact of microRNA 5†²-isoforms in four model species. Nucleic Acids Res. 2013;1–15. doi:10.1093/nar/gkt967.
33. Mohammed J, Flynt AS, Siepel A, Lai EC. The impact of age, biogenesis, and genomic clustering on *Drosophila* microRNA evolution. RNA. 2013;19(9):1295–308.
34. O'Connell RM, Rao DS, Chaudhuri AA, Baltimore D. Physiological and pathological roles for microRNAs in the immune system. Nat Rev Immunol. 2010;10(2):111–22.
35. Fullaondo A, Lee SY. Identification of putative miRNA involved in *Drosophila melanogaster* immune response. Dev Comp Immunol. 2012;36(2):267–73.
36. Wienholds E, Kloosterman WP, Miska E, Alvarez-Saavedra E, Berezikov E, de Bruijn E, Horvitz HR, Kauppinen S, Plasterk RH. MicroRNA expression in zebrafish embryonic development. Science. 2005;309(5732):310–1.
37. Murchison EP, Partridge JF, Tam OH, Cheloufi S, Hannon GJ. Characterization of Dicer-deficient murine embryonic stem cells. Proc Natl Acad Sci U S A. 2005;102(34):12135–40.
38. Jovanovic M, Hengartner MO. miRNAs and apoptosis: RNAs to die for. Oncogene. 2006;25(46):6176–87.
39. Baltimore D, Boldin MP, O'Connell RM, Rao DS, Taganov KD. MicroRNAs: new regulators of immune cell development and function. Nat Immunol. 2008;9(8):839–45.
40. Dumortier O, Hinault C, Van Obberghen E. MicroRNAs and metabolism crosstalk in energy homeostasis. Cell Metab. 2013;18(3):312–24.
41. Mathieu J, Ruohola-Baker H. Regulation of stem cell populations by microRNAs. Adv Exp Med Biol. 2013;786:329–51.
42. Di Leva G, Croce CM. miRNA profiling of cancer. Curr Opin Genet Dev. 2013;23(1):3–11.
43. Sun K, Lai EC. Adult-specific functions of animal microRNAs. Nat Rev Genet. 2013;4(8):535–48.
44. He L, Thomson JM, Hemann MT, Hernando-Monge E, Mu D, Goodson S, Powers S, Cordon-Cardo C, Lowe SW, Hannon GJ, Hammond SM. A microRNA polycistron as a potential human oncogene. Nature. 2005;435(7043):828–33.
45. O'Donnell KA, Wentzel EA, Zeller KI, Dang CV, Mendell JT. c-Myc regulated microRNAs modulate E2F1 expression. Nature. 2005;435(7043):839–43.
46. Lin H. Cell biology of stem cells: an enigma of asymmetry and self-renewal. J Cell Biol. 2008;180(2):257–60.
47. Singh SR. Stem cell niche in tissue homeostasis, aging and cancer. Curr Med Chem. 2012;19(35):5965–74.
48. Förstemann K, Tomari Y, Du T, Vagin VV, Denli AM, Bratu DP, Klattenhoff C, Theurkauf WE, Zamore PD. Normal microRNA maturation and germ-line stem cell maintenance requires Loquacious, a double-stranded RNA-binding domain protein. PLoS Biol. 2005;3(7):e236.
49. Shcherbata HR, Hatfield S, Ward EJ, Reynolds S, Fischer KA, Ruohola-Baker H. The MicroRNA pathway plays a regulatory role in stem cell division. Cell Cycle. 2006;5(2):172–5.

50. Hatfield S, Ruohola-Baker H. microRNA and stem cell function. Cell Tissue Res. 2008;331(1):57–66.
51. Stadler BM, Ruohola-Baker H. Small RNAs: keeping stem cells in line. Cell. 2008;132(4): 563–6.
52. Li Q, Gregory RI. MicroRNA regulation of stem cell fate. Cell Stem Cell. 2008;2(3):195–6.
53. Park JK, Liu X, Strauss TJ, McKearin DM, Liu Q. The miRNA pathway intrinsically controls self-renewal of Drosophila germline stem cells. Curr Biol. 2007;17(6):533–8.
54. Wang Y, Baskerville S, Shenoy A, Babiarz JE, Baehner L, Blelloch R. Embryonic stem cell-specific microRNAs regulate the G1-S transition and promote rapid proliferation. Nat Genet. 2008;40(12):1478–83.
55. Murashov AK. A brief introduction to RNAi and microRNAs in stem cells. Methods Mol Biol. 2010;650:15–25.
56. Huang XA, Lin H. The microRNA regulation of stem cells. Wiley Interdiscip Rev Dev Biol. 2012;1(1):83–95.
57. Hatfield SD, Shcherbata HR, Fischer KA, Nakahara K, Carthew RW, Ruohola-Baker H. Stem cell division is regulated by the microRNA pathway. Nature. 2005;435(7044):974–8.
58. Jiang F, Ye X, Liu X, Fincher L, McKearin D, Liu Q. Dicer-1 and R3D1-L catalyze microRNA maturation in Drosophila. Genes Dev. 2005, 19(14):1674–9.
59. Jin Z, Xie T. Dcr-1 maintains *Drosophila* ovarian stem cells. Curr Biol. 2007;17(6):539–44.
60. Shcherbata HR, Ward EJ, Fischer KA, Yu JY, Reynolds SH, Chen CH, Xu P, Hay BA, Ruohola-Baker H. Stage-specific differences in the requirements for germline stem cell maintenance in the Drosophila ovary. Cell Stem Cell. 2007;1(6):698–709.
61. Yang L, Chen D, Duan R, Xia L, Wang J, Qurashi A, Jin P, Chen D. Argonaute 1 regulates the fate of germline stem cells in Drosophila. Development. 2007;134(23):4265–72.
62. Yang L, Duan R, Chen D, Wang J, Chen D, Jin P. Fragile X mental retardation protein modulates the fate of germline stem cells in Drosophila. Hum Mol Genet. 2007;16(15):1814–20.
63. Neumüller RA, Betschinger J, Fischer A, Bushati N, Poernbacher I, Mechtler K, Cohen SM, Knoblich JA. Mei-P26 regulates microRNAs and cell growth in the *Drosophila* ovarian stem cell lineage. Nature. 2008;454(7201):241–5.
64. Yang Y, Xu S, Xia L, Wang J, Wen S, Jin P, Chen D. The bantam microRNA is associated with *Drosophila* fragile X mental retardation protein and regulates the fate of germline stem cells. PLoS Genet. 2009;5(4):e1000444.
65. Yu JY, Reynolds SH, Hatfield SD, Shcherbata HR, Fischer KA, Ward EJ, Long D, Ding Y, Ruohola-Baker H. Dicer-1-dependent Dacapo suppression acts downstream of Insulin receptor in regulating cell division of Drosophila germline stem cells. Development. 2009;136(9):1497–507.
66. Azzam G, Smibert P, Lai EC, Liu JL. *Drosophila* Argonaute 1 and its miRNA biogenesis partners are required for oocyte formation and germline cell division. Dev Biol. 2012;365(2):384–94.
67. Liu N, Han H, Lasko P. Vasa promotes Drosophila germline stem cell differentiation by activating mei-P26 translation by directly interacting with a (U)-rich motif in its 3' UTR. Genes Dev. 2009;23(23):2742–52.
68. Pek JW, Lim AK, Kai T. *Drosophila* maelstrom ensures proper germline stem cell lineage differentiation by repressing microRNA-7. Dev Cell. 2009;17(3):417–24.
69. Iovino N, Pane A, Gaul U. miR-184 has multiple roles in Drosophila female germline development. Dev Cell. 2009;17(1):123–33.
70. Wang H, Mu Y, Chen D. Effective gene silencing in *Drosophila* ovarian germline by artificial microRNAs. Cell Res. 2011;21(4):700–3.
71. Li Y, Maines JZ, Tastan OY, McKearin DM, Buszczak M. Mei-P26 regulates the maintenance of ovarian germline stem cells by promoting BMP signaling. Development. 2012;139(9):1547–56.
72. Toledano H, D'Alterio C, Czech B, Levine E, Jones DL. The let-7-Imp axis regulates ageing of the *Drosophila* testis stem-cell niche. Nature. 2012;485(7400):605–10.

73. Eun SH, Stoiber PM, Wright HJ, McMurdie KE, Choi CH, Gan Q, Lim C, Chen X. MicroRNAs downregulate Bag of marbles to ensure proper terminal differentiation in the *Drosophila* male germline. Development. 2013;140(1):23–30.
74. Pancratov R, Peng F, Smibert P, Yang S Jr, Olson ER, Guha-Gilford C, Kapoor AJ, Liang FX, Lai EC, Flaherty MS, DasgGupta R. The miR-310/13 cluster antagonizes β-catenin function in the regulation of germ and somatic cell differentiation in the Drosophila testis. Development. 2013;140(14):2904–16.
75. Joly W, Chartier A, Rojas-Rios P, Busseau I, Simonelig M. The CCR4 deadenylase acts with nanos and pumilio in the fine-tuning of Mei-P26 expression to promote germline stem cell self-renewal. Stem Cell Reports. 2013;1(5):411–24.
76. Song X, Zhu CH, Doan C, Xie T. Germline stem cells anchored by adherens junctions in the *Drosophila* ovary niches. Science. 2002;296(5574):1855–7.
77. Singh SR, Chen X, Hou SX. JAK/STAT signaling regulates tissue outgrowth and male germline stem cell fate in *Drosophila*. Cell Res. 2005;15(1):1–5.
78. Singh SR, Zheng Z, Wang H, Oh SW, Chen X, Hou SX. Competitiveness for the niche and mutual dependence of the germline and somatic stem cells in the Drosophila testis are regulated by the JAK/STAT signaling. J Cell Physiol. 2010;223(2):500–10.
79. Wang H, Singh SR, Zheng Z, Oh SW, Chen X, Edwards K, Hou SX. Rap-GEF signaling controls stem cell anchoring to their niche through regulating DE-cadherin-mediated cell adhesion in the *Drosophila* testis. Dev Cell. 2006;10(1):117–26.
80. Singh SR, Zhen W, Zheng Z, Wang H, Oh SW, Liu W, Zbar B, Schmidt LS, Hou SX. The *Drosophila* homolog of the human tumor suppressor gene BHD interacts with the JAK-STAT and Dpp signaling pathways in regulating male germline stem cell maintenance. Oncogene. 2006;25(44):5933–41.
81. Singh SR, Liu Y, Kango-Singh M, Nevo E. Genetic, immunofluorescence labeling, and in situ hybridization techniques in identification of stem cells in male and female germline niches. Methods Mol Biol. 2013;1035:9–23.
82. Fuller MT, Spradling AC. Male and female *Drosophila* germline stem cells: two versions of immortality. Science. 2007 Apr 20;316(5823):402–4.
83. Yamashita YM, Mahowald AP, Perlin JR, Fuller MT. Asymmetric inheritance of mother versus daughter centrosome in stem cell division. Science. 2007;315(5811):518–21.
84. Matunis EL, Stine RR, de Cuevas M. Recent advances in *Drosophila* male germline stem cell biology. Spermatogenesis. 2012;2(3):137–44.
85. Gönczy P, DiNardo S. The germ line regulates somatic cyst cell proliferation and fate during *Drosophila* spermatogenesis. Development. 1996;122(8):2437–47.
86. Voog J, D'Alterio C, Jones DL. Multipotent somatic stem cells contribute to the stem cell niche in the *Drosophila* testis. Nature. 2008;454(7208):1132–6.
87. Weng R, Cohen SM. Drosophila miR-124 regulates neuroblast proliferation through its target anachronism. Development. 2012;139(8):1427–34.
88. Sun K, Westholm JO, Tsurudome K, Hagen JW, Lu Y, Kohwi M, Betel D, Gao FB, Haghighi AP, Doe CQ, Lai EC. Neurophysiological defects and neuronal gene deregulation in *Drosophila* mir-124 mutants. PLoS Genet. 2012;8(2):e1002515.
89. Kucherenko MM, Barth J, Fiala A, Shcherbata HR. Steroid-induced microRNA let-7 acts as a spatio-temporal code for neuronal cell fate in the developing *Drosophila* brain. EMBO J. 2012;31(24):4511–23.
90. Morante J, Vallejo DM, Desplan C, Dominguez M. Conserved miR-8/miR-200 defines a glial niche that controls neuroepithelial expansion and neuroblast transition. Dev Cell. 2013;27(2):174–87.
91. Huang H, Li J, Hu L, Ge L, Ji H, Zhao Y, Zhang L. Bantam is essential for *Drosophila* intestinal stem cell proliferation in response to Hippo signaling. Dev Biol. 2014;385(2):211–9. doi:10.1016/j.ydbio.2013.11.008.
92. Tokusumi T, Tokusumi Y, Hopkins DW, Shoue DA, Corona L, Schulz RA. Germ line differentiation factor Bag of Marbles is a regulator of hematopoietic progenitor maintenance during *Drosophila* hematopoiesis. Development. 2011;138(18):3879–84.

93. Esquela-Kerscher A, Slack FJ. Oncomirs—microRNAs with a role in cancer. Nat Rev Cancer. 2006;6(4):259–69.
94. Wang D, Qiu C, Zhang H, Wang J, Cui Q, Yin Y. Human microRNA oncogenes and tumor suppressors show significantly different biological patterns: from functions to targets. PLoS One. 2010;5(9):e13067.
95. Calin GA, Croce CM. MicroRNA signatures in human cancers. Nat Rev Cancer. 2006;6(11):857–66.
96. Croce CM. Causes and consequences of microRNA dysregulation in cancer. Nat Rev Genet. 2009;10(10):704–14.
97. Shenouda SK, Alahari SK. MicroRNA function in cancer: oncogene or a tumor suppressor? Cancer Metastasis Rev. 2009;28(3–4):369–78.
98. Costa PM, Pedroso de Lima MC. MicroRNAs as molecular targets for cancer therapy: on the modulation of microRNA expression. Pharmaceuticals. 2013;6(10):1195–220.
99. Cheng AM, Byrom MW, Shelton J, et al. Antisense inhibition of human miRNAs and indications for an involvement of miRNA in cell growth and apoptosis. Nucleic Acids Res. 2005;33:1290–7.
100. Kumar MS, Lu J, Mercer KL, et al. Impaired microRNA processing enhances cellular transformation and tumorigenesis. Nat Genet. 2007;39:673–7.
101. Rubin GM, Hong L, Brokstein P, Evans-Holm M, Frise E, Stapleton M, Harvey DA. A *Drosophila* complementary DNA resource. Science. 2000;287(5461):2222–4.
102. Hombría JC, Serras F. Why should we care about fly tumors? The case of JAK-STAT and EGFR cooperation in oncogenesis. JAKSTAT. 2013;2(2):e23203.
103. Miles WO, Dyson NJ, Walker JA. Modeling tumor invasion and metastasis in *Drosophila*. Dis Model Mech. 2011;4(6):753–61.
104. Stefanatos RK, Vidal M. Tumor invasion and metastasis in *Drosophila*: a bold past, a bright future. J Genet Genomics. 2011;38(10):431–8.
105. Polesello C, Roch F, Gobert V, Haenlin M, Waltzer L. Modeling cancers in *Drosophila*. Prog Mol Biol Transl Sci. 2011;100:51–82.
106. Rudrapatna VA, Cagan RL, Das TK. Drosophila cancer models. Dev Dyn. 2012;241(1):107–18.
107. Brennecke J, Hipfner DR, Stark A, Russell RB, Cohen SM. bantam encodes a developmentally regulated microRNA that controls cell proliferation and regulates the proapoptotic gene hid in *Drosophila*. Cell. 2003;113(1):25–36.
108. Nolo R, Morrison CM, Tao C, Zhang X, Halder G. The bantam microRNA is a target of the hippo tumor-suppressor pathway. Curr Biol. 2006;16(19):1895–904.
109. Nairz K, Rottig C, Rintelen F, Zdobnov E, Moser M, Hafen E. Overgrowth caused by misexpression of a microRNA with dispensable wild-type function. Dev Biol. 2006;291(2):314–24.
110. Vallejo DM, Caparros E, Dominguez M. Targeting Notch signalling by the conserved miR-8/200 microRNA family in development and cancer cells. EMBO J. 2011;30(4):756–69.
111. Oh H, Irvine KD. Cooperative regulation of growth by Yorkie and Mad through bantam. Dev Cell. 2011;20(1):109–22.
112. Herranz H, Hong X, Hung NT, Voorhoeve PM, Cohen SM. Oncogenic cooperation between SOCS family proteins and EGFR identified using a *Drosophila* epithelial transformation model. Genes Dev 2012, 26:1602–11.
113. Da Ros VG, Gutierrez-Perez I, Ferres-Marco D, Dominguez M. Dampening the signals transduced through hedgehog via microRNA miR-7 facilitates notch-induced tumourigenesis. PLoS Biol. 2013;11(5):e1001554.
114. Zhang Y, Lai ZC. Mob as tumor suppressor is regulated by bantam microRNA through a feedback loop for tissue growth control. Biochem Biophys Res Commun. 2013;439(4):438–42.
115. Thompson BJ, Cohen SM. The Hippo pathway regulates the bantam microRNA to control cell proliferation and apoptosis in *Drosophila*. Cell. 2006;126(4):767–74.
116. Peng HW, Slattery M, Mann RS. Transcription factor choice in the Hippo signaling pathway: homothorax and yorkie regulation of the microRNA bantam in the progenitor domain of the *Drosophila* eye imaginal disc. Genes Dev. 2009;23(19):2307–19.

Chapter 5
MicroRNAs in Stem Cells and Cancer Stem Cells

Ryan D. Mohan, Bernadette Bibber, Garima Sinha, Shyam A. Patel and Pranela Rameshwar

Abstract Cancer is currently a leading cause of death worldwide, taking millions of lives yearly. Estimates indicate that this tragic trend will continue and even increase over time (GLOBOCAN 2008 v2.0, http://globocan.iarc.fr). Currently, lung, liver, colon, and breast cancers are responsible for the majority of cancer deaths (Int J Cancer 132:1133–1145, 2013) and so a concerted effort is being made to find the mechanisms underlying carcinogenesis, disease progression, and chemotherapeutic resistance in order to uncover novel pathways that may be targeted by anticancer therapies. Evidence increasingly supports the hypothesis that cancers arise from stem cells as much as healthy tissues do. However, cancer stem cells are able to evade the growth programs imposed on normal stem cells, leading to disease. Specifically targeting cancer stem cells might prove to be an effective therapeutic strategy since this would eliminate the source of new cancer cells. Unfortunately, cancer stem cells are able to enter a dormant state in which they are resistant to standard therapies. Recent findings have indicated that microRNAs are critical regulators of many aspects of stem cell biology including entry and maintenance of the dormant state. Here, we discuss evidence supporting the cancer stem cell theory and how targeting of microRNA-dependent pathways might be used to coax cancer stem cells out of dormancy and into the path of chemotherapeutics.

Keywords Stem cell · Cancer stem cell · MicroRNA · Breast cancer · Bone marrow · Dormancy · Connexins · Stroma · Metastasis

1 Introduction

The long-term goal for any cancer treatment strategy, after elimination of the primary tumor mass, is to eradicate the dormant cancer cells in the primary and secondary sites. This is paramount for the eradication of cancer because the dormant cancer

P. Rameshwar (✉) · B. Bibber · G. Sinha · S. A. Patel
Graduate School of Biomedical Sciences and New Jersey Medical School, Rutgers, 185 South Orange Ave, MSB, Room E-579, Newark, NJ 07103, USA
e-mail: rameshwa@njms.rutgers.edu

R. D. Mohan
Stowers Institute for Medical Research, Kansas City, MO, USA

S. R. Singh, P. Rameshwar (eds.), *MicroRNA in Development and in the Progression of Cancer,* DOI 10.1007/978-1-4899-8065-6_5,

cells may eventually exit the quiescence phase and give rise to new tumors, leading to relapse and recurrence. In the case of breast cancer (BC), the bone marrow appears to be the site of dormancy. There has been an intensive research effort directed toward understanding the molecular mechanisms underlying dormancy and applying the findings to reverse dormancy through targeted treatment. Dormancy is a state in which cells are fully transformed yet bear some non-tumorigenic properties conferring resistance to anticancer agents. Clinical dormancy has been defined as the time (5–25 yrs) between removing the primary tumor and relapse [3]. This chapter adds to this definition by including the existence of dormant cancer cells in the primary and distant sites long before clinical detection of the tumor [4].

BC remains a clinical challenge despite education and aggressive intervention [5]. Mammograms with human compliance have led to early detection and improvement in overall survival [6]. This discussion will focus on dormancy of BC cells in the bone marrow and discuss the role of microRNAs (miRNAs) in maintaining dormancy. The bone marrow was found to be a source of BC-initiating cells, or cancer stem cells (CSCs), in several cases of cancer resurgence [7, 8]. This chapter will refer to tumor-initiating cells as CSCs. Micrometastasis of BC to the bone marrow results in poor prognosis. This type of metastasis is perhaps worse than metastasis to sentinel lymph nodes [9–11]. BC cells in bone marrow interact with the supporting stromal cells located close to the endosteum to facilitate the cancer cells' ability to maintain pluripotency and to retain quiescence [12–16].

The mechanisms by which stroma facilitates the existence of the BC stem cells remain a subject of research. However, the experimental evidence provides insights into the mechanisms. The cycling quiescence of BC cells can be partly explained by the changes in cytokine production and their formation of gap junctional intercellular communication (GJIC) between cancer cells and stroma [14, 17, 18]. miRNAs can be exchanged between the cancer cells and stroma through gap junctions [17, 18]. Included among the transferred material, which passes from stroma to the cancer cells, are miRNAs that target CXCL12, to induce quiescence [19]. We have reported on a working hierarchy of BC cells and have identified the stem cell subset as those that form GJIC with stroma [19].

The bone marrow has been identified as a source of CSCs in a significant number of BC recurrences [8, 20, 21]. Resurgence of cancer can occur with cells from the bone marrow even after 15 yrs of remission [7]. This not only strongly supports the bone marrow as the "home" for dormant CSCs but also raises questions on how these CSCs are able to survive in bone marrow for such long periods without interrupting normal hematopoietic activity [18]. A better understanding of how CSCs survive in bone marrow and other organs might lead to more effective therapeutics which target dormant cancer cells within the bone marrow without toxicity to the hematopoietic system [22–27]. Bone marrow stroma is also relevant to bone metastasis due to the cells' role in facilitating bone invasion of the cancer cells [16].

The significance of understanding how CSCs exist in bone marrow is underscored by past clinical failures of autologous bone marrow transplantation for BC patients. The transplantation protocol entailed temporary removal of hematopoietic stem cells (HSCs) and their progenitors from the bone marrow of BC patients for high doses of chemotherapy to eradicate cancer cells. After this, the patients were reinfused with the autologous bone marrow cells. The method resulted in poor outcome, indicating that the cancer cells survived the high-dose chemotherapy. The current literature explains this failure at least partly, by showing GJIC between the cancer cells and stroma, and chemoresistance of BC stem cells [14, 19].

The relationship between BC stem cells and bone marrow could be extrapolated to other organs, including circulating CSCs in the blood. Cells with properties similar to BC stem cells have been identified in the blood of BC patients [19, 28]. Since CSCs are expected at low frequency, the functions of these cells, in vivo, will be technically challenging for live imaging in animal models [29]. Research studies going forward will need to use innovative methods to dissect the complex processes involved in the quiescence of CSCs in the bone marrow and other organs.

2 Stem Cells: Overview

2.1 Discovery of Stem Cells

The concept of stem cells was first reported in the nineteenth century as certain tissues were found to have different regenerative properties along with the discovery of cells with self-renewal ability [30]. As studies moved along in the 1950s, Dr. Ernest A. McCulloch and Dr. James Till introduced the idea of colony-forming units (CFUs). The CFUs were shown to form colonies on the spleen of irradiated mice and contained cells from the hematopoietic lineage, thus demonstrating the existence of self-renewing multipotent blood cells [31]. These CFUs have since come to be known as HSCs and are the most characterized adult stem cells [32].

2.2 Defining Stem Cells

Since their discovery, the definition of what constitutes a stem cell has been the subject of intense investigation. In general, stem cells are defined as undifferentiated cells with the ability to self-renew and differentiate into cells of one or more specific cell type [33]. This creates a population of long-lasting cells that have slow-cycling rates and self-renewal abilities which allow the stem cells to maintain their population while assisting in tissue renewal and regeneration [34]. Therefore, any dysfunction or dysregulation in a stem cell can cause issues in all of the differentiated cells derived from that stem cell [35].

2.3 Sources of Stem Cells

There are several types of stem cells. Embryonic stem cells (ESCs) are derived from the inner cell mass (ICM) of the blastocyst of fertilized oocytes. ESCs were originally found through studies on embryonic carcinoma (EC) cells derived from teratocarcinoma, which showed stem-likeness and were highly pluripotent [36]. Adult stem cells are found throughout the human body in stem cell-specific niches and include HSCs and mesenchymal stem cells (MSCs) [37]. Induced pluripotent stem cells (iPSCs) are differentiated cells that have been dedifferentiated into stem-like cells. In 2006, Kazutoshi Takahashi and Shinya Yamanaka dedifferentiated murine fibroblasts into stem cells in vitro by co-overexpressing the transcription factors *Oct3/4, Sox2, c-Myc,* and *Klf4* in murine fibroblasts isolated from adult mice [37]. The efficiency of this procedure, however, is very low and the reason for this is still under investigation.

2.4 Niche in Stem Cell Regulation

Stem cells reside in a niche, a region that supplies the stem cells with factors that dictate self-renewal, homing, division, and cell-survival [34, 38]. Within this microenvironment, stem cells receive signals from neighboring differentiated cells via intercellular interactions and cell signaling [39]. The signals help to maintain homeostasis and react to the need for tissue regeneration [40]. Protein secretions or cytokines help maintain the stem cell niche, but a disruption in their production or activity can lead to malignancies [41]. For example, Wnt proteins are intercellular signaling molecules that have been found in the bone marrow, a niche for HSCs, and when dysregulated can cause cancer [35, 41].

2.5 Self-Renewal and Differentiation of Stem Cells

By definition, stem cells have the ability to self-renew and produce differentiated cells. In order to do this, stem cells use asymmetric or symmetric division. The former method produces one daughter cell that is an exact copy of itself and one differentiated daughter cell [42]. This prevents depletion of the initial stem cell population over time as differentiated cells are produced for tissue repair and renewal. In case the stem cell population is diminished, stem cells are also able to use symmetric division in order to expand the number of stem cells in a population [42]. The decision to enter asymmetric or symmetric division is determined through communication with developmental and environmental signals in the niche [42].

2.6 Stem Cell Quiescence

The ability for stem cells to be quiescent and to reenter the cell cycle is very important to the survival of organisms. Twenty billion cells are lost daily in the human body and stem cells are responsible for their replacement. During pregnancy, the female body has to form extensive glands and branches and undergo differentiation of cells to accommodate milk production [38]. At times of injury, the body has an increased need for tissue regeneration. All of these needs cause an increased demand on the stem cells. Adult stem cells are usually found in a quiescent state, but both active and quiescent states have been found to coexist [34]. While in a quiescent state, stem cells are slow cycling and therefore experience less exhaustion [34, 43]. This creates a subpopulation that is available to replace and take some of the burden from the actively cycling stem cell population [34].

3 Cancer Stem Cell

3.1 The CSC hypothesis

The idea that tumor biology and stem cell biology are linked was proposed many decades ago, but the first definitive evidence of a stem cell basis for malignancy emerged only in the late twentieth century. In 1994 and 1997, the group of John Dick analyzed leukemic cells and demonstrated a hierarchical organization of these cells [44, 45]. This group demonstrated that the hematopoietic stem cell (HSC), a multipotent stem cell that gives rise to all lineages of the blood, was the target of transformation. The leukemic stem cell was shown to initiate acute myeloid leukemia in immune-deficient mice, and evidence of the in vivo repopulation ability of this cell spurred numerous investigations into the stem cell basis for other types of cancers [45].

The cancer stem cell (CSC) hypothesis holds that only a small subset of cells within every tumor harbors tumor-initiating properties [46]. The theory postulates that tumors are composed of heterogeneous populations of cells, and only the stem-like subset can give rise to all other components of the tumor. The stem-like cells drive initial tumor formation, maintenance, metastasis, and relapse after years of disease-free survival [47]. The bulk of the tumor is composed of differentiated cells with no self-renewal ability. Clinically, tumor-initiating cells are chemoresistant and radioresistant due to their slow-cycling ability and their drug-efflux properties [46]. The mechanisms of chemoresistance include expression of adenosine triphosphate (ATP)-binding cassette transporters, which allow for active export of toxic agents, and aldehyde dehydrogenase 1 (ALDH1), which is a detoxifying enzyme typically expressed in HSCs [48]. A formal definition of a CSC has not been established but functionally includes (1) the ability to undergo limitless self-renewal, (2) the ability to demonstrate multi-lineage differentiation, and (3) the ability to repopulate all cellular components of a tumor [49]. Although much has been discovered about unique genetic and molecular properties of CSCs of various tissues, complete characterization has been difficult to achieve due to the high degree of heterogeneity [50].

The CSC theory has largely replaced the stochastic or clonal selection model for cancer, which holds that all cells within a tumor can equally initiate and maintain tumors upon acquisition of mutations [48]. As evidence of the CSC theory surpasses

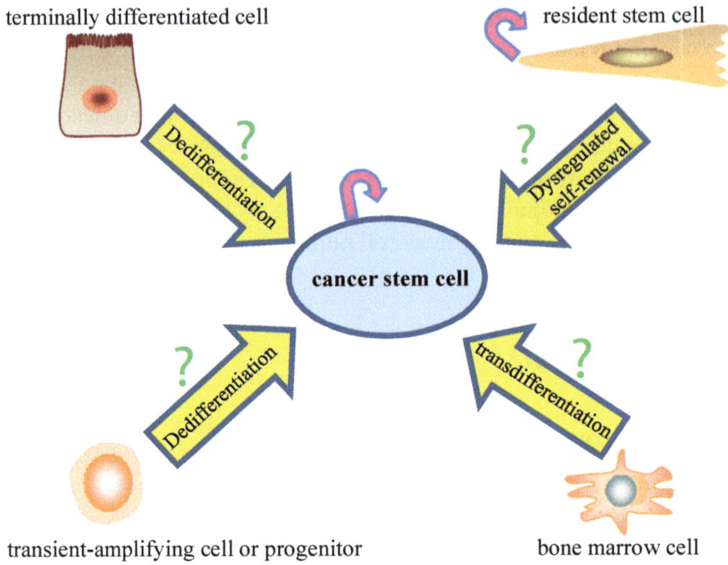

Fig. 5.1 The elusive origin of the cancer stem cell. Much controversy exists over the origin of the cancer stem cell. It is unknown whether these cells arise from dedifferentiation of terminally differentiated cells (*top left*), normal resident tissue stem cells that have acquired mutations that lead to dysregulation of self-renewal (*top right*), dedifferentiation of transient-amplifying cells or progenitors (*bottom left*), or transdifferentiation from bone marrow cells (*bottom right*) [52–54]

evidence for the stochastic model, investigations are beginning to focus on how the primitive subset of cells can be targeted. However, even within the CSC population, molecular heterogeneity exists [51]. Due to the high level of heterogeneity both between different tumors and among cells of the same tumor, the origins of CSCs remain unclear. Some groups postulate that these cells arise from resident stem cells and undergo dysregulation of self-renewal mechanisms, while other groups postulate that the cells arise from terminally differentiated malignant cells that undergo dedifferentiation programs, thereby acquiring self-renewal ability [52–54]. One cannot exclude the possibility that a transient-amplifying cell or a progenitor is the cell of origin [53, 54]. A final possibility is that transdifferentiation of bone marrow cells may lead to genesis of cancerous cells, a phenomenon that has been demonstrated in hepatocellular carcinoma [54] (Fig. 5.1). By identifying the cell of origin for each tissue-specific CSC, one may be able to focus investigations on this cell type and develop targeted therapies.

3.2 Emerging Evidence on the CSC Hypothesis

Since the identification of leukemic stem cells in the 1990s, the stem cell theory on cancer has been shown to involve many tissue types and many stem cell signaling pathways. In prostate cancer, for example, tumors in which the ESC transcription factor Sox2 is misregulated have been shown to have worse prognosis [55]. Interaction between androgen receptor signaling and Sox2 signaling contributes to castration-resistant prostate cancer [55]. In glioma, malignant astrocytes are thought to originate from neural stem cells or progenitors [53]. CD133, a HSC marker, has been suggested to be a marker for glioma stem cells [56]. To date, cancers of nearly every tissue type have shown a stem cell basis for malignancy. Signaling pathways that are common to stem cells, such as Notch, Wnt, and Hedgehog, have been shown to be instrumental in maintaining the viability of cancer cells [57].

In 2003, studies on the cellular heterogeneity of BC cells led to identification of a putative population of tumor-initiating cells in the breast [58]. A small subset of human BC cells obtained from primary sites and malignant pleural effusions was able to form new tumors in immune-deficient mice. Importantly, these cells could be serially passaged, suggesting a capacity for self-renewal. The phenotype was identified as $CD44^{+}/CD24^{-/low}/lin^{-}$. A minimum tumor burden of 100 cells of the $CD44^{+}/CD24^{-/low}/lin^{-}$ phenotype was able to initiate tumors, compared to the requirement of thousands of heterogeneous cells to initiate tumors [58]. For many years, the $CD44^{+}/CD24^{-/low}/lin^{-}$ phenotype was accepted as the best BC stem cell marker, but recent investigations have revealed that *Oct4* may more reliably indicate stem cell properties [59].

3.3 Targeted Therapy for Stem-Like Cancer Cells

As information on tumor-initiating cells unravels, novel phenotypic markers are being identified. Cell surface markers appear to be the most readily targetable and have therefore received the most attention. For example, anti-CD44 therapy has received attention recently due to the role of CD44 in cell migration and proliferation as well as the identification of CD44 as a putative marker of CSCs [58, 60]. CD44 has also been associated with cisplatin resistance in head and neck squamous cell carcinoma [60]. Signaling via CD44 has been shown to activate epidermal growth factor receptor (EGFR) signaling, and inhibition of this process can result in reduction in tumor growth [60]. EGFR disruption has been shown to chemosensitize stem cells from leiomyosarcoma [47]. As another example, melanoma stem cells have been shown to be $CD271^{+}$, and targeting this phenotype may deplete the stem cell population [61]. Continued identification of phenotypic markers is extremely useful for development of targeted therapy, since the majority of successful targeted therapy is based on depletion of cell populations expressing particular markers.

Aside from targeting cell surface markers, kinase inhibitors have also been evaluated for their ability to eliminate CSCs. Sorafenib, a multi-kinase inhibitor used in the treatment of renal cell carcinoma and hepatocellular carcinoma, has been shown to target glioma stem cells via induction of apoptosis and depletion of survival factors [62]. Sorafenib has been shown to inhibit tumorigenic potential in vivo via downregulation of key stem cell molecules, including nestin and Sox2 [62]. These findings are highly important clinically because the current therapy for glioma frequently targets the bulk of tumor cells while enriching the population of chemoresistant cells. Combination therapy employing sorafenib and temozolomide, for example, may be beneficial for optimal therapy. Perifosine, an inhibitor of Akt or protein kinase B, works synergistically with tumor necrosis factor-related apoptosis-inducing ligand (TRAIL) to induce apoptosis in myelogenous leukemia cells [63].

In addition to cell surface markers, cytokine networks are instrumental in maintaining the CSC population. In glioma stem cells, for example, inhibition of JAK2/STAT3 signaling can reduce tumor viability [51]. In BC, coculture experiments with MSCs have revealed that IL-6 and CXCL7 mediate bidirectional cross talk between BC cells and MSCs [57]. Targeting cytokine networks may disrupt communication between CSCs and components of their microenvironment, allowing for eradication of tumors.

Differentiation therapy is another modality of treatment that has been proposed for elimination of tumor-initiating cells. Since their primitive nature appears to drive tumor growth, promoting terminal differentiation of these cells may hamper their self-renewal capacity. The prospect of differentiation therapy arose from the use of all-*trans* retinoic acid for acute promyelocytic leukemia [64]. Inhibition of cyclin-dependent kinase 4 (CDK4) has been shown to favor differentiation of BC cells [65]. Although numerous mechanisms underlying CSC biology are known, much remains to be discovered about the behavior of these cells in order to achieve successful therapy in the clinic.

4 MicroRNA

Twenty years ago, the first miRNA was discovered when the lin-4 miRNA was found to repress translation of the lin-14 gene, controlling postembryonic cell lineage patterning in *Caenorhabditis elegans* [66]. Seven years passed before another miRNA was described, but the pace of discovery has been rapid since, with more than 3,000 miRNAs discovered between mouse and human and more than 21,000 entries in the miRBase database [67, 68]. miRNAs are 19–23-nucleotide-long, noncoding RNAs [69]. The primary miRNA is processed by the ribonuclease Drosha to generate precursor miRNAs (pre-miRs) [69]. The pre-miRs are then processed by Dicer to generate mature single-stranded miRNAs, which associate with the RNA-induced silencing complex (RISC) complex for loading onto the 3′ untranslated region (UTR) of mRNAs to repress translation [69, 70]. Specificity of mRNA targeting is achieved through complementary interactions between a short seed sequence located between nucleotides 2 and 8 of the 5′ end of the miRNA and sequences within the 3′ UTR of the targeted mRNA.

miRNAs have been broadly implicated in the regulation of diverse biological processes including development, aging, tissue homeostasis, and carcinogenesis as well as cardiovascular, neurological, and metabolic diseases [71]. Importantly, miRNAs are expressed in a tissue-specific manner in mouse and human [72].

Furthermore, some miRNAs are associated with specific cell states from pluripotency [73] to senescence and aging, suggesting that miRNA-targeting drugs may also function in a tissue-specific and state-specific manner [71].

5 Connexins

Cells constantly communicate with surrounding tissues and the local microenvironment for coordinated function within multicellular organisms. Breakdown of this back-and-forth communication can lead to many diseases including cancer. Direct intercellular communication between adjacent cells can occur through gap junctions, which connects the cytoplasm of the cells. This process is also known as "gap junctional intercellular communication." These small open channels allow for free exchange of inorganic ions such as Na^+, K^+, Ca^{2+}, as well as molecules of size ranging between 1 and 2 kDa such as cyclic adenosine monophosphate (cAMP), IP3, and miRNA, between connected cells.

Connexins (Cx) are the major structural component of gap junction channels. The Cxs comprise a large family of four-pass transmembrane proteins [74]. In humans and mice, 21 and 20 Cx genes have been identified, respectively [75–77]. Most commonly, Cxs are named by their molecular weight in kDa. For example, Cx isoforms include Cx26, Cx30, Cx32, Cx37, Cx43, Cx45, and Cx47. Interestingly, these isoforms are differentially expressed. The Cxs usually oligomerize as they move from the endoplasmic reticulum (ER) to the *trans* Golgi apparatus [78], although in some cases monomeric units may be retained in the *trans* Golgi region [79]. Cx26 is an exception to this as it is directly expressed on the plasma membrane without crossing the Golgi apparatus [80–82]. Cxs have a conserved amino-terminal domain and two extracellular loops (EL-1 and EL-2) along with highly divergent carboxy-terminal domains and three cytoplasmic loops [83, 84]. They form gap junctions through connection of hexameric hemichannel/connexons via extracellular loops which extend outward from each of the cells being connected [74, 83]. The hemichannels can be either homotypic (identical Cxs) or heterotypic (nonidentical Cxs) with different properties. The half-life of Cxs is only a few hours, making Cx formation and degradation a dynamic process. Interestingly, Cx regulation has been found to be altered during tumor growth and metastasis.

The process known as epithelial–mesenchymal transition (EMT) results in invasive and metastatic cancer by allowing cells to migrate from the primary tumor site to other organs in the body. Bone marrow has been shown to be a site where BC cells can migrate and survive primary treatment and also hide from the immune response [85]. It is thought that entry into quiescence is a major mechanism which allows these cells to escape death since many therapies target dividing cells. Formation of gap junctions in bone marrow through pairing of the BC connexins Cx26, Cx32, and Cx43 with Cx43/Cx32 on stroma promotes entry into quiescence by allowing GJIC with local stromal cells [19, 77, 80, 86–90]. Transfer of miRNAs through the GJIC is integral to this process [17].

During pregnancy, the expression level of Cx26 (gap junction beta 2 (GJB2)) increases and is at its peak during the lactation period in mammary glands. Cx26 may be involved in differentiation of the mammary gland as it is present in non-pregnant women [91, 92]. Earlier, Cx26 was thought to be a tumor suppressor and recent reports support such a role [93]. Methylation of cytosines within promoter CpGs has been shown to silence gene expression of tumor suppressors in some cancers [94, 95]. In BC, differential CpG methylation was observed at many locations, including frequent methylation of Sp1 transcription factor-binding sites important for Cx26 gene expression, resulting in reduced Cx26 expression. This indicates that Cx26 is indeed a tumor suppressor when active and is usually downregulated in invasive BC cells. The miRNA Cx32 (GJB1), however, is highly expressed in the lactating mammary gland (basolateral region of luminal cells), but is not detectable in noncancerous human breast tissue. Interestingly, it is expressed in the BC cell line MDA-MB-435, suggesting that it may act as an oncogene [90, 91, 96–98].

Cx43 (GJA1) is widely expressed and can be activated by kinases like protein kinase A (PKA), protein kinase C (PKC), p34cdc2 ($p34^{cdc2}$/cyclin B kinase), protein kinase CK1, mitogen-activated protein kinase (MAPK), and Src ($pp60^{Src}$ kinase) [99–121]. They phosphorylate Cx43, opening and activating Cx43-containing gap junctions. Cx43 is important in growth and differentiation of myoepithelial cells in the mammary gland and is important for lactation [122]. By facilitating formation of GJIC, Cx43 can also promote metastasis in a process called "homing" through which mobile cancer cells increase their ability to adhere to tissues and form secondary tumors. Accordingly, Cx43 has been shown to be upregulated along contact sites between tumor cells and endothelial cells [123]. BC cells with increased Cx43 expression tend to metastasize to the brain (124). Interestingly, CXCL12 can activate PKC in BC cells, leading to phosphorylation and activation of Cx43, opening the gap junction and promoting GJIC after adhesion [59].

According to recent findings, the expression of different miRNAs can control gap junction formation by regulating translation of Cx transcripts. For example, miR-1/206 can inhibit translation of Cx 43, thereby regulating gap junction formation and altering GJIC [125]. Recently, in glioblastoma cells, miR 221/222 has been found to downregulate expression of Cx43, leading to suppression of cell proliferation and invasion [126]. These observations suggest that miRNAs play an important role in cancer progression through regulation of Cxs and GJIC.

Cxs are associated with various other proteins such as tight and adherent junctional proteins, phosphatase, kinases, and cytoskeletal proteins. Cxs can also interact with microtubules to transport the newly formed Cx proteins from ER to plasma membrane, where the interaction with the actin cytoskeleton stabilizes the gap junction [75]. The permeability of tight junctions is also regulated by Cxs. To some extent, the interaction between the adherens and Cx regulates cell proliferation and growth. Since N-cadherin and β-catenin also assemble in ER and Golgi apparatus, there is evidence that they regulate the expression of Cxs such as Cx43 [122]. Hence, these Cxs might prove to be good targets for therapy in the future.

6 Stromal Cell-Derived Factor 1/CXCL12

6.1 Overview

Stromal cell-derived factor 1 (SDF-1), also known as CXCL12, is a chemokine ligand of the CXC family that binds to the seven-transmembrane G protein-coupled receptor CXCR4 [127]. CXCL12 is one of the major players in leukocyte and stem cell trafficking through interaction with CXCR4 [128]. It also plays critical roles in hematopoiesis and angiogenesis [129]. CXCR4 expression on HSCs allows for stem cell homing to the bone marrow, which is a major source of CXCL12. CXCR4 is expressed on hematopoietic cells of varying stages of differentiation, including long- and short-term repopulating stem cells as well as hematopoietic progenitor cells [130]. Thus, the interaction between CXCL12 and CXCR4 is vital to maintenance of the hematopoietic niche [131]. CXCL12 is also expressed in the endothelial niche and osteoblastic niche [130].

Recent studies have revealed that CXCR7, a G protein-independent receptor, is another receptor that binds CXCL12. CXCR4 and CXCR7 jointly form a functional receptor complex [132, 133]. Upon ligand binding, CXCL12 can lead to activation of intracellular signal transduction pathways such as Erk1/2 and Akt [133]. The pathways may explain the ability of CXCL12 to induce proliferation of cells to which the chemokine binds. Regulation of CXCL12 occurs through multiple mechanisms, including cross talk with tackykinins and cleavage by matrix metalloproteases (MMPs) [134]. MMP-8 proteolytically cleaves CXCL12, a process that inhibits HSC and progenitor cell migration [128].

6.2 Emerging Role of CXCL12 in Cancer Biology

The CXCL12–CXCR4 axis has been strongly implicated in tumorigenesis [127, 133]. Many transformed cells express high levels of CXCR4, permitting chemotaxis and therefore establishment of sites of metastasis in tissues that express high levels of CXCL12. For example, BC cells express CXCR4 and preferentially migrate to the bone marrow, where CXCL12 is abundant. Other distant metastatic sites, such as the liver and lung, have been shown to have high CXCL12 levels, facilitating the migration of cancer cells into these tissues. Upon hijacking the CXCL12–CXCR4 axis to metastasize to distant sites, cancer cells can establish unique niches within the target microenvironment that allow for them to survive and flourish. For these reasons, pharmacological targeting of the CXCL12–CXCR4 interaction has become an area of high interest. Blockade of this interaction using AMD3100 can uncouple stromal–epithelial interactions [135]. Plerixafor, another CXCL12–CXCR4 antagonist that transiently and reversibly interferes with the interaction, has been used in the clinic for HSC mobilization. This therapy has a similar effect as granulocyte colony-stimulating factor (G-CSF), which is used for stem cell mobilization for bone marrow transplantation. These agents act in synergy to induce mobilization [136].

In addition to the chemotactic function of CXCL12, CXCL12 binding to CXCR4 transduces signaling via Erk1/2 and Akt pathways [127, 133]. These signaling pathways contribute to dysregulated cell cycle control and tumor cell proliferation. Thus, the dynamic role of CXCL12 in both chemotaxis and proliferation makes this signaling axis an important therapeutic target.

6.3 CXCL12–CXCR4 Signaling Loops in Cancer

In recent years, the role of CXCL12 in cancer progression has become clearer. CXCL12 is involved in the interaction not only between cancer cells and surrounding stroma but also within a single cancer cell itself. Thus, besides the role of CXCL12 in chemotaxis of cancer cells to target organs, paracrine and autocrine signaling loops drive cancer progression and dormancy. One example of the paracrine effect of CXCL12 in cancer was demonstrated upon studying cellular interactions between BC cells and MSCs. Tumor-conditioned media from BC cells can induce MSCs to adopt a carcinoma-associated fibroblast-like phenotype and release CXCL12 [137]. Carcinoma-associated fibroblasts, in turn, promote tumor growth via secretion of CXCL12 [137].

In addition to the paracrine influence of CXCL12 in cancer, an autocrine signaling loop allows cancer cells to acquire an invasive phenotype and proliferate [127]. Many BC cell lines, such as MDA-MB-231, express CXCR4, and some cell lines also express CXCL12 [129]. When CXCL12(-) MDA-MB-231 cells are experimentally manipulated to express CXCL12, they demonstrate increased migratory and invasive capability [129]. In lung cancer cells, CXCL12–CXCR4 autocrine signaling increased the ability of the cells to invade via upregulation of MMP-2 and MMP-9 [127]. These autocrine loops also promote increased proliferation of lung cancer cells [127]. These mechanistic studies point to the critical role of CXCL12 in initially supporting cancer invasion into distant sites of metastasis and then supporting growth of the tumor [127]. Finally, in oral squamous cell carcinoma, the autocrine signaling loop promotes anchorage-independent growth and motility. Squamous cells expressing CXCL12 had increased metastatic potential and resulted in decreased survival [138]. Importantly, CXCR4 inhibitor AMD3100/Plerixafor abrogated this effect [138]. An autocrine signaling phenomenon has also been shown in Ewing's sarcoma. Ewing's sarcoma cells express both CXCR4 and CXCL12, and CXCR4 expression is positively associated with tumor size [139]. CXCL12 promotes proliferation of sarcoma cells via interaction with CXCR4 [139]. This mechanism may rely on Erk1/2 and/or Akt pathways, since these pathways are involved in cell proliferation.

Aside from the well-established role of CXCR4 in cancer biology, there is increasing evidence of the role of the CXCL12–CXCR7 axis in cancer. CXCR7 has been shown to be expressed in a variety of neural tumors, including neurilemmomas and meningiomas [140]. The role of CXCR7 in the context of CXCL12 signaling remains to be further explored.

6.4 CXCL12 in Tumor Dormancy: Pharmacological Targeting

The specific mechanisms of how CXCL12 is involved in BC progression are becoming clearer. Bone marrow stromal cells have been shown to transmit CXCL12-specific miRNAs through gap junctions into BC cells [17]. Posttranscriptional suppression of CXCL12 in BC cells by stromal-derived miRNAs allows BC cells to enter G_0 phase of the cell cycle and maintain dormancy [17]. The interaction between CXCL12 and gap junctions was further elucidated in studies demonstrating the inverse link between CXCL12 expression and the establishment of gap junctions between BC cells and stromal cells [59]. The mechanism of interaction between CXCL12 and gap junction formation involves protein kinase C-mediated phosphorylation of Cx43 [59]. A complete understanding of these interactions may pave the way for targeted therapy for BC.

In addition to BC, other malignancies have been shown to employ the CXCL12–CXCR4 axis for establishment of dormancy. Leukemic blasts interact with bone marrow stroma through this chemokine–cytokine pair [136]. A phase I and II clinical trial on the use of plerixafor for acute myeloid leukemia suggested that uncoupling of the CXCL12–CXCR4 interaction renders leukemic blasts susceptible to chemotherapy. The basis for this effect is that plerixafor disrupts interactions between blasts and bone marrow microenvironment, allowing the cells to exit dormancy and reenter the cell cycle [136]. These studies point to the significant role that chemokine-based targeted therapy will play in the future of cancer treatment.

7 miRNA and CSC

Although cancer is a disease involving heterogeneous cell types, miRNAs have shown some promise as predictive markers of cancer outcome and progression [141]. Accordingly, the microenvironment has a major role in tumor behavior. The literature on CSCs, although limited on the specifics of markers useful for identifying this population, is clear on their potent tumor-initiating properties. Similar to other stem cells, CSCs are not expected to be metabolically active and are small in number. Since miRNAs are involved in developmental processes such as stem cell maturation it is intuitively understandable that some effort is being made to evaluate their potential in regulating CSC function [142, 143].

Pluripotency genes, such as Nanog and Oct4, can be regulated by miRNAs [144]. Since CSCs express the same pluripotency genes [19], they might be regulated by similar miRNAs. However, there is no clear evidence that genes linked to pluripotency are also involved in self-renewal. Independent studies using robust array approaches have linked specific miRNAs such as let-7 in self-renewal and malignancy in BC [145]. Methylation of the miRNA 34C gene is required to maintain self-renewal and EMT capability in CSCs [146]. Although other miRNAs such as miR21 can also induce EMT, it is unclear if this occurs through DNA methylation or the suppression of other genes [147].

miRNAs also have properties in addition to maintaining self-renewal ability. miRNA 30 not only maintains self-renewal but also prevents apoptosis [148]. miRNAs such as miR200 can negatively regulate EMT [149]. CSCs generally evade treatment with chemotherapy and chemo resistance has been linked to specific miRs. miR-128 facilitates chemoresistance through Bmi-1 and the multiple-drug resistance gene [150]. Mir-451 imparts chemoresistance of BC cells to doxorubicin [151]. The question is how the information on miRNA can be exploited to target CSCs. This question will require side-by-side studies with mammary epithelial cells to understand how miRNAs are involved in normal development and to understand if the same miRNA becomes dysfunctional to transition the normal stem cells to EMT [152].

8 Conclusion

Drugs, including small molecules, are available for rapid translation of research to target CSCs. As examples, antagonists are available for neurokinin receptors, γ-secretase inhibitors can target Notch and N-cadherin, and AMD3100 can inhibit CXCL12 receptors. In addition, specific knockout mouse models are available to investigate the function of these inhibitors. It is envisioned that small molecule drugs like these can potentially reverse dormancy and/or induce differentiation of the CSCs to make them chemosensitive. In the case of bone marrow, it will be important to eliminate all subsets of cancer cells without toxicity to hematopoiesis.

New classes of drugs targeting miRNAs are currently under development and evaluation. Targeting of disease-related miRNAs by antisense anti-miRs (antagomirs) which act to inhibit miRNA function has been shown to be effective in treating cardiovascular disease in a mouse model [153, 154]. Another approach, using "locked nucleic acid" inhibitors to target miR-122 was shown to be effective in treating hepatitis in a phase II clinical trial performed in nonhuman primates (NCT01200420) [155]. These successes indicate that miRNA-targeted therapeutic interventions might also be designed to treat cancer as well.

There is still much to learn about miRNA function. For example, the seed sequences of many co-expressed miRNAs are very similar. In human pluripotent stem cells, a number of pluripotency-associated miRNAs are expressed which share six or even seven out of seven of the same nucleotides at the seed site [156]. This redundancy not only suggests an important regulatory role for these miRNAs in regulating stem cell function but also hints at undiscovered modes of regulation beyond the seed sequence. Basic research into this and other mechanisms of miRNA regulation should provide new avenues for drug design in the future.

References

1. Ferlay JSH, Bray F, Forman D, Mathers C, Parkin DM. GLOBOCAN 2008 v2.0, Cancer Incidence and Mortality Worldwide: IARC CancerBase No. 10. 2010 (http://globocan.iarc.fr).
2. Bray F, Ren JS, Masuyer E, Ferlay J. Global estimates of cancer prevalence for 27 sites in the adult population in 2008. Int J Cancer. 2013;132:1133–45.

3. Uhr JW, Pantel K. Controversies in clinical cancer dormancy. Proc Natl Acad Sci U S A. 2011;108:12396–400.
4. Talmadge JE. Clonal selection of metastasis within the life history of a tumor. Cancer Res. 2007;67:11471–5.
5. Jemal A, Siegel R, Xu J, Ward E. Cancer statistics, 2010. CA Cancer J Clin. 2010;60:277–300.
6. Green BB, Taplin SH. Breast cancer screening controversies. J Am Board Fam Pract. 2003;16:233–41.
7. Mansi JL, Berger U, McDonnell T, Pople A, Rayter Z, Gazet JC, et al. The fate of bone marrow micrometastases in patients with primary breast cancer. J Clin Oncol. 1989;7:445–9.
8. Habeck M. Bone-marrow analysis predicts breast-cancer recurrence. Mol Med Today. 2000;6:256–7.
9. Braun S, Auer D, Marth C. The prognostic impact of bone marrow micrometastases in women with breast cancer. Cancer Invest. 2009;27:598–603.
10. Banys M, Hartkopf AD, Krawczyk N, Becker S, Fehm T. Clinical implications of the detection of circulating tumor cells in breast cancer patients. Biomark Med. 2012;6:109–18.
11. Giuliano AE, Hunt KK, Ballman KV, Beitsch PD, Whitworth PW, Blumencranz PW, et al. Axillary dissection vs no axillary dissection in women with invasive breast cancer and sentinel node metastasis: a randomized clinical trial. JAMA. 2011;305:569–75.
12. Oh HS, Moharita A, Potian JG, Whitehead IP, Livingston JC, Castro TA, et al. Bone marrow stroma influences transforming growth factor-beta production in breast cancer cells to regulate c-myc activation of the preprotachykinin-I gene in breast cancer cells. Cancer Res. 2004;64:6327–36.
13. Ramkissoon SH, Patel PS, Taborga M, Rameshwar P. Nuclear factor-kappaB is central to the expression of truncated neurokinin-1 receptor in breast cancer: implication for breast cancer cell quiescence within bone marrow stroma. Cancer Res. 2007;67:1653–9.
14. Rao G, Patel PS, Idler SP, Maloof P, Gascon P, Potian JA, et al. Facilitating role of preprotachykinin-I gene in the integration of breast cancer cells within the stromal compartment of the bone marrow: a model of early cancer progression. Cancer Res. 2004;64:2874–81.
15. Reddy BY, Greco SJ, Patel PS, Trzaska KA, Rameshwar P. RE-1-silencing transcription factor shows tumor-suppressor functions and negatively regulates the oncogenic TAC1 in breast cancer cells. Proc Natl Acad Sci U S A. 2009;106:4408–13.
16. Lu X, Wang Q, Hu G, Van Poznak C, Fleisher M, Reiss M, et al. ADAMTS1 and MMP1 proteolytically engage EGF-like ligands in an osteolytic signaling cascade for bone metastasis. Genes Dev. 2009;23:1882–94.
17. Lim PK, Bliss SA, Patel SA, Taborga M, Dave MA, Gregory LA, et al. Gap junction-mediated import of microRNA from bone marrow stromal cells can elicit cell cycle quiescence in breast cancer cells. Cancer Res. 2011;71:1550–60.
18. Moharita AL, Taborga M, Corcoran KE, Bryan M, Patel PS, Rameshwar P. SDF-1alpha regulation in breast cancer cells contacting bone marrow stroma is critical for normal hematopoiesis. Blood. 2006;108:3245–52.
19. Patel SA, Ramkissoon SH, Bryan M, Pliner LF, Dontu G, Patel PS, et al. Delineation of breast cancer cell hierarchy identifies the subset responsible for dormancy. Sci Rep. 2012;2:906.
20. Naume B, Zhao X, Synnestvedt M, Borgen E, Russnes HG, Lingjaerde OC, et al. Presence of bone marrow micrometastasis is associated with different recurrence risk within molecular subtypes of breast cancer. Mol Oncol. 2007;1:160–71.
21. Riethdorf S, Wikman H, Pantel K. Review: Biological relevance of disseminated tumor cells in cancer patients. Int J Cancer. 2008;123:1991–2006.
22. Zon LI. Intrinsic and extrinsic control of haematopoietic stem-cell self-renewal. Nature. 2008;453:306–13.
23. Miller JS, McCullar V, Punzel M, Lemischka IR, Moore KA. Single adult human CD34(+)/Lin-/CD38(-) progenitors give rise to natural killer cells, B-lineage cells, dendritic cells, and myeloid cells. Blood. 1999;93:96–106.
24. Weissman IL. Stem cells: units of development, units of regeneration, and units in evolution. Cell. 2000;100:157–68.

25. Liu K, Castillo MD, Murthy RG, Patel N, Rameshwar P. Tachykinins and hematopoiesis. Clin Chim Acta. 2007;385:28–34.
26. Broxmeyer HE. Chemokines in hematopoiesis. Curr Opin Hematol. 2008;15:49–58.
27. Quelle FW. Cytokine signaling to the cell cycle. Immunol Res. 2007;39:173–84.
28. Meng S, Tripathy D, Frenkel EP, Shete S, Naftalis EZ, Huth JF, et al. Circulating tumor cells in patients with breast cancer dormancy. Clin Cancer Res. 2004;10:8152–62.
29. Bos PD, Nguyen DX, Massague J. Modeling metastasis in the mouse. Curr Opin Pharmacol. 2010;10(5):571–7.
30. Robey PG. Series Introduction: Stem cells near the century mark. J Clin Invest. 2000;105:1489–91.
31. Eaves C, Eaves A. A fond farewell to Ernest A. McCulloch-A beacon who inspired generations of scientists and framed the foundation of modern stem cell biology. Exp Hematol. 2011;39:602–3.
32. Zon LI. Intrinsic and extrinsic control of haematopoietic stem-cell self-renewal. Nature. 2008;453:306–13.
33. Wallenfang MR, Matunis E. Orienting stem cells. Science. 2003;301:1490–1.
34. Li L, Clevers H. Coexistence of quiescent and active adult stem cells in mammals. Science. 2010;327:542–5.
35. Reya T, Morrison SJ, Clarke MF, Weissman IL. Stem cells, cancer, and cancer stem cells. Nature. 2001;414:105–11.
36. Knoepfler PS. Deconstructing stem cell tumorigenicity: A roadmap to safe regenerative medicine. Stem Cells. 2009;27:1050–6.
37. Shevde N. Stem Cells: flexible friends. Nature. 2012;483:S22–S6.
38. Fuchs E. The tortoise and the hair: slow-cycling cells in the stem cell race. Cell. 2009;137:811–9.
39. Spradling A, Drummond-Barbosa D, Kai T. Stem cells find their niche. Nature. 2001; 414:98–104.
40. Chen T, Heller E, Beronja S, Oshimori N, Stokes N, Fuchs E. An RNA interference screen uncovers a new molecule in stem cell self-renewal and long-term regeneration. Nature. 2012;485:104–8.
41. Medema JP, Vermeulen L. Microenvironmental regulation of stem cells in intestinal homeostasis and cancer. Nature. 2011;474:318–26.
42. Morrison SJ, Kimble J. Asymmetric and symmetric stem-cell divisions in development and cancer. Nature. 2006;441:1068–74.
43. Beachy PA, Karhadkar SS, Berman DM. Tissue repair and stem cell renewal in carcinogenesis. Nature. 2004;432:324–31.
44. Lapidot T, Sirard C, Vormoor J, Murdoch B, Hoang T, Caceres-Cortes J, et al. A cell initiating human acute myeloid leukaemia after transplantation into SCID mice. Nature. 1994;367:645–8.
45. Bonnet D, Dick JE. Human acute myeloid leukemia is organized as a hierarchy that originates from a primitive hematopoietic cell. Nat Med. 1997;3:730–7.
46. Lorico A, Rappa G. Phenotypic heterogeneity of breast cancer stem cells. J Oncol. 2011;2011:135039.
47. Sette G, Salvati V, Memeo L, Fecchi K, Colarossi C, Di Matteo P, et al. EGFR inhibition abrogates leiomyosarcoma cell chemoresistance through inactivation of survival pathways and impairment of CSC potential. PLoS One. 2012;7:e46891.
48. Alison MR, Lin WR, Lim SM, Nicholson LJ. Cancer stem cells: in the line of fire. Cancer Treat Rev. 2012;38:589–98.
49. Valent P, Bonnet D, De Maria R, Lapidot T, Copland M, Melo JV, et al. Cancer stem cell definitions and terminology: the devil is in the details. Nat Rev Can. 2012;12:767–75.
50. Gibbs KD Jr, Jager A, Crespo O, Goltsev Y, Trejo A, Richard CE, et al. Decoupling of tumor-initiating activity from stable immunophenotype in HoxA9-Meis1-driven AML. Cell Stem Cell. 2012;10:210–7.
51. Stechishin OD, Luchman HA, Ruan Y, Blough MD, Nguyen SA, Kelly JJ, et al. On-target JAK2/STAT3 inhibition slows disease progression in orthotopic xenografts of human glioblastoma brain tumor stem cells. Neuro Oncol. 2013;15:198–207.

52. van Deurzen CH, Lee AH, Gill MS, Menke-Pluijmers MB, Jager A, Ellis IO, et al. Metaplastic breast carcinoma: tumour histogenesis or dedifferentiation? J Pathol. 2011;224:434–7.
53. Dufour C, Cadusseau J, Varlet P, Surena AL, de Faria GP, Dias-Morais A, et al. Astrocytes reverted to a neural progenitor-like state with transforming growth factor alpha are sensitized to cancerous transformation. Stem Cells. 2009;27:2373–82.
54. Wu XZ, Chen D. Origin of hepatocellular carcinoma: role of stem cells. J Gastroenterol Hepatol. 2006;21:1093–8.
55. Kregel S, Kiriluk KJ, Rosen AM, Cai Y, Reyes EE, Otto KB, et al. Sox2 is an androgen receptor-repressed gene that promotes castration-resistant prostate cancer. PloS One. 2013;8:e53701.
56. Jamal M, Rath BH, Tsang PS, Camphausen K, Tofilon PJ. The brain microenvironment preferentially enhances the radioresistance of CD133(+) glioblastoma stem-like cells. Neoplasia. 2012;14:150–8.
57. Liu S, Ginestier C, Ou SJ, Clouthier SG, Patel SH, Monville F, et al. Breast cancer stem cells are regulated by mesenchymal stem cells through cytokine networks. Cancer Res. 2011;71:614–24.
58. Al-Hajj M, Wicha MS, Benito-Hernandez A, Morrison SJ, Clarke MF. Prospective identification of tumorigenic breast cancer cells. Proc Natl Acad Sci U S A. 2003;100:3983–8.
59. Park JM, Munoz JL, Won BW, Bliss SA, Greco SJ, Patel SA, et al. Exogenous CXCL12 activates protein kinase C to phosphorylate connexin 43 for gap junctional intercellular communication among confluent breast cancer cells. Cancer Lett. 2013;331:84–91.
60. Perez A, Neskey DM, Wen J, Pereira L, Reategui EP, Goodwin WJ, et al. CD44 interacts with EGFR and promotes head and neck squamous cell carcinoma initiation and progression. Oral Oncol. 2013;49:306–13.
61. Boiko AD. Isolation of melanoma tumor-initiating cells from surgical tissues. Methods Mol Biol. 2013;961:253–9.
62. Carra E, Barbieri F, Marubbi D, Pattarozzi A, Favoni RE, Florio T, et al. Sorafenib selectively depletes human glioblastoma tumor-initiating cells from primary cultures. Cell cycle. 2013;12:491–500.
63. Tazzari PL, Tabellini G, Ricci F, Papa V, Bortul R, Chiarini F, et al. Synergistic proapoptotic activity of recombinant TRAIL plus the Akt inhibitor Perifosine in acute myelogenous leukemia cells. Cancer Res. 2008;68:9394–403.
64. Miftakhova R, Sandberg T, Hedblom A, Nevzorova T, Persson JL, Bredberg A. DNA methylation in ATRA-treated leukemia cell lines lacking a PML-RAR chromosome translocation. Anticancer Res. 2012;32:4715–22.
65. Han YK, Lee JH, Park GY, Chun SH, Han JY, Kim SD, et al. A possible usage of a CDK4 inhibitor for breast cancer stem cell-targeted therapy. Biochem Biophys Res Commun. 2013;430:1329–33.
66. Lee RC, Feinbaum RL, Ambros V. The C. elegans heterochronic gene lin-4 encodes small RNAs with antisense complementarity to lin-14. Cell. 1993;75:843–54.
67. Kozomara A, Griffiths-Jones S. miRBase: integrating microRNA annotation and deep-sequencing data. Nucleic Acids Res. 2011;39:D152–7.
68. Reinhart BJ, Slack FJ, Basson M, Pasquinelli AE, Bettinger JC, Rougvie AE, et al. The 21-nucleotide let-7 RNA regulates developmental timing in Caenorhabditis elegans. Nature. 2000;403:901–6.
69. Scherr M, Eder M. Gene silencing by small regulatory RNAs in mammalian cells. Cell cycle. 2007;6:444–9.
70. Johnson C, Sundaresan V. Regulatory small RNAs in plants. EXS. 2007;97:99–113.
71. Dimmeler S, Nicotera P. MicroRNAs in age-related diseases. EMBO Mol Med. 2013;5:180–90.
72. Lagos-Quintana M, Rauhut R, Yalcin A, Meyer J, Lendeckel W, Tuschl T. Identification of tissue-specific microRNAs from mouse. Curr Biol. 2002;12:735–9.
73. Leonardo TR, Schultheisz HL, Loring JF, Laurent LC. The functions of microRNAs in pluripotency and reprogramming. Nat Cell Biol. 2012;14:1114–21.
74. Kumar NM, Gilula NB. The gap junction communication channel. Cell. 1996;84:381–8.

75. Harris AL. Emerging issues of connexin channels: biophysics fills the gap. Q Rev Biophys. 2001;34:325–472.
76. Saez JC, Berthoud VM, Branes MC, Martinez AD, Beyer EC. Plasma membrane channels formed by connexins: their regulation and functions. Physiol Rev. 2003;83:1359–400.
77. Willecke K, Eiberger J, Degen J, Eckardt D, Romualdi A, Guldenagel M, et al. Structural and functional diversity of connexin genes in the mouse and human genome. Biol Chem. 2002;383:725–37.
78. Musil LS, Goodenough DA. Multisubunit assembly of an integral plasma membrane channel protein, gap junction connexin43, occurs after exit from the ER. Cell. 1993;74:1065–77.
79. Koval M, Harley JE, Hick E, Steinberg TH. Connexin46 is retained as monomers in a trans-Golgi compartment of osteoblastic cells. J Cell Biol. 1997;137:847–57.
80. Zhang JT, Chen M, Foote CI, Nicholson BJ. Membrane integration of in vitro-translated gap junctional proteins: co- and post-translational mechanisms. Mol Biol Cell. 1996;7:471–82.
81. Ahmad S, Evans WH. Post-translational integration and oligomerization of connexin 26 in plasma membranes and evidence of formation of membrane pores: implications for the assembly of gap junctions. Biochem J. 2002;365:693–9.
82. George CH, Kendall JM, Evans WH. Intracellular trafficking pathways in the assembly of connexins into gap junctions. J Biol Chem. 1999;274:8678–85.
83. Wei CJ, Xu X, Lo CW. Connexins and cell signaling in development and disease. Annu Rev Cell Dev Biol. 2004;20:811–38.
84. Evans WH, Martin PEM. Lighting up gap junction channels in a flash. Bioessays. 2002;24:876–80.
85. van der Pluijm G. Epithelial plasticity, cancer stem cells and bone metastasis formation. Bone. 2011;48:37–43.
86. Wilgenbus KK, Kirkpatrick CJ, Knuechel R, Willecke K, Traub O. Expression of Cx26, Cx32 and Cx43 gap junction proteins in normal and neoplastic human tissues. Int J Cancer. 1992;51:522–9.
87. Laird DW, Fistouris P, Batist G, Alpert L, Huynh HT, Carystinos GD, et al. Deficiency of connexin43 gap junctions is an independent marker for breast tumors. Cancer Res. 1999;59:4104–10.
88. Hirschi KK, Xu CE, Tsukamoto T, Sager R. Gap junction genes Cx26 and Cx43 individually suppress the cancer phenotype of human mammary carcinoma cells and restore differentiation potential. Cell Growth Differ. 1996;7:861–70.
89. Jamieson S, Going JJ, D'Arcy R, George WD. Expression of gap junction proteins connexin 26 and connexin 43 in normal human breast and in breast tumours. J Pathol. 1998;184:37–43.
90. Monaghan P, Moss D. Connexin expression and gap junctions in the mammary gland. Cell Biol Int. 1996;20:121–5.
91. Monaghan P, Perusinghe N, Carlile G, Evans WH. Rapid modulation of gap junction expression in mouse mammary gland during pregnancy, lactation, and involution. J Histochem Cytochem. 1994;42:931–8.
92. Talhouk RS, Elble RC, Bassam R, Daher M, Sfeir A, Mosleh LA, et al. Developmental expression patterns and regulation of connexins in the mouse mammary gland: expression of connexin30 in lactogenesis. Cell Tissue Res. 2005;319:49–59.
93. Kanczuga-Koda L, Sulkowski S, Lenczewski A, Koda M, Wincewicz A, Baltaziak M, et al. Increased expression of connexins 26 and 43 in lymph node metastases of breast cancer. J Clin Pathol. 2006;59:429–33.
94. Jones PA, Laird PW. Cancer epigenetics comes of age. Nat Genet. 1999;21:163–7.
95. Baylin SB, Herman JG, Graff JR, Vertino PM, Issa JP. Alterations in DNA methylation: a fundamental aspect of neoplasia. Adv Cancer Res. 1998;72:141–96.
96. Monaghan P, Clarke C, Perusinghe NP, Moss DW, Chen XY, Evans WH. Gap junction distribution and connexin expression in human breast. Exp Cell Res. 1996;223:29–38.
97. Saunders MM, Seraj MJ, Li Z, Zhou Z, Winter CR, Welch DR, et al. Breast cancer metastatic potential correlates with a breakdown in homospecific and heterospecific gap junctional intercellular communication. Cancer Res. 2001;61:1765–7.

98. Locke D, Perusinghe N, Newman T, Jayatilake H, Evans WH, Monaghan P. Developmental expression and assembly of connexins into homomeric and heteromeric gap junction hemichannels in the mouse mammary gland. J Cell Physiol. 2000;183:228–37.
99. Lau AF, Kanemitsu MY, Kurata WE, Danesh S, Boynton AL. Epidermal growth factor disrupts gap-junctional communication and induces phosphorylation of connexin43 on serine. Mol Biol Cell. 1992;3:865–74.
100. Kanemitsu MY, Lau AF. Epidermal growth factor stimulates the disruption of gap junctional communication and connexin43 phosphorylation independent of 12-0-tetradecanoylphorbol 13-acetate-sensitive protein kinase C: the possible involvement of mitogen-activated protein kinase. Mol Biol Cell. 1993;4:837–48.
101. Warn-Cramer BJ, Cottrell GT, Burt JM, Lau AF. Regulation of connexin-43 gap junctional intercellular communication by mitogen-activated protein kinase. J Biol Chem. 1998;273:9188–96.
102. Polontchouk L, Ebelt B, Jackels M, Dhein S. Chronic effects of endothelin 1 and angiotensin II on gap junctions and intercellular communication in cardiac cells. FASEB J. 2002;16:87–9.
103. Petrich BG, Gong X, Lerner DL, Wang X, Brown JH, Saffitz JE, et al. c-Jun N-terminal kinase activation mediates downregulation of connexin43 in cardiomyocytes. Circ Res. 2002;91:640–7.
104. Cameron SJ, Malik S, Akaike M, Lerner-Marmarosh N, Yan C, Lee JD, et al. Regulation of epidermal growth factor-induced connexin 43 gap junction communication by big mitogen-activated protein kinase1/ERK5 but not ERK1/2 kinase activation. J Biol Chem. 2003;278:18682–8.
105. Cooper CD, Lampe PD. Casein kinase 1 regulates connexin-43 gap junction assembly. J Biol Chem. 2002;277:44962–8.
106. Kanemitsu MY, Jiang W, Eckhart W. Cdc2-mediated phosphorylation of the gap junction protein, connexin43, during mitosis. Cell Growth Differ. 1998;9:13–21.
107. Lampe PD, Kurata WE, Warn-Cramer BJ, Lau AF. Formation of a distinct connexin43 phosphoisoform in mitotic cells is dependent upon p34cdc2 kinase. J Cell Sci. 1998; 111:833–41.
108. Xie H, Laird DW, Chang TH, Hu VW. A mitosis-specific phosphorylation of the gap junction protein connexin43 in human vascular cells: biochemical characterization and localization. J Cell Biol. 1997;137:203–10.
109. Berthoud VM, Ledbetter ML, Hertzberg EL, Saez JC. Connexin43 in MDCK cells: regulation by a tumor-promoting phorbol ester and Ca2+. Eur J Cell Biol. 1992;57:40–50.
110. Berthoud VM, Rook MB, Traub O, Hertzberg EL, Saez JC. On the mechanisms of cell uncoupling induced by a tumor promoter phorbol ester in clone 9 cells, a rat liver epithelial cell line. Eur J Cell Biol. 1993;62:384–96.
111. Brissette JL, Kumar NM, Gilula NB, Dotto GP. The tumor promoter 12-O-tetradecanoylphorbol-13-acetate and the ras oncogene modulate expression and phosphorylation of gap junction proteins. Mol Cell Biol. 1991;11:5364–71.
112. Lampe PD. Analyzing phorbol ester effects on gap junctional communication: a dramatic inhibition of assembly. J Cell Biol. 1994;127:1895–905.
113. Reynhout JK, Lampe PD, Johnson RG. An activator of protein kinase C inhibits gap junction communication between cultured bovine lens cells. Exp Cell Res. 1992;198:337–42.
114. Paulson AF, Lampe PD, Meyer RA, TenBroek E, Atkinson MM, Walseth TF, et al. Cyclic AMP and LDL trigger a rapid enhancement in gap junction assembly through a stimulation of connexin trafficking. J Cell Sci. 2000;113:3037–49.
115. Atkinson MM, Lampe PD, Lin HH, Kollander R, Li XR, Kiang DT. Cyclic AMP modifies the cellular distribution of connexin43 and induces a persistent increase in the junctional permeability of mouse mammary tumor cells. J Cell Sci. 1995;108:3079–90.
116. Burghardt RC, Barhoumi R, Sewall TC, Bowen JA. Cyclic AMP induces rapid increases in gap junction permeability and changes in the cellular distribution of connexin43. J Membr Biol. 1995;148:243–53.

117. Yogo K, Ogawa T, Akiyama M, Ishida-Kitagawa N, Sasada H, Sato E, et al. PKA implicated in the phosphorylation of Cx43 induced by stimulation with FSH in rat granulosa cells. J Reprod Dev. 2006;52:321–8.
118. Loo LW, Berestecky JM, Kanemitsu MY, Lau AF. pp60src-mediated phosphorylation of connexin 43, a gap junction protein. J Biol Chem. 1995;270:12751–61.
119. Crow DS, Kurata WE, Lau AF. Phosphorylation of connexin43 in cells containing mutant src oncogenes. Oncogene. 1992;7:999–1003.
120. Swenson KI, Piwnica-Worms H, McNamee H, Paul DL. Tyrosine phosphorylation of the gap junction protein connexin43 is required for the pp60v-src-induced inhibition of communication. Cell Reg. 1990;1:989–1002.
121. Crow DS, Beyer EC, Paul DL, Kobe SS, Lau AF. Phosphorylation of connexin43 gap junction protein in uninfected and Rous sarcoma virus-transformed mammalian fibroblasts. Mol Cell Biol. 1990;10:1754–63.
122. Dbouk HA, Mroue RM, El-Sabban ME, Talhouk RS. Connexins: a myriad of functions extending beyond assembly of gap junction channels. Cell commun signal. 2009;7:4.
123. Elzarrad MK, Haroon A, Willecke K, Dobrowolski R, Gillespie MN, Al-Mehdi AB. Connexin-43 upregulation in micrometastases and tumor vasculature and its role in tumor cell attachment to pulmonary endothelium. BMC Med. 2008;6:20.
124. Bos PD, Zhang XH, Nadal C, Shu W, Gomis RR, Nguyen DX, et al. Genes that mediate breast cancer metastasis to the brain. Nature. 2009;459:1005–9.
125. Klotz LO. Posttranscriptional regulation of connexin-43 expression. Arch Biochem Biophys. 2012;524:23–9.
126. Zhang CZ, Zhang JX, Zhang AL, Shi ZD, Han L, Jia ZF, et al. MiR-221 and miR-222 target PUMA to induce cell survival in glioblastoma. Mol Cancer. 2010;9:229.
127. Dai X, Mao Z, Huang J, Xie S, Zhang H. The CXCL12/CXCR4 autocrine loop increases the metastatic potential of non-small cell lung cancer in vitro. Oncol Lett. 2013;5:277–82.
128. Steinl C, Essl M, Schreiber TD, Geiger K, Prokop L, Stevanovic S, et al. Release of matrix metalloproteinase-8 during physiological trafficking and induced mobilization of human hematopoietic stem cells. Stem Cells Dev. 2013;22:1307-18.
129. Kang H, Watkins G, Parr C, Douglas-Jones A, Mansel RE, Jiang WG. Stromal cell derived factor-1: its influence on invasiveness and migration of breast cancer cells in vitro, and its association with prognosis and survival in human breast cancer. Breast Cancer Res. 2005;7:R402–10.
130. Ratajczak MZ, Kim C, Janowska-Wieczorek A, Ratajczak J. The expanding family of bone marrow homing factors for hematopoietic stem cells: stromal derived factor 1 is not the only player in the game. ScientificWorldJournal. 2012;2012:758512.
131. Chou SH, Ko BS, Chiou JS, Hsu YC, Tsai MH, Chiu YC, et al. A knock-in npm1 mutation in mice results in myeloproliferation and implies a perturbation in hematopoietic microenvironment. PloS one. 2012;7:e49769.
132. Hattermann K, Mentlein R. An infernal trio: the chemokine CXCL12 and its receptors CXCR4 and CXCR7 in tumor biology. Ann Anat. 2013;195:103–10.
133. Lipfert J, Odemis V, Wagner DC, Boltze J, Engele J. CXCR4 and CXCR7 form a functional receptor unit for SDF-1/CXCL12 in primary rodent microglia. Neuropathol Appl Neurobiol. 2013;39:667–80.
134. Corcoran KE, Trzaska KA, Fernandes H, Bryan M, Taborga M, Srinivas V, et al. Mesenchymal stem cells in early entry of breast cancer into bone marrow. PloS One. 2008;3:e2563.
135. Greco SJ, Patel SA, Bryan M, Pliner LF, Banerjee D, Rameshwar P. AMD3100-mediated production of interleukin-1 from mesenchymal stem cells is key to chemosensitivity of breast cancer cells. Am J Cancer Res. 2011;1:701–15.
136. Uy GL, Rettig MP, Motabi IH, McFarland K, Trinkaus KM, Hladnik LM, et al. A phase 1/2 study of chemosensitization with the CXCR4 antagonist plerixafor in relapsed or refractory acute myeloid leukemia. Blood. 2012;119:3917–24.

137. Mishra PJ, Mishra PJ, Humeniuk R, Medina DJ, Alexe G, Mesirov JP, et al. Carcinoma-associated fibroblast-like differentiation of human mesenchymal stem cells. Cancer Res. 2008;68:4331–9.
138. Uchida D, Onoue T, Tomizuka Y, Begum NM, Miwa Y, Yoshida H, et al. Involvement of an autocrine stromal cell derived factor-1/CXCR4 system on the distant metastasis of human oral squamous cell carcinoma. Mol Cancer Res. 2007;5:685–94.
139. Berghuis D, Schilham MW, Santos SJ, Savola S, Knowles HJ, Dirksen U, et al. The CX-CR4-CXCL12 axis in Ewing sarcoma: promotion of tumor growth rather than metastatic disease. Clinical Sarcoma Res. 2012;2:24.
140. Tang T, Xia QJ, Chen JB, Xi MR, Lei D. Expression of the CXCL12/SDF-1 Chemokine Receptor CXCR7 in Human Brain Tumours. Asian Pac J Cancer Prev. 2012;13:5281–6.
141. Nair VS, Maeda LS, Ioannidis JP. Clinical outcome prediction by microRNAs in human cancer: a systematic review. J Natl Cancer Inst. 2012;104:528–40.
142. Iorio MV, Croce CM. Causes and consequences of microRNA dysregulation. Cancer J. 2012;18:215–22.
143. Liu C, Tang DG. MicroRNA regulation of cancer stem cells. Cancer Res. 2011;71:5950–4.
144. Tay Y, Zhang J, Thomson AM, Lim B, Rigoutsos I. MicroRNAs to Nanog, Oct4 and Sox2 coding regions modulate embryonic stem cell differentiation. Nature. 2008;455:1124–8.
145. Yu F, Yao H, Zhu P, Zhang X, Pan Q, Gong C, et al. let-7 regulates self renewal and tumorigenicity of breast cancer cells. Cell. 2007;131:1109–23.
146. Yu F, Jiao Y, Zhu Y, Wang Y, Zhu J, Cui X, et al. MicroRNA 34c gene down-regulation via DNA methylation promotes self-renewal and epithelial-mesenchymal transition in breast tumor-initiating cells. J Biol Chem. 2012;287:465–73.
147. Han M, Liu M, Wang Y, Mo Z, Bi X, Liu Z, et al. Re-expression of miR-21 contributes to migration and invasion by inducing epithelial-mesenchymal transition consistent with cancer stem cell characteristics in MCF-7 cells. Mol Cell Biochem. 2012;363:427–36.
148. Yu F, Deng H, Yao H, Liu Q, Su F, Song E. Mir-30 reduction maintains self-renewal and inhibits apoptosis in breast tumor-initiating cells. Oncogene. 2010;29:4194–204.
149. Guttilla IK, Adams BD, White BA. ERalpha, microRNAs, and the epithelial-mesenchymal transition in breast cancer. Trends Endocrinol Metab. 2012;23:73–82.
150. Zhu Y, Yu F, Jiao Y, Feng J, Tang W, Yao H, et al. Reduced miR-128 in breast tumor-initiating cells induces chemotherapeutic resistance via Bmi-1 and ABCC5. Clin Cancer Res. 2011;17:7105–15.
151. Kovalchuk O, Filkowski J, Meservy J, Ilnytskyy Y, Tryndyak VP, Chekhun VF, et al. Involvement of microRNA-451 in resistance of the MCF-7 breast cancer cells to chemotherapeutic drug doxorubicin. Mol Cancer Ther. 2008;7:2152–9.
152. Liu H. MicroRNAs in breast cancer initiation and progression. Cell Mol Life Sci. 2012;69:3587–99.
153. Thum T. MicroRNA therapeutics in cardiovascular medicine. EMBO Mol Med. 2012;4:3–14.
154. van Rooij E, Marshall WS, Olson EN. Toward microRNA-based therapeutics for heart disease: the sense in antisense. Circ Res. 2008;103:919–28.
155. Lindow M, Kauppinen S. Discovering the first microRNA-targeted drug. J Cell Biol. 2012;199:407–12.
156. Laurent LC, Chen J, Ulitsky I, Mueller FJ, Lu C, Shamir R, et al. Comprehensive microRNA profiling reveals a unique human embryonic stem cell signature dominated by a single seed sequence. Stem Cells. 2008;26:1506–16.

Chapter 6
MicroRNAs in Epithelial Mesenchymal Transition and Breast Cancer Progression

George R. Nahas, Bernadette M. Bibber and Pranela Rameshwar

Abstract Breast cancer remains a serious public health issue despite early diagnosis and aggressive treatment. This chapter discusses the molecular mechanism by which epithelial mesenchymal transition (EMT) occurs and its implication for metastasis. The chapter also discusses how the different subsets, including cancer stem cells, contribute to cancer evasion and resistance, through dormancy. Included in the discussion are studies on mesenchymal stem cells as protection for the cancer cells from immune clearance. We reviewed the growing information on microRNA (miRNA) in the cellular mechanisms of EMT and its role in facilitating metastasis and/or dormancy of the cancer cells. Overall, this chapter provides a "snapshot" of EMT and miRNA in breast cancer dormancy and metastasis.

Keywords Breast cancer · Epithelial mesenchymal transition · miRNA · Bone marrow · Cytokine · Chemoresistance · Cancer stem cells

1 Introduction: Breast Cancer

Breast cancer (BC) remains a major public health challenge in the USA [20]. Mortality of BC is mostly due to metastatic disease [10]. In this regard, it is important to understand metastasis, which will lead to the development of new targets. Despite education and compliance for screening by mammogram, the incidence of BC deaths remains the same [32]. However, there are advances in screening, such as molecular breast imaging, which has been determined to be efficient in diagnosing women with dense breast tissue to increase the chance of early detection [56]. This population generally showed indeterminate results from traditional mammography [56].

This underscores the importance for further research and open-mindedness for "out-of-the-box" hypotheses to understand the development of tumors and to evaluate how tumor cells exist at an early stage. Approximately a century ago,

Grant support: This work was funded by the Department of Defense.

P. Rameshwar (✉) · G. R. Nahas · B. M. Bibber
Graduate School of Biomedical Sciences and New Jersey Medical School, Rutgers,
185 South Orange Ave, MSB, Room E-579, Newark, NJ 07103, USA
e-mail: rameshwa@njms.rutgers.edu

S. R. Singh, P. Rameshwar (eds.), *MicroRNA in Development and in the Progression of Cancer*, DOI 10.1007/978-1-4899-8065-6_6,

Stephen Paget introduced his "seed and soil" hypothesis that described a necessary compatibility between the tumor cell and the microenvironment for successful metastasis. It is now clear that the microenvironment or niche is important for the survival and function of the cancer cells.

BC has metastatic predilection for the bone marrow, suggesting that cells within the marrow support the BC cells (BCCs) as the "soil." The role of the "soil" appears to be multifunctional, supporting dormancy as well as cancer growth [10]. The dormant phase could be extremely long, based on clinical evidence of > 10 years in remission in bone marrow, and the length of time before BC can be clinically detected [8, 53, 63]. This chapter discusses the role of microRNAs (miRNAs) in cancer progression, as this will be relevant to understanding cancer resurgence from bone marrow and other organs. Since epithelial mesenchymal transition (EMT) can be involved in tumor metastasis and invasion, these topics are also reviewed. The relevance to chemoresistance and to prognosis is discussed.

2 Dormancy

Synergism appears to exist between EMT and cancer stem cells (CSCs). Interactions between cancer cells and their niche resulted in the cancer cells undergoing an EMT. At the time of EMT, the cancer cells undergo a change to a mesenchymal phenotype with a decrease in E-cadherin and an increase in N-cadherin, vimentin, and α-smooth muscle actin [67]. Dormant BCCs are mostly in the growth 1 (G1) phase of the cell cycle, which could make the cells refractory to anti-cycling therapy [69]. Following the extravasation of cancer cells, solitary tumor cells can be detected in the bone marrow years before the development of overt metastasis [9].

There could be several rate-limiting steps during the process of metastasis by the primary tumor. For example, the inability of angiogenesis at the metastatic site may be a rate-limiting step, similar to the rate-limiting step at the primary tumor site, which requires hypoxia-linked factors such as vascular endothelial growth factor (VEGF) for angiogenesis to occur [28]. Despite the information on the rate-limiting step, this might not be an overall mechanism given that the cancer cells at sites of metastasis seem to be heterogeneous. For example, dormant cells extracted from the bone marrow of different types of tumors were shown to have different proliferative potentials [61]. This raised the question of equilibrium. It is possible that the heterogeneity with regard to cell proliferation might be aimed at maintaining the tumor size through an equilibrium between apoptosis and proliferation [29]. In order to fully eradicate tumor cells, rather than putting the individual in remission, the mechanism of dormancy must be understood in order to lead to the reversal of the quiescence phase for effective targeted treatment.

It appears that dormant BCCs can survive in the bone marrow, close to the endosteum, in the form of gap junctional intercellular communication (GJIC) with hematopoietic-supporting cells such as stoma [37]. miRNA can be exchanged though GJIC to cause cycling quiescence. In addition to cellular communication through gap junctions, cytokines are a particular topic of interest with regard to BC.

C-X-C motif chemokine 12 (CXCL12), normally constitutively expressed and a known regulator in hematopoiesis, is downregulated when contact is made between the same stroma and the BCCs; it was discovered that decreased levels correlated with decreased proliferation of BC [37, 43]. In search of possible mechanisms for decreased levels, it was found that particular miRNAs that cross GJICs between BC and stromal cells specifically reduce CXCL12 levels and transition the BCCs to the G0 phase [37]. The importance of CXCL12 is outlined in its interaction with C-X-C chemokine receptor type 4 (CXCR4) to facilitate the protection of BCCs from the immune system [13, 50]. This contact, inducing a variety of different states including but not limited to immunosuppression, may form the beginning of a microenvironment conducive to the survival of BC, including those in bone marrow. These mechanisms may outline specific targets for future treatments and therapeutic interventions.

Dormancy is achieved through a variety of factors, including factors intrinsic to the cell such as ATP-binding cassette (ABC) drug transporters and GJICs, and those that are extrinsic to the cells including mesenchymal stem cell (MSC)-derived oncoprotection, among other immune system interactions. While many mechanisms are understood and others are currently being researched and explored, significant problems and questions remain to be answered in order to further understand the mechanisms underlying BC metastasis and dormancy. Many current treatment regimens take a classical route and are primarily concerned with killing cells that are rapidly dividing, but treatment modalities must evolve and embrace the concept that CSCs drive tumorigenesis.

3 MicroRNA: Overview

miRNAs are small, noncoding RNAs that constitute ~1–2% of mammalian genes [4, 63]. Since their discovery in 2000, miRNAs have been implicated in a number of biological processes including neural patterning, neurodevelopment, and oncogenesis. Despite their relatively small size, miRNAs are stable in normal circulation, thus enabling their utilization as a biomarker in oncology [75].

miRNAs represent a diverse family of RNA molecules that can be found inter- and intragenetically (intronic and exonic). Primary miRNAs are transcribed by RNA polymerase II, and then cleaved by Drosha, an RNase type III. Exportin 5 transports the precursor miRNA in a RAN-GTP-dependent manner from the nuclei to the cytoplasm where RNase type III, Dicer, cleaves the precursor miRNA into mature 18–22 double-stranded nucleotides. It is believed that the more stable of the two stands recruits the RNA-induced silencing complex (RISC) to degrade the complementary strand. After integration to the RISC, the miRNA may bind to the 3′ untranslated region (UTR) of its target mRNA. Depending on complementarity, the target mRNA may be degraded by RISC-associated argonaute (Ago) proteins or the translation of the targeted mRNA could be suppressed.

In general, miRNAs are considered posttranscriptional regulators of genes [44]. This occurs through interactions with the 3′ UTR of transcripts (mRNA) to suppress translation. Interestingly, miRNAs and their targets are generally conserved, suggesting their critical functions in development [4]. A single miRNA has even been reported to suppress target genes. In most instances, outcomes of a single miRNA are generally modest. The efficiency of miRNAs in regulating gene expression appears to require one or more miRNAs in clusters at multiple sites within the transcript [2]. Although the role of miRNAs in development has been exhaustively studied, it is yet to be determined if the miRNAs are involved in "fine-tuning" the developmental processes or if they mediate a central role in development.

4 Breast Cancer Stem Cells

The idea that each cancer cell can initiate a tumor is now considered a concept of the past. The expansion of stem cell biology has led to a growth of information that was extrapolated to research on CSCs to support CSCs as the initiator of tumor [55, 75]. The existence of CSCs is further supported by the inability to eradicate cancer, regardless of the treatment [41]. Other than CSCs as the subset that resist drug treatment, other mechanisms have been linked as the cause of chemoresistance. These include increased expressions of anti-apoptotic proteins and the multidrug resistance gene [62, 70].

The presence of normal stem cells in adult organs poses a significant challenge to targeting the CSCs without bystander effect on the normal stem cells. An understanding of how miRNAs are involved in protecting CSCs is fundamental to the development of drugs to eradicate cancer. An understanding of breast cancer stem cells (BCSCs) is better explained by first describing the mammary stem cells and lineage development. The mutation of mammary epithelial cells has been thought to be the source of tumor initiation. Although incomplete, phenotypically, mammary stem cells have been identified as $CD29^{high}/CD24^{+}$ or $CD49F^{high}/CD24^{+}$. This subset has been shown to express the stem cell marker aldehyde dehydrogenase 1 and can form a complete and functional mammary gland when transplanted into a cleared mouse mammary fat pad [59]. BCSCs are phenotypically distinct from normal mammary stem cells and are identified as $CD44^{+}CD24^{-/low}$ [1].

5 Epithelial Mesenchymal Transition

EMT is defined as the alteration of the architectural support for normal epithelial cells, which are strongly attached to the basement membrane. EMT causes a loss in polarity with the attainment of a mesenchymal-like or fibroblastoid phenotype at the primary site of the tumor in cells that show an increase in motility and invasion [11]. EMT causes the cells to lose intercellular junctions and to show a decrease in E-cadherin but increases in N-cadherin, matrix metalloprotienases, vimentin, and α-smooth muscle [68]. EMT is the beginning of tumor metastasis, which is defined as the spread of tumor cells from the primary tumor site to distant sites. First proposed by Stephen Paget in 1889, successful metastasis is a result of compatibility between tumor cells and the environment. Paget's "seed and soil" hypothesis states that metastasis depends on both the characteristics of selected cancer cells (the seeds) and their compatibility with certain microenvironments (the soil) [47].

A recent study linked specific miRNA to EMT and CSCs [24]. The authors indicated that miR-21 cancer reverses EMT to differentiate CSC phenotype. This occurs through targeting of phosphatase and tensin homolog (PTEN), resulting in the inactivation of the tumor-promoting signaling pathways involving AKT and extracellular signal-regulated kinase (ERK)1/2.

6 Supporting Factors in EMT

Successful metastasis occurs when there are growth signals, insensitivity to growth-inhibitory (antigrowth) signals, evasion of programmed cell death (apoptosis), limitless replicative potential, sustained angiogenesis, and tissue invasion [25]. These properties are linked to two categories, those that are intrinsic and those that are extrinsic to the tumor cell. Both intrinsic and extrinsic factors resulted in the control of cell cycle regulation. However, extrinsic factors involve those that influence the cell from the microenvironment, such as extracellular matrix components, basement membranes, reactive oxygen species (ROS), and factors from immune cells.

An example of an intrinsic trigger of other functions is the mutation of p53, which is linked to apoptosis and cell cycle arrest [35]. Mutations in p53 have been reported in almost all human tumors, which result in poor prognosis [15, 71]. Normally found at relatively low levels, p53 is increased by stressors, such as oncogenic activation, DNA damage, and hypoxia [38].

Other genes such as Wnt-β-catenin pathway, Notch, and Hedgehog are also linked to tumor metastasis [54, 74]. Dysregulation of the Wnt-β-catenin and Notch-mediated pathways have been implicated in BC [48, 57]. The continued activation of the Hedgehog pathway is associated with unfavorable prognosis and progression of BC [66].

The tumor cells are required to detach for metastasis. It is prudent, therefore, to explore how this occurs. The experimental evidence suggested that defects in adhesion molecules and altered functions within the microenvironment can lead to cancer progression and metastasis [31]. Extensive studies were done with β1-integrin, which interacts with collagen, fibronectin, E-cadherin, among other matrices [31]. In endometrial cancer, mutant p53 binds to the regulatory region of miR130b to promote EMT [5]. Although we focus on BC, this report might be critical to understanding how mutant p53 is involved in EMT.

ROS can be considered as an extrinsic factor, which is produced by the mitochondria under hypoxic conditions [23]. ROS induces hypoxia-inducible factor 1 (HIF-1α) in cancer cells [18]. HIF-1 activates the expression of growth survival genes such as VEGF and erythropoietin [28]. In conclusion, a hypoxic microenvironment leads to elevated HIF-1α, which can promote cancer metastasis and progression through HIF-1α-induced VEGF expression.

7 miRNA: Cancer Niche

The debate continues whether circulating tumor cells (CTCs) "seed" at the first site of contact. In an experimental model of melanoma, it was shown that the cells selectively metastasize to specific organs [26]. Insights were gained in support of a nonrandom process for metastasis in patients with peritoneovenous shunts, which is a palliative intervention for patients with ascites [64]. Despite viable cancer cells being introduced into the venous system, there was no evidence of hematogenous metastasis [10, 65]. These human studies support a strong role for the niche in supporting the "behavior" of the cancer. We discuss a possible role of miRNAs in the interaction between the niche and the cancer cells.

In order to fully understand miRNAs as mediators of the BCCs and the niche, the discussion reverts to the heterogeneity of BCCs. A broad statement cannot be made about miRNA since the interaction with different subsets is likely to differ. This brings up a confound of the field since a comprehensive hierarchy of BCCs has not been identified. The different subsets might acquire metastatic properties over time, allowing only those cells to metastasize.

There are different models to study tumor metastasis and each might address a different question on the role of miRNAs, which would provide insights on EMT. The clonal evolution theory describes single cells accumulating a broad spectrum of changes over time, thus creating cells with variable phenotypic and behavioral characteristics, which includes the ability to metastasize [63]. The transition of the clone into a heterogeneous subset of cancer cells has not been established as a hierarchy, making it a significant obstacle when studies are intended to understand how miRNAs are involved in a particular BCC subset. There are several studies on miRNA with unsorted/heterogeneous BCCs. While the findings from these studies can be discussed, there is still a question whether all subsets would behave similarly.

Regardless of the above discussion, the clonal development of a cancer cell does attest to the fact that given the right circumstances any cell has the potential to transform into a cancer cell with metastatic capabilities. On the other hand, the stem cell cancer theory states that a specific subset of multipotent cells that compose the CSC compartment are initiated for subsequent metastasis [63]. This model defends the idea that only a particular subset of cells within a tumor can have the potential for metastasis. Both models, nevertheless, describe a scenario in which only cells that have acquired those abilities to metastasize do so and proceed to form secondary tumors at distant sites [17].

A recent model proposed six characteristics that are imperative to malignant tumor growth: self sufficiency in growth signals, insensitivity to growth-inhibitory (antigrowth) signals, evasion of programmed cell death (apoptosis), limitless replicative potential, sustained angiogenesis, and tissue invasion and metastasis [25]. Despite the necessity for these characteristics, cancer remains a clinical dilemma. Thus, it is not sufficient to understand the properties at each step, butthe molecular mechanisms, including the role of miRNAs, need to be determined.

The evolution processes of metastasis do not necessarily end with the growth of tumor cells at a distant site. At the new/distant site, the metastatic cells can develop as heterogeneous cells with different metastatic profiles than the original metastatic cells [63]. This concept was demonstrated with BC in which the degree of tumor cell heterogeneity in bone marrow caused a negative impact on prognosis. The theory was further developed to include the ability of metastatic cells to reseed the primary tumor site, which further resulted in a diverse tumor [45]. This mechanism adds another layer to the diversity of metastatic capabilities within the primary tumor site [63].

8 Tumor Niche: Cellular and Humoral Components

Regardless of the molecular mechanisms discussed above, the central role in the process involves the protection of the tumors from the immune system. This provides immune cells as a key cellular component of the tumor microenvironment. In the absence of the niche, tumor cells can elicit an immune response but in vivo, the tumor antigens can be modulated by the tumor microenvironment [46]. The mechanism by which this occurs is complex. A dysregulated immune system can promote tumor progression. T-helper 17 (Th17) cells can promote the production of proinflammatory cytokines, generally linked to an autoimmune-type response, such asinterleukin-6 (IL-6), transforming growth factor-beta (TGF-β), and IL-17 [40]. The cytokines produced by Th17 cells, combined with angiogenic factors, such as VEGF, and other inflammatory mediators, such as prostaglandin E2, contribute to the humoral component of the tumor microenvironment [34].

In contrast to Th17 responses, which provide an autoimmune-type response, protection of the tumor can occur through negative immune responses by regulatory T cells (T_{regs}) [14]. Patients with epidermal growth factor receptor 2 (HER2)-expressing BC when treated with Herceptin showed an inverse switch in the ratio of Th17 to T_{regs} [30]. The tumors can induce their own protection through soluble factors. Tumor-derived soluble factor, phosphatidylserine, can induce the production of IL-10 and TGF-β and inhibit the response from dendritic and T cells [34].

It is difficult to discuss the cellular component of tumors without including the role of MSCs and their recent association with Snail 1, which is considered as a key factor in EMT [5]. Although a transcriptional factor, its ability to support the MSCs could be significant if it can maintain the "stemness" of MSCs. This will explain how MSCs could protect the tumor from the immune system by inducing the differentiation and expansion of T_{regs} [50].

MSCs are ubiquitously expressed in the adult and fetal organs and are abundant in the adult bone marrow and adipose tissue [36]. The anatomical location of MSCs in the bone marrow suggests that they could be the first set of cells that the cancer cells will encounter upon entry into the cavity [7]. MSCs appear to serve a bimodal role in BC progression as they not only can inhibit BCC proliferation but can also promote its progression at the primary and metastatic sites [13, 42, 50]. As a support, MSCs can differentiate into tumor-associated fibroblasts to produce factors such as CXCL12 [21, 42]. Exosomes are also involved in the biology of cancer by facilitating the movement of RNA, including miRNA from one cell to the other

[37]. Exosomes and membrane vesicles, which contain RNA and protein, are able to convert MSCs into tumor-associated myofibroblast and to increase the expression of tumor-promoting factors [12]. The evidence that MSCs support cancer metastasis is still a subject of investigation. The co-injection of BCCs and MSCs in immunodeficient mice resulted in enhanced lung and liver metastasis [33]. The experimental evidence suggested that MSC-derived fibroblasts expressed genes that regulate matrices and angiogenesis to support the stromal microenvironment to facilitate metastasis [19].

The immune suppressive properties of MSCs seem to be used by tumors for survival, collectively known as oncoprotection [49]. As part of the stromal compartment of tumors, MSCs can suppress the immune response, including natural killer and cytotoxic T cells [50]. MSCs can also suppress the proliferation of T cells, partly through the production of TGF-β and hepatocyte growth factor and induced apoptosis of T cells [49, 52]. These functions are consistent with the other properties of MSCs, discussed above.

9 Bone Metastasis

EMT, which describes metastasis, is directly linked to bone invasion of cancer cells. Bone homeostasis is accomplished through a delicate balance between bone-resorbing osteoclasts and bone-forming osteoblasts. Osteoblasts are derived from MSCs, whereas osteoclasts are members of the monocyte/macrophage family. The differentiation of monocytes into osteoclasts relies on the activation of the receptor for receptor activator of nuclear factor kappa-B ligand (RANKL) and macrophage colony-stimulating factor (M-CSF). Parathyroid hormone-related protein (PTHrP) stimulates the expression of RANKL. BCC-induced osteolysis was blunted by neutralizing antibodies against PTHrP [22]. Osteoblasts express a decoy receptor for RANK, osteoprotegrin (OPG), which could counteract osteoclast formation. By understanding these processes, insights can be gained to understand the transition of BC metastasis to bone invasion.

Bone metastasis may be either osteolytic or osteoblastic. The osteolytic process is initiated by the release of osteoclastic agents in the bone microenvironment, such as BC-derived IL-8, which induces osteoclastic formation, independent of RANKL [6]. Although metastasis of BC to bone tends to be osteolytic, regardless of whether metastasis is predominantly osteolytic or osteoblastic, each has components of the other as a primarily osteolytic period may be followed by an osteoblastic period [51]. This may provide clues to the worth of the paradoxical treatment of bone resorption-inhibiting agents in an osteoblastic lesion.

Other factors such as cytokines have a significant role in bone metastasis. TGF-β, which is stored in the bone, can signal through Smad2 and Smad3, which then interact with Smad4 for nuclear translocation and transcription of genes [60]. The TGF-β1–Smad2 pathway controls cell growth and mediates EMT [60]. The vascularity of the marrow and the production of angiogenic factors, such as VEGF

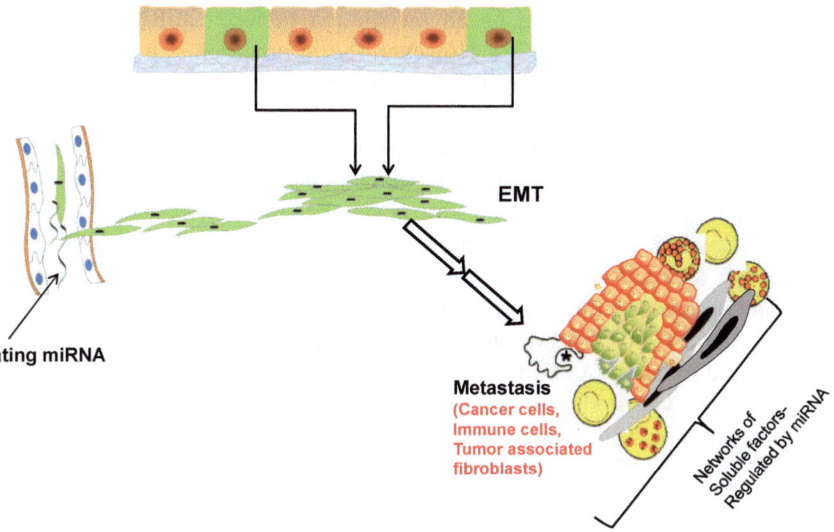

Fig. 6.1 Shown is the mutation of epithelial cells, perhaps the stem cells. The mutants undergo EMT and then migrate to the blood vessels and metastasize at a distant site. miRNAs are produced and remain in the circulation, which can be used as prognostic markers. The tumor mass is supported by immune cells and associated fibroblasts. The complex interactions between the cancer cells and supporting cells are regulated by miRNA

and bone-reacting cytokines, are elevated as compared to soft tissue resulting in an environment that facilitates EMT. The bone itself can provide a nourishing environment for tumor cells by providing TGF-β and other factors such as insulin-like growth factor, fibroblast growth factor, platelet-derived growth factors, and bone morphogenetic protein [27]. The mechanisms of metastasis, particularly to the bone, in addition to the description of a microenvironment conducive to overall tumor survival, are important for overall tumorigenicity. Whether or not treatment and research are tailored to thwart any number of the pathways, BC dormancy continues to be a clinical problem.

10 miRNA in Drug Resistance

CSCs and EMT cells exert drug resistance [73]. As discussed above, the interactions of these cancer cells with the microenvironment as well as the molecular mechanisms of drug resistance will be critical to make cancer cells sensitive to drugs. miRNAs are important for drug resistance [58]. However, the role of miRNA in facilitating chemoresistance of CSCs and EMT is considered a “new” field. The dissection of miRNA in CSCs and EMT cells is not mutually exclusive to the microenvironment because miRNA can be shared between cells through GJIC and exosomes [37].

The data are developing on an association between miRNA expression and drug-resistant tumor cells [73]. However, the information is not always linked to the intracellular biology of miRNA in the chemoresistance of the cell. Circulating miR125b has been shown to predict the chemoresistance of BCCs [72]. Lin28, which has been shown to be involved in the chemoresistance of tumors, was targeted by Let-7 miRNA following paclitaxel treatment [39]. In other cases, the same miRNAs that are involved in EMT are also linked to chemoresistance [16]. miRNA mimics could be a viable treatment. A recent review article discusses the possibility of therapy with miRNA34 as a mimic due to its ability to antagonize tumor promotion and its role as a tumor suppressor [3].

In summary, this brief discussion attempts to address how EMT and CSCs are maintained by miRNA and to provide insights into the mechanisms by which this occurs. However, it should be remembered that there are subsets of cancer cells and the function of the cancer cells could be influenced by the microenvironment. Figure 6.1 summarizes this chapter.

References

1. Al-Hajj M, Wicha MS, Benito-Hernandez A, et al. Prospective identification of tumorigenic breast cancer cells. Proc Natl Acad Sci. 2003;100:3983–8.
2. Arvey A, Larsson E, Sander C, et al. Target mRNA abundance dilutes microRNA and siRNA activity. Mol Syst Biol 2010; 6:363, doi: 10.1038/msb.2010.24.
3. Bader AG. miR-34: A microRNA replacement therapy is headed to the clinic. Frontiers in Genetics 2012; 3:120. doi: 10.3389/fgene.2012.00120.
4. Bartel DP. MicroRNAs: Target recognition and regulatory functions. Cell. 2009;136:215–33.
5. Batlle R, Alba-Castellon L, Loubat-Casanovas J, et al. Snail1 controls TGF-[beta] responsiveness and differentiation of mesenchymal stem cells. Oncogene (In press).
6. Bendre MS, Margulies AG, Walser B, et al. Tumor-derived interleukin-8 stimulates osteolysis independent of the receptor activator of nuclear factor-+¦B ligand pathway. Cancer Res. 2005;65:11001–9.
7. Binaco P. Minireview: The Stem Cell Next Door: Skeletal and Hematopoietic Stem Cell 'Niches' in Bone. Endocrinol. 2011;152:2957–62.
8. Braun S, Vogl FD, Naume B, et al. A pooled analysis of bone marrow micrometastasis in breast cancer. N Engl J Med. 2005;353:793–802.
9. Braun S, Auer D, Marth C. The prognostic impact of bone marrow micrometastases in women with breast cancer. Cancer Invest. 2009;27:598–603.
10. Castano Z, Tracy K, McAllister SS. The tumor macroenvironment and systemic regulation of breast cancer progression. Int J Dev Biol. 2011;55:889–97.
11. Chaffer CL, Thompson EW, Williams ED. Mesenchymal to epithelial transition in development and disease. Cells Tissues Organs. 2007;185(1–3):7–19.
12. Cho JA, Park H, Lim EH, et al. Exosomes from breast cancer cells can convert adipose tissue-derived mesenchymal stem cells into myofibroblast-like cells. Int J Oncol. 2012;40:130–8.
13. Corcoran KE, Trzaska KA, Fernandes H, et al. Mesenchymal stem cells in early entry of breast cancer into bone marrow. PLoS ONE. 2008;3:e2563.
14. Decker T, Fischer G, Bucke W, et al. Increased number of regulatory T cells (T-regs) in the peripheral blood of patients with Her-2/neu-positive early breast cancer. J Cancer Res Clin Oncol (In press).
15. Deng G, Chen LC, Schott DR, et al. Loss of heterozygosity and p53 gene mutations in breast cancer. Cancer Res. 1994;54:499–505.

16. Feng B, Wang R, Chen LB. Review of MiR-200b and cancer chemosensitivity. Biomed Pharmacother (In press).
17. Fidler IJ. The pathogenesis of cancer metastasis: the 'seed and soil' hypothesis revisited. Nat Rev Cancer. 2003;3:453–8.
18. Flugel D, Gorlach A, Michiels C, et al. Glycogen synthase kinase 3 phosphorylates hypoxia-inducible factor 1alpha and mediates its destabilization in a VHL-independent manner. Mol Cell Biol. 2007;27:3253–65.
19. Goldstein RH, Reagan MR, Anderson K, et al. Human bone marrow-derived MSCs can home to orthotopic breast cancer tumors and promote bone metastasis. Cancer Res. 2010;70:10044–50.
20. Gompel A. Breast cancer incidence rates in US women are no longer declining. Climacteric. 2011;14:690–1.
21. Grisendi G, Bussolari R, Veronesi E, et al. Understanding tumor-stroma interplays for targeted therapies by armed mesenchymal stromal progenitors: the Mesenkillers. Am J Cancer Res. 2011;1:787–805.
22. Guise TA, Yin JJ, Taylor SD, et al. Evidence for a causal role of parathyroid hormone-related protein in the pathogenesis of human breast cancer-mediated osteolysis. J Clin Invest. 1996;98:1544–9.
23. Guzy RD, Schumacker PT. Oxygen sensing by mitochondria at complex III: the paradox of increased reactive oxygen species during hypoxia. Exp Physiol. 2006;91:807–19.
24. Han M, Liu M, Wang Y, et al. Antagonism of miR-21 reverses epithelial-mesenchymal transition and cancer stem cell phenotype through AKT/ERK1/2 inactivation by targeting PTEN. PLoS ONE. 2012;7:e39520.
25. Hanahan D, Weinberg R. Hallmarks of cancer: the next generation. Cell. 2011;144:646–74.
26. Hart IR, Fidler IJ. Role of organ selectivity in the determination of metastatic patterns of B16 melanoma. Cancer Res. 1980;40:2281–7.
27. Hauschka PV, Mavrakos AE, Iafrati MD, et al. Growth factors in bone matrix. Isolation of multiple types by affinity chromatography on heparin-Sepharose. J Biol Chem. 1986;261:12665–74.
28. Hockel M, Vaupel P. Biological consequences of tumor hypoxia. Semin Oncol. 2001;28:36–41.
29. Holmgren L, O'Reilly MS, Folkman J. Dormancy of micrometastases: balanced proliferation and apoptosis in the presence of angiogenesis suppression. Nat Med. 1995;1:149–53.
30. Horlock C, Stott B, Dyson PJ, et al. The effects of trastuzumab on the CD4+CD25+FoxP3+and CD4+IL17A+T-cell axis in patients with breast cancer. Br J Cancer. 2009;100:1061–7.
31. Hynes RO. Integrins: bidirectional, allosteric signaling machines. Cell. 2002;110:673–87.
32. Jorgensen K. Is the tide turning against breast screening? Breast Cancer Res. 2012;14(4):107.
33. Karnoub AE, Dash AB, Vo AP, et al. Mesenchymal stem cells within tumour stroma promote breast cancer metastasis. Nature. 2007;449:557–63.
34. Kim R, Emi M, Tanabe K. Cancer cell immune escape and tumor progression by exploitation of anti-inflammatory and pro-inflammatory responses. Cancer Biol Ther. 2005;4(9):924–33.
35. Ko LJ, Prives C. p53: puzzle and paradigm. Genes Dev. 1996;10:1054–72.
36. Le Blanc K, Mougiakakos D. Multipotent mesenchymal stromal cells and the innate immune system. Nat Rev Immunol. 2012;12:383–96.
37. Lim PK, Bliss SA, Patel SA, et al. Gap junction-mediated import of MicroRNA from bone marrow stromal cells can elicit cell cycle quiescence in breast cancer cells. Cancer Res. 2011;71:1550–60.
38. Lowe SW, Ruley HE. Stabilization of the p53 tumor suppressor is induced by adenovirus 5 E1A and accompanies apoptosis. Genes Dev. 1993;7:535–45.
39. Lv K, Liu L, Wang L, et al. Lin28 mediates paclitaxel resistance by modulating p21, Rb and Let-7a miRNA in breast cancer cells. PLoS ONE. 2012;7:e40008.
40. Manel N, Unutmaz D, Littman DR. The differentiation of human TH-17 cells requires transforming growth factor-[beta] and induction of the nuclear receptor ROR[gamma]t. Nat Immunol. 2008;9:641–9.
41. Mathiesen RR, Borgen E, Renolen A, et al. Persistence of disseminated tumor cells after neoadjuvant treatment for locally advanced breast cancer predicts poor survival. Breast Cancer Res. 2012;14:R117.

42. Mishra PJ, Mishra PJ, Glod JW, et al. Mesenchymal stem cells: flip side of the coin. Cancer Res. 2009;69:1255–8.
43. Moharita AL, Taborga M, Corcoran KE, et al. SDF-1alpha regulation in breast cancer cells contacting bone marrow stroma is critical for normal hematopoiesis. Blood. 2006;108:3245–52.
44. Mukherji S, Ebert MS, Zheng GXY, et al. MicroRNAs can generate thresholds in target gene expression. Nat Genet. 2011;43:854–9.
45. Norton L, Massague J. Is cancer a disease of self-seeding? Nat Med. 2006;12:875–8.
46. Ochsenbein AF, Klenerman P, Karrer U, et al. Immune surveillance against a solid tumor fails because of immunological ignorance. Proc Natl Acad Sci. 1999;96:2233–8.
47. Paget S. The distribution of secondary growths in cancer of the breast. 1889. Cancer Metastasis Rev. 1989;8:98–101.
48. Parr C, Watkins G, Jiang WG. The possible correlation of Notch-1 and Notch-2 with clinical outcome and tumour clinicopathological parameters in human breast cancer. Int J Mol Med. 2004;14:779–86.
49. Patel SA, Meyer JR, Greco SJ, et al. Mesenchymal stem cells protect breast cancer cells through regulatory T cells: role of mesenchymal stem cell-derived TGF-beta. J Immunol. 2010;184:5885–94.
50. Patel SA, Dave MA, Murthy RG, et al. Metastatic breast cancer cells in the bone marrow microenvironment: novel insights into oncoprotection. Oncol Rev. 2011;5:93–102.
51. Pelger RCM, Hamdy NAT, Zwinderman AH, et al. Effects of the bisphosphonate olpadronate in patients with carcinoma of the prostate metastatic to the skeleton. Bone. 1998;22:403–8.
52. Plumas J, Chaperot L, Richard MJ, et al. Mesenchymal stem cells induce apoptosis of activated T cells. Leukemia. 2005;19:1597–604.
53. Rao G, Patel PS, Idler SP, et al. Facilitating role of preprotachykinin-I gene in the integration of breast cancer cells within the stromal compartment of the bone marrow. Cancer Res. 2004;64:2874–81.
54. Reya T, Clevers H. Wnt signalling in stem cells and cancer. Nature. 2005;434:843–50.
55. Reya T, Morrison SJ, Clarke MF, Weissman IL. Stem cells, cancer, and cancer stem cells. Nature. 2001;414:105–11.
56. Rhodes DJ, O'Connor MK, Phillips SW, et al. Molecular breast imaging: a new technique using technetium Tc 99m scintimammography to detect small tumors of the breast. Mayo Clinic Proc. 2005;80:24–30.
57. Ryo A, Nakamura M, Wulf G, et al. Pin1 regulates turnover and subcellular localization of [beta]-catenin by inhibiting its interaction with APC. Nat Cell Biol. 2001;3:793–801.
58. Sarkar FH, Li Y, Wang Z, et al. Implication of microRNAs in drug resistance for designing novel cancer therapy. Drug Resist Updates. 2010;13:57–66.
59. Shackleton M, Vaillant F, Simpson KJ, Stingl J, Smyth GK, Asselin-Labat ML, Wu L, Lindeman GJ, Visvader JE. Generation of a functional mammary gland from a single stem cell. Nature. 2006;439:84–8.
60. Shi Y, Massague J. Mechanisms of TGF-beta signaling from cell membrane to the nucleus. Cell. 2003;113:685–700.
61. Solakoglu O, Maierhofer C, Lahr G, et al. Heterogeneous proliferative potential of occult metastatic cells in bone marrow of patients with solid epithelial tumors. Proc Natl Acad Sci. 2002;99:2246–51.
62. Subramaniam K, Hirpara JL, Tucker-Kellogg L, et al. FLIP: A flop for execution signals. Cancer Lett (In press).
63. Talmadge JE. Clonal selection of metastasis within the life history of a tumor. Cancer Res. 2007;67:11471–5.
64. Tarin D, Price JE, Kettlewell MG, et al. Clinicopathological observations on metastasis in man studied in patients treated with peritoneovenous shunts. Br Med J. 1984a;288:749–51.
65. Tarin D, Price JE, Kettlewell M, G W, et al. Mechanisms of human tumor metastasis studied in patients with peritoneovenous shunts. Cancer Res. 1984b;44:3584–92.
66. Ten HA, Bektas N, von SS, et al. Expression of the glioma-associated oncogene homolog (GLI) 1 in human breast cancer is associated with unfavourable overall survival. BMC Cancer. 2009;9:298.

67. Thiery JP, Sleeman JP. Complex networks orchestrate epithelial-mesenchymal transitions. Nat Rev Mol Cell Biol. 2006;7:131–42.
68. Thiery JP, Acloque H, Huang RYJ, et al. Epithelial-Mesenchymal transitions in development and disease. Cell. 2009;139:871–90.
69. Uhr JW, Pantel K. Controversies in clinical cancer dormancy. Proc Natl Acad Sci. 2011;108:12396–400.
70. van DA, Naaijkens BA, Jurgens WJ, et al. The multidrug resistance protein breast cancer resistance protein (BCRP) protects adipose-derived stem cells against ischemic damage. Cell Biol Toxicol (In press).
71. Vogelstein B, Lane D, Levine AJ. Surfing the p53 network. Nature. 2000;408:307–10.
72. Wang Z, Li Y, Ahmad A, Azmi AS, Kong D, Banerjee S, Sarkar FH. Targeting miRNAs involved in cancer stem cell and EMT regulation: an emerging concept in overcoming drug resistance. Drug Resist Updates. 2010;13:109–18.
73. Wang H, Tan G, Dong L, Cheng L, Li K, Wang Z, Luo H. Circulating MiR-125b as a marker predicting chemoresistance in breast cancer. PLoS ONE. 2012;7:e34210.
74. Zardawi SJ, O'Toole SA, Sutherland RL, Musgrove EA. Dysregulation of Hedgehog, Wnt and Notch signalling pathways in breast cancer. Histol Histopathol. 2009;24:385–98.
75. Zen K, Zhang CY. Circulating MicroRNAs: a novel class of biomarkers to diagnose and monitor human cancers. Med Res Rev. 2012;32:326–48.

Chapter 7
MicroRNAs in Development and Progression of Breast Cancer

Muthusawamy Thangraju and Ashok Jain

Abstract Over the past several years, it has become clear that alterations in the expression of microRNA (miRNA) genes contribute to the pathogenesis of most human malignancies. These alterations can be caused by various mechanisms, including deletions, amplifications, or mutations involving miRNA loci, epigenetic silencing, or the dysregulation of transcription factors that target specific miRNAs. Further, every cellular process is likely to be regulated by miRNAs and an aberrant miRNA expression signature is a hallmark of several diseases, including breast cancer. miRNA expression profiling has provided evidence of the association of these molecules with tumor development and progression. An increasing number of evidences have demonstrated that miRNAs can function as either potential oncogenes or tumor suppressor genes, depending on the cellular context and on the target genes they regulate. Here, we review our current knowledge about the involvement of miRNAs in breast cancer and their potential as diagnostic, prognostic, and therapeutic tools.

Keywords Breast cancer · miRNA · Oncogene · Tumor suppression · Stem cells · Cancer stem cell · Breast cancer treatment

1 Introduction

Cancer is a complex genetic disease involving structural and expression abnormalities of both coding and noncoding genes. For the past several decades, it has been thought that cancer is caused by genetic and/or epigenetic alterations to protein-coding oncogenes and tumor suppressor genes. These alterations result mainly from somatic genetic events that occur over long periods of time, particularly in solid malignancies like lung, breast, prostate, and gastrointestinal cancers [1]. These findings have enlightened the development of novel therapies, which are based on the specific genetic alterations that are involved in cancer pathogenesis. Although it

A. Jain (✉)
Department of Natural Sciences, Albany State University, Albany GA, USA
e-mail: ashok.jain@asurams.edu

M. Thangraju
Cancer Research Center, Georgia Health Science University, Augusta, GA, USA

S. R. Singh, P. Rameshwar (eds.), *MicroRNA in Development and in the Progression of Cancer*, DOI 10.1007/978-1-4899-8065-6_7,

is clear that tumors are initiated by somatic genetic alterations, it is also evident that many tumors show alterations in the expression of tumor suppressor genes because of epigenetic alterations, such as the methylation of CpG islands in their promoters, which lead to loss of function [2]. Therefore, therapies directed toward the reversal of the epigenetic changes that occur in various malignancies have also been developed. Even though significant progress has been made in identifying the genetic and epigenetic causes of cancer and identifying targets for therapy, several challenges still remain. For example, targeted therapies that are based on the identification of oncogenic mutations with causal roles in cancer have been developed, but the treatment of malignancies that are initiated through the loss of function of tumor suppressor genes is more difficult, because the lost gene function must be replaced in all the cancer cells and this should be a very challenging task to achieve. Thus, the identification of additional alterations that cause or contribute to malignancy is a high priority.

In 1993, Victor Ambros and colleagues discovered a gene, *lin-4,* that affected development in *Caenorhabditis elegans* and found that its product was a small noncoding RNA [3]. After these seminal findings, the cloning and characterization of small, 20–22-nucleotide-long members of the noncoding RNA family, called microRNAs (miRNAs), have led to the identification of ~1,000 miRNAs. After this initial discovery, the field of miRNAs has undergone a long period of silence and it took several more years to realize that these small RNA molecules are actually expressed in several organisms, including Homo sapiens. miRNAs are highly conserved across different species and are highly specific for tissue and developmental stages. In the past few years, miRNAs have taken their place in the complex circuitry of cell biology, revealing a key role as regulators of gene expression. miRNA genes represent approximately 1% of the genome of different species, and each of them has hundreds of different conserved or non-conserved targets and it has been estimated that approximately 30% of the genes are regulated by, at least, one miRNA [4]. Interestingly, more than 30% of miRNAs have roles in the regulation of fundamental processes such as development, differentiation, cell proliferation, apoptosis, and stress responses in a range of organisms, including *C. elegans*, plants, *Drosophila melanogaster,* and mammals, including humans [5, 6].

miRNAs are transcribed for the most part by RNA polymerase II as long primary transcripts characterized by hairpin structures (pri-miRNAs) and processed into the nucleus by RNAse III Drosha into 70–100-nt-long pre-miRNAs. These precursor molecules are exported by an exportin-5 mediated mechanism to the cytoplasm, where an additional step mediated by the RNAse III Dicer generates a double-stranded RNA (dsRNA) of approximately 22 nts, named matured miRNA. The mature single-stranded miRNA product is then incorporated in the complex known as miRNA-containing ribonucleoprotein complex (miRNP) or miRNA-containing RNA-induced silencing complex (miRISC), whereas the other strand is likely subjected to degradation. In this complex, the mature miRNA is able to regulate gene expression at the posttranscriptional level, binding through partial complementarity for the most part to the 3′ untranslated region UTR of target messenger RNAs (mRNAs), and leading at the same time to some degree of mRNA degradation and translation inhibition (Fig. 7.1) [4, 7].

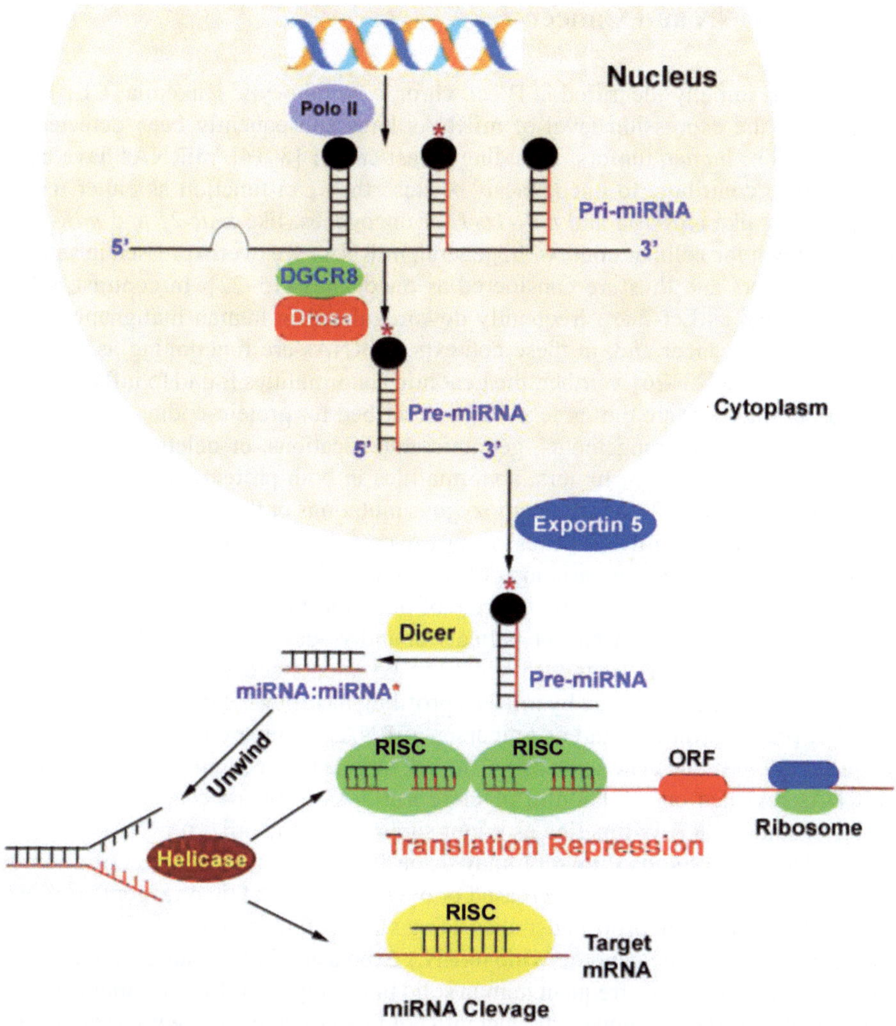

Fig. 7.1 Processing of pri-microRNA and maturation of microRNAs (miRs). miRs are transcribed mainly by RNA polymerase II as long primary transcripts characterized by hairpin structures (pri-miRs) and processed in the nucleus by RNAse III Drosha in a 70-nucleotide-long pre-miR. This precursor molecule is exported by the exportin 5 to the cytoplasm, where RNAse III Dicer generates a dsRNA of approximately 22 nucleotides, named miR:miR*. The mature miRNA product is then incorporated in the complex known as miRNA-induced silencing complex (miRISC), whereas the other strand is likely subjected to degradation. As part of this complex, the mature miRNA is able to regulate gene expression binding through partial homology of the 3′UTR of target mRNAs and leading to mRNA degradation or translation inhibition

2 MicroRNA and Cancer

miRNA was initially identified in B cell chronic lymphocytic leukemia (CLL) and changes in the expression level of miRNAs have subsequently been detected in many types of human tumors, including breast cancer [8–14]. miRNAs have been proposed to contribute to oncogenesis because they can function as either tumor suppressors, like *miR-15*a and *miR-16-1,* or oncogenes, like *miR-21* and *miR-155,* depending on the cellular context. In general, miRNAs are overexpressed in several human cancers and thus are considered as oncogenes [15–22]. In contrast, other miRNAs such as Let-7 are frequently downregulated in human malignancies including breast cancer and, in these contexts, miRNAs are functioning as a tumor suppressor gene [23–26]. Further, the genomic abnormalities found to influence the activity of miRNAs are the same as those described for protein-coding genes, such as chromosomal rearrangements, genomic amplifications, or deletions and mutations. In a specific set of tumors, abnormalities in both protein-coding genes and miRNAs can be identified [10]. Homozygous mutations or the combination of deletion plus mutation in miRNA genes is a rare event, and the functional consequences of heterozygous sequence variations of miRNAs in human cancers have not been identified [27, 28]. Furthermore, the role of polymorphisms in the complementary sites of target mRNAs in cancer patients or individuals with a predisposition to other hereditary diseases has also started to be understood [29, 30]. In addition, every type of tumor analyzed by miRNA profiling has shown significantly different miRNA profiles (mature and/or precursor miRNAs) compared with normal cells from the same tissue. The expression studies targeted to recognize the function of miRNAs revealed only a handful of these miRNAs in breast cancer. As in other cancers, some miRNAs function as tumor suppressors and other miRNAs as oncogenes. Therefore, tumor formation may occur by reduction or deletion of a tumor suppressor miRNA and/or by increased or overexpression of an oncogenic miRNA (Table 7.1). Two large profiling studies using various tumors and two distinct technologies to investigate genome-wide miRNA expression in 540 samples, including 363 from six of the most frequent human solid tumor types, and 177 normal controls found that cancer cells showed distinct miRNA profiles compared with normal cells that have been demonstrated [31, 32].

Further, studies have shown that miR-155 is upregulated in Burkitt's lymphoma, diffuse large B cell lymphoma, primary mediastinal B cell lymphoma, and Hodgkin's lymphoma [63, 64]. Mice overexpressing miR-155 in B-lymphocytes develop polyclonal preleukemic pre-B cell proliferation followed by a full-blown B cell malignancy [37]. More recently, two knockout mice models have demonstrated a critical role of miR-155 in immunity by showing that BIC/miR-155$^{-/-}$ have defective dendritic cell functions, impaired cytokine secretion, and T_H cells intrinsically biased toward T_H2 differentiation [65, 66]. Moreover, miR-155 could represent the connection between inflammation, immunity, and cancer, because its expression can be induced by mediators of inflammation and this is involved in response to endotoxic shock [67]. In contrast, members of the miR-29 family have been shown to be downregulated in aggressive CLL, invasive breast cancer, lung cancer, and

Table 7.1 MicroRNAs that function as oncogene or tumor suppressor in human cancers

MicroRNA	Dysregulation	Function	Target genes	Oncogene or tumor suppressor (Ref.)
miR-21	Upregulated in breast, colon, pancreatic, lung, prostate, liver, and stomach cancers	Inhibits apoptosis and increases tumorigenicity	PTEN, PDCD4, TPM1 and TIMP3	Oncogene [33–35]
miR-155	Upregulated in breast, lung, colon, and CLL cancers	Induces cell proliferation and leukemia or lymphoma in mice	MAF and SHIP1	Oncogene [36, 37]
miR-373	Upregulated in breast, prostate, and testicular cancers	Stimulates invasion and metastasis of breast and prostate cancers	LSAST, p53, CD44	Oncogene [38–40]
miR-17-92	Upregulated in breast, lung, colon, stomach, and pancreatic cancers	Induces proliferation	E2F1, BIM and PTEN	Oncogene [41–43]
miR-10b	Highly expressed in metastatic cancer cells	Promotes cell migration and invasion in vitro and initiates tumor invasion and metastasis in vivo	HOXD10	Oncogene [44]
miR-145	Loss in breast cancer	Inhibits proliferation and induces apoptosis of breast cancer cells	ERG	Tumor suppressor [16]
miR-34	Loss in breast, pancreatic, colon, and liver cancers	Induces apoptosis	CDK4, CDK6, cyclin E2, EZF3 and MET	Tumor suppressor [45-48]
miR-29	Loss in breast cancers and cholangiocarcinoma	Induces apoptosis and inhibits tumorigenicity Reactivates silenced tumor suppressor genes	TCL1, MCL1 and DNMTs	Tumor suppressor [49–51]
Let-7	Loss in breast, lung and in various solid and hematopoietic malignancies	Induces apoptosis and inhibits tumorigenesis	RAS, MYC, and HMGA2	Tumor suppressor [52, 53]
miR-206	Upregulated in estrogen receptor (ER)α-negative breast cancers	Inhibits the expression of ESR1 mRNA and induces cell cycle arrest to inhibit estrogen induced proliferation	ESR1	Tumor suppressor [54, 55]
miR-17-5p [also known as miR-91]	Undergoes loss of heterozygosity in multiple cancers, including breast cancer	Represses the translation of A1B1 thereby inhibiting the E2F1 and estrogen-stimulated proliferation and estrogen/ER-independent breast cancer cell proliferation	*AIB1*, *CCND1*, *E2F1*	Tumor suppressor [56, 57]

Table 7.1 (continued)

MicroRNA	Dysregulation	Function	Target genes	Oncogene or tumor suppressor (Ref.)
miR-125a and miR-125b	These miRNAs are downregulated in HER2-amplified and HER2-overexpressing breast cancers	Overexpression of these miRNAs suppresses HER2 and HER3 mRNA and protein leading to a reduction in anchorage-dependent growth and cell motality	HER2, HER3	Tumor suppressor [58, 59]
miR-200 family	Downregulated expression in breast cancer is associated with highly aggressive, metaplastic breast tumor	Suppresses the expression of the EMT inducers ZEB1 and ZEB2. Transforming growth factor (TGF)-β signaling	*BMI1*, *ZEB1*, *ZEB2*	Tumor suppressor [23, 60, 61]
miR-31	miR-31 expression decreased or undetectable in metastatic breast cancer cells	Prevents metastasis by inhibiting the expression of prometastatic genes	*FZD3*, *ITGA5*, *M-RIP*, *MMP16*, *RDX*, *RHOA*	Tumor suppressor [62]

ERG estrogen response genes, *FZD3* frizzled 3, *ITGA5* integrin α-5, *MMP16* matrix metalloproteinase 16, *RDX* radixin, *M-RIP* myosin phosphatase Rho-interacting protein, *CDK* cyclin-dependent kinases, *MAF* *M*usculo*a*poneurotic *f*ibrosarcoma, *SHIP1* *S*rc *H*omology 2-Containing *I*nositol *P*hosphatase-*1*, *LATS* *La*rge *t*umor *s*uppressor kinase, *CD44* *C*luster of *D*ifferentiation *44*, *E2F1* *E2* Transcription *F*actor *1*, *BIM* *B*cl2-*i*nteracting *m*ediator gamma, *ERG* *E*TS-*r*elated *g*ene, *EZF3* Kruppel-like zinc finger protein 3, *MET* *met* proto-oncogene, *RAS* *Ra*t *S*arcoma, *TCL1* *T* *C*ell *l*ymphoma breakpoint *1*, *BMI1* *B* lymphoma Mo-*M*LV *i*nsertion region *1*, *RHOA* *R*as *ho*molog gene family member A cholangiocarcinoma [49, 51, 68, 69]. The transfection of miR-29b induces apoptosis in cholangiocarcinoma cell lines and reduces the tumorigenicity of lung cancer cells in nude mice. Further, it has been shown that rhabdomyosarcoma loses miR-29 expression because of an elevation of nuclear factor κB (NFκB) and YY1 levels, and introduction of miR-29s into the tumor delays rhabdomyosarcoma progression in mice [70]. miR-29s was also found to directly target myeloid cell leukemia 1 (Mcl-1) [49], an oncogene overexpressed in acute myeloid leukemia (AML), and the *de novo* DNA methyltransferases (DNMT)-3A and DNMT-3B and the maintenance of DNMT1 [51, 71]. Thus, the loss of miR-29 family member results in the constitutive overexpression of Mcl-1 and of DNMT, causing epigenetic changes characteristic of AML. These recent results suggest that the loss of miR-29s may be important, perhaps critical, for the pathogenesis of a major group of myelodysplastic syndromes and AMLs.

Breast Cancer One of the first solid tumors to be profiled for miRNA expression was breast cancer. Iorio et al. [69] described the first miRNA signature for breast carcinoma and identified 13 miRNAs that discriminate breast tumors from normal tissues with 100 % accuracy. One of the most significant miRNAs differentially expressed between normal and breast tumor is miR-21, which is overexpressed in breast carcinoma. This miR-21 has been demonstrated to play a crucial role in regulation of cell survival and proliferation by directly targeting the tumor suppressor genes phosphatase and tensin homolog (PTEN), programmed cell death 4 (PDCD4), and tropomyosin 1 (TPM1). Further, this miR-21 has been associated with advanced clinical stage, lymph node metastasis, and poor patient prognosis [72, 73]. miR-21, is one of the first cancer miRNAs described, has been found overexpressed in a variety of other malignancies like glioblastoma, ovarian cancer, lung cancer, etc. [74–78]. Further, miR-21 overexpression is also associated with poor survival and poor therapeutic outcome in colorectal, pancreatic, endocrine, and exocrine tumors [79–81]. Conversely, downregulated miRNAs such as miR-125b and miR-205 regulate oncogenes like tyrosine kinase receptors human epidermal growth factor receptor (HER)-2 and HER-3 [82, 83]. Ectopic expression of miR-205 in a breast cancer cell line decreases proliferation and improves the responsiveness to tyrosine kinase inhibitors like gefitinib and lapatinib [84]. miRNA expression is also related to some histopathologic features of breast carcinoma, such as estrogen receptor (ER) and progesterone receptor (PR) expression, grade and stage, and presence of invasion [69]. Further studies have shown the correlation between miRNA expression and the classification in different subtypes of breast cancer [85]. In addition, studies have shown the correlation between miRNAs and ER status in human breast cancer. For example, miR-206 directly targets ER-α, and miR-221 and miR-222 confer tamoxifen resistance by regulating p27 and ER-α [86, 87]. Further, studies have also shown that there is a regulatory loop between ER-α and miR-221 and miR-222: The two miRNAs are able to directly target ER-α receptor, which in turn negatively regulates their expression, binding estrogen-responsive elements on their promoter region [88].

3 MicroRNAs are Regulated by Transcription Factors in Breast Cancer

Breast cancer is the second leading cause of death among women in the Western world and its molecular pathogenesis is still not fully understood. Several lines of evidence indicate that ERα-negative breast tumors, which are highly aggressive and unresponsive to hormonal therapy, arise from ERα-positive precursors through different molecular pathways. High levels of miR-221 and miR-222 were found in ERα-negative cells and in primary breast tumor samples. Overexpression of miR-221 and miR-222 in ERα-positive cells suppressed ERα protein and luciferase assays confirmed that ERα is a target of miR-221 and miR-222. ERα was also found to negatively regulate the expression of miR-221 and miR-222 by promoter binding [88]. Therefore, silencing ERα, either by methylation or by dysregulating miR-221 and miR-222, results in the constitutive activation of miR-221 and miR-222 and the subsequent inhibition of the tumor suppressors like p27, p57, PTEN, and tissue inhibitor of metalloproteinases-3 (TIMP3), which in turn contribute to the development of the invasive phenotype [45]. This regulatory feedback loop seems to be involved in the development of ERα-negative breast cancers [89]. Further, studies have shown that the family of miR-34, which consists of miR-34a, miR-34b, and miR-34c, is induced directly by the p53 tumor suppressor and suggested that some p53 effects could be mediated by these miRNAs [46–48]. By using various cellular models, these authors compared miRNA expression in cells with high or low levels of p53 expression and found that the expression levels of p53 correlated with the levels of expression of miR-34 family members [46]. In addition, chromatin immunoprecipitation experiments showed that p53 binds to the miR-34 promoters [47, 48]. The introduction of miR-34 family members into cells that had lost miR-34 expression resulted in cell cycle arrest.

Therefore, miRNAs could be dysregulated by transcription factors and, therefore, genetic or epigenetic alterations that result in the dysregulation of transcription factors can cause miRNA dysregulation, which contributes to malignant transformation. Further, Volinia and coworkers have defined miRNA expression signatures, which distinguish cancerous tissues from normal tissues [32]. Interestingly, they observed that some miRNA genes were dysregulated not just in one tumor type but also in many tumors, suggesting that these miRNAs may be downstream targets of pathways that are commonly dysregulated in cancer. Thus, if miRNAs are downstream targets of pathways that are commonly dysregulated in human cancer they become excellent targets for therapeutic intervention. It is known that proliferative signals lead to the activation of the c-Myc transcription factor (that integrates the cell cycle machinery), which positively and negatively regulates the expression of many miRNA genes, including miR-17-92 cluster (c-Myc regulates positively) and miR-15a and miR-16-1 (c-Myc regulates negatively). Therefore, it seems that the mechanism of action of the activated oncogene is, at least partly, miRNA dysregulation and thereby it causes the downregulation of tumor suppressors and upregulation of oncogenes. Therefore, genetic or epigenetic alterations in protein-coding cancer genes or in miRNA genes may have similar consequences. Until now, it has not been possible to target the overexpression of c-Myc in tumors with drugs, but it is possible to target the miRNAs that are dysregulated c-Myc by treating the cancer cells with anti-miR-17-92 or with miR-15a and miR-16-1. Recently, Bonci et al. have shown that the miR-15a-miR-16-1 cluster can control prostate cancer by targeting multiple oncogenic activities [90]. In general, in tumors with alterations in protein-coding cancer genes, it should be possible to induce tumor regression using miRNAs and/or anti-miRNAs. In fact, recent studies have shown that c-Myc-induced hepatocellular carcinomas in mice are regressed by targeting miRNA that is induced by c-Myc [91].

4 Role of Cancer Epigenetics in Regulation of miRNAs

The most-studied epigenetic changes in cancer cells are the methylation of cytosines in the dinucleotide CpG in DNA [92]. Such "methylable" sites, known as CpG islands, are preferentially located in the 5′ region (which consists of the promoter, 5′ UTR, and exon 1) of many genes and are non-methylated in normal cells but are transcribed in the presence of the appropriate transcription factors. Methylation of the CpG islands of tumor suppressors results in their silencing and contributes to malignant transformation [92]. miRNAs are affected by genetic changes, such as deletion, gene amplification, and mutation, and by transcription factors. In addition, the expression of miRNAs is affected by epigenetic changes, such as methylation of the CpG islands of their promoters. Saito et al. [93] reported that miR-127 is silenced by promoter methylation in bladder tumors and that its expression could be restored by demethylating agents such as 5-azacitidine. miR-127 targets B cell lymphoma 6 (Bcl6), an oncogene that is involved in the regulation of apoptosis signaling and in the development of B cell lymphoma. Therefore, silencing of miR-127 may lead to the overexpression of Bcl6, which leads to malignant transformation. Further, miRNAs can also regulate enzymes that are involved in the methylation of the CpG islands of tumor suppressor genes. For example, the overexpression of DNMTs is a poor prognostic indicator in several human malignancies, and the miR-29 family targets de novo DNMT3A and DNMT3B [94]. However, recent studies have also shown that miR-29 not only targets DNMT3A and DNMT3B but also targets indirectly DNMT1 [71]. Interestingly, the introduction of miR-29 into cancer cell lines caused demethylation of the CpG islands in the promoter regions of tumor suppressor genes, which allowed their reactivation and resulted in the loss of tumorigenicity [71]. The introduction of miR-29 into cancer cells resulted in the loss of expression of the oncogene Mcl1 and the three DNMTs with reactivation of the p16 tumor suppressor [71]. Further, the loss of miR-29 cluster in Dicer-deficient mouse embryonic stem cells leads to downregulation of DNMT1, DNMT3a, and DNMT3b through the modulation of their repressor, RBL-2, a proven target of miR-290 [95, 96]. Therefore, treatment of the malignancies with miR-29 family members caused the reversion of epigenetic changes that contribute to malignant transformation in human cancers. Further, miRNAs also target histone deacetylases (HDACs) and other proteins that are involved in chromatin structure and function [97]. Therefore, all these observations indicate that alterations in the expression of miRNAs could be responsible for some of the epigenetic changes that are observed in cancer cells, and that such miRNAs could provide novel targets for cancer therapy.

5 Role of MicroRNAs in Regulation of Cancer-Initiating Stem Cells in Breast Cancer

Cancer stem cells (CSC) are a small subpopulation of cells capable of self-renewal, differentiation, and tumor initiation. The prevailing view is that CSCs are the root of cancer origin and recurrence. Although not all cancers have been found to contain CSC populations, to date, the data on the roles of miRNAs in CSCs are consistent with the existence of CSCs in breast, prostate, lung, pancreatic, and liver cancers. It has been well established that miRNAs regulate tumor development, prognosis, and metastasis either as oncogenes or as tumor suppressors [98, 99]. In addition, emerging evidence suggests that miRNAs also play essential roles in stem cell self-renewal and differentiation by negatively regulating the expression of key stem cell-regulating genes [100]. Evidence for a role of miRNAs in stem cell maintenance and differentiation is accumulating from analysis of mutations in key RNA interference (RNAi) components. For example, Dicer-mutant mice die early in development with a loss of Oct4-positive multipotent stem cells [101]. Further, abnormal miRNA expression may result in dysregulation of self-renewal in CSCs during cancer progression [102, 103]. Silber et al. reported that miR-124 and miR-137 induce differentiation of neural and glioblastoma stem cells and induce cell cycle arrest [104]. These results suggest that the targeted delivery of miR-124 and miR-137 to glioblastoma cells may be therapeutically efficacious for the glioblastoma treatment. Further, studies have shown that prostate cancer stem and/or progenitor cell populations have lower levels of miR-34a and Let-7b compared to bulk tumor cells [105]. In addition, studies have also shown that miR34a targets CD44, resulting in impaired tumor growth and decreased metastases in mouse models of prostate cancer. The increased survival of mice treated with systemically delivered miR34a suggests a novel strategy to target prostate CSCs, thereby inhibiting tumor growth and metastasis [105]. Further, a number of studies describing the role of miRNA in the regulation of normal and malignant breast stem cells have been conducted. Recent studies have shown that both miR-205 and miR-22 are highly expressed in mouse mammary stem/progenitor cells, whereas miR-93 and Let-7 are depleted in this population [106]. Further, studies have also reported that miR-205 overexpression in mouse mammary cells led to an expansion of the progenitor cell population, decreased cell size, and increased cellular proliferation [107]. More recent studies have shown that overexpression of miR-200c reduced the clonogenic and tumor-initiation activities of breast cancer stem cells (BCSCs) and suppressed mammary duct formation by normal mammary stem cells. This occurred through the downregulation of the polycomb gene Bmi-1, a target of miR-200c. This work demonstrated a molecular link between normal breast stem cells and BCSCs [23]. Yu et al. showed that Let-7 is decreased in BCSCs and that overexpression of Let-7 inhibits the cell proliferation, mammosphere formation, BCSC self-renewal and differentiation, and tumor formation and metastasis in nonobese diabetic/severe combined immunodeficiency (NOD/SCID) mice [108]. These effects were shown to be mediated through downregulation of the Let-7 targets H-Ras and HMGA2 [108]. This group also demonstrated that expression of miR-30 markedly reduced BCSCs by targeting ubiquitin-conjugating enzyme 9 (UBC9) and integrin b3 (ITGB3). A more complete inhibition of self-renewal and mammosphere formation of BCSCs was observed when both Let-7 and miR-30 were simultaneously introduced compared to each miRNA individually [108]. The ability of these miRNAs to target BCSCs suggests that they may have significant therapeutic potential. All these studies show that miRNA plays a crucial role in the regulation of stem cells and progenitor and cancer-initiating stem cells. Thus, the most effective cancer therapy must be directed against both the small, quiescent pool of CSCs as well as the more actively proliferating bulk tumor mass. This may be possible if specific CSC signals are inhibited using molecularly targeted therapy, while simultaneously attacking proliferating cells by conventional therapies like chemotherapy and radiotherapy. To attain this goal, developments in miRNA regulation present exciting new prospects.

6 MicroRNAs in Breast Cancer Invasion, Angiogenesis, and Metastasis

The leading cause of death in breast cancer (BC) patients is not the primary tumor in the breast per se, but metastasis to distant organs. Metastasis accounts for more than 90 % of the deaths in BC patients. Millions of cells are released from the primary tumor into the blood circulation, but only a small portion of these cells survive and colonize on distant organs. The development of cancer metastases depends on multiple factors, including miRNAs [109–112], epithelial–mesenchymal transition (EMT) [113–116], CSCs [117, 118], etc. miRNAs have been implicated not only in the development of primary tumors, but also in affecting progression and in the metastatic phase of the disease. Indeed, several lines of evidence show how miRNAs are involved in the regulation of biologic processes leading to the acquisition of metastatic potential, as adhesion, migration and invasion, and angiogenesis. Further, the connection between miRNAs and EMT has also been well established. Numerous studies have revealed that miRNAs are dysregulated in cancer versus normal tissue, and in noninvasive versus invasive forms [119, 120]. Discordant miRNA expression between normal and breast tumor tissues and between ERα-positive and ERα-negative tumors is also well established [69]. In general, the levels of the most mature (processed) miRNAs are lower in cancer versus normal tissue, and in ERα-negative versus ERα-positive breast cancer, and both Drosha and Dicer levels are subnormal in several cancers [121, 122]. Interestingly, estrogen (E2) induces Dicer expression in MCF-7 cells, suggesting that the loss of ERα and estrogen signaling in breast cancers may contribute to the decreased expression of Dicer and consequently lower levels of miRNA expression [123]. Some evidence supports a role of miRNAs in dampening Dicer expression. Martello et al. [124] showed that miR-103/107 targets Dicer mRNA and inhibits Dicer expression, thereby promoting EMT in breast cancer cell lines and the metastatic spread in mice. These investigators also provided evidence that the miR-103/107-induced decrease in Dicer leads to compromised processing of pre-miRNAs belonging to the EMT-inhibitory miR-200 family. Cochrane et al. [121] demonstrated that miR-221, miR-222, and miR-29a, which are elevated in ERα-negative breast cancer cell lines, target Dicer. Upregulation of miR-200c increased Dicer in two ERα-negative cell lines. Although the majority of miRNAs are decreased after neoplastic transformation, some miRNAs clearly show an increase [119].

To perform miRNA effector functions, miRNAs must be incorporated into Argonaute (AGO)-containing complexes [125]. The expression of the miRNA effector AGO proteins is also altered in cancer. The AGO1 gene is frequently deleted in several cancers, including breast cancer [126]. However, studies have also shown that both AGO1 and AGO2 were elevated in ERα-negative versus ERα-positive breast cancer [127).]. Further, forced expression of AGO2 reduced E-cadherin expression and enhanced motility in epithelial breast cancer cell lines [128]. Additional work is needed to examine whether increased AGO2 expression induces a complete EMT in breast cancer. Specific miRNAs have been demonstrated to promote EMT and metastasis. One of the seminal studies by the Weinberg Group [129] showed that Twist (but not Snail) induced expression of miR-10b. They observed that miR-10b was downmodulated in all the breast carcinomas from metastasis-free patients, but surprisingly, 50% of metastasis-positive patients had elevated miR-10b levels in their primary tumors. miR-10b targets and represses the homeobox D10 (HOXD10) expression, thereby relieving transcriptional inhibition of the prometastatic Ras homolog gene family member C (RHOC) and thus leading to tumor cell invasion and metastasis. Further, as a functional screen that aimed to discover miRNAs that promote cell migration in vitro, Huang et al. [38] identified miR-373 and validated its metastatic potential in tumor transplantation experiments using breast cancer cells. More recently, several "metastamiRs" have been characterized [130]. Conversely, several miRNAs have been characterized as suppressors of metastasis [130]. In particular, members of the miR-200 family of miRNAs and miR-205 have been shown to reduce cell migration and invasiveness by targeting zinc-finger E-box binding (ZEB) transcription factors, which are known inducers of EMT [131, 132]. Further, the oncogenic miR-21 stimulates invasion, extravasation, and metastasis in different tumor types, including breast cancer, whereas oncosuppressor miR-205 has opposite effects, reducing invasion in vitro and suppressing lung metastasis in vivo [133, 84]. With the same aim of searching for regulators of breast cancer metastasis, Tavazoie et al. [134] identified miR-126 and miR-335 as metastasis suppressors: Reduced levels of the two miRNAs are associated with poor metastasis-free survival of patients with breast cancer, whereas their re-expression inhibits metastasis in a cell transplantation model. Interestingly, it has been recently observed that primary tumors and metastasis from the same tissue show a similar pattern of miRNAs expression [135]. Being a more accurate classifier than mRNA expression studies, miRNA profiling has thus revealed the potential to solve one of the most demanding issues in cancer diagnostics: the origin of metastasis of unknown primary tumors.

7 MicroRNAs Are a New Clinical Tool

As active players in important oncogenic signaling pathways, miRNAs should affect cancer diagnosis and prognosis.

miRNA Profiling as a Diagnostic Tool Metastatic cancer of unknown primary site (CUP) is one of the ten most frequent cancer diagnoses worldwide, and constitutes 3–5 % of all human malignancies [136]. Patients with CUP present with metastases (late-stage disease) without an established primary tumor (a site at which the tumor has initially developed and from which it has metastasized). The study by Lu et al. [31] produces an important advance in the diagnosis of this peculiar type of cancer. Analyzing 17 poorly differentiated tumors with nondiagnostic histological appearance, they showed that the miRNA-based classifier was much better at establishing the correct diagnosis of the samples than the mRNA classifier [31]. This result is exciting because profiling a few hundred miRNAs has a much better predictive power for CUP diagnosis than profiling several tens of thousands of mRNAs. As miRNA expression changes with differentiation, the poorly differentiated tumors have lower global expression levels of miRNAs compared with well-differentiated tumors from control groups [31]. The reduced expression levels of miRNAs in poorly differentiated tumors reveal why miRNA profiling is effective in the diagnosis of CUP.

miRNA Profiling as a Prognostic Tool Breast cancer is the leading cause of death from cancer in women worldwide. Therefore, the identification of new prognostic markers (markers that correlate with disease evolution) could be a significant advance for the identification of patients that would benefit from more aggressive therapy. In univariate analyses, the expression of both miR-155 (high levels) and Let-7a-2 (low levels) has been shown to correlate with poor survival in 104 patients; in multivariate analyses, the expression of miR-155 also correlated with a poor prognosis when all clinical variables were considered together.

8 MicroRNAs and Anti-MicroRNA in Cancer Treatment

The evidence collected to date demonstrate how microRNAs could represent valid diagnostic, prognostic, and predictive markers in cancer. Indeed, the aberrant miRNA expression is correlated with specific biopathologic features, disease outcome, and response to specific therapies in different tumor types. Considering the importance of miRNAs in development, progression, and treatment of cancer, the potential usefulness of an miRNA-based therapy in cancer is now being exploited, with the attempt to modulate their expression, reintroducing miRNAs lost in cancer, or inhibiting oncogenic miRNAs by using anti-miRNA oligonucleotides. For example, the transfection of miR-15a/16–1 induces apoptosis in leukemic MEG01 cells and inhibits tumor growth in vivo in a xenograft model [137], whereas the inhibition of miR-21 with antisense oligonucleotides generates a proapoptotic and antiproliferative response in vitro in different cellular models and reduces tumor development and metastatic potential in vivo [138].

Moreover, miRNAs involved in specific networks, such as the apoptotic, proliferation, or receptor-driven pathways, could likely influence the response to targeted therapies or to chemotherapy: Inhibition of miR-21 and miR-200b enhances sensitivity to gemcitabine, probably by modulation of CLOCK, PTEN, and PTPN12, whereas reintroduction of miR-205 in breast cancer cells can improve the responsiveness to tyrosine kinase inhibitors through HER-3 silencing [139, 83]. Besides targeted therapies and chemotherapy, miRNAs could also alter the sensitivity to radiotherapy; the Let-7 family of miRNAs can suppress the resistance to anticancer radiation therapy, probably through RAS regulation [140]. Evidence described to date represents the experimental bases for the use of miRNAs as both targets and tools in anticancer therapy, but there are at least two primary issues to address to translate these fundamental research advances into medical practice: the development of engineered animal models to study cancer-associated miRNAs and the improvement of the efficiency of miRNAs/anti-miRNAs delivery in vivo. To this aim, modified miRNA molecules with longer half-lives and efficiency have been developed, such as anti-miRNA oligonucleotides, locked nucleic acid-modified oligonucleotides, and cholesterol-conjugated antagomirs [141–143]. Interestingly, Ebert et al. [144] have recently described a new approach to inhibit miRNAs function: Synthetic mRNAs containing multiple binding sites for a specific miRNA, called miRNA sponges, are able to bind up the miRNA, preventing its association with endogenous targets. To improve the in vivo delivery of either miRNAs or anti-miRNAs, the methods that have been tested in preclinical studies over the past decades for short-interfering RNAs (siRNA) or short heteroduplex RNA (shRNA) could be applied also to miRNAs. Moreover, the advantage of miRNAs over siRNA/shRNA is their ability to affect multiple targets with a single hit, thus regulating a whole network of interacting molecules.

9 Future Perspective

It has been unequivocally proven that miRNA dysregulation occurs in many human malignancies including the most common human malignancies like lung, breast, prostate, and gastrointestinal cancers. Such dysregulation, like the dysregulation of oncogenes and tumor suppressor genes, can be caused by multiple mechanisms, such as deletion, amplification, mutation, transcriptional dysregulation, and epigenetic changes. As miRNAs have multiple targets, their function in tumorigenesis could be due to their regulation of a few specific targets, possibly even one, or many targets. A future challenge would be to identify all of the targets of the miRNAs involved in cancer and establish their contribution to malignant transformation. An additional challenge would be the identification of all of the miRNAs that are dysregulated by pathways consistently dysregulated in various types of human cancers. If these miRNA targets are crucial for the expression of the malignant phenotype and the cancer cells depend on their dysregulation for proliferation and survival, we can expect that the use of miRNAs or anti-miRNAs will result in tumor regression. Over the past several years, we have observed a shift from conventional chemotherapy to targeted therapies, and miRNAs and anti-miRNAs could contribute extensively in the forthcoming years.

References

1. Croce CM. Oncogenes and cancer. N Engl J Med. 2008;358:502–11.
2. Jones PA, Baylin SB. The epigenomics of cancer. Cell. 2007;128:683–92.
3. Lee RC, Feinbaum RL, Ambros V. The C. elegans heterochronic gene lin-4 encodes small RNAs with anti-sense complementarity to lin-14. Cell. 1993;75:843–54.
4. Bartel DP. MicroRNAs: genomics, biogenesis, mechanism, and function. Cell. 2004;116:281–97.
5. Lagos-Quintana M, Rauhut R, Lendeckel W, Tuschl T. Identification of novel genes coding for small expressed RNAs. Science. 2001;294:853–8.
6. Calin GA, Croce CM. MicroRNA signatures in human cancers. Nature Rev Cancer. 2006;6:857–66.
7. Gregory RI, Shiekhattar R. MicroRNA biogenesis and cancer. Cancer Res. 2005;65:3509–12.
8. Calin GA, Dumitry CD, Shimizu M, Bichi R, et al. Frequent deletions and downregulation of micro-RNA genes miR15 and miR16 at 13q14 in chronic lymphocytic leukemia. Proc Natl Acad Sci U S A. 2002;99:15524–9.
9. Caldas C, Brenton JD. Sizing up miRNAs as cancer genes. Nature Med. 2005;11:712–4.
10. Calin GA, Croce CM. MicroRNA-cancer connection: the beginning of a new tale. Cancer Res. 2006;66:7390–4.
11. Chen CZ. MicroRNAs as oncogenes and tumor suppressors. N Engl J Med. 2005;353:1768–71.
12. Hwang HW, Mendell JT. MicroRNAs in cell proliferation, cell death, and tumorigenesis. Br J Cancer. 2006;94:776–80.
13. Hammond SM. MicroRNAs as oncogenes. Curr Opin Genet Dev. 2006;16:4–9.
14. Esquela-Kerscher A, Slack FJ. Oncomirs-microRNAs with a role in cancer. Nature Rev Cancer. 2006;6:259–69.
15. Petrocca F, Vecchione A, Croce CM. Emerging role of miR-106b-25/miR-17-92 clusters in the control of transforming growth factor beta signaling. Cancer Res. 2008;68:8191–4.
16. Hayashita Y, Osada H, Tatematsu Y, Yamada H, et al. A polycistronic microRNA cluster, miR-17-92, is overexpressed in human lung cancers and enhances cell proliferation. Cancer Res. 2005;65:9628–32.
17. Mendell JT. miRiad roles for the miR-17-92 cluster in development and disease. Cell. 2008;133:217–22.
18. Bandres E, Agirre X, Ramirez N, Zarate R, Garcia-Foncillas J. MicroRNAs as cancer players: potential clinical and biological effects. DNA Cell Biol. 2007;26:273–82.
19. Hernando E. microRNAs and cancer: role in tumorigenesis, patient classification and therapy. Clin Transl Oncol. 2007;9:155–60.

20. Negrini M, Ferracin M, Sabbioni S, Croce CM. MicroRNAs in human cancer: from research to therapy. J Cell Sci. 2007;120:1833–40.
21. Wiemer EA. The role of microRNAs in cancer: no small matter. Eur J Cancer. 2007;43:1529–44.
22. Wijnhoven BP, Michael MZ, Watson DI. MicroRNAs and cancer. Br J Surg. 2007;94:23–30.
23. Yu F, Yao H, Zhu P, Zhang X, et al. let-7 regulates self-renewal and tumorigenicity of breast cancer cells. Cell. 2007;131:1109–23.
24. Shimono Y, Zabala M, Cho RW, Lobo N, et al. Downregulation of miRNA-200c links breast cancer stem cells with normal stem cells. Cell. 2009;138:592–603.
25. Esquela-Kerscher A, Trang P, Wiggins JF, Patrawala L, et al. The let-7 microRNA reduces tumor growth in mouse models of lung cancer. Cell Cycle. 2008;7:759–64.
26. Cimmino A, Calin GA, Fabbri M, Iorio MV, et al. miR-15 and miR-16 induce apoptosis by targeting BCL2. Proc Natl Acad Sci U S A. 2006;102:13944–9.
27. Calin GA, Farracin M, Cimmino A, Di Lava G, et al. A unique microRNA signature associated with prognostic factors and disease progression in B cell chronic lymphocytic leukemia. N Engl J Med. 2005;352:1667–76.
28. Diederichs S, Haber DA. Sequence variations of microRNAs in human cancer: alterations in predicted secondary structure do not affect processing. Cancer Res. 2006;66:6097–104.
29. He H, Jazdzewki K, Li W, Liyanarachchi S, et al. The role of microRNA genes in papillary thyroid carcinoma. Proc Natl Acad Sci U S A. 2005;102:19075–80.
30. Abelson JF, Kwan KY, O'Roak BJ, Baek DY, et al. Sequence variants in SLITRK1 are associated with Tourette's syndrome. Science. 2006;310:317–20.
31. Lu J, Getz G, Miska EA, Alvarez-Saavedra E, et al. MicroRNA expression profiles classify human cancers. Nature. 2005;435:834–8.
32. Volinia S, Calin GA, Liu CG, Ambs S, et al. A microRNA expression signature of human solid tumors define cancer gene targets. Proc Natl Acad Sci U S A. 2006;103:2257–61.
33. Meng F, Henson R, Wehbe-Janek H, Ghoshal K et al. MicroRNA-21 regulates expression of the PTEN tumor suppressor gene in human hepatocellular cancer. Gastroenterology. 2007;133:647–58.
34. Frankel LB, Chrisoffersen NR, Jacobsen A, Lindow M et al. Programmed cell death 4 (PDCD4) is an important functional target of the microRNA miR-21 in breast cancer cells. J Biol Chem. 2008;283:1026–33.
35. Zhu S, Si M-L, Win H, Mo Y-Y. MicroRNA-21 targets the tumor suppressor gene tropomyosin 1 (TPM1). J Biol Chem. 2007;282:14328–36.
36. Eis PS, Tam W, Sun L, Chadburn AZ et al. Accumulation of miR-155 and BIC RNA in human B cell lymphomas. Proc Natl Acad Sci U S A. 2005;102:3627–32.
37. Costinean S, Zanesi N, Pekarsh Y, Tili E et al. Pre-B cell proliferation and lymphoblastic leukemia/high grade lymphoma in Eμ-miR155 transgenic mice. Proc Natl Acad Sci U S A. 2006;103:7024–9.
38. Huang Q, Gumireddy K, Schrier M, le Sage C et al. The microRNAs miR-373 and miR-520c promote tumor invasion and metastasis. Nat Cell Biol. 2008;10:202–10.
39. Yan GR, Xu SH, Tan ZL, Liu L, He QY. Global identification of miR-373-regulated genes in breast cancer by quantitative proteomics. Proteomics. 2011;11:912–20.
40. Voorhoeve PM, le Sage C, Schrier M, Gillis AJ et al. A genetic screen implicates miRNA-372 and miRNA-373 as oncogenes in testicular germ cell tumors. Cell. 2006;124:1169–81.
41. Ventura AG, Young AG, Winslow MM, Meissner A, et al. Targeted deletion reveals essential and overlapping functions of the miR-17 through 92 family of miRNA clusters. Cell. 2008;132:875–86.
42. Xiao C, Srinivasan L, Calado DP, Patterson HC, et al. Lymphoproliferative disease and autoimmunity in mice with increased miR-17-92 expression in lymphocytes. Nature Immunol. 2008;9:405–14.
43. Ota A, Tagawa H, Karnan S, Tsuzuki S, et al. Identification and characterization of a novel gene, C13orf25, as a target for 13q31-q32 amplification in malignant lymphoma. Cancer Res. 2004;64:3087–95.

44. Ma L, Teruya-Feldstein J, Weinberg RA. Tumour invasion and metastasis initiated by microRNA-10b in breast cancer. Nature. 2007;449:682–8.
45. Garofalo M, Di Leva G, Romano G, Suh SS, et al. MiR-221 and 222 regulate TRAIL resistance and enhance tumorigenicity through PTEN and TIMP3 downregulation. Cancer Cell. 2009;16:498–509.
46. He L, He X, Lim LP, de Stanchina E, et al. A microRNA component of the p53 tumour suppressor network. Nature. 2007;447:1130–4.
47. Raver-Shapira N, Marciano E, Meiri E, Spectpr Y, et al. Transcriptional activation of miR-34a contributes to p53-mediated apoptosis. Mol Cell. 2007;26:731–43.
48. Chang TC, Wentzel EA, Kent OA, Ramachandran K, et al. Transactivation of miR-34a by p53 broadly influences gene expression and promotes apoptosis. Mol Cell. 2007;26:745–52.
49. Mott JL, Kobayashi S, Bronk SF, Gores GJ. miR-29 regulates Mcl-1 protein expression and apoptosis. Oncogene. 2007;26:6133–40.
50. Pekarsky Y, Santanam U, Cimmino A, Palamarchuk A, et al. Tcl1 expression in chronic lymphocytic leukemia is regulated by miR-29 and miR-181. Cancer Res. 2006;66:11590–3.
51. Fabbri M, Garzon R, Cimmino A, Liu Z, et al. MicroRNA family reverts aberrant methylation in lung cancer by targeting DNA methyltransferases 3A and 3B. Proc Natl Acad Sci U S A. 2007;104:15805–10. (2007.
52. Johnson SM, Grosshans H, Shingara J, Byrom M, et al. RAS is regulated by the let-7 microRNA family. Cell. 2005;120:635–47.
53. Lee YA, Dutta A. The tumor suppressor microRNA let-7 represses the HMGA2 oncogene. Genes Dev. 2007;21:1025–30.
54. Iorio MV, Ferracin M, Liu CG, Veronese A, Spizzo R, Sabbioni S, Magri E, Pedriali M, Fabbri M, Campiglio M, Menard S, Palazzo JP, Rosenberg A, Musiani P, Volinia S, Nenci I, Calin GA, Querzoli P, Negrini M, Croce CM. MicroRNA gene expression deregulation in human breast cancer. Cancer Res. 2005;65:7065–70.
55. Adams BD, Furneaux H, White BA. The micro-ribonucleic acid (miRNA) miR-206 targets the human estrogen receptor-alpha (ERalpha) and represses ERalpha messenger RNA and protein expression in breast cancer cell lines. Mol Endocrinol. 2007;21:1132–47.
56. Hossain A, Kuo MT, Saunders GF. Mir-17-5p regulates breast cancer cell proliferation by inhibiting translation of AIB1 mRNA. Mol Cell Biol. 2006;26:8191–201.
57. Yu Z, Wang C, Wang M, Li Z, Casimiro MC, Liu M, Wu K, Whittle J, Ju X, Hyslop T, McCue P, Pestell RG. A cyclin D1/microRNA 17/20 regulatory feedback loop in control of breast cancer cell proliferation. J Cell Biol. 2008;182:509–17.
58. Mattie MD, Benz CC, Bowers J, Sensinger K, Wong L, Scott GK, Fedele V, Ginzinger D, Getts R, Haqq C. Optimized high-throughput microRNA expression profi ling provides novel biomarker assessment of clinical prostate and breast cancer biopsies. Mol Cancer. 2006;5:24.
59. Scott GK, Goga A, Bhaumik D, Berger CE, Sullivan CS, Benz CC. Coordinate suppression of ERBB2 and ERBB3 by enforced expression of micro-RNA miR-125a or miR-125b. J Biol Chem. 2007;282:1479–86.
60. Gregory PA, Bert AG, Paterson EL, Barry SC, Tsykin A, Farshid G, Vadas MA, Khew-Goodall Y, Goodall GJ. The miR-200 family and miR-205 regulate epithelial to mesenchymal transition by targeting ZEB1 and SIP1. Nat Cell Biol. 2008;10:593–601.
61. Dykxhoorn DM, Wu Y, Xie H, Yu F, Lal A, Petrocca F, Martinvalet D, Song E, Lim B, Lieberman J. miR-200 enhances mouse breast cancer cell colonization to form distant metastases. PLoS One. 2009;4:e7181.
62. Valastyan S, Reinhardt F, Benaich N, Calogrias D, Szasz AM, Wang ZC, Brock JE, Richardson AL, Weinberg RA. A pleiotropically acting microRNA, miR-31, inhibits breast cancer metastasis. Cell. 2009;137:1032–46.
63. Metzler M, Wilda M, Busch K, et al. High expression of precursor microRNA-155/BIC RNA in children with Burkitt lymphoma. Genes Chromosomes Cancer. 2004;39:167–9.
64. Kluiver J, Poppema S, de JD, et al. BIC and miR-155 are highly expressed in Hodgkin, primary mediastinal and diffuse large B cell lymphomas. J Pathol. 2005;207:243–9.

65. Thai TH, Calado DP, Casola S, et al. Regulation of the germinal center response by microRNA-155. Science. 2007;316:604–8.
66. Rodriguez A, Vigorito E, Clare S, et al. Requirement of bic/microRNA-155 for normal immune function. Science. 2007;316:608–11.
67. Tili E, Michaille JJ, Cimino A, et al. Modulation of miR-155 and miR-125b levels following lipopolysaccharide/TNF-alpha stimulation and their possible roles in regulating the response to endotoxin shock. J Immunol. 2007;179:5082–9.
68. Calin GA, Ferracin M, Cimmino A, et al. A MicroRNA signature associated with prognosis and progression in chronic lymphocytic leukemia. N Engl J Med. 2005;353:1793–801.
69. Iorio MV, Ferracin M, Liu CG, et al. MicroRNA gene expression deregulation in human breast cancer. Cancer Res. 2005;65:7065–70.
70. Wang H, Garzon R, Sun H, et al. NF-kappaB YY1-miR-29 regulatory circuitry in skeletal myogenesis and rhabdomyosarcoma. Cancer Cell. 2008;14:369–81.
71. Garzon R, Liu S, Fabbri M, et al. MicroRNA-29b induces global DNA hypomethylation and tumor suppressor gene re-expression in acute myeloid leukemia by targeting directly DNMT3A and 3B and indirectly DNMT1. Blood. 2009;113:6411–8.
72. Yan LX, Huang XF, Shao Q, et al. MicroRNA miR-21 overexpression in human breast cancer is associated with advanced clinical stage, lymph node metastasis and patient poor prognosis. RNA. 2008;14:2348–60.
73. Qian B, Katsaros D, Lu L, et al. High miR-21 expression in breast cancer associated with poor disease-free survival in early stage disease and high TGF-beta1. Breast Cancer Res Treat. 2009;117:131–40.
74. Chan JA, Krichevsky AM, Kosik KS. MicroRNA-21 is an antiapoptotic factor in human glioblastoma cells. Cancer Res. 2005;65:6029–33.
75. Ciafre SA, Galardi S, Mangiola A, et al. Extensive modulation of a set of microRNAs in primary glioblastoma. Biochem Biophys Res Commun. 2005;334:1351–8.
76. Iorio MV, Visone R, Di LG, et al. MicroRNA signatures in human ovarian cancer. Cancer Res. 2007;67:8699–707.
77. Yanaihara N, Caplen N, Bowman E, Seike M, et al. Unique microRNA molecular profiles in lung cancer diagnosis and prognosis. Cancer Cell. 2006;9:189–98.
78. Markou A, Tsaroucha EG, Kaklamanis L, Fotinou M, et al. Prognostic value of mature microRNA-21 and microRNA-205 overexpression in non-small cell lung cancer by quantitative real-time RT-PCR. Clin Chem. 2008;54:1696–704.
79. Schetter AJ, Leung SY, Sohn JJ, Zanetti KA, et al. MicroRNA expression profiles associated with prognosis and therapeutic outcome in colon adenocarcinoma. JAMA. 2008;299:425–36.
80. Roldo C, Missiaglia E, Hagan JP, Falconi M, et al. MicroRNA expression abnormalities in pancreatic endocrine and acinar tumors are associated with distinctive pathologic features and clinical behavior. J Clin Oncol. 2006;24:4677–84.
81. Bloomston M, Frankel WL, Petrocca F, Volinia S, et al. MicroRNA expression patterns to differentiate pancreatic adenocarcinoma from normal pancreas and chronic pancreatitis. JAMA. 2007;297:1901–8.
82. Scott GK, Goga A, Bhaumik D, Berger CE, et al. Coordinate suppression of ERBB2 and ERBB3 by enforced expression of micro-RNA miR-125a or miR-125b. J Biol Chem. 2007;282:1479–86.
83. Iorio MV, Casalini P, Piovan C, Di Leva G, et al. MicroRNA-205 regulates HER3 in human breast cancer. Cancer Res. 2009;69:2195–200.
84. Wu H, Zhu S, Mo YY. Suppression of cell growth and invasion by miR-205 in breast cancer. Cell Res. 2009;19:439–48.
85. Blenkiron C, Goldstein LD, Thorne NP, Spiteri I, et al. MicroRNA expression profiling of human breast cancer identifies new markers of tumor subtype. Genome Biol. 2007;8:R214.
86. Miller TE, Ghoshal K, Ramaswamy B, Roy S, et al. MicroRNA-221/222 confers tamoxifen resistance in breast cancer by targeting p27Kip1. J Biol Chem. 2008;283:29897–903.

87. Zhao JJ, Lin J, Yang H, Kong W, et al. MicroRNA-221/222 negatively regulates estrogen receptor alpha and is associated with tamoxifen resistance in breast cancer. J Biol Chem. 2008;283:31079–86.
88. Di Leva G, Gasparini P, Piovan C, Ngankeu A, et al. MicroRNA cluster 221-222 and estrogen receptor alpha interactions in breast cancer. J Natl Cancer Inst. 2010;102:706–21.
89. Derynck R, Akhurst RJ, Balmain A. TGF-β signaling in tumor suppression and cancer progression. Nature Genet. 2001;29:117–29.
90. Bonci D, Coppola V, Musumeci M, Adddario A, et al. The miR-15a-miR-16-1 cluster controls prostate cancer by targeting multiple oncogenic activities. Nature Med. 2008;14:1271–7.
91. Kota J, Chivukula RR, O'Donnell KA, Wentzel EA, et al. Therapeutic microRNA delivery suppresses tumorigenesis in a murine liver cancer model. Cell. 2009;137:1005–17.
92. Esteller M. Cancer epigenomics: DNA methylomes and histone-modification maps. Nature Rev Genet. 2007;8:286–98.
93. Saito Y, Liang G, Egger G, Friedman JM, et al. Specific activation of microRNA-127 with downregulation of the proto-oncogene BCL6 by chromatin-modifying drugs in human cancer cells. Cancer Cell. 2006;9:435–43.
94. Fabbri M, Garzon R, Cimmino A, Liu Z, et al. MicroRNA family reverts aberrant methylation in lung cancer by targeting DNA methyltransferases 3A and 3B. Proc Natl Acad Sci U S A. 2007;104:15805–10.
95. Benetti R, Gonzalo S, Jaco I, Munoz P, et al. A mammalian microRNA cluster controls DNA methylation and telomere recombination via Rbl2-dependent regulation of DNA methyltransferases. Nat Struct Mol Biol. 2008;15:268–79.
96. Sinkkonen L, Hugenschmidt T, Berninger P, Gaidatzis D, et al. MicroRNAs control de novo DNA methylation through regulation of transcriptional repressors in mouse embryonic stem cells. Nat Struct Mol Biol. 2008;15:259–67.
97. Noonan EJ, Place RF, Pookot D, Basak S, et al. miR-449a targets HDAC-1 and induces growth arrest in prostate cancer. Oncogene. 2009;28:1714–24.
98. Lowery AJ, Miller N, McNeill RE, Kerin MJ. MicroRNAs as prognostic indicators and therapeutic targets: potential effect on breast cancer management. Clin Cancer Res. 2008;14:360–5.
99. Wiemer EA. The role of microRNAs in cancer: no small matter. Eur J Cancer. 2007;43:1529–44.
100. Hatfield S, Ruohola-Baker H. microRNA and stem cell function. Cell Tissue Res. 2008;331:57–66.
101. Bernstein E, Kim SY, Carmell MA, Murchison EP, et al. Dicer is essential for mouse development. Nat Genet. 2003;35:215–7.
102. Croce CM, Calin GA. miRNAs, cancer, and stem cell division. Cell. 2005;122:6–7.
103. Yu F, Yao H, Zhu P, Zhang X, et al. let-7 regulates self renewal and tumorigenicity of breast cancer cells. Cell. 2007;131:1109–23.
104. Silber J, Lim DA, Petritsch C, Persson AI, et al. miR-124 and miR-137 inhibit proliferation of glioblastoma multiforme cells and induce differentiation of brain tumor stem cells. BMC Med. 2008;6:14–31.
105. Liu C, Kelnar K, Liu B, Chen X, et al. The microRNA miR-34a inhibits prostate cancer stem cells and metastasis by directly repressing CD44. Nat Med. 2011;17:211–5.
106. Ibarra I, Erlich Y, Muthuswamy SK, Sachidanandam R, et al. A role for microRNAs in maintenance of mouse mammary epithelial progenitor cells. Genes Dev. 2007;21:3238–43.
107. Greene SB, Gunaratne PH, Hammond SM, Rosen JM. A putative role for microRNA-205 in mammary epithelial cell progenitors. J Cell Sci. 2010;123:606–18.
108. Yu F, Deng H, Yao H, Liu Q, et al. Mir-30 reduction maintains self-renewal and inhibits apoptosis in breast tumor initiating cells. Oncogene. 2010;29:4194–204.
109. Olson P, Lu J, Zhang H, Shai A, et al. MicroRNA dynamics in the stages of tumorigenesis correlate with hallmark capabilities of cancer. Genes Dev. 2009;23:2152–65.
110. Farazi TA, Horlings HM, Ten Hoeve JJ, Mihailovic A, et al. MicroRNA sequence and expression analysis in breast tumors by deep sequencing. Cancer Res. 2011;71:4443–53.

111. Baranwal S, Alahari SK. miRNA control of tumor cell invasion and metastasis. Int J Cancer. 2010;126:1283–90.
112. Ceppi P, Mudduluru G, Kumarswamy R, Rapa I, et al. Loss of miR-200c expression induces an aggressive, invasive, and chemoresistant phenotype in non-small cell lung cancer. Mol Cancer Res. 2010;8:1207–16.
113. Thiery JP, Sleeman JP. Complex networks orchestrate epithelial-mesenchymal transitions. Natl Rev Mol Cell Biol. 2006;7:131–42.
114. Thiery JP. Epithelial-mesenchymal transitions in tumor progression. Nat Rev. Cancer. 2002;2:442–54.
115. Brabletz T, Jung A, Spaderna S, Hlubek F, et al. Opinion: migrating cancer stem cells-an integrated concept of malignant tumor progression. Nat Rev Cancer. 2005;5:744–9.
116. Lee JM, Dedhar S, Kalluri R, Thompson EW. The epithelial- mesenchymal transition: new insights in signaling, development, and disease. J Cell Biol. 2006;172:973–81.
117. Hurt EM, Farrar WL. Cancer stem cells: the seeds of metastasis? Mol Interv. 2008;8:140–2.
118. Visvader JE, Lindeman GJ. Cancer stem cells in solid tumors: accumulating evidence and unresolved questions. Nat Rev Cancer. 2008;8:755–68.
119. Garofalo M, Croce CM. MicroRNAs: master regulators as potential therapeutics in cancer. Annu Rev Pharmacol Toxicol. 2011;51:25–43.
120. Adams BD, Guttilla IK, White BA. Involvement of microRNAs in breast cancer. Semin Reprod Med. 2008;26:522–36.
121. Cochrane DR, Cittelly DM, Howe EN, Spoelstra NS, et al. MicroRNAs link estrogen receptor alpha status and Dicer levels in breast cancer. Horm Cancer. 2010;1:306–19.
122. Krol J et al. The widespread regulation of microRNAs biogenesis, function and decay. Nat Rev Genet. 2010;11:597–610.
123. Bhat-Nakshatri P, Wang G, Collins NR, Thomson MJ, et al. Estradiol-regulated microRNAs control estradiol response in breast cancer cells. Nucleic Acids Res. 2009;37:4850–61.
124. Martello G, Rosato A, Ferrari F, Manfrin A, et al. A MicroRNA targeting dicer for metastasis control. Cell. 2010;141:1195–207.
125. Ender C, Meister G. Argonaute proteins at a glance. J Cell Sci. 2010;123:1819–23.
126. Davis-Dusenbery BN, Hata A. Mechanisms of control of microRNA biogenesis. J Biochem. 2010;148:381–92.
127. Cheng C, Fu X, Alves P, Gerstein M. mRNA expression profiles show differential regulatory effects of microRNAs between estrogen receptor-positive and estrogen receptor-negative breast cancer. Genome Biol. 2009;10:R90.
128. Adams BD, Claffery KP, White BA. Argonaute-2 expression is regulated by epidermal growth factor receptor and mitogen-activated protein kinase signaling and correlates with a transformed phenotype in breast cancer cells. Endocrinology. 2009;150:14–23.
129. Ma L, Teruya-Feldstein J, Weinberg RA. Tumour invasion and metastasis initiated by microRNA-10b in breast cancer. Nature. 2007;449:682–8.
130. Wang L, Wang J. MicroRNA-mediated breast cancer metastasis: from primary site to distant organs. Oncogene. 2011;31:2499–511.
131. Gregory PA, Bracken CP, Bert AG, Goodall GJ. MicroRNAs as regulators of epithelial-mesenchymal transition. Cell Cycle. 2008;7:3112–8.
132. Park SM, Gaur AB, Lengyel E, Peter ME. The miR-200 family determines the epithelial phenotype of cancer cells by targeting the E-cadherin repressors ZEB1 and ZEB2. Genes Dev. 2008;22:894–907.
133. Zhu S, Wu H, Wu F, Nie D, et al. MicroRNA-21 targets tumor suppressor genes in invasion and metastasis. Cell Res. 2008;18:350–9.
134. Tavazoie SF, Alarcon C, Oskarsson T, Padua D, et al. Endogenous human microRNAs that suppress breast cancer metastasis. Nature. 2008;451:147–52.
135. Rosenfeld N, Aharonov R, Meiri E, Rosenwald S, et al. MicroRNAs accurately identify cancer tissue origin. Nat Biotechnol. 2008;26:462–9.
136. Pavlidisa N, Fizazib K. Cancer of unknown primary (CUP). Crit Rev Oncol Hematol. 2008;54:243–50.

137. Calin GA, Cimmino A, Fabbri M, Ferracin M, et al. MiR-15a and miR-16-1 cluster functions in human leukemia. Proc Natl Acad Sci U S A. 2008;105:5166–71.
138. Si ML, Zhu S, Wu H, Lu Z, et al. MiR-21-mediated tumor growth. Oncogene. 2007;26: 2799–803.
139. Meng F, Henson R, Lang M, Wehbe H, et al. Involvement of human micro-RNA in growth and response to chemotherapy in human cholangiocarcinoma cell lines. Gastroenterology. 2006;130:2113–29.
140. Weidhaas JB, Babar I, Nallur SM, Trang P, et al. MicroRNAs as potential agents to alter resistance to cytotoxic anticancer therapy. Cancer Res. 2007;67:11111–6.
141. Weiler J, Hunziker J, Hall J. Anti-miRNA oligonucleotides (AMOs): Ammunition to target miRNAs implicated in human disease? Gene Ther. 2006;13:496–502.
142. Orom UA, Kauppinen S, Lund AH. LNA modified oligonucleotides mediate specific inhibition of microRNA function. Gene. 2006;372:137–41.
143. Krutzfeldt J, Rajewsky N, Braich R, Rajeev KG, et al. Silencing of microRNAs in vivo with 'antagomirs'. Nature 2005;438:685–689.
144. Ebert MS, Neilson JR, Sharp PA. MicroRNA sponges: competitive inhibitors of small RNAs in mammalian cells. Nat Methods. 2007;4:721–6.

Chapter 8
The Role of MicroRNAs in Hematopoietic Stem Cells and Leukemia Development

Safak Yalcin, Stephen S. Chung and Christopher Y. Park

Abstract Hematopoietic stem cells (HSCs) are multipotent cells capable of self-renewal as well as differentiation into all mature blood cell types to sustain lifelong hematopoiesis. Leukemias are thought to be initiated and maintained by leukemic stem cells (LSCs), which also have the capacity for self-renewal, but they are also characterized by varying levels of impaired differentiation and increased proliferation which give rise to disease phenotypes. LSCs and HSCs share the common ability to self-renew, and they may rely on similar pathways for this unique function. Thus, it is critical to understand the molecular mechanisms shared between these two cell populations, as these are likely to represent key features that drive leukemic transformation of HSCs and/or their downstream progeny. Recently, microRNAs (miRNAs) have been implicated as important regulators of self-renewal and differentiation in the hematopoietic system. Profiling of normal and malignant hematopoietic cells, corroborated with a limited but growing number of functional studies, has demonstrated that miRNAs are critical regulators of HSC function, are dysregulated in leukemias, and likely play an important role in leukemogenesis. Herein, we will review these studies and discuss their contributions toward our understanding of the importance of miRNAs in normal and malignant stem cell function in the hematopoietic system.

Keywords MicroRNAs · Hematopoietic stem cells · Leukemia stem cells · Self-renewal · Leukemia · Acute myeloid leukemia (AML)

C. Y. Park (✉)
Departments of Pathology and Laboratory Medicine, 1275 York Ave, Box 20,
New York, NY 10065, USA
e-mail: parkc@mskcc.org

S. Yalcin · S. S. Chung · C. Y. Park
Human Oncology and Pathogenesis Program, 1275 York Ave, Box 20,
New York, NY 10065, USA

S. S. Chung
Department of Medicine and Leukemia Service,
Memorial Sloan-Kettering Cancer Center, New York, NY, USA

S. R. Singh, P. Rameshwar (eds.), *MicroRNA in Development and in the Progression of Cancer*, DOI 10.1007/978-1-4899-8065-6_8,

1 Introduction

Normal hematopoiesis represents a dynamic orchestration of tightly regulated cellular events, reflecting an integration of signals that promote or restrain proliferation and differentiation across a hierarchy of cells with varying capacities for self-renewal, proliferation, and lineage fate. Hematopoietic stem cells (HSCs) reside at the top of this hierarchy and are characterized by the unique ability to self-renew and give rise to all mature blood cell types [1, 2]. In acute myeloid leukemia (AML), the counterpart of the HSC appears to be the leukemic stem cell (LSC), with seminal human leukemia xenograft studies demonstrating that AMLs include a population of cells that can self-renew and give rise to all the components of the leukemia, all while retaining their hierarchically developmental relationships [3, 4]. Only a subpopulation of $CD34^{+}CD38^{-}$ leukemic cells—similar to the immunophenotype of normal HSCs—was capable of engrafting human AML in nonobese diabetic/severe combined immunodeficiency (NOD/SCID) mice, with no engraftment potential in the $CD34CD38^{+}$ or $CD34^{-}$ fractions. Importantly, transplantation of $CD34^{+}CD38^{-}$ cells resulted in engraftment of leukemia-containing cells exhibiting all three immunophenotypes, thereby recapitulating the immunophenotypic diversity of the original leukemia [3]. As such, AML blasts appears to be organized in a hierarchy akin to that seen in the normal hematopoietic system, with self-renewing LSCs giving rise to more differentiated non-self-renewing leukemic cells. Given these parallels, it is not surprising that several signaling pathways shown to be highly relevant in HSCs (*Hox, Notch, Wnt,* and *Sonic-hedgehog* [5–9] have also been found to regulate LSCs. These findings likely reflect a shared dependence of these two cell populations on many of the same molecular pathways that regulate self-renewal and differentiation. Identifying these pathways through direct comparisons of HSCs and LSCs represents a potentially powerful approach to identifying novel regulators of leukemogenesis.

Micro-ribonucleic acids (miRNAs) are evolutionarily conserved small (20–22-nucleotide, nt) noncoding RNAs, first described in the 1990s [10, 11], that regulate gene expression at the posttranscriptional level by binding the 3′-untranslated region (3′-UTR) of messenger RNAs (mRNAs), leading to the inhibition of translation or mRNA degradation [12]. miRNAs have been shown to regulate a variety of cellular processes including differentiation, metabolism, apoptosis, and proliferation [13]. miRNAs are first transcribed by RNA polymerase II in the nucleus as large primary transcripts (pri-miRNAs) [14], followed by processing by Drosha, an RNAse III-type endonuclease (RN3), into ~70 nt pre-miRNAs containing a hairpin structure. Pre-miRNAs are then transported into the cytoplasm by the guanosine triphosphate (GTP)-dependent RAN transporter Exportin-5, where they are further processed by the endonuclease Dicer into mature ~20 bp double-stranded miRNAs. One of the miRNA strands is then incorporated into a protein–RNA complex called the RNA-induced silencing complex (RISC), thereby allowing miRNAs to bind to their mRNA targets [15, 16]. miRNAs interact with their targets via 6–8-nt "seed sequences," which have sequence complementarity to "seed match" sequences in the 3′-UTR of target mRNAs. In most situations, such base pairing is imperfect, resulting in translational suppression, but in the presence of perfect base pairing, endonucleolytic mRNA cleavage is triggered [17]. Through these mechanisms, each particular miRNA may target many mRNAs, and in total approximately 30 % of human protein-coding genes are predicted to be regulated by miRNAs [12]. As developmental transitions in hematopoiesis depend on the coordinated and rapid modulation of many transcription factors, the promiscuous nature of miRNA:mRNA interactions are ideally suited to regulate such changes. Accordingly, there are increasing experimental data to support roles of miRNAs as key regulators of differentiation and self-renewal in normal and malignant stem cells.

In this chapter, we will provide an overview of the known biological functions of miRNAs in HSCs and LSCs. While a relatively small number of studies have evaluated the functional role of specific miRNAs in these cell populations, a large number of miRNA profiling studies have also been performed, providing numerous additional potential candidates for experimental verification. Thus, we will also discuss the results of these studies and their implications for future research and potential therapeutic strategies.

2 miRNA Expression in the Hematopoietic Stem Cell

The functional importance of miRNAs in hematopoiesis has been identified primarily through miRNome-wide expression profiling and in silico bioinformatic prediction studies. The majority of such studies have been performed using normal HSCs and hematopoietic progenitors from different stages of lineage differentiation. As summarized in Table 8.1, miRNA expression profiling studies using a variety of techniques have identified that miRNAs are highly and commonly expressed at high levels in both mouse and human HSCs.

One of the earliest efforts to identify biologically relevant miRNAs in HSCs and progenitor cells was performed using human $CD34^+$ cells. While $CD34^+$cells are enriched for immature hematopoietic cells, HSCs comprise only ~ 1 %, with the vast majority of cells representing non-self-renewing progenitors [18, 19]. This first large-scale profiling of miRNA expression identified 33 miRNAs that were differentially expressed in $CD34^+$ hematopoietic stem-progenitor cells (HSPCs) when compared to normal human bone marrow (BM) and mobilized human peripheral blood $CD34^+$ cells [20]. The investigators then combined these data with mRNA expression data generated from human $CD34^+$ cells and used an in silico algorithm to predict the interactions between HSPC-expressed miRNAs and mRNAs, particularly those previously shown to regulate lineage fate decisions. Based on pairing of HSPC-expressed miRNAs with their putative mRNA targets, they formulated a model in which many of the genes associated with hematopoietic differentiation are expressed at an early time point by undiffrentiated HSPCs, but are supressed by a small subset of miRNAs until differentiation occurs. While the predicted interactions between miRNAs and their targets were, for the most part, not functionally

Table 8.1 MicroRNAs identified in mouse and human HSCs by profiling assays

Surface markers for HSC	miRNAs	Reference
Lin-c-Kit+Sca-1+CD150+CD48- (mouse)	*miR-99a, -125a/b, -155 miR-126, -196b, -130a, -181c miR-193b, -542-5p, let-7e*	O'Connell et al. (2010)
Lin-c-Kit+Sca-1+CD34-Flk2- (mouse)	*miR-99a/b, -125a/b, let-7e, miR-10a/b, -130a, -31, -18a, miR-15b, -100, -146, -425, 422b*	Guo et al. (2010)
ES150+(CD45+EPCR+CD48-CD150+) ES150-(CD45+EPCR+CD48-CD150-) SLAM HSC-(CD150+CD48-CD45+) LSK (Lin-c-Kit+Sca-1+) (mouse)	*miR-29a (in ES150+/150-) miR-125b miR-196a/b, -130a miR-148b, -351, and let-7d*	Petriv et al. (2010)
CD34+CD38- (human)	*miR-452, -127, -526b*, -520h, predicted-miR-209, -105, -149, predicted-miR-100*	Liao et al. (2008)
CD133+ (human)	*miR-10a, -125b, -146a, -125a, miR-551b, -99a, -29a/b/c, -24 miR-23a/b, -146b*	Bissels et al. (2011)

validated, this study represented an important first attempt to describe the function of miRNAs in early human hematopoiesis.

O'Connell et al. measured the expression of 137 miRNAs in HSPC-enriched $Lin^{-}c\text{-}Kit^{+}Sca\text{-}1^{+}$(LSK) cells, as well as unfractionated total BM from C57BL/6 mice [21]. Of these 137 miRNAs, 11 were found to be enriched in LSK cells as compared with total BM (Table 8.1). Compellingly, the same set of miRNAs was found to be enriched in a similar analysis comparing $CD34^{+}$ human cord blood (CB) cells to $CD34^{-}$CB cells, suggesting high evolutionary conservation and likely important functions of these miRNAs in primitive hematopoietic cells. Subsequently, Petriv et al. developed a high-throughput microfluidics-based real-time quantitative polymersase chain reaction (RT-qPCR) approach in order to study global miRNA expression in 27 phenotypically distinct cell populations isolated from normal adult mouse hematopoietic tissue [22]. A phylogenetic analysis of their study has revealed that 27 cell populations could be grouped into six clusters including stem and progenitor cells, lymphoid cells, and four distinct branches of myeloid lineage cells. Interestingly, expression of a subset of miRNAs in single hematopoietic cells further demonstrated that progenitors such as granulocyte–macrophage progenitors (GMPs), thought to be relatively homogeneous with respect to function, showed less cell-to-cell miRNA expression variation compared to functionally heterogeneous populations such as $CD45^{+}CD48^{+}$ cells. These data suggest that miRNA expression variability in single cells may underlie the functional heterogeneity observed in such populations, and thus miRNA profiling of individual cells may provide a powerful approach to evaluate the purity of cell populations and to identify new functional subsets of cells. Nonetheless, while these initial studies identified many promising candidate miRNAs that likely regulate different stages of hematopoiesis and HSC function, they did not make any attempts at functional validation.

Subsequent studies have aimed to define the functional importance of many of these candidate miRNAs in hematopoiesis by overexpression or selective targeting of specific miRNAs in mice. The mouse system is well suited to such studies since mouse HSPCs are well defined immunophenotypically, and the function of highly purified populations can be precisely assessed in vivo using well-established congenic transplantation models [23].

The first experimental evidence that miRNAs regulate hematopoiesis came from studies of mice in which *Ago2,* a gene encoding a protein component of the RISC, was conditionally inactivated, leading to severe hematopoietic defects such as impaired B cell and erythroid cell differentiation [24]. Additional evidence supporting a role of miRNAs in hematopoiesis came from a mouse model in which *Dicer*was conditionally deleted in the hematopoietic system, resulting in a cell-autonomous impairment in HSPC function and increased apoptosis, indicating that miRNAs are required to maintain the immature HSPC pool. *Dicer*-null HSCs were able to engraft lethally irradiated recipient mice, but they showed a significant reduction in stably engrafted LSK and mature populations, particularly myeloid cells [25].

Guo and colleagues also measured miRNA expression in mouse HSPCs and identified a number of highly expressed miRNAs, including the miRNA polycistron that contains *miR-125a, miR-99b,*and *let-7e.* To investigate the role of this miRNA cluster in hematopoiesis, either the entire cluster or individual members were ectopically overexpressed in unfractionated mouse BM cells and transplanted into lethally irradiated recipient mice. Overexpression of *miR-125a* alone resulted in a more than eightfold expansion of donor-derived reconstituting HSCs as well as all major lineages, and this was accompanied by a reduction in apoptosis in primitive hematopoietic cells. The anti-apoptotic effect of *miR-125a* was found to be associated with its ability to suppress the expression of its direct target, the pro-apoptotic protein *Bak1* [25]. Interestingly, *Bak1*$^{-/-}$ mice did not show any alterations in HSC number or function, suggesting that *miR-125a* induces HSC expansion by targeting multiple genes [25]. In addition, since a single miRNA may have multiple target mRNAs, and a single mRNA can be targeted by many miRNAs, it is perhaps not surprising that loss of a single target, even if physiologically relevant, may not be sufficient to induce a dramatic phenotype [26].

MiR-125b, another member of miR-125 family, is also highly expressed in HSCs and HSC-enriched populations, suggesting a role in HSC regulation. We have confirmed *miR-125b*'s role in HSC self-renewal by demonstrating that *miR-125b*-transduced mouse HSCs (Lin$^-$cKit$^+$Sca-1$^+$CD150$^+$CD34$^-$) exhibit increased engraftment capability in serial transplantation assays [21, 27]. In these studies, overexpression of *miR-125b* in HSCs also resulted in a preferential expansion of lymphoid-biased and lymphoid-balanced (intermediate myeloid and lymphoid lineage bias) HSCs [27]. Similar to *miR-125a,* the *miR-125b*-induced enhancement in HSC function was also associated with induction of anti-apoptotic pathways, likely through inhibition of *Bmf* and *KLF13* mRNA expression. O'Connell and colleagues also identified *miR-125b* among a set of miRNAs highly expressed in

HSC-enriched populations (LSK CD150$^+$CD48–; Table 8.1). Interestingly, in their studies, overexpression of *miR-125b* in unfractionated mouse BM cells not only induced an expansion of HSCs, but also caused a dose-dependent myeloproliferative disorder that eventually progressed to AML [21]. We will discuss the relevance of *miR-125b* to human AML in a later section.

*MiR-29a*has also been identified to be highly expressed in mouse and human HSCs, as well as human LSCs [28], and overexpression of *miR-29a* in mouse HSCs and progenitors results in a myeloproliferative disorder with biased myeloid differentiation that progresses to AML upon serial transplantation. Moreover, *miR-29a* overexpression induced aberrant self-renewal in committed myeloid progenitors almost immediately following the establishment of grafts, indicating that it is likely an important regulator of HSC and LSC self-renewal as well. Indeed, mice deficient in *miR-29a/b* exhibit a significant decrease in HSC self-renewal (Hu and Park, unpublished observations). Although this study did not investigate the presence of possible cooperating lesions at the time of leukemic progression, it was the first to show that a single miRNA may function as an oncogene. These studies demonstrating the ability of HSC-associated miRNAs to induce leukemia are consistent with a model in which HSCs and LSCs exhibit a common dependence on particular miRNAs which confer shared properties such as self-renewal, increased quiescence, and impaired apoptosis.

Most recently, *miR-126* was shown to be highly expressed and functionally active in human and mouse HSCs [29]. Lechman et al. showed that overexpression of *miR-126* in lineage negative HSPC-enriched mouse or human CB cells results in decreased cell division, whereas knockdown of *miR-126* (using miRNA "sponges") induces HSC expansion without exhaustion (as assessed through tertiary transplants) or induction of hematological malignancies. *MiR-126*knockdown promoted increased proliferation only in the stem-cell-enriched LSK compartment but not immediate downstream progenitors, suggesting that *miR-126*'s function in HSC expansion is cell context specific. Gene expression profiling of human CD34$^+$CD38-lin-CB transduced to overexpress or knockdown *miR-126*demonstrated enrichment for multiple genes in the PI3K/AKT/GSK3β pathway with *miR-126*knockdown. By regulating signaling through this pathway, *miR-126*appears to set a cell-signaling input threshold for HSC activation and limit the HSC pool size through cell cycle regulation [30]. Given *miR-126*'s functional properties, the authors suggested that manipulation of *miR-126* expression in HSPCs might represent a promising approach to promote ex vivo expansion of human and mouse HSPC cells.

Together, these studies demonstrate major roles of miRNAs in HSC biology and suggest that modulation of miRNA expression levels may represent a promising strategy by which to expand HSCs for regenerative medicine approaches (HSC-related miRNAs are summarized in Table 8.2). In addition, overexpression of some HSC-associated miRNAs such as *miR-29a* and *miR-125* enhances HSC or progenitor self-renewal capacity and is sufficient to induce AML in mice. Thus, these studies suggest that some miRNAs that regulate HSC function may also contribute to myeloid leukemogenesis, perhaps by regulating self-renewal pathways and lineage choice.

Table 8.2 MicroRNAs with characterized roles in normal and/or leukemia stem cells function

miRNA	Function/phenotype	Affected cells	Targets	Reference
miR-125a	Overexpression in mouse bone marrow cells leads to an increase in HSC pool size	HSC	*Bak1*	Guo et al. 2010
miR-125b	Overexpression in mouse bone marrow cells leads to a myeloproliferative disease, progressing to AML Overexpression in CD34+ CB cells increases engraftment in xenotransplants	N/A	N/A	O'Connell et al. 2010
miR-125b	Overexpression in mouse fetal liver cells in a mouse transplantation model results in a myeloproliferative disorder and B-/T-ALL	N/A	*Bak1*, *Bmf*, and *p53*	Bousquet et al. 2010
miR-125b	Overexpression in mouse HSCs leads to expansion of lymphoid-balanced and lymphoid-biased HSCs	HSC	*Klf13* and *Bmf*	Ooi et al. 2010
miR-125b	Overexpression in HSPC-enriched mouse bone marrow cells and CD34+ human bone marrow cells induces myeloid-biased differentiation Overexpression in mouse fetal liver megakaryocyte and erythroid progenitors (MkP and MEP) leads to increased proliferation and self-renewal	MkP/MEP	*Dicer* and *ST18*	Klusmann et al. 2010
miR-29a	Overexpression in mouse HSPC-enriched bone marrow cells causes a myeloproliferative disease that progresses to AML	MPP, CMP, GMP as well as leukemic blasts	*Hbp1*	Han et al. 2010
miR-155	Overexpression in HSC-enriched mouse bone marrow expands granulocytes/monocytes with the eventual development of myeloid neoplasia	N/A	*SHIP1*, *PU.1*, *Picalm*, *Cutl1*, *Csf1r*	O'Connell et al. 2010 and 2009
miR-126	Overexpression in mouse or human CB Lin-bone marrow cells promotes HSC quiescence and decreases the hematopoietic contribution from these cells Knockdown in mouse or human CB Lin-bone marrow cells results in the expansion of HSC pool by increasing proliferation	HSC	Multiple targets in the PI3K/AKT/GSK3 pathway	Lechman et al. 2012

3 The Role of MicroRNAs in Acute Myeloid Leukemia

Given the importance of miRNAs in HSC self-renewal and differentiation, it is perhaps not surprising that miRNAs have also been implicated in the initiation, progression, and/or maintenance of human hematologic malignancies. In fact, the first convincing evidence of a pathogenic role of miRNAs in human cancer was described in chronic lymphocytic leukemia (CLL). The observation that *miR-15a*and *miR-16–1*are located on 13q4, which is deleted in 68% of CLL patients, suggested their role as tumor suppressors [31]. Subsequent studies demonstrated a global downregulation of mature miRNAs encoding tumor suppressors, often accompanied by upregulation of specific subsets of oncogenic miRNAs [32–34]. Spontaneous mouse models of CLL also exhibit loss of these miRNAs, further supporting their major roles as tumor suppressors [35].

A large number of studies have measured miRNA expression in human leukemias including AML, CLL, acute lymphocytic leukemia (ALL), and Hodgkin's and non-Hodgkin's lymphomas [31, 36–39, 41, 42]. Variable findings between different studies may be in part due to heterogeneity in sample selection and varying methods of miRNA expression profiling. Furthermore, in many studies, leukemic cells were not purified prior to the measurement of miRNA expression, increasing the likelihood of contamination of the miRNA signature by non-leukemic components of the BM. Nevertheless, a large number of dysregulated miRNAs were identified in these studies, but the number of functional studies performed to verify their pathogenic roles has been limited to only a few miRNAs (e.g., *miR-125b, miR-155, miR-196b*). Moreover, understanding the role of these miRNAs in the biology of LSCs has been limited by the lack of appropriate and easily applicable in vivo and in vitro assays to study miRNA biology in primary leukemic cells. Nevertheless, for the purposes of this chapter, we will focus on the role of miRNAs in LSCs in AML, since AML is a disorder arising from HSPCs, and the LSC has been functionally defined most stringently in AML. We propose that miRNAs with important functions in normal HSCs may be co-opted by LSCs to promote essential cellular properties such as self-renewal, quiescence, and resistance to exogenous genotoxic insults.

AML is thought to arise from HSPCs [43]. While it is driven by acquired and inherited somatic mutations and epigenetic alterations [44–48], in some cases, the molecular basis of the disease is not clear even though blasts exhibit dramatic changes in gene expression. A large number of studies have evaluated mRNA expression profiles in hematologic malignancies and have shown that dysregulated transcript expression can be utilized to predict disease progression, prognosis, and response to therapy [49, 50]. Similar expression profiling efforts have been performed for miRNAs and will be described below [33, 51].

Lu et al. performed one of the first studies to evaluate miRNA dysregulation in human cancer using a bead-based flow cytometric platform and systematically analyzed the expression of 217 miRNAs from 334 samples that included multiple human solid and liquid tumors [33]. Somewhat surprisingly, miRNA signatures not only accurately separated tumor from non-tumor samples but they were also quite effective at identifying the tissue origin of tumors (even when not apparent morphologically) and appeared to be more effective for these analyses than mRNA profiles incorporating many more transcripts (e.g., 277 miRNAs generated more robust clustering than > 16,000 mRNAs). This highlights one of the potential advantages of using miRNAs as biomarkers in cancer diagnosis. Subsequently, numerous miRNA profiling studies have identified distinct miRNA signatures that correlate with different AML subtypes, and specific miRNAs have been described as predictors of clinical outcome, as summarized in Table 8.3 [34, 37, 52–56]. For example, high levels of *miR-181a* expression correlated with specific AML morphological subtypes [56, 57] and upregulation of *miR-181* was also associated with cytogenetic abnormalities in AML (including t(11q23), isolated trisomy 8, and *FLT3-ITD* mutations). In addition, upregulation of *miR-181a*and *miR-335*was identified in AML patients carrying *C/EBPα* mutations [53, 55]. Interestingly, samples characterized by t(8;21) and inv(16) abnormalities have similar miRNA expression profiles, supporting the long-held notion that they represent core-binding factor (CBF) AMLs with similar pathogenic features [58]. Other distinct biologic subtypes of AML were characterized by unique overexpression of particular miRNAs, such as *miR-127, miR-224, miR-323, miR-154, miR-370,* and *miR-382,* in t(15;17)-associated acute promyelocytic leukemia, and the *miR-17–92* cluster in mixed-lineage leukemia (*MLL*)-rearranged AML [58]. Patients with high expression of *miR-191* and *miR-199a* have been described to have a worse disease-free and overall survival, and a signature composed of [52] 12 miRNAs has been shown to predict event-free survival in cytogenetically normal AML [55]. The latter miRNA signature included five members of the *miR-181* family, overexpression of which was shown to inversely correlate with event-free survival, as well as decreased expression of many predicted targets participating in mechanisms of innate immunity controlled by Toll-like receptors and nucleotide-binding oligomerization domain (NOD)-like receptors. Through these mechanisms, it was proposed that low *miR-181*expression may contribute to disease aggressiveness by activating pathways controlled by Toll-like receptors and interleukin-1β (IL-1β). Although studies functionally validating *miR*-181's role in AML pathogenesis remain lacking, other miRNAs such as *miR-145/miR-146a* have been demonstrated to repress innate immune mediators (e.g., TIRAP/TRAF6), with downregulation of these miRNAs leading to a myelodysplastic syndrome (MDS)-like phenotype that progresses to AML [59]. Additional evidence implicating dysregulated miRNAs in human AML in disease pathogenesis remains largely circumstantial with few exceptions.

Accumulating evidence suggests that the Hox subgroup of homeobox proteins serve a critical role in normal and malignant hematopoiesis. Expression of *Hox* genes is associated with self-renewal in both HSCs and LSCs [60], and overexpression of *HoxA10*in mouse BM cells leads to perturbed myeloid differentiation and progression to AML [61]. Both *HoxA9* overexpression alone and in combination with *Meis1* lead to myeloproliferative disorders that eventually progress to AML [62]. Specific subtypes of AML are associated with high *Hox*gene expression (e.g., *NPM1c*63,64). Upregulation of *miR-10a, -10b,*and *-196a* has been found to be a distinct feature of patients with *NPM1* mutations [34, 53], and, interestingly, these

Table 8.3 miRNA identified by expression profiling in AML subtypes

Sample type	Upregulated miRNAs	Downregulated miRNAs	Reference
t(11q23)ALL versus AML	miR-130, -1, -210, miR-128a/b	miR-223, -125a, -221, miR-222, -23a/b, -26, miR-199, -21, -23, -24, -27 let-7a/b/c/e	Mi et al. 2007
AML versus normal CD34+	N/A	miR-130a, -135, -126, miR-93, -146, -106a, -95, miR-92, -125a, -155, -25, miR-96, -124a, -18, -20, miR-101, -338, -371, -7d, miR-29b, -199b, -301	Garzon et al. 2008a
T (11q23) versus other AMLs	miR-324, -339, -301, -99b miR-328, -326, -219, -194	miR-102, -34b, -15a, -30a miR-29b/c, -372, -196a, miR-331, -229, -193	Garzon et al. 2008a
FLT3-mutated AML versus FLT3 wild-type AML	miR-155, -10a/b	N/A	Garzon et al. 2008a
FLT3-mutated AML versus FLT3 wild-type AML	miR-155, -302, -133a	N/A	Garzon et al. 2008b
NPM1-mutated AML versus NPM1 wild-type AML	miR-10a/b, -100, -21, -9 miR-16a/b, -19a/b, -18a miR-29a/b/c, -16-1, -24 miR-20, -17, -369, -106, miR-16-2, 155, -195, -102 miR-152, -142, -378, -15a miR-374, -98, let-7a-1/2/3, let-7c/d/f/g	miR-22, -192, -128a, -383, miR-373, -324, -373, -127, miR-373*, -139, -193b, miR-145, -498, -135a, -299 miR-493, -326, -429, -204, miR-198, -486	Garzon et al. 2008b
t(8;21), inv(16), t(16;16) AML versus other AMLs	miR-126/126*, -130a	miR-17-5p, -17-3p, -18a, miR-19a/b, -20a, -92, -miR-196b	Li et al. 2008
t(15;17) versus other AMLs	miR-181a/b/c/d, -100, -125b miR-224, -368, -382, -424	miR-126/126*, -422b, -10, miR-150, -124a, -17-5p, miR-20a	Li et al. 2008
t(11q23) versus other AMLs	miR-10a/b, -124a, -196b, miR-20a, -19a/b, -92, -18a, miR-17-5p, -17-3p	miR-126/126*, -130, -224 miR-146a, -181a/b/c/d, miR-382, -368, -424	Li et al. 2008
Inv(16) versus other AMLs	miR-424, -199b, -365, -335 miR-511, -193a	miR-192, -296, -155, -148a miR-218, -135a/b, -196b, miR-10b, -127, -let-7b	Jongen-Lavrencic et al. 2008

Table 8.3 (continued)

Sample type	Upregulated miRNAs	Downregulated miRNAs	Reference
t(8;21) versus other AMLs	miR-126*	miR-19a, -221, -107, -188, miR-342, -338, -20b, -187, miR-501, -339, -502, -210, miR-500, -182, -152, -148a miR-135a, -100, -125b, -99a miR-1, -133a/b, -224, miR-10a/b/, -196a,/b, let-7b, let-7c,miR-9	Jongen-Lavrencic et al. 2008
t(15;17) versus other AMLs	miR-130a/b, -335, -148a, miR-222, -146a, -181a/b/d miR-193a, -450, -213, -199 miR-496, -409-5p, -497 miR-496, -154, -125b, -365 miR-369-5p, -99a, -203, miR-433, -323-494, -100 miR-370, -432, -224, -127 miR-452, -299-5p, -376a, miR-134, -485-5p, -382, miR-379, -193b	miR-196a/b,-151,-10b, let-7c	Jongen-Lavrencic et al. 2008
t(11q23) versus other AMLs	miR-9, -429	miR-213,-146a	Jongen-Lavrencic et al. 2008
NPM1-mutated AML versus other AMLs	miR-10a/b, -135a, -196a/b, miR-152, let-7b	miR-99b, -143, 146a, -323 miR-497, -320, -511, -450 miR-151, -494, -193b, -365 miR-203, -335, -130a, -433 miR-126*, -485-5p, -451 miR-299-5p, 134, -370, miR-379, -432, -224, -382 miR-376a, -424, -127	Jongen-Lavrencic et al. 2008
FLT3-mutated AML versus other AMLs	miR-511, -155, -10b, -135a	miR-30a-3p, -203, -130a, miR-214, -338, -143, -145, miR-182	Jongen-Lavrencic et al. 2008

miRNAs are located in genomic clusters with Hox genes. The correlation between expression of these miRNAs and *Hox* genes in the same AML subtypes suggests a common transcriptional regulatory mechanism. Furthermore, Garzon and colleagues reported downregulation of miR-204 and -128a as additional features of the NPM1-mutant AML miRNA signature and showed that miR-204 directly targets HoxA10 and Meis1 in cell-line experiments [34]. These data suggest that miRNAs may both negatively regulate and be co-expressed with *Hox*genes in *NPM1*-mutated AML.

Several other miRNAs have been shown to be involved in the regulation of AML-related self-renewal pathways. For example, high levels of *miR-21* expression characterize *MLL*-rearranged AML samples [34] and *miR-21* has been shown to target the tumor suppressor *Pten* [65]. *Pten* has been implicated in HSC function, and loss of *Pten* in mice results in the development of a myeloproliferative disorder that progresses to acute leukemia [66, 67]. Thus, *miR-21*overexpression may, in part, promote MLL-rearranged AML by suppressing phosphatase and tensin homolog (PTEN) expression. *MiR-21*has also been recently described to be overexpressed in primary MDS specimens as well as a transforming growth factor beta (TGF-β)-driven transgenic MDS mouse model characterized by anemia and thrombocytopenia [68]. Inhibition of *miR-21*in this model led to improvements in anemia and erythroid colony formation, suggesting that *miR-21* overexpression in AML may promote anemia independent of direct effects on leukemic blast growth. Again, experiments functionally validating the role of *miR-21*in AML remain lacking.

Li and colleagues performed miRNA expression profiling of AML samples harboring CBF abnormalities (e.g., t(8;21) and inv(16)), t(15;17), or *MLL* rearrangements, identifying overexpression of *miR-126/mir126** as a distinguishing feature of CBF AML. Overexpression of *miR-126*in AML cell lines led to decreased apoptosis and increased cell viability, while *miR-126*-transduced mouse normal BM progenitor cells exhibited enhanced proliferation and colony-forming activity, particularly when transduced in combination with the t(8;21) fusion gene (*AML-ETO*). Thus, *miR-126*may cooperate with other genetic lesions to foster leukemogenesis, suggesting that other miRNAs may cooperate with well-described oncogenes in a similar manner [58, 69].

MiR-155 has been found to be upregulated in AML patients with FLT3-ITD mutations [52] as well as myelomonocytic and monocytic subtypes of AML [70]. Overexpression of *miR-155* alone in mature mouse BM cells resulted in a myeloproliferative phenotype characterized by the expansion of granulocytes and monocytes, suggesting that *miR-155* contributes to the physiologic expansions of these cell types seen in inflammatory states [70]. While *miR-155*overexpression alone was not sufficient to induce AML, these data strongly suggest that *miR-155* plays an important function during myeloid leukemogenesis, either by determining lineage fate or by regulating proliferation and/or self-renewal of myeloid-committed progenitors [70].

Together, the existing miRNA expression profiling studies in AML demonstrate that miRNAs have a robust ability to classify AML subtypes, with differentially expressed miRNAs also likely playing a significant role in myeloid leukemogenesis. Unfortunately, few of these miRNAs have been rigorously investigated with respect to their functional significance in AML. Moreover, miRNA profiling has not yet been adopted as an adjunct to the standard molecular workup in AML, most likely because the information gained from such studies has not yet been rigorously shown to provide actionable clinical information independent of established cytogenetic and molecular genetic abnormalities. Thus, validation of prognostic miRNAs or miRNA signatures in independent studies incorporating cytogenetic and molecular genetic data will be required to fully explore the potential of miRNA expression profiling in the prognostication of AML. miRNAs may become clinical utility in the prospective identification of clinical responders to therapy, similar to what has been shown for *DNMT3a* and *MLL* abnormalities in the context of daunorubicin intensification [47]. For example, Blum and colleagues reported that *miR-29b* overexpression was a robust predictor of clinical responses to the hypomethylating agent decitabine in AML [71]. While an independent study was unable to confirm this finding [72], it used a different hypomethylating agent (5-azacytidine) in combination with a histone deacetylase inhibitor and all-*trans* retinoic acid, both potentially confounding factors. Furthermore, in the latter study, *miR-29b*expression was measured in samples taken from the peripheral blood, as opposed to the BM as in the study by Blum and colleagues. This may underscore the importance of cell source (e.g., BM vs. peripheral blood) and purity (e.g., total BM or peripheral blood vs. purified leukemic blasts or LSCs) when attempting to identify clinically relevant miRNAs. With further studies assessing miRNA expression prospectively in a uniformly treated and well-annotated cohort, we anticipate that miRNA expression will become a more prominently featured test in the clinical setting in the future.

4 MicroRNA Regulation of Leukemia Stem Cells

Similar to normal HSCs, LSCs are defined by their capability to self-renew and differentiate, giving rise to both themselves and non-self-renewing leukemic progenitors and blasts. Assessment for LSC function requires the use of serial transplantation models to demonstrate both the tumor-initiating and self-renewal potential of this population [73]. The studies reviewed above strongly suggest that miRNAs serve as critical regulators of molecular networks involved in leukemogenesis. It thus follows that such miRNAs may also confer essential cellular properties of the LSC. While studies comprehensively assessing the role of miRNAs in LSC function remain lacking, in part due to technical challenges with xenotransplantation models and limited primary sample availability, some studies have indirectly assessed the role of miRNAs in LSC function by characterizing their ability to induce increased self-renewal in normal HSPCs or transplantable AML or myeloid disorders in mice.

Popovic and colleagues reported that *miR-196b* is overexpressed specifically in AMLs harboring *MLL* rearrangements [74]. *MLL* leukemias are characterized by overexpression of *Hox* genes that are critical for *MLL*-induced leukemogenesis, including *HoxA9* and *HoxA10*. Binding of *MLL* to specific clusters of CpG islands in *HoxA9* protects these clusters from DNA methylation [75, 76]. *Mir-196b*lies adjacent to these clusters and has been shown to be regulated by *MLL* in the same manner. *MiR-196b* overexpression in c-Kit+ mouse BM cells increases replating potential in methylcellulose and partially blocks differentiation [74], suggesting that *miR-196* may enhance self-renewal in HSPCs. However, *miR-196* has been shown to target both the oncogenic genes *HoxA9*and *Meis1,* as well as the pro-apoptotic gene *Fas* during progression of oncogenic *MLL*-fusion-induced AML, indicating that *miR-196* may be bifunctional, operating as both a tumor suppressor and an oncogene in *MLL*-fusion-induced leukemia [77]. Whether *miR-196*is required for the enhanced self-renewal properties of the LSCs in these leukemias remains an open question, and further studies will be required to fully define the role of *miR-196* in leukemogenesis.

The *miR-17–92* polycistronic cluster is frequently amplified or aberrantly expressed in various human cancers and it has been found to be highly expressed in *MLL*-fusion-associated AMLs [78]. A recent model of *MLL-AF10*-induced AML demonstrated preferential expression of this polycistron in LSC-enriched c-Kit+ cell populations. Overexpression of all *miR-17–92* polycistron members together except for *miR-92b* along with *MLL-AF10*led to a more aggressive leukemia and an increased frequency of "LSCs" as determined by in vitro colony formation. *MiR-17–92*overexpression appeared to increase cell cycle entry and impair differentiation in these "LSCs" via suppression of *CDKN1A* (*p21*), whose deletion phenocopied the *MiR-17–92*overexpression phenotype. Unfortunately, serial transplantation of *miR-17–92/MLL-AF10*leukemias was not performed, so a definitive assessment of functional LSC frequency could not be made in this study [79].

MiR-146 expression has been found to be consistently downregulated across multiple subtypes of AML [52], and *miR-145 and miR-146a*are also notable for their presence on the commonly deleted region of chromosome five in the 5q-subtype of MDS [59]. Coordinate knockdown of *miR-145* and *miR-146a* in mouse HSPCs using "miRNA sponges" followed by transplantation into lethally irradiated recipients leads to an MDS phenotype characterized by peripheral cytopenias and eventual progression to acute leukemia. As discussed previously in this chapter, *miR-145*and *miR-146a* appear to regulate components of the innate immune response such as *TIRAP*and *TRAF6,*respectively, and overexpression of *TRAF6*phenocopies the *miR-145/miR-146a*model in an IL-6-dependent manner. Interestingly, *TRAF6*-overexpressing mice that are $IL6^{-/-}$do not develop cytopenias but still eventually develop AML, suggesting that *TRAF6*overexpression contributes to the transformation of a self-renewing LSC independent of the induction of cytopenias. Thus, it might follow that *miR-146*downregulation in AML may contribute to enhanced LSC function, and it would be interesting to test this hypothesis by rescuing decreased *miR-146*expression in primary AML samples and/or cell lines.

As discussed previously, *miR-125a* and *miR-125b* are highly expressed in normal HSCs as well as in several AML subtypes (*MLL*-fusion associated, t(8;21), and t(15;17)). Overexpression of *miR-125b* (on average, >1,000-fold overexpression) in unfractionated mouse BM followed by transplantation into lethally irradiated recipient mice leads to the development of AML in a dose-dependent manner [21]. Overexpression of *miR-125b* in unfractionated mouse fetal liver cells at somewhat lower levels (~700-fold overexpression) followed by transplantation leads to the development of a B-/T-acute lymphoid leukemia in the majority of animals, with a low frequency of AML observed as well [80]. Overexpression of *miR-125b* in purified HSCs at much lower levels (~35-fold overexpression) leads to an expansion of lymphoid-balanced and lymphoid-biased HSCs, with a concomitant expansion of mature lymphoid cells in mice [27]. While the oncogenic effects of *miR-125b*are thought to be at least partially mediated by direct effects on apoptotic regulators such as *Bak, Bmf,* and *KLF13* [25, 27], more recent studies in myeloid cell lines show that *miR-125* can also block myeloid differentiation in part by targeting *CBFB* and apoptosis through downregulation of multiple genes downstream of *p53. MiR-125b*may also confer a proliferative advantage in part by targeting *ABTB1* [81]. The relevance of each of these *miR-125b*targets in each particular overexpression model remains to be determined.

Together, these data indicate that the molecular pathways dysregulated by *miR-125b* overexpression are sufficient to drive leukemic transformation, but that this leukemogenic effect is highly dependent on cell context and level of expression. Very high levels of overexpression likely drive transformation of progenitor populations, with lineage bias dependent on the cell type utilized and age of the hematopoietic tissue (e.g., the increased myeloid lineage bias of HSPCs with age [82] may account for the predilection for myeloid leukemias to arise from adult BM). However, such high expression levels are less likely to be physiologically relevant, and the ~35-fold overexpression used in the study by Ooi and colleagues may more accurately reflect overexpression levels seen in primary human AML. Whether the enhanced self-renewal and lymphoid lineage bias conferred by *miR-125b*in normal HSCs is also relevant to LSCs remains to be tested. A compelling model might be one in which *miR-125b*overexpression at relatively low levels is sufficient to expand HSCs without the accompanying exhaustion that is seen with overexpression or deletion of strong oncogenes or tumor suppressors such as *K-ras* [83] and *PTEN* [66]. This might allow for the accumulation of "pre-leukemic HSCs" that can accumulate and propagate cooperating leukemogenic lesions through continued self-renewal and expansion. Overexpression or knockdown of *miR-125b* in primary AML xenografts or transgenic mouse leukemia models will be necessary to determine the relative contribution *of miR-125b* to transformation and maintenance of the LSC.

While perturbing the expression of a number of specific miRNAs is sufficient to initiate leukemia, relatively few of the models emanating from this approach have been evaluated rigorously with respect to the biologic events occurring during early disease pathogenesis. Han et al. transduced mouse HSPCs to overexpress *miR-29a* followed by transplantation into lethally irradiated recipients, with a resultant myeloproliferative disorder that progressed to AML. A key finding in this study was the induction of aberrant self-renewal in committed myeloid progenitors (CMP and GMP), as assayed by serial transplantation. Acquisition of aberrant self-renewal in these committed progenitors occurred as early as 4 weeks after transduction and transplantation of HSPCs, suggesting that this is an early event in disease initiation that does not require additional cooperating genetic lesions. Induction of self-renewal in the transformed progenitors was accompanied by increased cell cycle progression, postulated to be due in part to downregulation of the cell cycle regulator (and predicted *miR-29a*target) *HBP1* [28]. While *miR-29a* overexpression induces AML in vivo, overexpression of *miR-29a/b* in both K562 and Kasumi-1 AML cell lines causes decreased proliferation and increased apoptosis [84]. Thus, it is likely

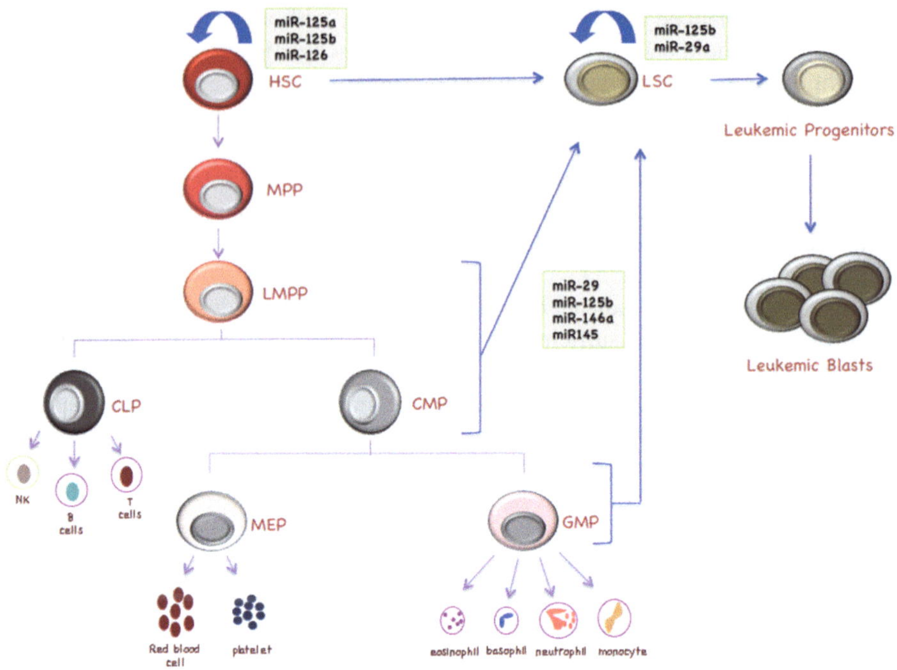

Fig. 8.1 The role of microRNAs in hematopoietic stem cell (HSC) and leukemic stem cell (LSC) function. HSC/progenitor cells (LMPP, CMP, and GMP) are hypothesized to undergo transformation to become leukemic stem cells. Each of the indicated microRNAs has been shown to be involved in the pathogenesis of acute myeloid leukemia in the mouse. See the text for details. *HSC* hematopoietic stem cell, *LSC* leukemic stem cell, *MPP* multipotent progenitor, *LMPP* lymphoid-primed multipotential progenitors, *CMP* common myeloid progenitor, *MEP* megakaryocyte-erythroid progenitor, *GMP* granulocyte-macrophage progenitor, *CLP* common lymphoid progenitor, *NK* natural killer cell

that the biologic effects of *miR-29* vary depending on the stage of leukemogenesis (i.e., *miR-29a*may be necessary for leukemia initiation but not maintenance), the cell population evaluated, and/or differential response to the presence of stromal factors. Differences in phenotypes induced by different *miR-29* family members may also be explained by differing subcellular localization, as *miR-29b* contains a nuclear localization signal and *miR-29a* does not [85]. Nevertheless, *miR-29a* was the first miRNA to be formally shown to confer self-renewal to HSPC populations that do not usually self-renew. This acquisition of aberrant self-renewal represents a critical step in development of the LSC, but other genetic lesions are likely required for full leukemic transformation (not assessed in this study; Fig. 8.1). Future studies are needed to identify genetic lesions that cooperate with *miR-29a*to induce AML as well to assess the ability of other AML-associated miRNAs to confer aberrant self-renewal to HSPCs. Together, such studies promise to provide a more robust understanding of miRNA function in LSCs.

5 Conclusions and Future Directions

Since the discovery of miRNAs in 1993 [10], tremendous progress has been made in identifying miRNAs expressed in normal and malignant hematopoiesis. Based on studies presented here, miRNAs clearly play an important regulatory role in normal HSC function as well as in the transformation of HSCs and/or progenitor cells to LSCs (e.g., *miR-125, miR-29,* and *miR-155*).

Large-scale miRNA expression profiling studies have largely correlated the expression of specific miRNAs to AML subtypes. Although there are several miRNAs that show promise in their ability to provide prognostic information in AML, it remains to be seen whether these will meaningfully add to existing clinical treatment algorithms that already incorporate cytogenetic and molecular genetic data. It is also important to note the significant limitations of existing profiling studies. In many of these studies, leukemic cells were not highly purified from normal hematopoietic cells prior to the measurement of miRNA expression, thereby potentially contaminating leukemic miRNA expression signatures with those of non-leukemic cells. The functional relevance of miRNAs identified in these profiling studies remains largely unclear, and since each miRNA has many different mRNA targets, it will be important to use bioinformatic prediction tools, paired miRNA:mRNA expression data sets from the same samples, and experimental validation to identify truly disease-relevant, functional miRNA:mRNA interactions. Such analyses promise to better identify more powerful disease biomarkers and bona fide therapeutic targets in AML. In addition, identification of highly relevant miRNAs may require integration of miRNA and mRNA profiling studies with various complementary approaches including CHIP-Seq and epigenome studies in different AML subtypes in order to understand how miRNAs may interact with and regulate these other networks within the context of disease.

While there is strong evidence that miRNAs are biologically relevant in primitive hematopoietic cell populations, it is unclear whether they can be efficiently targeted for therapeutic benefit. Recent studies have shown that miRNAs can be effectively targeted in vivo with antagomir therapeutics to treat hepatic tumors [86] or mammary tumors [87] but it remains to be proven that such strategies can penetrate the BM sufficiently to treat leukemia.

Unfortunately, none of the currently published studies has rigorously characterized the specific function(s) of miRNAs in LSCs. Existing studies are largely limited to overexpression and knockdown of miRNAs in HSPCs in serial replating or in vivo transplantation assays, and a truly quantitative assessment of the effect of miRNA dysregulation on LSC function and early disease pathogenesis remains lacking. A more rigorous study of the effects of miRNA dysregulation in existing mouse models is necessary, along with the development of refined techniques to xenograft and perturb miRNA expression in primary human AML specimens. Such experimental progress will help to define the relative contributions of miRNAs in LSC development and potentially identify promising candidates for therapeutic approaches.

References

1. Kondo M, Wagers AJ, Manz MG, et al. Annu Rev immunol. 2003;21:759.
2. Weissman IL. Cell. 2000;100(1):157.
3. Bonnet D, Dick JE. Nat Med. 1997;3(7):730.
4. Lapidot T, Sirard C, Vormoor J, et al. Nature. 1994;367(6464):645.
5. Sauvageau G, Lansdorp PM, Eaves CJ, et al. Proc Natl Acad Sci USA. 1994;91(25):12223.
6. Thorsteinsdottir U, Sauvageau G, Humphries RK. Blood. 1999;94(8):2605.
7. Varnum-Finney B, Brashem-Stein C, Flowers D, et al. Blood. 2000;96(11):64A.
8. Reya T, Duncan AW, Ailles L, et al. Nature. 2003;423(6938):409.
9. Bhardwaj G, Murdoch B, Wu D, et al. Nat Immunol. 2001;2(2):172.
10. Lee RC, Feinbaum RL, Ambros V. Cell. 1993;75(5):843.
11. Wightman B, Ha I, Ruvkun G. Cell. 1993;75(5):855.
12. Filipowicz W, Bhattacharyya SN, Sonenberg N. Nat Rev Genet. 2008;9(2):102.
13. Bartel DP. Cell. 2004;116(2):281.
14. Kim VN. Mol Cells. 2005;19(1):1.
15. Rana TM. Nature reviews. Mol Cell Biol. 2007;8(1):23.
16. Bushati N, Cohen SM. Annu Rev cell Dev Biol. 2007;23:175.
17. Jones-Rhoades MW, Bartel DP, Bartel B. Annu Rev Plant Biol. 2006;57:19.
18. Majeti R, Becker MW, Tian Q, et al. Proc Natl Acad Sci USA. 2009;106(9):3396.
19. Peault B, Weissman IL, Buckle AM, et al. Nouvelle revue francaise d'hematologie. 1993;35(1):91.
20. Georgantas RW, Hildreth R, Morisot S. Proc Natl Acad Sci USA. 2007;104(8):2750 .
21. O'Connell RM, Chaudhuri AA, Rao DS, et al. Proc Natl Acad Sci USA. 2010;107(32):14235.
22. Petriv OI, Kuchenbauer F, Delaney AD, et al. Proc Natl Acad Sci USA. 2010;107(35):15443.
23. Wilson A, Oser GM, Jaworski M, et al. Ann Ny Acad Sci. 2007;1106:64.
24. O'Carroll D, Mecklenbrauker I, Das PP, et al. Genes Dev. 2007;21(16):1999.
25. Guo SQ, Lu J, Schlanger R, et al. Proc Natl Acad Sci USA. 2010;107(32):14229.
26. Esquela-Kerscher A, Slack FJ. Nat Rev Cancer. 2006;6(4):259.
27. Ooi AG, Sahoo D, Adorno M, et al. Proc Natl Acad Sci USA. 2010;107(50):21505.
28. Han YC, Park CY, Bhagat G, et al. J Exp Med. 2010;207(3):475.
29. Gentner B, Visigalli I, Hiramatsu H, et al. Sci Transl Med. 2010;2(58):58ra84.
30. Lechman ER, Gentner B, Galen P van, et al. Cell stem cell. 2012;11(6):799.
31. Calin GA, Dumitru CD, Shimizu M, et al. Proc Natl Acad Sci USA. 2002;99(24):15524.
32. Croce CM, Calin GA. Nat Rev Cancer. 2006;6(11):857.
33. Lu J, Getz G, Miska EA, et al. Nature. 2005;435(7043):834.
34. Garzon R, Garofalo M, Martelli MP, et al. Proc Natl Acad Sci USA. 2008:105(10):3945.
35. Raveche ES, Salerno E, Scaglione BJ, et al. Blood. 2007;109(12):5079. Scaglione BJ, Salerno E, Balan M, et al. Br J Haematol. 2007;139(5):645.
36. Schotte D, Menezes RX de, Moqadam FA, et al. Blood. 2010;116(21):1326.
37. Mi SL, Lu J, Sun M, et al. Proc Natl Acad Sci USA. 2007;104(50):19971.
38. Wang YG, Li ZJ, He CJ, et al. Blood Cell Mol Dis. 2010;44(3):191.
39. Kaddar T, Chien WW, Bertrand Y, et al. Leukemia Res. 2009;33(9):1217.
40. He L, Thomson JM, Hemann MT et al. Nature. 2005;435(7043):828.
41. Eis PS, Tam W, Sun LP, et al. Proc Natl Acad Sci USA. 2005;102(10):3627.
42. Kluiver J, Poppema S, Jong D de, et al. J Pathol. 2005;207(2):243.
43. Passegue E, Jamieson CHM, Ailles LE, et al. Proc Natl Acad Sci USA. 2003;100:11842.
44. Figueroa ME, Abdel-Wahab O, Lu C, et al. Cancer Cell. 2010;18(6):553.
45. Figueroa ME, Lugthart S, Li YS, et al. Cancer Cell. 2010;17(1):13.
46. Figueroa ME, Skrabanek L, Li YS, et al. Blood. 2009;114(16):3448.
47. Patel JP, Gonen M, Figueroa ME, et al. New Engl J Med. 2012;366(12):1079.
48. Abdel-Wahab O, Kilpivaara O, Patel J, et al. Leukemia. 2010;24(9):1656.
49. Valk PJ, Verhaak RG, Beijen MA, et al. N Engl J Med. 2004;350(16):1617.

50. Wilson CS, Davidson GS, Martin SB, et al. Blood. 2006;108(2):685.
51. Calin GA, Liu CG, Sevignani C, et al. Proc Natl Acad Sci USA. 2004;101(32):11755.
52. Garzon R, Volinia S, Liu CG, et al. Blood. 2008;111(6):3183.
53. Jongen-Lavrencic M, Sun SM, Dijkstra MK, et al. Blood. 2008;111(10):5078.
54. Marcucci G, Maharry K, Radmacher MD, et al. J Clin oncol. 2008;26(31):5078.
55. Marcucci G, Radmacher MD, Maharry K, et al. N Engl J Med. 2008;358(18):1919.
56. Dixon-McIver A, East P, Mein CA, et al. PloS one. 2008;3(5):e2141.
57. Debernardi S, Skoulakis S, Molloy G, et al. Leukemia. 2007;21(5):912.
58. Li Z, Lu J, Sun M, et al. Proc Natl Acad Sci USA. 2008;105(40):15535.
59. Starczynowski DT, Kuchenbauer F, Argiropoulos B, et al. Nat Med. 2010;16(1):49.
60. Argiropoulos B, Humphries RK. Oncogene. 2007;26(47):6766.
61. Thorsteinsdottir U, Sauvageau G, Hough MR, et al. Mol cell Biol. 1997;17(1):495.
62. Kroon E, Krosl J, Thorsteinsdottir U, et al. EMBO J. 1998;17(13):3714.
63. Verhaak RG, Goudswaard CS, Putten W van, et al. Blood. 2005;106(12):3747.
64. Alcalay M, Tiacci E, Bergomas R, et al. Blood. 2005;106(3):899.
65. Meng F, Henson R, Wehbe-Janek H, et al. Gastroenterology. 2007;133(2):647.
66. Yilmaz OH, Valdez R, Theisen BK, et al. Nature. 2006;441(7092):475.
67. Zhang JW, Grindley JC, Yin T, et al. Nature. 2006;441(7092):518.
68. Bhagat TD, Zhou L, et al. Blood. 2013;Feb 6.
69. Li Z, Chen J. Methods Mol Biol. 2011;676:185.
70. O'connell RM, Chaudhuri AA, Rao DS, et al. Proc Natl Acad Sci USA. 2009;106(17):7113.
71. Blum W, Garzon R, Klisovic RB, et al. Proc Natl Acad Sci USA. 2010;107(16):7473.
72. Yang H, Fang ZH, Wei Y, et al. Blood. 2010;116(21):713.
73. Clarke MF, Dick JE, Dirks PB, et al. Cancer Res. 2006;66(19):9339.
74. Popovic R, Riesbeck LE, Velu CS, et al. Blood. 2009;113(14):3314.
75. Rozovskaia T, Feinstein E, Mor O, et al. Oncogene. 2001;20(7):874.
76. Erfurth FE, Popovic R, Grembecka J, et al. Proc Natl Acad Sci USA. 2008;105(21):7517.
77. Li ZJ, Huang H, Chen P, et al. Nat Commun. 2012;3.
78. Mi S, Li Z, Chen P, et al. Proc Natl Acad Sci USA. 2010;107(8):3710.
79. Wong P, Iwasaki M, Somervaille TC, et al. Cancer Res. 2010;70(9):3833.
80. Bousquet M, Harris MH, Zhou BY, et al. Proc Natl Acad Sci USA. 2010;107(50):21558.
81. Bousquet M, Nguyen D, Chen C, et al. Haematologica. 2012;97(11):1713.
82. Pang WW, Price EA, Sahoo D, et al. Proc Natl Acad Sci USA. 2011;108(50):20012.
83. Tuveson DA, Braun BS. Proc Natl Acad Sci USA. 2004;101(2):597.
84. Garzon R, Heaphy CE, Havelange V, et al. Blood. 2009;114(26):5331.
85. Hwang HW, Wentzel EA, Mendell JT. Science. 2007;315(5808):97.
86. Rajewsky N, Krutzfeldt J, et al. Nature. 2005;438(7068):685.
87. Ma L, Reinhardt F, Pan E, et al. Nat Biotechnol. 2010;28(4):341.

Chapter 9
MicroRNAs and Chronic Lymphocytic Leukemia

Heba Degheidy, Siddha Kasar, Chingiz Underbayev, Yao Yuan, Smruti Mehta, Marilyn Lightfoote, Gerald Marti and Elizabeth Raveche

Abstract MicroRNA abnormalities are involved in the development and progression of chronic lymphocytic leukemia (CLL). In the majority of CLL malignant cluster of differentiation CD5+ (B-1) clones, the expression of miR-15a/16 is decreased relative to polyclonal CD5− (B-2) cells. Levels of the anti-apoptotic protein, bcl-2, correlate with the deletion of the Dleu2 region on chromosome 13q14 containing the *mir-15a/16-1* loci. The expansion of CD5+ B cells and development of B-1 clones usually precedes CLL disease and is referred to as monoclonal B lymphocytosis (MBL). An early event in MBL may decrease miR-15a/16 expression. In the New Zealand black mouse model of CLL, there is a mutation in the *mir-15a/16-1* loci resulting in decreased expression of miR-15a/16 due to a processing defect. Since this is a germline mutation, induced pluripotent stem cells and hematopoietic stem cells were employed to determine the role of decreased miR-15a/16 on B-1 lineage development. Enforced overexpression of miR-15a/16 leads to almost exclusive B-2 development. These results suggest that decreased miR-15a/16 is critical for the development of B-1 cells and the progression to malignant B-1 clonal expansion.

Keywords Chronic lymphocytic leukemia · miR-15a/16 · Monoclonal B lymphocytosis · Murine models of CLL · NZB mouse model of CLL · Dleu2 mutations/deletions · Induced pluripotent stem cells · B-1 lineage development

---

Support: This work was supported by NCI R01CA129826 (ER) and an NSF SIR award (ER)

E. Raveche (✉) · S. Kasar · C. Underbayev · Y. Yuan · S. Mehta
Pathology and Laboratory Medicine, New Jersey Medical School,
MSB C512, 185 S Orange Avenue, Newark, NJ 07103, USA
e-mail: raveches@njms.rutgers.edu

H. Degheidy · M. Lightfoote · G. Marti
OSEL, CDRH, FDA, White Oak, MD, USA

S. R. Singh, P. Rameshwar (eds.), *MicroRNA in Development and in the Progression of Cancer*, DOI 10.1007/978-1-4899-8065-6_9,

1 CLL and MicroRNAs

RNAs, in addition to encoding messenger RNA (mRNA) that can be translated into proteins, also serve as regulatory elements. MicroRNAs (miRNAs) are small (20–22 nt), evolutionarily conserved, noncoding single-stranded RNAs discovered in the 1990s [53, 103], mainly functioning to target the 3′ untranslated region (UTR) of mRNAs in antisense sequence-specific way and regulate genes posttranscriptionally for degradation or translation suppression [31]. The majority of miRNAs are first transcribed by RNA polymerase II in the nucleus as large primary transcript (pri-miRNA) [47] and further processed into ~70 nt pre-miRNA with a hairpin structure by Drosha, an RNase-III type endonuclease (RN3) in the nucleus. In the cytoplasm, ~20 bp miRNA/miRNA* duplex are generated by Dicer and one of the miRNA strands is incorporated into the RNA-induced silencing complex (RISC) [5, 23, 57, 64, 80]. The RISC is composed of Argonaute (Ago) proteins and GW182 proteins which localize in the cytoplasmic foci called processing bodies (P bodies or GW bodies), wherein mRNA may be sequestered from being translated [18, 25, 26, 72]. miRNAs are critical regulators of cell programming, proliferation, activation, and apoptosis, which are all critical for cancer development. Abnormal miRNA expression has been found in chronic lymphocytic leukemia (CLL) (reviewed in [69]).

In CLL, there is dysregulation of several miRNAs including decreased miR-15a/16, miR-34 cluster, miR-181a/b, and miR-125b [7–9, 89, 99, 106]. The CLL malignant B-1 clones have chromosomal alterations and the regions involved contain miRNAs. The 11q23 deletions contain the miR-34b/c cluster and the 13q14 deletions contain the *mir-15a/16-1* region [20, 89]. The resultant elevated miRNA target gene expression usually involves proliferation and anti-apoptotic pathways. Decreased miR-181, which normally targets Tcl1, results in highly elevated Tcl1, which is found in aggressive CLL [74]. Decreased miR-34 expression, which normally targets Zap70, would result in increased Zap70 found in aggressive CLL [27]. The miRNA profile of CLL cells is similar to the microRNA levels observed in activated B cells [54]. The miR-29 expression is decreased in aggressive CLL, while it is increased in indolent CLL as compared to normal volunteers [74, 92]. Thus, the same miRNA can function as both an oncogene and a tumor suppressor in CLL. miR-155, miR-150, and miR-21 expression is increased in B-chronic lymphocytic leukemia (B-CLL) cells as compared to normal B cells [9, 32]. miR-155 overexpression has been associated with CLL [17] and been shown to target DNA repair mechanisms, particularly mismatch repair [100, 101]. While miRNA abnormalities are present in CLL, this chapter focuses on decreased expression of miR-15a/16 as a critical initial event in CLL development. Decreased miR-15a/16 and deletions involving this region are found in other malignancies in addition to CLL [28, 95], suggesting that the loss of *mir-15a/16-1* loci, which normally acts as a tumor suppressor, is a critical event in the development of malignancies.

CLL is characterized by the accumulation of malignant cluster of differentiation CD5+ B cells (B-1; $CD5^{+}CD19^{+}CD20^{dull}CD23^{+}IgM^{dull}$) in peripheral lymphoid organs, bone marrow (BM), and peripheral blood [40]. It accounts for 30 % of all leukemia in the western world, making it the most common lymphoid malignancy.

This is an age-associated leukemia with a peak onset in the sixth decade. CLL is preceded by a precursor state, monoclonal B lymphocytosis, MBL [50, 51, 61, 83, 86]. CLL is broadly classified into aggressive ($Zap70^{hi}$ unmutated IgH) and indolent ($Zap70^{low}$ mutated IgH) [37]. CLL cells have genomic instability, chromosomal alterations, and several characteristic genetic abnormalities. These include trisomy 12 and deletions of 11q, 13q14, and 17p [19]. The deletions on the 13q14 region of patients with CLL are heterogeneous and studies have indicated that the minimal deleted region (MDR) encompasses the *DLEU2* gene, which contains the *mir-15a/16-1* region within an intronic region of the host gene, *DLEU2* [14, 67, 87]. Recent reports have found rare cases of 13q14 deletions, which did not involve the DLEU2 region containing the *mir-15a/16-1* loci but only the DLEU1 [24]. In some CLL patients, a larger region of chromosome 13q is involved and a common deleted region (CDR) has been identified that includes *DLUE1, DLEU7,*and *RNASEH2B* with only decreased *DLEU7* transcript associated with CLL, and considered to be an additional tumor suppressor gene in addition to miR-15a/16 [41]. Studies have indicated that in the precursor states of CLL and MBL, one of the consistent features is decreased miR-15a/16 and in the majority of CLL patients, levels of miR-15a/16 are decreased [29]. Decreased levels of miR-15a/16 are critical for the development of the B-1 clone, which eventually can progress to CLL.

Decreased miR-15a/16 confers a growth advantage as these miRNAs target key cell cycle regulatory and anti-apoptotic proteins such as cyclin D1 and Bcl2 [6, 13]. In this chapter, the role of decreased miR-15a/16 in the commitment to become B-1 lineage cells (the cells that give rise to CLL), the development of the precursor state MBL, and the progression to the malignant state, CLL, will be the focus. While much more information about the expression of these miRNAs in isolated subpopulations of cells is needed, insights can be drawn from the literature and preliminary studies in a de novo mouse model of CLL, the New Zealand black (NZB) mouse model [63, 78, 81, 82, 91, 93].

1.1 The Precursor State: Monoclonal B Lymphocytosis

There is a precursor state of CLL, referred to as monoclonal B lymphocytosis (MBL) [33, 34, 60, 61, 68, 96]. This is defined by lymphocytosis with a count of less than 5×10^9/L and no clinical features of CLL [60]. MBL with B cell expansions of cells with the surface markers consistent with CLL (CLL-like-MBL) is the major focus of this chapter. In addition to the MBL cases identified with an increase in absolute number of lymphocytes, there are additional types of MBL including low-count MBL in which monoclonal CLL-like lymphocytes are found in an environment of polyclonal B lymphocytes [85], and a typical MBL which can be CD20bright. Interestingly, MBL has been found to increase with age and be more prevalent in males than in females (both features of CLL). MBL is also increased in relatives of patients with CLL [33, 62]. Since only a small percentage of MBL progresses to CLL (1–5 % per year) [33, 84, 88], identification of individuals with clones that will progress to CLL would give insight into the mechanism of disease development.

Some MBLs may represent the expansion of several different clones [49], and the development of CLL may start from an oligoclonal state with one clone progressing and eventually becoming the dominant clone [15]. One of the questions is, what determines the dominant clone and what is the mechanism that drives the B-1 clone to expand to become the CLL state?

In terms of the role of decreased miR-15a/16 in the development of MBL, individuals with MBL have a high rate of 13q14 deletions (resulting in decreased miR-15a/16). This is true for CLL-like MBL as well as low-count MBL. Since MBL is the precursor state of CLL, the findings of 13q14 deletions in MBL support that decreased levels of miR-15a/16 may be critical for B-1 clonal development. While 13q14 deletions are consistently found in MBL, not all the cytogenetic abnormalities found in CLL patients are found in MBL cases, and 17p and trisomy 12 are infrequent relative to the occurrence of these abnormalities in CLL [29].

Insights from this precursor state of CLL suggest that the development of long-lived clonal populations of CLL cells is a "natural" occurrence and may depend upon reduced miR-15a/16 to support this clonal expansion. A recent article found that in non-cancer patients over the age of 50, there was an increase in cells with a deletion in the 13q14 region often resulting in deleted *mir-15a/16-1* [52]. Other events (additional mutations, etc.) that eventually lead to CLL development may occur and be required, but what emerges from these studies of MBL is that decreased miR-15a/16 may be a critical first step toward CLL progression. Since a major target of miR-15a/16 is the anti-apoptotic gene bcl-2 transcript, increased bcl-2 is associated with 13q14 (containing the *mir-15a/16-1* loci) deletions in humans. The increased bcl-2 in the MBL clone is most consistent with decreased expression of miR-15a/16 in the MBL clone relative to the miR-15a/16 expression in the polyclonal B cells. While decreased miR-15a/16 expression as the sole abnormality may lead to indolent CLL, this most likely is also responsible for the existence of the precursor state, MBL. These B-1 clones in MBL with decreased miR-15a/16 most likely would eventually develop into CLL, given enough time to accumulate due to the failure to undergo apoptosis or if additional alterations are acquired.

1.2 CLL and miR-15a/16 Levels

In CLL, 70% of patients have at least monoallelic 13q14 deletions (del13qX1) and 19% have biallelic 13q14 deletions (del13qX2) (DeWald 2003). Nevertheless, as CLL progresses, the increase in the percentage of cells with 13q14 deletions as well as an increase in biallelic deletions in the 13q14 region occur. Malignant B-1 cells from CLL patients with biallelic 13q14 deletions (del13qX2) have significantly faster growth kinetics than CLL patients with monoallelic 13q14 deletions (del13qX1) (Pfeifer D 2007). In CLL patients, the miR15a and miR16 levels are lower in patients with monoallelic deletions of 13q14 (del13qX1) when compared to patients with no 13q14 deletions (Smonskey 2012). In the CLL patient shown in Fig. 9.1, peripheral blood mononucleated cell (PBMC) obtained from two time points were subjected to fluorescence in situ hybridization (FISH) analysis for

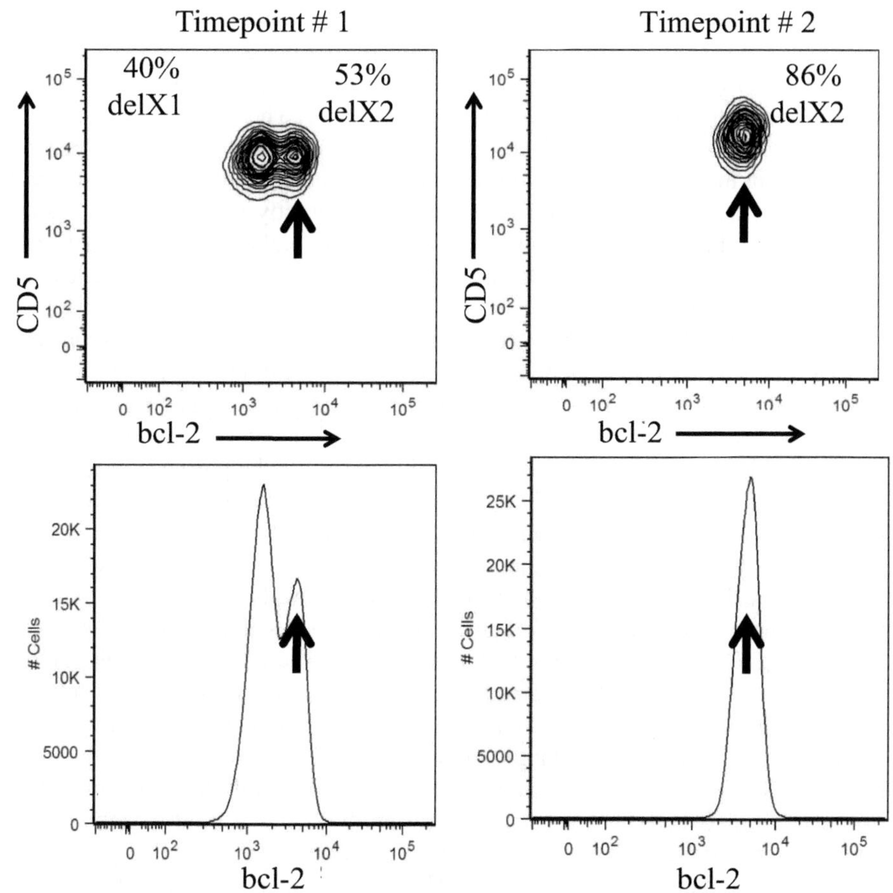

Fig. 9.1 Progressive loss of 13q14 (*mir-15a/16-1*loci) found in CLL: Decreased miR-15a and miR-16 favors CLL progression. CLL patient peripheral blood cells were stained with CD5, CD20, CD19, CD3, and intracellular bcl-2. B clonal cells were obtained by gating on CD20dimCD5+ cells. Bcl-2 was evaluated on the gated clonal cells. *Upper* two color contour plots show bcl-2 versus CD5 obtained from the same patient at two different time points. *Lower* single histograms show bimodal and unimodal bcl-2 expression at the two time points. At time point #1, the 13q14 status was 40% 13q14+/− (del13qX1), 53% 13q14−/− (del13qX2). At the later time point #2, 86% of the cells had deleted both alleles (del13qX2) and were 13q14−/−. Decreased miR-15a/16 levels correlated with increased bcl-2 levels. At time point #1, there are two populations of B-1 cells based on bcl-2 levels, the population that represents cells with one allele remaining of the 13q14 loci had lower levels of bcl-2 than the second population of malignant B-1 cells with higher levels of bcl-2 associated with the loss of both alleles of the 13q14 loci (indicated by *arrow*). With disease progression, the majority of the CLL cells have decreased miR-15a/16 as a result of biallelic deletions (del13qX2) resulting in only the bright bcl-2 malignant cell population remaining

13q14 status. At the later time point, the percentage of malignant B-1 cells that were homozygous for the 13q14 deletion (del13qX2) increased. This indicates that the malignant clonal progressed from monoallelic deletion to biallelic deletion.

Previous studies have also shown progression from monoallelic to biallelic deletions [12]. The B cells were analyzed for intracellular bcl-2, an important target of miR-15a/16 (Fig. 9.1). The homozygous 13q14–/– (del13qX2) cells can be detected as an increase in cells expressing the target gene, bcl-2. The biallelic deletion of the 13q14 region results in further decreased expression of miR-15a/16 that targets bcl-2 resulting in increased bcl-2 protein levels. While these results are consistent with the previous findings that miR-15a/16 levels are directly related to 13q14 status (and decrease during progression from monoallelic to biallelic deletions of 13q14), other epigenetic factors serve to further decrease the levels of miR-15a/16 levels in the monoallelic state [66].

Clonal heterogeneity has been found in CLL. This may represent the presence of two independent clones or more commonly represent subclones derived from an original clone. These subclones differ in the expression of surface markers as well as the level of miR-15a/16 expression and 13q14 status. As shown in Fig. 9.2, peripheral blood analysis of a CLL patient demonstrated the presence of three subclones, which differed in light chain expression and CD20 expression. The kappa-expressing clone had lower levels of bcl-2 expression. This is consistent with the miR-15a/16 status, since in the kappa expression subclone, the majority of the malignant B-1 cells had only deleted one of the 13q14 loci. The mechanism that leads to interclonal variability in CLL is unclear. Recent studies employing IgH sequence analysis suggest that most of the subclones in CLL are derived from an initial clone. Many of these subclones involve single IgH mutations that may lead to a subclone without diminished capacity to expand relative to the initial clone, or vice versa.

Based on the limited immunoglobulin variable (IgVH) repertoire of both CLL patients and the NZB mouse model of CLL, it is most likely that CLL precursors originate from B cells that produce natural antibodies that bind to antigens displayed on apoptotic cells and self-like microbial peptides [21]. The presence of subsets of CLL patients based on the IgVH mutational status suggests that different maturational stages can be converted to CLL cells. Recent studies have suggested that the production of activation-induced deaminase (AID), which induces DNA double-strand breaks and is responsible for IgVH mutations, is found in actively dividing CLL cells, particularly in the unmutated CLL subset. However, the AID in this subset may be inhibited from leading to the accumulation of IgVH mutations. Nevertheless, the expression of AID may be responsible for clonal evolution as well as the accumulation of additional genomic alterations due to the off-target effects of AID [73].

1.3 NZB As a Murine Model of CLL

There are several murine models of CLL (reviewed in [93]). Several of these models which do not directly involve decreased miR-15a/16 result in increased expression of targets of miR-15a/16. The tumor necrosis factor receptor-associated factor (TRAF) DN/bcl-2 model [76, 104] which has a clonal B-1 expansion and

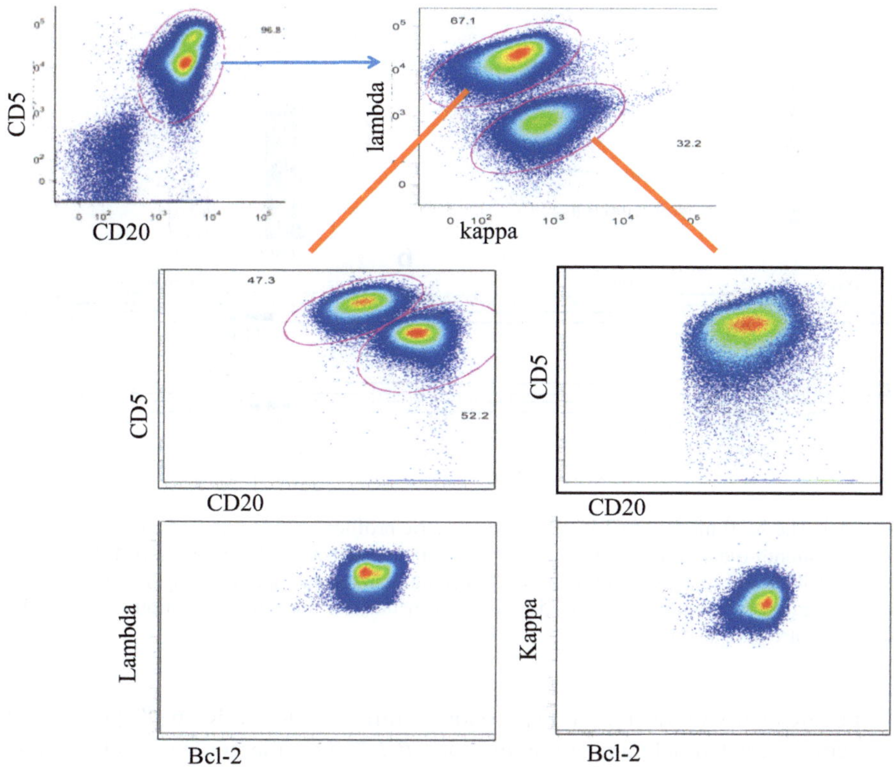

Fig. 9.2 Clonal heterogeneity in CLL patient. PBMC cells from a single CLL patient were stained with CD45, CD3, CD5, CD20, kappa, lambda, and intracellular bcl-2. CD45+ cells were gated on CD3− and further gated into CD5+, CD20+ (B-1 cells). This population was found to contain two subclones distinguished on the basis of kappa and lambda expression. The lambda expression cells could be further distinguished into two subpopulations based upon CD20 expression. The lambda-expressing cells were sorted and analyzed for 13q14 status and found to be mainly (69 %) 13q14−/− (del13qX2), while the kappa-expressing population was mainly 13q14+/− (del13qX1). This correlated with the bcl-2 levels along with the lambda-expressing malignant B-1 subclone having higher bcl-2 levels than the kappa-expressing malignant B-1 subclone

splenomegaly involves increased expression of the anti-apoptotic gene, bcl-2, which normally is targeted by miR-15a/16, suggesting that increasing the miR-15a/16 target bypasses the need to have decreased miR-15a/16 expression. Focusing on mouse models with decreased miR-15a/16, there are several models linked to the murine chromosome 14, which has synteny with human 13q14. We have found that the NZB mice spontaneously develop B-1 malignancies as they age and are a de novo murine model of CLL [78, 79, 82] (Fig. 9.3). The NZB mice initially have an oligoclonal expansion of the B-1 population and, in the majority of year-old NZB mice, an eventual single clone dominates. This has similarities to a subpopulation of human MBL cases in which more than one B-1 clone is present. In addition, we have linked the development of CLL in NZB mice to a mutation in the *mir-15a/16-1*

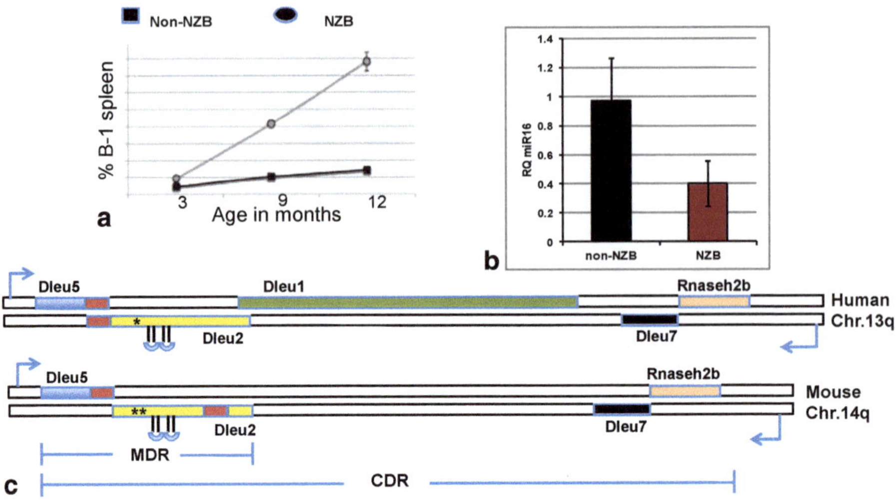

Fig. 9.3 The NZB model of CLL. **a** Flow cytometric profiles of B-1 cells (indicated by *red box*) from 11-month-old control strain and NZB mice spleen stained for CD5 and IgM. **b** Percentage of B-1 cells in the spleens of NZB and C57 mice at the indicated ages of study. **c** Scheme of the genetic regions containing the *mir-15a/16-1* loci in humans and mice. Locations of identified mutations are indicated by *asterisk*

loci associated with decreased expression of miR-15a and miR-16 [82] (Fig. 9.3). Moreover targeted deletion of the *mir-15a/16-1* loci or a larger surrounding MDR led to the development of CLL in mice, further confirming the tumor suppressor function of this locus [48].

2 MicroRNAs in Hematopoietic Stem Cells in CLL and the NZB Mouse Model

Multiple evidences suggest that miRNAs play a significant role in the posttranscriptional genetic regulation in stem and progenitor cells. miR profiling is critical in order to distinguish stem cells of different origins, developmental stages, and genetic conditions [3]. Recent studies have demonstrated a causative role for miRNAs in malignant disease development in the hematopoietic system. For instance, overexpression of miR-155 or miR-29a in the mouse hematopoietic system leads to a myeloproliferative disorder [71] or leukemia [42], respectively. In another study on human umbilical cord blood, two particular miRs, miR-520h and miR-526b* levels appeared to be elevated. Interestingly, adenosine triphosphate-binding cassette subfamily G member 2 (ABCG2), an important factor of stem cells' maintenance, is a known target of hsa-miR-520h [55].

In CLL, insight into the possibility that the hematopoietic stem cells (HSCs) may already be abnormal has come from studies of xenogenic transplants in which

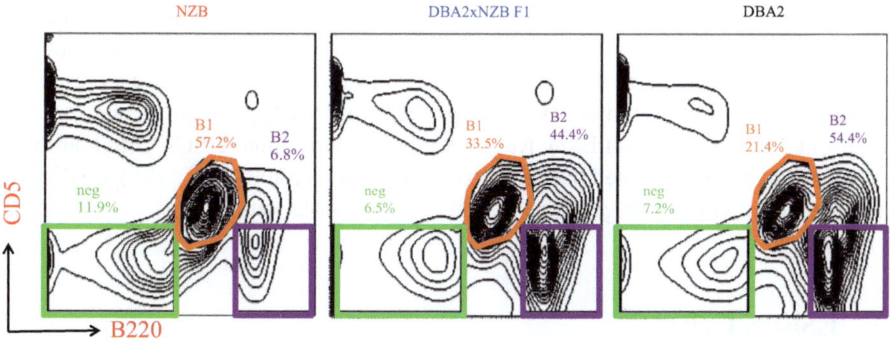

Fig. 9.4 Loss of repopulation of B-2 and increase in B-1 repopulation in NOD/SCID injected with BM HSCs from NZB mice which have germline mutation in *mir-15a/16-1* loci. HSCs (lin-, CD105+, c-kit+, Sca-1+) were obtained from the BM of three different strains: the inbred NZB, the inbred DBA/2, and the F1 intercross (NZB × DBA/2) F1. NOD/SCID recipients were sublethally irradiated (450 rads) followed by an injection of 2×10^5 HSCs (purified by flow cytometric sorting). Post-repopulation analysis (day 46) flow cytometric data are shown for the peritoneal washout cells (PWCs)

CLL HSCs are transplanted into nonobese diabetic/severe combined immunodeficiency (NOD/SCID) recipients [46]. These resultant chimeras demonstrate that the HSCs (not the malignant B-1 cells) from CLL patients have the ability to give rise to monoclonal and oligoclonal B-1 cells. Analysis of the IgH has indicated that these B-1 clones that develop in the NOD/SCIDs, while clonal, are not the same as the malignant clone of B-1 cells present in the original patient. This suggests that clonal development of B-1 cells is abnormal in CLL cells. The question remains if decreased miR-15a/16 levels are a prerequisite for the aberrant HSC propensity to give rise to B-1 clonal development of B-1 cells. Previous data have suggested that there is gradual epigenetic silencing of the region containing the *mir-15a/16-1* loci which may supersede the deletions of 13q14 region [66]. Thus, the HSCs from CLL patients may already have abnormal regulation of miR15a/16. While the HSC cell does not express miR-15a/16, commitment to the B lineage cells does require expression of miR-15a/16. The level of expression may determine the type of B cell that develops.

Our studies employing the NZB mouse model of CLL have shed insight into the ability of HSCs to give rise to B-1 cells. Purified HSC cells (lin-, CD105+) were obtained by flow cytometric sorting and injected into sublethally irradiated NOD/SCID recipients. The NZB stem cells gave rise to more B-1 lineage cells than did the non-NZB donors (Fig. 9.4). This was in part related to the levels of miR-15a/16 since the (NZB × DBA/2 ((DBA/2))) F1 donor (which has intermediate levels of miR-15a/16). Given that commitment of the multipotential progenitor cell to the B lineage is a stochastic process rather than a deterministic process [70], expression of moderate levels of miR-15a/16 may be critical for commitment to the B-1 lineage and higher levels of miR-15a/16 may lead to B-2 commitment. Recent studies have demonstrated that an important target of miR-15a/16 is RNF4 [43], which has been shown to drive HSC differentiation away from the erythroid and myeloid lineages. These data support the concept that commitment to the B-1 lineage is abnormal as early as the lineage commitment of the HSC differentiation and that decreased but not absent miR-15a/16 may have a stronger effect on the type of B-lineage commitment (i.e., B-1 higher than B-2). Recent studies of individual HSCs from adult murine BM have demonstrated that HSCs that give rise to the B-1a (CD5+) are distinct and rarer than HSCs that give rise to other B subpopulations [35].

3 ES/iPS Cells

To further determine if abnormalities that give rise to CLL are present in very early B lineage precursor cells or even in HSCs, we evaluated the pluripotent stem cell. In 2006, direct reprogramming of fibroblasts to pluripotent stem cells was accomplished by introducing four transcription factors, *Oct4, Sox3, Klf4,*and *cMyc,*that are important for self-renewal of embryonic stem cells (ESs) and have been shown to reprogram both mouse and human somatic cells into ES-like pluripotent cells referred to induced pluripotent stem cells (iPSs) [97]. With iPS technology, genetic alterations can be introduced to explore disease mechanisms and points of intervention both in vitro and in vivo. miRNAs are involved in the regulation of many biological processes, including the stem cells' self-renewal and pluripotency [65, 98]. It is known that the expression of ES-specific miRNAs, such as miR-294, promotes iPS cell induction from somatic cells [59]. Recently, it has even been demonstrated that expression of the miR-302/367 cluster can directly reprogram somatic cells to a pluripotent stem-cell state in the absence of the commonly used reprogramming factors [2]. Alternatively, specific miRNAs need to be expressed in order for ES/iPS differentiation. MiR-145 has been shown to induce ES cell differentiation by inhibiting the expression of Sox2, Oct4, Klf4, and c-Myc, key reprogramming factors, and led to an increase of HSC number in vivo by more than eightfold [4, 38, 90].

We have generated iPSs from NZB spleen stromal fibroblasts by means of lentiviral delivery of three factors Oct4, Sox2, and Klf4 in a single polycistronic vector. After 4 weeks in culture medium supplemented with small molecules (glycogen synthase kinase (GSK)-3b and mitogen activated protein kinase kinase (MEK) inhibitors), we were able to see ES-like colonies which stained positive for alkaline phosphatase and stage-specific embryonic antigen (SSEA)-1 surface antigen. Colonies were picked and expanded for further analysis. Reverse transcription polymerase chain reaction (RT-PCR) assay showed the expression of basic endogenous pluripotency genes such as Oct4, Sox2, Klf4, cMyc, and Nanog. The pluripotent capacity of NZB iPS cells was further successfully confirmed by the teratoma formation assay in NOD/SCID recipient mice. These iPS cells were then efficiently transduced with lentiviral vector expressing wild-type miR-15a/16 gene (Fig. 9.5). Increased expression of miR-15a/16 in NZB iPS cells did not alter the expression of pluripotency markers Nanog and Oct4. Control (wild-type/C57Bl/6) and NZB-derived iPS cells; both NZB iPS and NZB iPS transduced with miR-15a/16, were allowed to form spheroid embryoid bodies (EBs; Fig. 9.5). The NZB iPS cells

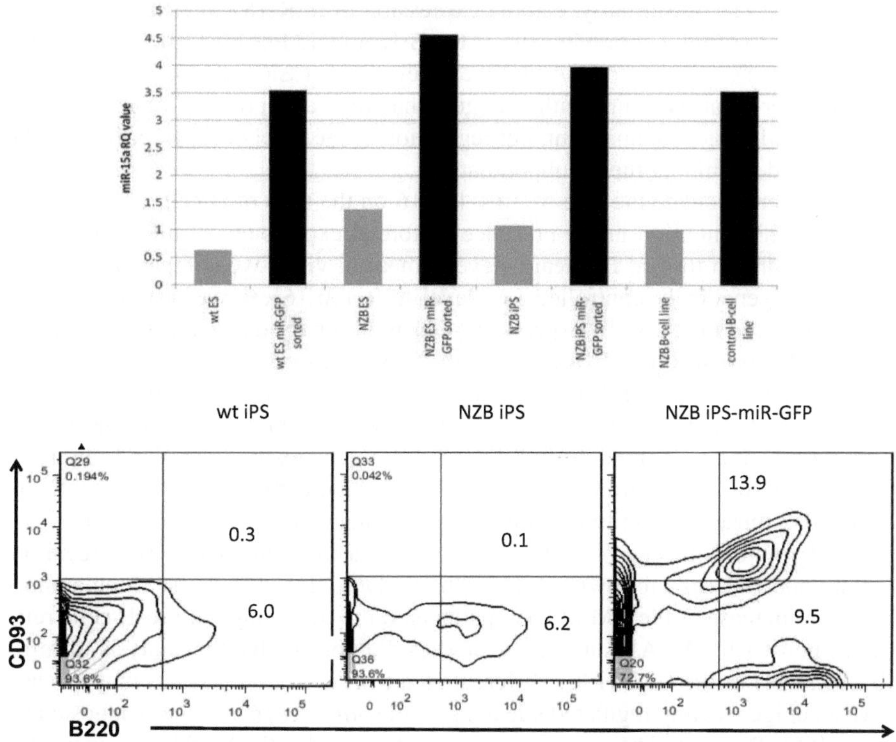

Fig. 9.5 Analysis of B lineage markers on EBs in vitro. The iPS cells were generated as previously described using lentivirus to deliver Oct-4, Sox-2, and c-kit. The iPS cells were maintained for several weeks in culture. A subclone of iPS cells was derived from the NZB by transfecting the iPS with a lentivirus containing green fluorescent protein (GFP) and miR-15a/16.To obtain a stably expressing NZB iPS miR-15a/16, GFP-positive cells were selected by flow cytometric sorting and maintained in culture. The miR-15a/16 levels (TaqMan RT-PCR relative expression) were measured in extracted RNA from the iPS cells or ESs (*upper panel*) and compared to the miR-15a/16 levels in an NZB-derived malignant B-1 cell line [75] or a control (non-NZB-derived B-cell line, A20). EBs were formed from iPS derived from either C57Bl6 tail-tip fibroblasts (WT-iPS) or NZB tail-tip fibroblasts and analyzed for surface marker expression (*lower panel*). Only NZB iPS cells which constitutively express high levels of miR-15a/16 (not regulated by differentiation factors) have an unusual pro-B cell population (Ueda Y 2007), which is CD93+, B220dull, and IgM–

formed EB bodies not yet committed to a lineage. The EB bodies with constitutive overexpression of miR-15a/16 had a population of cells already committed to the B-lineage. The NZB iPS with enforced miR-15a/16 expression formed embryoid bodies with a subpopulation which expressed elevated expression of the B-lineage marker, B220 (B-cell isoform of CD45), suggesting that high early expression of miR-15a/16 led to B-lineage commitment. This is consistent with the findings of others that miRNAs play an important role in the regulation of B cell development [30]. However, the timing of the expression of miRNAs is critical in determining the fate of HSC. For instance, miR-150 is sharply elevated in immature B cells but enforced overexpression of miR-150 in HSCs leads to impairment of B-lineage development [105]. Similarly, ectopic expression of miR-181 in HSCs resulted in more than a twofold increase in cells of the B-lymphoid lineage [10]. These results suggest that the type of miRNA and the level of expression play a critical role in B cell differentiation. Our results suggest that forced expression of miR-15a/16 increases B-lineage commitment, but regulation of the level of miR-15a/16 may be critical for determining the B subpopulation.

To further elucidate the role of miR-15a/16 on the type of B cell, which results from iPS differentiation in the presence of forced expression of miR-15a/16, iPS cells were driven toward B lineage development in vitro. We compared NZB iPS cells (with genetically controlled low-level miR-15a/16 expression) versus NZB iPS with forced miR-15a/16 expression. Our preliminary data suggest that exogenously delivered miR-15a/16 affects B cell differentiation resulting in expression of higher levels of B220 (CD45R) and low to negative levels of CD5 surface marker, suggesting a loss of B-1 lineage cells in the presence of high levels of miR-15a/16 (Fig. 9.6). In the NZB model of CLL, there is a genetically determined low level of miR-15a/16, which strongly supports B-1 lineage commitment and negatively affects B-2 lineage commitment (Fig. 9.6). In human CLL, BM-derived HSCs displayed lymphoid lineage priming and produced a high number of B progenitors in vitro and B-1 cells following transfer to NOD/SCIDs [46]. These same authors found an increase in the BM of CLL patients in the percentage of pro-B cells relative to non-CLL BM. Although these studies did not involve the analysis of miR-15a/16 levels as the CLL HSCs differentiated to B cells, these HSCs may already have an epigenetically regulated silencing of the *mir-15a/16-1* loci as suggested by others [66]. This would be consistent with our finding that the NZB has an increased differentiation into B-1 cells and forced increased expression of miR-15a/16 drives the B differentiation toward the B-2 lineage and not the B-1.

4 Side Population Cells in CLL

In many cancers, the presence of a cancer progenitor or cancer stem cell has been identified [11, 16, 22]. This is a critical finding since these cells are most often chemoresistant and, despite successive depletion of the cancer, the cancer stem cells may eventually give rise to new cancer cells. The search for a cancer stem cell in CLL has not elucidated this population, but studies of xenogenic transplantation of HSCs from CLL patients have suggested that even at the HSC level the propensity to develop B-1 clones [46]. Another population, referred to as side population (SP), has been shown to contain immature and undifferentiated cells, many of which have stem-cell markers [45]. In CLL, there is an increase in this stem-cell-like population, the SP [36, 39, 44], which is identified by flow cytometry due to their ability to extrude the dye, Hoechst 33342. Verification of SP phenotype is accomplished by the loss of this population when verapamil is added [94]. The SP cells in the BM are enriched in HSC population and, for some solid tumors, cancer stem cells are enriched in the SP cells [56]. SP cells have the ability to extrude Hoechst dye and chemotherapeutic agents because of the elevated expression of the ABC transporter

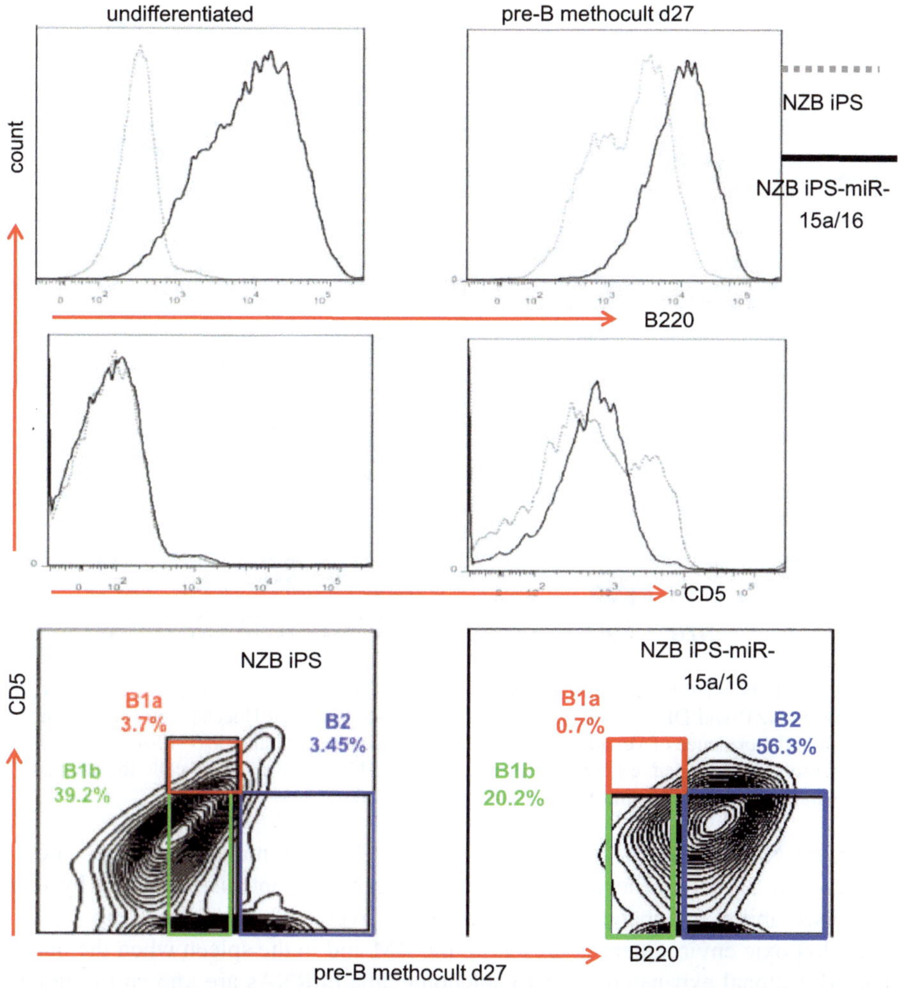

Fig. 9.6 In vitro differentiation of NZB iPS cells toward B cell lineage. Two types of NZB iPS cells were grown for 27 days in pre-B methocult to allow differentiation of B cells. NZB iPS cells have a germline-encoded mutation in the *mir-15a/16-1*loci leading to a processing error, which results in decreased ability to generate mature miR-15a/16 from the precursor molecules. NZB iPS cells were transfected with a lentiviral construct, sorted on the basis of GFP expression and also subjected to culture conditions to induce differentiation to B lineage cells. The cells were stained with antibodies that recognize B220 (*upper panel*) and CD5 (*middle panel*) before and after differentiation in pre-B methocult for 27 days. The NZB iPS cells expressing high levels of miR-15a/16 were induced to express high levels of B220 (even prior to expansion in B-cell differentiation media). The NZB iPS cells gave rise to mainly B-1a and B-1b cells and very few $B220^{bright}$ CD5− (B-2) cells (*lower left*). In contrast, NZB iPS miR-15a/16 gave rise to predominantly CD5− $B220^{bright}$ B-2 cells

family members including the multidrug resistance gene 1 (MDR1) and ABCG2 [77, 94]. Despite these studies, other studies have not found HSCs or increased expression of ABCG2 in SP cells [1].

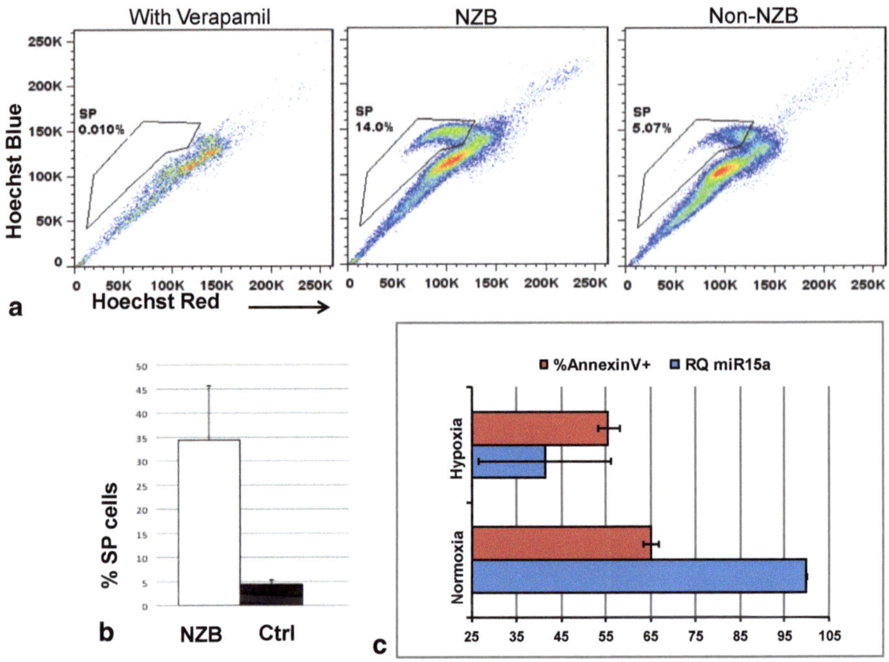

Fig. 9.7 Side population in a murine model of CLL. **a** The presence of the side population was detected in NZB and DBA/2 mice by gating on cells that excluded Hoechst and this population was lost in the presence of Verapamil. **b** Mean percent of side population in 9–12-month NZB and control (non-NZB) spleen. **c** Percent apoptosis and miR-15a/16 levels in SP cells following 48 h culture in the presence of normoxia (5 %) or hypoxia (2 %)

In the NZB mice, as the mice age they develop B-1 malignant clones and an increase in SP cells. These SP cells are resistant to apoptosis as shown by their increased ability to expand in the presence of a hypoxic environment (Fig. 9.7). In vivo, hypoxic environment is present in the BM and in the spleen when the malignant B-1 clonal expansion leads to splenomegaly. miRNAs are known to regulate normal stem cells [102] and miRNA profiles of SP cells have been shown to be different from non-SP cells [58]. Sorted SP cells were analyzed for expression levels of miR-15a/16 as well as the ability to survive a hypoxic environment (2 % CO_2, 48 h). Hypoxia induced a further decrease in miR-15a/16 levels suggesting that decreased miR-15a/16 is a survival factor following hypoxia.

5 Conclusions

In this chapter, the role of miR-15a/16 in the development of CLL, a malignancy of B-1 cells, was the major focus. Decreased miR-15a/16 levels are found in majority of patients with CLL. In addition, findings suggest that even before deletion of the 13q14 region, which contains the *mir-15a/16-1* loci, epigenetic regulation may

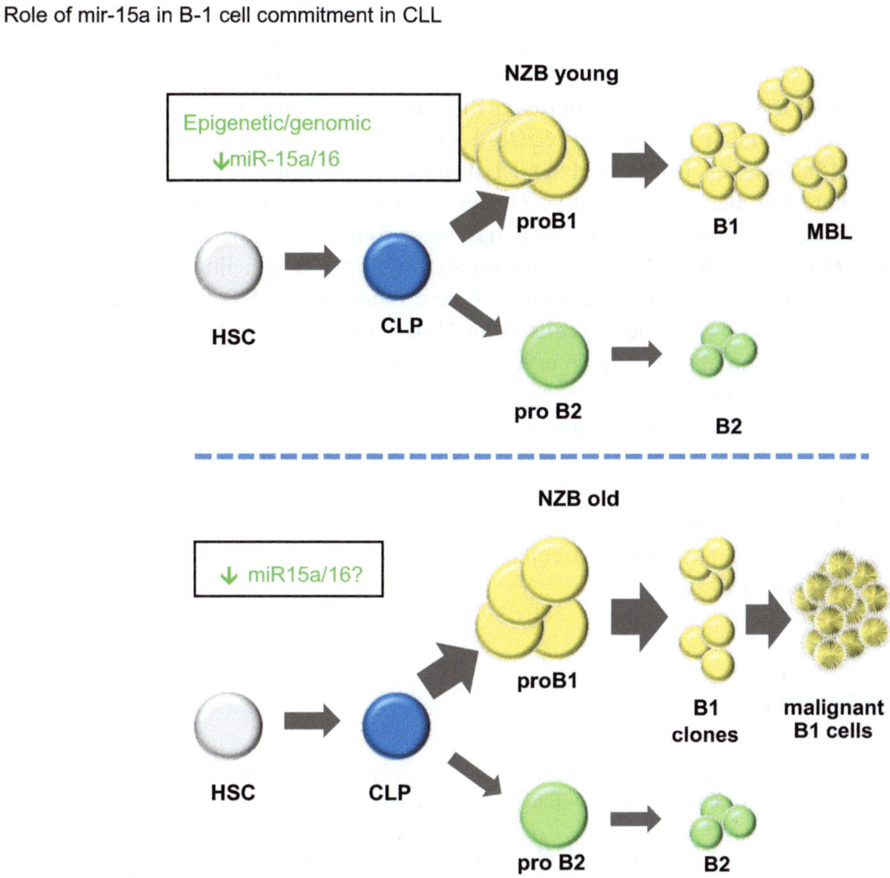

Fig. 9.8 Schematic diagram of the role of miR-15a/16 levels in the development of B-1 cells. The HSCs give rise to committed lymphoid progenitors (CLP) that differentiate into CD5+ (B-1) or CD5− (B-2) lineage cells. In young NZB (*top*), both epigenetic silencing of the host gene, Dleu2 and germline-encoded mutation in the *mir15a/16-1* loci, which together lead to decreased miR-15a/16, play a negative role in differentiation to B-2 and increase the B-1 lineage cells. This increase in B-1 cells results in MBL. As NZB mice age (*bottom*), the development of CLL clones occurs

lead to decreased miR-15a/16 in CLL. Recent data have also found that the precursor state of CLL, MBL, has decreased expression of miR-15a/16. Taken together, decreased miR-15a/16 is critical for the development of CLL. In several mouse models, deletion of the loci containing *mir-15a/16-1* led to a CLL-like disease. In the NZB de novo mouse model of CLL, there is a decrease in miR-15a/16 levels due to mutation in the mouse syntenic region of human 13q14. HSCs from NZB mice gave rise to more B-1 cells in vivo. In addition, iPS cells from NZB with enforced overexpression of miR-15a/16 gave rise almost exclusively to B-2 lineage cells. In contrast, NZB mice with the genetically determined low miR-15a/16 levels, gave rise to mainly the B-1 lineage. Taken together, decreased miR-15a/16 is a critical factor in the development and progression of human CLL (Fig. 9.8).

It is clear that acquiring of additional genetic alterations, including alterations in additional miRNAs, is required to progress from indolent to aggressive CLL. While deletion of the 13q14 region (resulting in decreased miR-15a/16) alone is associated with low risk, as CLL progresses there is often an increase in cells which have deleted both alleles of the *mir-15a/16-1* loci. This suggests that further reduction of miR-15a/16 levels is involved in CLL disease progression.

There has been no definitive cancer stem cells identified in CLL. However, increased B-1 clonal development has been found in xenogenic transplantation studies of HSCs from CLL, suggesting abnormalities in B-1 lineage development. Decreased miR-15a/16 expression may be a fundamental defect that increases the development of B-1 lineage cells from HSCs. In studies in NZB mice, enforced high expression of miR-15a/16 reversed the type of B cells which developed from a skewed B-1 toward exclusively a B-2 differentiation. Based on the literature, HSCs may compromise a heterogeneous group of stem cells with a unique B-1 lineage stem cell, which, by genetic alterations or epigenetic silencing, may express lower levels of miR-15a/16 than do the HSCs that give rise to B-2 cells.

Overall, decreased miR-15a/16 levels play a role in differentiation toward the B-1 lineage, expansion of B-1 cells, development of premalignant B-1 clones, and progression to CLL.

References

1. Alt R, Wilhelm F, et al. ABCG2 expression is correlated neither to side population nor to hematopoietic progenitor function in human umbilical cord blood. Exp Hematol. 2009;37(2):294–301.
2. Anokye-Danso F, Trivedi CM, et al. Highly efficient miRNA-mediated reprogramming of mouse and human somatic cells to pluripotency. Cell Stem Cell. 2011;8(4):376–88.
3. Arnold CP, Tan R, et al. MicroRNA programs in normal and aberrant stem and progenitor cells. Genome Res. 2011;21(5):798–810.
4. Bissels U, Bosio A, et al. MicroRNAs are shaping the hematopoietic landscape. Haematologica. 2012;97(2):160–7.
5. Bushati N, Cohen SM. microRNA functions. Annu Rev Cell Dev Biol. 2007;23:175–205.
6. Calin GA, Cimmino A, et al. MiR-15a and miR-16-1 cluster functions in human leukemia. Proc Natl Acad Sci USA. 2008;105(13):5166–71.
7. Calin GA, Croce CM. Chronic lymphocytic leukemia: interplay between noncoding RNAs and protein-coding genes. Blood. 2009;114(23):4761–70.
8. Calin GA, Dumitru CD, et al. Frequent deletions and down-regulation of micro- RNA genes miR15 and miR16 at 13q14 in chronic lymphocytic leukemia. Proc Natl Acad Sci USA. 2002;99(24):15524–9.
9. Calin GA, Liu CG, et al. MicroRNA profiling reveals distinct signatures in B cell chronic lymphocytic leukemias. Proc Natl Acad Sci USA. 2004;101(32):11755–60.
10. Chen CZ, Li L, et al. MicroRNAs modulate hematopoietic lineage differentiation. Science. 2004;303(5654):83–6.
11. Chen SY, Huang YC, et al. An overview of concepts for cancer stem cells. Cell Transplant. 2011;20(1):113–20.
12. Chena C, Avalos JS, et al. Biallelic deletion 13q14.3 in patients with chronic lymphocytic leukemia: cytogenetic, FISH and clinical studies. Eur J Haematol. 2008;81(2):94–9.

13. Cimmino A, Calin GA, et al. miR-15 and miR-16 induce apoptosis by targeting BCL2. Proc Natl Acad Sci USA. 2005;102(39):13944–9.
14. Corcoran MM, Rasool O, et al. Detailed molecular delineation of 13q14.3 loss in B-cell chronic lymphocytic leukemia. Blood. 1998;91(4):1382–90.
15. Dagklis A, Fazi C, et al. The immunoglobulin gene repertoire of low-count chronic lymphocytic leukemia (CLL)-like monoclonal B lymphocytosis is different from CLL: diagnostic implications for clinical monitoring. Blood. 2009;114(1):26–32.
16. Dalerba P, Clarke MF. Cancer stem cells and tumor metastasis: first steps into uncharted territory. Cell Stem Cell. 2007;1(3):241–2.
17. di Iasio MG, Norcio A, et al. SOCS1 is significantly up-regulated in Nutlin-3-treated p53(wild-type) B chronic lymphocytic leukemia (B-CLL) samples and shows an inverse correlation with miR-155. Invest New Drugs. 2012;30(6):2403–6.
18. Ding L, Han M. GW182 family proteins are crucial for microRNA-mediated gene silencing. Trends Cell Biol. 2007;17(8):411–6.
19. Dohner H. The use of molecular markers in selecting therapy for CLL. Clin Adv Hematol Oncol. 2005;3(2):103–4.
20. Dohner H, Stilgenbauer S, et al. Genomic aberrations and survival in chronic lymphocytic leukemia. N Engl J Med. 2000;343(26):1910–6.
21. Dorshkind K, Montecino-Rodriguez E. Fetal B-cell lymphopoiesis and the emergence of B-1-cell potential. Nat Rev Immunol. 2007;7(3):213–9.
22. Driessens G, Beck B, et al. Defining the mode of tumour growth by clonal analysis. Nature. 2012;488(7412):527–30.
23. Du T, Zamore PD. microPrimer: the biogenesis and function of microRNA. Development. 2005;132(21):4645–52.
24. Edelmann J, Holzmann K, et al. High-resolution genomic profiling of chronic lymphocytic leukemia reveals new recurrent genomic alterations. Blood.2012;120(24):4783–94.
25. Eulalio A, Behm-Ansmant I, et al. P bodies: at the crossroads of post-transcriptional pathways. Nat Rev Mol Cell Biol. 2007;8(1):9–22.
26. Eystathioy T, Jakymiw A, et al. The GW182 protein colocalizes with mRNA degradation associated proteins hDcp1 and hLSm4 in cytoplasmic GW bodies. RNA. 2003;9(10):1171–3.
27. Fabbri M, Bottoni A, et al. Association of a microRNA/TP53 feedback circuitry with pathogenesis and outcome of B-cell chronic lymphocytic leukemia. JAMA. 2011;305(1):59–67.
28. Facon T, Avet-Loiseau H, et al. Chromosome 13 abnormalities identified by FISH analysis and serum beta2-microglobulin produce a powerful myeloma staging system for patients receiving high-dose therapy. Blood. 2001;97(6):1566–71.
29. Fazi C, Scarfo L, et al. General population low-count CLL-like MBL persists over time without clinical progression, although carrying the same cytogenetic abnormalities of CLL. Blood. 2011;118(25):6618–25.
30. Fernando TR, Rodriguez-Malave NI, et al. MicroRNAs in B cell development and malignancy. J Hematol Oncol. 2012;5:7.
31. Filipowicz W, Bhattacharyya SN, et al. Mechanisms of post-transcriptional regulation by microRNAs: are the answers in sight? Nat Rev Genet. 2008;9(2):102–14.
32. Fulci V, Chiaretti S, et al. Quantitative technologies establish a novel microRNA profile of chronic lymphocytic leukemia. Blood. 2007;109(11):4944–51.
33. Ghia P, Caligaris-Cappio F. Monoclonal B-cell lymphocytosis: right track or red herring? Blood. 2012;119(19):4358–62.
34. Ghia P, Prato G, et al. Monoclonal CD5+ and CD5− B-lymphocyte expansions are frequent in the peripheral blood of the elderly. Blood. 2004;103(6):2337–42.
35. Ghosn EE, Yamamoto R, et al. Distinct B-cell lineage commitment distinguishes adult bone marrow hematopoietic stem cells. Proc Natl Acad Sci USA. 2012;109(14):5394–8.
36. Goodell MA, Brose K, et al. Isolation and functional properties of murine hematopoietic stem cells that are replicating in vivo. J Exp Med. 1996;183(4):1797–806.
37. Gribben JG, O'Brien S. Update on therapy of chronic lymphocytic leukemia. J Clin Oncol. 2011;29(5):544–50.

38. Guo S, Lu J, et al. MicroRNA miR-125a controls hematopoietic stem cell number. Proc Natl Acad Sci USA. 2010;107(32):14229–34.
39. Hadnagy A, Gaboury L, et al. SP analysis may be used to identify cancer stem cell populations. Exp Cell Res. 2006;312(19):3701–10.
40. Hallek M, Cheson BD, et al. Guidelines for the diagnosis and treatment of chronic lymphocytic leukemia: a report from the International Workshop on Chronic Lymphocytic Leukemia updating the National Cancer Institute-Working Group 1996 guidelines. Blood. 2008;111(12):5446–56.
41. Hammarsund M, Corcoran MM, et al. Characterization of a novel B-CLL candidate gene-DLEU7-located in the 13q14 tumor suppressor locus. FEBS Lett. 2004;556(1–3):75–80.
42. Han YC, Park CY, et al. microRNA-29a induces aberrant self-renewal capacity in hematopoietic progenitors, biased myeloid development, and acute myeloid leukemia. J Exp Med. 2010;207(3):475–89.
43. Hanlon K, Rudin CE, et al. Investigating the targets of MIR-15a and MIR-16-1 in patients with chronic lymphocytic leukemia (CLL). PLoS One. 2009;4(9):e7169.
44. Hirschmann-Jax C, Foster AE, et al. A distinct "side population" of cells with high drug efflux capacity in human tumor cells. Proc Natl Acad Sci USA. 2004;101(39):14228–33.
45. Johnnidis JB, Camargo FD. Isolation and functional characterization of side population stem cells. Methods Mol Biol. 2008;430:183–93.
46. Kikushige Y, Ishikawa F, et al. Self-renewing hematopoietic stem cell is the primary target in pathogenesis of human chronic lymphocytic leukemia. Cancer Cell. 2011;20(2):246–59.
47. Kim VN. MicroRNA biogenesis: coordinated cropping and dicing. Nat Rev Mol Cell Biol. 2005;6(5):376–85.
48. Klein U, Lia M, et al. The DLEU2/miR-15a/16-1 cluster controls B cell proliferation and its deletion leads to chronic lymphocytic leukemia. Cancer Cell. 2010;17(1):28–40.
49. Lanasa MC, Allgood SD, et al. Single-cell analysis reveals oligoclonality among 'low-count' monoclonal B-cell lymphocytosis. Leukemia. 2010;24(1):133–40.
50. Landgren O, Albitar M, et al. B-cell clones as early markers for chronic lymphocytic leukemia. N Engl J Med. 2009;360(7):659–67.
51. Landgren O, Tilly H. Epidemiology, pathology and treatment of non-follicular indolent lymphomas. Leuk Lymphoma. 2008;49(Suppl. 1):35–42.
52. Laurie CC, Laurie CA, et al. Detectable clonal mosaicism from birth to old age and its relationship to cancer. Nat Genet. 2012;44(6):642–50.
53. Lee RC, Feinbaum RL, et al. The C. elegans heterochronic gene lin-4 encodes small RNAs with antisense complementarity to lin-14. Cell. 1993;75(5):843–54.
54. Li S, Moffett HF, et al. MicroRNA expression profiling identifies activated B cell status in chronic lymphocytic leukemia cells. PLoS One. 2011;6(3):e16956.
55. Liao R, Sun J, et al. MicroRNAs play a role in the development of human hematopoietic stem cells. J Cell Biochem. 2008;104(3):805–17.
56. Lin KK, Goodell MA. Purification of hematopoietic stem cells using the side population. Methods Enzymol. 2006;420:255–64.
57. Liu J, Carmell MA, et al. Argonaute2 is the catalytic engine of mammalian RNAi. Science. 2004;305(5689):1437–41.
58. Liu WH, Tao KS, et al. Differences in the properties and mirna expression profiles between side populations from hepatic cancer cells and normal liver cells. PLoS One. 2011;6(8):e23311.
59. Mallanna SK, Rizzino A. Emerging roles of microRNAs in the control of embryonic stem cells and the generation of induced pluripotent stem cells. Dev Biol. 2010;344(1):16–25.
60. Marti G, Abbasi F, et al. Overview of monoclonal B-cell lymphocytosis. Br J Haematol. 2007;139(5):701–8.
61. Marti GE. MBL: mostly benign lymphocytes, but…. Blood. 2011;118(25):6480–1.
62. Marti GE, Carter P, et al. B-cell monoclonal lymphocytosis and B-cell abnormalities in the setting of familial B-cell chronic lymphocytic leukemia. Cytometry B Clin Cytom. 2003;52(1):1–12.
63. Marti GE, Metcalf RA, et al. The natural history of a lymphoproliferative disorder in aged NZB mice. Curr Top Microbiol Immunol. 1995;194:117–26.

64. Meister G, Landthaler M, et al. Human Argonaute2 mediates RNA cleavage targeted by miRNAs and siRNAs. Mol Cell. 2004;15(2):185–97.
65. Melton C, Judson RL, et al. Opposing microRNA families regulate self-renewal in mouse embryonic stem cells. Nature. 2010;463(7281):621–6.
66. Mertens D, Philippen A, et al. Chronic lymphocytic leukemia and 13q14: miRs and more. Leuk Lymphoma. 2009;50(3):502–5.
67. Migliazza A, Bosch F, et al. Nucleotide sequence, transcription map, and mutation analysis of the 13q14 chromosomal region deleted in B-cell chronic lymphocytic leukemia. Blood. 2001;97(7):2098–104.
68. Mowery YM, Lanasa MC. Clinical aspects of monoclonal B-cell lymphocytosis. Cancer Control. 2012;19(1):8–17.
69. Nana-Sinkam SP, Croce CM. MicroRNA in chronic lymphocytic leukemia: transitioning from laboratory-based investigation to clinical application. Cancer Genet Cytogenet. 2010;203(2):127–33.
70. Nutt SL, Busslinger M. Monoallelic expression of Pax5: a paradigm for the haploinsufficiency of mammalian Pax genes? Biol Chem. 1999;380(6):601–11.
71. O'Connell RM, Rao DS, et al. Sustained expression of microRNA-155 in hematopoietic stem cells causes a myeloproliferative disorder. J Exp Med. 2008;205(3):585–94.
72. Parker R, Sheth U. P bodies and the control of mRNA translation and degradation. Mol Cell. 2007;25(5):635–46.
73. Patten PE, Chu CC, et al. IGHV unmutated and mutated chronic lymphocytic leukemia cells produce activation-induced deaminase protein with a full range of biologic functions. Blood. 2012;120(24):4802–11.
74. Pekarsky Y, Santanam U, et al. Tcl1 expression in chronic lymphocytic leukemia is regulated by miR-29 and miR-181. Cancer Res. 2006;66(24):11590–3.
75. Peng B, Sherr DH, et al. A cultured malignant B-1 line serves as a model for Richter's syndrome. J Immunol. 1994;153(4):1869–80.
76. Perez-Chacon G, Llobet D, et al. TNFR-associated factor 2 deficiency in B lymphocytes predisposes to chronic lymphocytic leukemia/small lymphocytic lymphoma in mice. J Immunol. 2012;189(2):1053–61.
77. Pfister O, Oikonomopoulos A, et al. Role of the ATP-binding cassette transporter Abcg2 in the phenotype and function of cardiac side population cells. Circ Res. 2008;103(8):825–35.
78. Phillips JA, Mehta K, et al. The NZB mouse as a model for chronic lymphocytic leukemia. Cancer Res. 1992;52(2):437–43.
79. Ramachandra S, Metcalf RA, et al. Requirement for increased IL-10 in the development of B-1 lymphoproliferative disease in a murine model of CLL. J Clin Invest. 1996;98(8):1788–93.
80. Rana TM. Illuminating the silence: understanding the structure and function of small RNAs. Nat Rev Mol Cell Biol. 2007;8(1):23–36.
81. Raveche E, Dang A, et al. Analysis of NZB derived B-1 Malignancies. In: Melcher F, Potter M, editors. Mechanisms of B cell Neoplasia. Basel, Switzerland: Editiones Roche; 1993. p. 137–148.
82. Raveche ES, Salerno E, et al. Abnormal microRNA-16 locus with synteny to human 13q14 linked to CLL in NZB mice. Blood. 2007;109(12):5079–86.
83. Rawstron AC. Prevalence and characteristics of monoclonal B-cell lymphocytosis (MBL) in healthy individuals and the relationship with clinical disease. J Biol Regul Homeost Agents. 2004;18(2):155–60.
84. Rawstron AC, Bennett FL, et al. Monoclonal B-cell lymphocytosis and chronic lymphocytic leukemia. N Engl J Med. 2008;359(6):575–83.
85. Rawstron AC, Shanafelt T, et al. Different biology and clinical outcome according to the absolute numbers of clonal B-cells in monoclonal B-cell lymphocytosis (MBL). Cytometry B Clin Cytom. 2010;78(Suppl. 1):19–23.
86. Rawstron AC, Shingles J, et al. Chronic lymphocytic leukaemia (CLL) and CLL-type monoclonal B-cell lymphocytosis (MBL) show differential expression of molecules involved in lymphoid tissue homing. Cytometry B Clin Cytom. 2010;78(Suppl. 1):42–6.

87. Rondeau G, Moreau I, et al. Comprehensive analysis of a large genomic sequence at the putative B-cell chronic lymphocytic leukaemia (B-CLL) tumour suppresser gene locus. Mutat Res. 2001;458(3–4):55–70.
88. Rossi D, Sozzi E, et al. The prognosis of clinical monoclonal B cell lymphocytosis differs from prognosis of Rai 0 chronic lymphocytic leukaemia and is recapitulated by biological risk factors. Br J Haematol. 2009;146(1):64–75.
89. Rossi S, Shimizu M, et al. microRNA fingerprinting of CLL patients with chromosome 17p deletion identify a miR-21 score that stratifies early survival. Blood. 2010;116(6):945–52.
90. Sachdeva M, Zhu S, et al. p53 represses c-Myc through induction of the tumor suppressor miR-145. Proc Natl Acad Sci USA. 2009;106(9):3207–12.
91. Salerno E, Yuan Y, et al. The New Zealand black mouse as a model for the development and progression of chronic lymphocytic leukemia. Cytometry B Clin Cytom. 2010;78(Suppl. 1):98–109.
92. Santanam U, Zanesi N, et al. Chronic lymphocytic leukemia modeled in mouse by targeted miR-29 expression. Proc Natl Acad Sci USA. 2010;107(27):12210–5.
93. Scaglione BJ, Salerno E, et al. Murine models of chronic lymphocytic leukaemia: role of microRNA-16 in the New Zealand Black mouse model. Br J Haematol. 2007;139(5):645–57.
94. Scharenberg CW, Harkey MA, et al. The ABCG2 transporter is an efficient Hoechst 33342 efflux pump and is preferentially expressed by immature human hematopoietic progenitors. Blood. 2002;99(2):507–12.
95. Schlade-Bartusiak K, Stembalska A, et al. Significant involvement of chromosome 13q deletions in progression of larynx cancer, detected by comparative genomic hybridization. J Appl Genet. 2005;46(4):407–13.
96. Shim YK, Middleton DC, et al. Prevalence of monoclonal B-cell lymphocytosis: a systematic review. Cytometry B Clin Cytom. 2010;78(Suppl. 1):10–8.
97. Takahashi K, Yamanaka S. Induction of pluripotent stem cells from mouse embryonic and adult fibroblast cultures by defined factors. Cell. 2006;126(4):663–76.
98. Tay Y, Zhang J, et al. MicroRNAs to Nanog, Oct4 and Sox2 coding regions modulate embryonic stem cell differentiation. Nature. 2008;455(7216):1124–8.
99. Tili E, Michaille JJ, et al. Mutator activity induced by microRNA-155 (miR-155) links inflammation and cancer. Proc Natl Acad Sci USA. 2011;108(12):4908–13.
100. Tili E, Michaille JJ, et al. The down-regulation of miR-125b in chronic lymphocytic leukemias leads to metabolic adaptation of cells to a transformed state. Blood. 2012;120(13):2631–8.
101. Volinia S, Galasso M, et al. Reprogramming of miRNA networks in cancer and leukemia. Genome Res. 2010;20(5):589–99.
102. Wang Y, Russell I, et al. MicroRNA and stem cell regulation. Curr Opin Mol Ther. 2009;11(3):292–8.
103. Wightman B, Ha I, et al. Posttranscriptional regulation of the heterochronic gene lin-14 by lin-4 mediates temporal pattern formation in C. elegans. Cell. 1993;75(5):855–62.
104. Zapata JM, Krajewska M, et al. TNF receptor-associated factor (TRAF) domain and Bcl-2 cooperate to induce small B cell lymphoma/chronic lymphocytic leukemia in transgenic mice. Proc Natl Acad Sci USA. 2004;101(47):16600–5.
105. Zhou B, Wang S, et al. miR-150, a microRNA expressed in mature B and T cells, blocks early B cell development when expressed prematurely. Proc Natl Acad Sci USA. 2007;104(17):7080–5.
106. Zhu DX, Zhu W, et al. miR-181a/b significantly enhances drug sensitivity in chronic lymphocytic leukemia cells via targeting multiple anti-apoptosis genes. Carcinogenesis. 2012;33(7):1294–301.

Chapter 10
The Role of miRNAs in the Development of Normal Pancreas and Pancreatic Cancer, and Their Roles in Tumor Progression

Yiwei Li, Dejuan Kong, Aamir Ahmad, Bin Bao and Fazlul H. Sarkar

Abstract The microRNAs (miRNAs) have been shown to play important roles in the control of many normal biological processes including cell differentiation and organogenesis. During the development of normal pancreas, several miRNAs including miR-375, miR-7, miR-124, etc. have been shown to regulate exocrine and endocrine cell differentiation. These regulations could in part be mediated through the miRNA-mediated deregulation of transforming growth factor-β, Notch, and Hedgehog signaling, which are the signal transduction pathways that are critically involved in organogenesis during normal development. However, deregulated expression of miRNAs could also lead to the development and progression of pancreatic cancer. Moreover, miRNAs are also known to regulate the development and maintenance of pancreatic cancer stem cells and epithelial-to-mesenchymal transition phenotypic cells, which are typically responsible for drug resistance, tumor recurrence, and metastasis. Therefore, targeting specific miRNAs by oligonucleotide/nanoparticle vector delivery or regulation of miRNAs by natural agents could become novel strategies for the treatment of pancreatic cancer with better treatment outcome in combination with conventional therapeutics.

Keywords miRNA · Pancreas development · Pancreatic cancer · TGF-β · Notch · Hedgehog · Cancer stem cell · EMT · Drug resistance · Natural agent

---

F. H. Sarkar (✉) · Y. Li · D. Kong · A. Ahmad · B. Bao
Department of Pathology, Barbara Ann Karmanos Cancer Institute, Wayne State University School of Medicine, 740 Hudson Webber Cancer Research Center, 4100 John R, Detroit, MI 48201, USA
e-mail: fsarkar@med.wayne.edu

F. H. Sarkar
Department of Oncology, Barbara Ann Karmanos Cancer Institute, Wayne State University School of Medicine, Detroit, MI, USA

S. R. Singh, P. Rameshwar (eds.), *MicroRNA in Development and in the Progression of Cancer,* DOI 10.1007/978-1-4899-8065-6_10,

1 Introduction

Recently, microRNAs (miRNAs) have received significant attention in the area of medical research, especially, cancer research. Each miRNA contains 19–25 nucleotides (nt) in a single strand. Most miRNAs have about 22 nt, much shorter than messenger RNAs (mRNAs). Unlike mRNAs, miRNAs do not code for any protein or peptide. However, they play important roles in the regulation of gene expression by degrading mRNAs or by inhibiting translational processes. Specific miRNA could imperfectly bind to the 3′-untranslated region (UTR) of the target mRNAs, resulting in target mRNA cleavage or translational repression. Therefore, by regulating gene expression, miRNAs are known to contribute to the control of cell development, differentiation, growth, apoptosis, and other biological processes.

The first miRNA lin-4 was identified and reported in 1993 [1]. It was found that a small RNA, named as miRNA lin-4, contained sequences complementary to the sequences in the 3′-UTR of lin-14 mRNA and that the miRNA lin-4 could suppress lin-14 mRNA translation through complementary RNA–RNA interaction. Another critical miRNA, let-7, was found afterward [2]. The miRNA let-7 is a 21 nt small RNA which could complementarily bind to the 3′-UTR of heterochronic genes, including lin-14, lin-28, lin-41, lin-42, and daf-12, and inhibit the expression of these genes. Functional analysis revealed that miRNA lin-4 and let-7 could alter the developmental timing by complementarily binding to the mRNA of heterochronic genes and inhibiting the expression of these genes, suggesting the biological regulation of lin-14 and let-7 during the process of normal development. Since this initial discovery, significant efforts have been made to discover newer miRNAs and characterizing their functions in human cells. Now, it is clear that these miRNAs play critical roles in the regulation of many physiological and pathological processes in humans [3, 4, 5]. Since miRNAs play important roles in the control of cellular development, differentiation, growth, and apoptosis, miRNAs are also critically involved in the development and progression of cancer. Emerging evidence suggests the role of many miRNAs in the normal development of pancreas as well as in the development and progression of pancreatic cancer.

It is well known that pancreas mainly contain two compartments in one system. One compartment is the exocrine acinar tissues, which produce and secrete digestive enzymes into digestive system. Another compartment is endocrine tissues (including the islets housing β cells and other endocrine cells), which secrete insulin and other hormones directly into the blood to regulate blood glucose metabolism and maintain the homeostasis of blood glucose. If β-cells are lost or malfunction during pancreas development or in adult, this could lead to the development of diabetes. Moreover, if the exocrine epithelial cells are dedifferentiated, it may contribute to the development of pancreatic cancer. Since miRNAs play critical roles in the control of development and differentiation, understanding the molecular effects of specific miRNAs on pancreas development and pancreatic cell differentiation is important and such knowledge may lead to the development of strategies for the treatment of diabetes and pancreatic cancer.

Pancreas is of endodermal origin because it forms from embryonic foregut during development. First, the ventral and dorsal anlages (or the so-called buds) are formed and connected to foregut by ducts. Then, rotation and fusion of the ventral and dorsal pancreatic buds occur, leading to the formation of pancreas. During these processes, the progenitor cells of pancreas differentiate by two different pathways, forming exocrine and endocrine tissues, respectively. By the stimulation of follistatin and fibroblast growth factors (FGFs), the progenitor cells of exocrine tissue in pancreas differentiate to mature exocrine pancreas in three successive stages (pre-differentiated, proto-differentiated, and differentiated stages) with different levels of digestive enzyme activity detected (from undetectable to low, and to high). At the same time, under the stimulation of neurogenin-3 and Isl-1, the progenitor cells of the endocrine tissue in pancreas differentiate to α-, β-, γ- and δ-cells. It is now known that some specific miRNAs contribute to the control of normal pancreas development and differentiation.

It has been widely accepted that stem cells are present in most adult organs including pancreas. In adult pancreas, cell turnover including cellular program death and replacement with new differentiated cells from stem cells controls the normal pancreas tissue homeostasis [6]. At the level of molecular regulation, various processes such as DNA repair and metabolism contribute to the preservation of normal anatomical structure and biological function [7]. However, upon DNA damage such as inherited or acquired mutation and if DNA repair fails, then this process could lead to the development of pancreatic cancer. In pancreatic cancer cells, the acquired DNA mutation could be caused by cigarette smoke, aging, chronic pancreatitis, or defects in DNA replication and copy number through cell division. If these DNA mutations are located in cancer-associated genes in a pancreatic cell, then the cell may undergo deregulated cell growth, which could eventually lead to the development of pancreatic cancer. In addition, diabetes and obesity, where disorders of metabolism occur, could also cause pancreatic cancer. It is known that miRNAs are critically involved in the control of development and differentiation of pancreas. The miRNAs also regulate DNA repair and molecule metabolism in pancreas. Moreover, emerging studies have shown that miRNAs could regulate the formation of cancer stem cells (CSCs) [8, 9], the acquisition of epithelial-to-mesenchymal transition (EMT) phenotype [10], and the development of drug resistance to conventional therapeutics [10, 11]. Therefore, it is now clear that miRNAs are critically involved in the development and progression of pancreatic cancer.

2 miRNAs in the Regulation of Pancreas Development

The investigations delineating the role of miRNAs in the regulation of pancreas development, largely focused on the endocrine compartment. Dicer is a critical molecule participating in the synthesis of miRNA. An animal study showed that conditional knockout of Dicer in the pancreas of mouse significantly damaged the development of all pancreatic endocrine cell lineages [12]. The damage of pancreas development

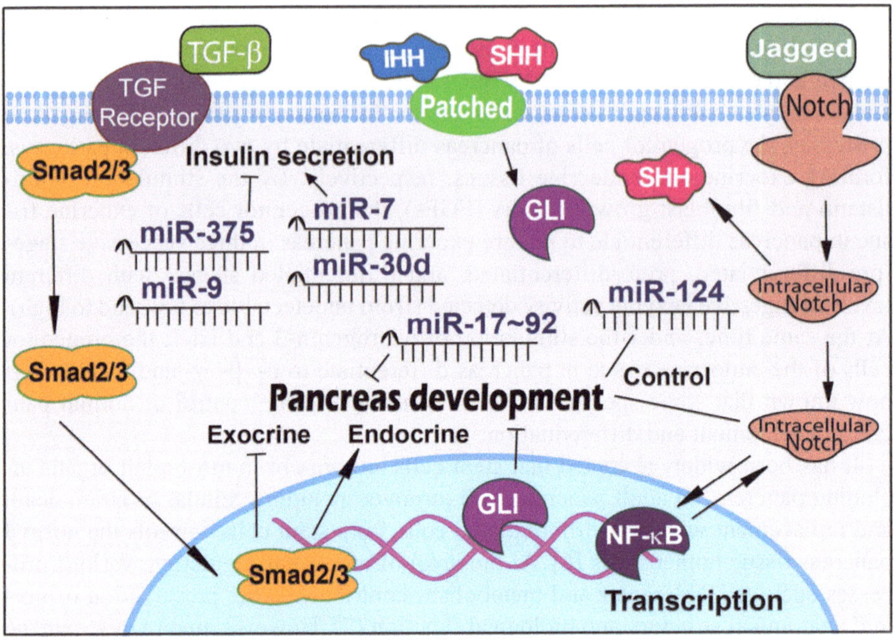

Fig. 10.1 miRNAs and signaling pathways in the regulation of pancreas development

by Dicer knockout was accompanied with reduced number of progenitor cells and activated Notch and apoptosis signaling in progenitor cells, suggesting that miRNAs could regulate progenitor cell proliferation and differentiation in the pancreas.

2.1 *miRNA-Regulated Signaling Pathways in Pancreas Development*

During pancreas development, the interactions among epithelial, midline mesoderm, and mesenchymal cells control pancreatic structure formation and cell differentiation. It has been shown that several signaling pathways are important for the control of cellular interactions in embryonic pancreas. These pathways include transforming growth factor (TGF), Notch, Hedgehog, FGF, and epidermal growth factor (EGF) signaling pathways (Fig. 10.1). All of these signaling pathways also contribute to the development and progression of pancreatic cancer. Importantly, the cross talk among these signaling is also mediated by miRNAs.

TGF protein, especially TGF-β, is a key molecule which controls the differentiation of both pancreatic endocrine and exocrine cells. TGF-β is expressed in embryonic pancreas [13]. Activation of TGF-β promotes the development of endocrine cells and inhibits the differentiation of exocrine cells. In contrast, downregulation of TGF-β signaling enhances differentiation of embryonic exocrine cells and inhibits the differentiation of endocrine cells. Recently, several miRNAs including miR-9, miR-30b, miR-30d, miR-375, and miR-494, have been correlated with the activity of TGF-β [14–16], suggesting that these miRNAs could be involved in the development of pancreas and cellular differentiation.

Notch signaling is known to contribute to the differentiation of pancreatic endocrine and exocrine cells. In pancreas development, expression of neurogenin genes is a critical process. In embryonic pancreas, epithelial cells and adjacent cells express ngn3, which is a member of the neurogenin family [17]. It has been found that the peak of ngn3 expression reached maximum just before the formation of islets and that the level of ngn3 was undetectable in adult pancreas, suggesting the critical role of ngn3 during pancreas development. Importantly, the expression of ngn3 has been shown to be correlated with the activity of Notch signaling. Animal experiments have shown that deficient Notch signals enhanced the differentiation of pancreatic epithelial cells which express ngn3, suggesting the role of Notch signaling in the control of pancreas development [18]. Notch-deficient mice also showed aberrant development of pancreatic endocrine cells [19]. However, inactivation of Notch1 could lead to the abnormal islet cellular distribution, exocrine cell hypoplasia, and β-cell hyperplasia although increased differentiation of pancreatic exocrine and endocrine cells has been usually observed [19, 20]. It is well known that miRNAs could regulate Notch signaling. Moreover, recent studies have also shown that Drosha and miR-20 regulates the expression of neurogenin [21, 22], suggesting the regulatory effects of miRNAs on pancreas development.

Hedgehog signaling is an important signaling pathway during normal development and differentiation. It is known that Sonic Hedgehog (Shh) is not detectable in embryonic pancreas and adult islets. However, the levels of Indian Hedgehog (Ihh), Desert Hedgehog (Dhh), and Patched (Hedgehog receptor) are detectable in developing pancreas and adult islets. It has been found that Ihh-deficient embryos contained increased number of pancreatic cells which express insulin and glucagons, and Shh-deficient embryo also had significantly increased number of endocrine cells [23], suggesting that Hedgehog signaling is required for limiting differentiation of pancreatic embryonic cells. Since embryonic pancreas does not express Shh, the Shh from adjacent organs could regulate pancreas development. Hedgehog signals such as Shh and Ihh bind to Patched and inhibit pancreas development. In addition, the activity of Hedgehog signaling has been found to be regulated by several miRNAs. Studies have shown that miR-212, miR-214, miR-196, and miR-17–92 cluster could regulate Hedgehog signaling [24–27], suggesting that these miRNAs could participate in the development of pancreas.

2.2 *miRNA Regulates Pancreatic Cell Differentiation and Pancreas Development*

It has been found that specific miRNAs are expressed at high levels during pancreas development and participate in the synthesis and secretion of insulin. Several studies have focused on the investigation of miRNA profiling during pancreas development [28–30]. The results showed that a number of miRNAs are involved in pancreas organogenesis. These miRNAs include miR-7, miR-9, miR-30d, miR-375, miR-376, miR-124, miR-15a, miR-15b, miR-16, miR-19b, etc. (Fig. 10.1). Among them, miR-375, miR-7, miR-124, and miR-9 are most investigated in pancreatic cell differentiation.

The miR-375 is expressed in pancreatic islet cells including α-, β-, and δ-cells. The miR-375 has inhibitory effect on glucose-induced insulin secretion by suppressing exocytosis; however, it does not directly regulate insulin synthesis [31]. The identified targets of miR-375 are myotrophin and phosphoinositide-dependent kinase 1 (PDK1) [32]. Knockdown of myotrophin caused similar effect on β-cells as miR-375. Another target of miR-375 is PDK1, which could regulate insulin synthesis. It is known that glucose promotes insulin expression by downregulation of miR-375 and upregulation of PDK1. Therefore, miR-375 exerts its inhibitory effect on insulin expression through the downregulation of PDK1. Animal studies have also demonstrated that miR-375 is required for keeping the balance of β and α cells during pancreas development.

The expression of miR-7 could be found in both adult and fetal endocrine cells of the pancreas. The miR-7 is islet-enriched miRNA and its expression is increased during pancreatic islet development [33], especially, β-cell differentiation. Experimental studies have shown that the inhibition of miR-7 could cause death of insulin-producing cells during developmental process. Animal studies have demonstrated that the inhibition of miR-7 by antisense miR-7 could lead to the inhibition of β-cell formation and insulin production in mouse embryos.

The miR-124a also plays a critical role in the regulation of pancreas development. The miR-124a exerts its effect on pancreas development through its target FOXA2 that is a transcription factor for early pancreas development. FOXA2 and its downstream targets including Pdx-1, Kir6.2, and sulfonylurea receptor 1, regulate the development and also maintain the function of the pancreas. It is believed that miR-124a has different functions in different periods of pancreas development. In early stages of pancreas development, miR-124a control differentiation by the regulation of FOXA2. However, in later stages of pancreas development and in the adult pancreas, miR-124a is critically involved in the maintenance of β-cell functions [34]. In neuronal development, miR-124a could regulate the expression of Hes1, which controls proliferation of ngn3-producing cell [35]. Since ngn3 plays critical roles in pancreas development, miR-124a could be an important molecule, which regulates Notch1 signaling in pancreas development. In addition to miR-124a, other miRNAs including miR-15a, miR-15b, miR-16, and miR-195 are also critically involved in the control of translation of ngn3, leading to a decreased number of hormone-producing cells during pancreas development.

The miR-9 is highly expressed during islet development. In pancreatic islet cells, miR-9 has been found to inhibit the expression of its target Onecut-2 and, in turn, increase the expression of granuphilin/Slp4, leading to the suppression of glucose-induced insulin secretion [36]. Other miRNAs also play their roles in the differentiation of pancreatic cells. The miR-19b is a member of the miR-17–92 cluster, which is highly expressed in pancreatic progenitor cells. The miR-19b could directly bind to the 3′-UTR of NeuroD1 mRNA and reduce the expression of NeuroD1 at mRNA and protein levels, leading to the inhibition of proliferation of pancreatic progenitor cells. The miR-30d has been found to increase the expression of insulin in a β-cell line [37] and miR-218 has been found to be expressed in early fetal pancreas of mice. The miR-218 has been found to control pancreas and liver development through the regulation of Onecut-2 [38]. In addition, miR-503 and miR-376a are expressed at high levels during islet development [33]. These miRNAs participate in the control of pancreatic cell differentiation during different stages of pancreas development. Therefore, the knowledge regarding the molecular function and regulation of these miRNAs is likely to be very important for the regeneration of pancreatic endocrine cells, which would be highly valuable to restore endocrine function of the pancreas.

3 miRNAs in the Regulation of Pancreatic Cancer Development and Progression

Since, miRNAs play important roles in normal biological processes during pancreas development, pancreatic cell differentiation, proliferation, apoptosis, and metabolism, it is not surprising that miRNAs are also critically involved in the development and progression of pancreatic cancer. In pancreatic cancer, the aberrant expression of specific miRNAs has been clinically linked with the disease stage, invasion, metastasis, and survival of patients diagnosed with pancreatic cancer [39, 40]. The miRNAs could act as tumor suppressors or oncogenes to regulate the biological behavior of pancreatic cancer cells. Importantly, miRNAs could also control the growth and differentiation of pancreatic CSCs and EMT-type cells, which contribute to drug resistance and tumor progression [41,–43]. Therefore, miRNAs could serve as molecular markers and targets for diagnosis, prognosis, and treatment of pancreatic cancer.

3.1 miRNAs that Inhibit Pancreatic Cancer Development and Progression

The deregulation of miRNAs in pancreatic cancer could be monitored by the expression profiling of miRNAs using chip assay or reverse transcriptase polymerase chain reaction (RT-PCR) technologies. Over the past years, several studies have been conducted to reveal the expression levels of miRNAs in normal pancreatic epithelial cells and pancreatic cancer cells. It has been found that a set of miRNAs are downregulated in pancreatic cancer or PanIN cells, compared to normal cells. The expression of miR-452, miR-126, miR-218, miR-125b, miR-127–3p, miR-139–3p, miR-139–5p, miR-216b, and miR-296–5p has been found to be downregulated in the samples with PanIN lesion [44]. It has also been found that the expression of

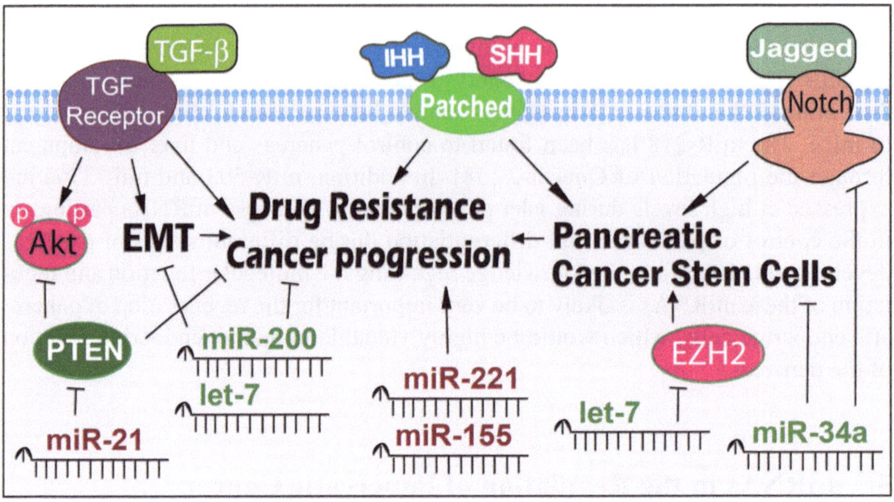

Fig. 10.2 miRNAs and signaling pathways in the progression of pancreatic cancer

miR-96, miR-34a, miR-130b, miR-139, miR-142, miR-148a, miR-148b, miR-216, miR-217, miR-345, and miR-375 was downregulated in pancreatic cancer cells [43, 45, 46]. These miRNAs have been considered as tumor suppressors, which inhibit pancreatic cancer development and progression. Among these tumor suppressive miRNAs, miR-216, miR-217, and miR-34a are mostly downregulated, suggesting their potent inhibitory effects on pancreatic cancer development and progression.

In a comprehensive miRNA expression profiling analysis, tissues from normal pancreas, chronic pancreatitis, pancreatic cancer, and non-pancreatic tissues were subjected to 377 miRNA array analysis. It was found that miR-216 and miR-217 were absent or only minimally expressed in pancreatic cancer cell lines and tissues, suggesting that miR-216 and miR-217 could serve as tumor suppressive miRNAs in pancreatic cancer [46]. Furthermore, the expression of miR-216 and miR-217 and the absence of miR-133a expression were found to be characteristic of pancreas tissue, suggesting the importance of these specific miRNAs in maintaining the normal function of the pancreas.

The miR-34 has also been found to be a tumor suppressive miRNA (Fig. 10.2). The miR-34a could be directly transactivated by p53 [47], leading to the induction of apoptosis and the inhibition of proliferation and angiogenesis. Studies have also shown that CpG methylation of the miR-34a promoter is prevalent in various cancers including pancreatic cancer (3/19; 15.7%) [48, 49]. These findings suggest that the low expression of miR-34a found in pancreatic cancer cells and tissues is due to epigenetically regulated miR-34a silencing. Furthermore, miR-34a has been found to inhibit CSCs through directly downregulating CD44 [50]; therefore, targeting epigenetic regulation of miR-34a could be a novel strategy for the treatment of pancreatic cancer by inhibition of pancreatic CSCs [43].

miR-150 is another miRNA functioning as tumor suppressive miRNA in pancreatic cancer. In pancreatic cancer tissues, much lower expression of miR-150 and high expression of MUC4 have been observed. Mechanistic studies have shown that the 3′-UTR of MUC4 contains a highly conserved miR-150-binding motif and that miR-150 downregulated the expression of MUC4 protein, suggesting that MUC4 is the target of miR-150 [51]. Moreover, the miR-150-mediated downregulation of MUC4 led to a decrease in the levels of EGFR2 and p-EGFR2, resulting in the inhibition of cell growth, clonogenicity, migration, and invasion in pancreatic cancer cells [51]. These findings demonstrate that miR-150 is a tumor suppressor miRNA in pancreatic cancer.

3.2 miRNAs that Promote Pancreatic Cancer Development and Progression

By miRNA expression profiling analysis in pancreatic cancer, a set of specific miRNAs has been found to be significantly upregulated in pancreatic cancers or PanIN compared to normal pancreas. The expression of miR-21, miR-18a, miR-182, miR-18b, miR-183, miR-422a, miR-190, miR-29b, miR-93, miR-101, miR-193a-3p, miR-135b, miR-320b, miR-222, miR-106b, and miR-196b was significantly upregulated in the tissues from PanIN-3 lesions compared to normal tissues [44]. In pancreatic cancer, the expression of miR-10, miR-15b, miR-21, miR-23, miR-24, miR-92, miR-95, miR-100, miR-107, miR-143, miR-145, miR-155, miR-181, miR-190, miR-194, miR-196, miR-205, miR-210, miR-212, miR-213, miR-220, miR-221, miR-222, miR-223, and miR-301 has been found to be significantly upregulated compared to normal tissues [45], suggesting that these miRNAs possess oncogenic activity. Among these miRNA, miR-21, miR-155, and miR-221 are more frequently upregulated in other cancers, and thus commonly accepted as oncogenic miRNAs.

The miR-21 is well known as oncogenic miRNA (Fig. 10.2). It has been found that miR-21 is upregulated in different types of cancers [52–54]. In early noninvasive intraductal papillary mucinous neoplasms which are noninvasive precursor lesions of pancreatic cancer, miR-21 is highly expressed with highest relative fold-changes [55]. By in situ hybridization, using the samples from patients, upregulated miR-21 expression was found in 79 % of pancreatic cancers; however, only 8 % of benign pancreas and 27 % of chronic pancreatitis expressed miR-21 [40], suggesting that miR-21 plays important roles in the development of pancreatic cancer. Moreover, the expression level of miR-21 was predictive of poorer outcome compared to absence or low expression of miR-21 in patients without lymph node metastasis [40], suggesting the critical role of miR-21 in pancreatic cancer progression. By in vitro studies, overexpression of miR-21 was observed in pancreatic cancer cell lines compared to nonmalignant cells [56]. Forced expression of miR-21 using miR-21 precursor significantly increased proliferation, invasion, and chemoresistance of pancreatic cancer cells. In contrast, knockdown of miR-21 inhibited cell proliferation and invasion and sensitized pancreatic cancer cells to chemotherapy [56]. The inhibition of cell proliferation, invasion, and drug resistance by knockdown of miR-21 could be mediated through the upregulation of phosphatase and tensin homolog (PTEN), one of the miR-21 targets, and downregulation of matrix metalloproteinase (MMP)-9, MMP-2, and vascular endothelial growth factor (VEGF) [56, 57].

The miR-221 is another known oncogenic miRNA (Fig. 10.2). High expression of miR-221 has been found in various cancers. In pancreas, it was found that the expression of miR-221 was significantly elevated in the mouse pancreas with PanIN lesions [58]. In human pancreatic cancer, the expression of miR-221 has been found to be significantly increased [59]. Silencing of miR-221 has been found to inhibit cell proliferation and invasion, and sensitize pancreatic cancer cells to chemotherapy [57, 58]. The oncogenic effect of miR-221 could be mediated by direct targeting of cell cycle inhibitor $p27^{kip1}$ [57], which is a key regulator of cell proliferation and invasion.

The miR-155 is another oncogenic miRNA (Fig. 10.2). By conducting miRNA expression profiling analysis, the elevated level of miR-155 expression has been observed in various cancers including pancreatic cancer [55, 60]. The miR-155 is one of the most highly expressed miRNAs in early noninvasive intraductal papillary mucinous neoplasms which are noninvasive precursor lesions of pancreatic cancer [55]. Moreover, elevated miR-155 expression has been found to be correlated with low overall survival ($p=0.005$) of patients diagnosed with pancreatic cancer [60]. The patients with increased miR-155 expression in pancreatic tumor tissue had a 6.2-fold increased risk of cancer-related death compared to patients who had low level of miR-155 expression [60], suggesting that miR-155 plays an important role in the progression of pancreatic cancer. The targets of miR-155 have been investigated and it has been found that the tumor protein 53-induced nuclear protein 1 (TP53INP1) is the direct target of miR-155 and that miR-155 could inhibit p53-induced apoptosis by targeting TP53INP1 [61].

3.3 miRNAs that Control Pancreatic EMT and Cancer Stem Cells

As described in the previous section, miRNAs could regulate pancreatic embryonic cell differentiation and pancreas development. Therefore, miRNAs could also control the formation and growth of CSCs and EMT-type cells because normal stem cells, CSCs, and EMT-type cells share many similar molecular characteristics. Indeed, growing evidence has shown that miRNAs could play important roles in the formation of CSCs and could lead to the acquisition of EMT phenotype [9, 62, 63]. By sphere-forming assay to enrich CSC population and by miRNA expression profiling, differentially expressed miRNAs including miR-99a, miR-100, miR-125b, miR-192, and miR-429 have been identified in pancreatic CSCs [64]. These miRNAs are critically involved in maintaining and regulating pancreatic stem cells. It has also been found that miR-34a participates in the regulation of pancreatic CSCs. Experimental studies have shown that CD44+/CD133+MiaPaCa2 pancreatic cancer cells are enriched in sphere-forming cells [65]. These tumor-initiating cells or cancer stem/progenitor cells showed high expression of Notch/Bcl-2 and loss of miR-34 expression. Forced overexpression of miR-34 in the pancreatic tumor-initiating cells has been found to significantly inhibit Bcl-2 and Notch1/2 expression [65]. The upregulation of miR-34a caused significant inhibition of sphere-forming capacity and up to 87 % reduction of tumor-initiating cell population. The miR-34a transfection also caused significant inhibition of tumor formation induced by these tumor-initiating cells in animals [65], suggesting the inhibitory effect of miR-34a on pancreatic CSCs (Fig. 10.2).

It has also been known that miRNAs could regulate the acquisition of EMT phenotype [66]. We have found that the expression of miR-200 family and let-7 family are downregulated in gemcitabine-resistant pancreatic cancer cells which showed EMT phenotype [11]. However, the reexpression of miR-200 family in the EMT-type cells caused the upregulation of epithelial marker E-cadherin and downregulation of mesenchymal marker ZEB1 and vimentin, which was consistent with the reversal of mesenchymal phenotype to epithelial phenotype [11]. The reexpression of miR-200 family also increased the drug sensitivity of pancreatic cancer cells. These results suggested the importance of miR-200 in the maintenance of epithelial identity and sensitivity to chemotherapy and that the loss of miR-200 could be responsible for drug resistance and tumor aggressiveness. We have also found high expression of miR-21 and low expression of miR-200 family and let-7 family and miR-146a could contribute to the maintenance of pancreatic EMT-type and stem cell phenotypes through the regulation of Notch1, FoxM1, and EZH2 [67–69]. These findings demonstrated that miRNAs could play key roles in the progression of pancreatic cancer through the regulation of pancreatic EMT and CSCs.

3.4 miRNAs as Prognostic Markers in Pancreatic Cancer

Since miRNAs are critically involved in the progression of various cancers, specific miRNAs could be useful as prognostic markers [70, 71]. Indeed, growing in vitro and in vivo evidences clearly suggest that the alterations in several miRNAs including miR-10b, miR-21, miR-17–5p, miR-150, miR-196a, and miR-200c are clearly seen in pancreatic cancer, and thus these miRNAs could be useful as diagnostic and prognostic markers for pancreatic cancer [51, 56, 72–77].

As pancreatic cancer commonly has a poor prognosis, identification of novel biomarkers for predicting prognosis of pancreatic cancer is urgently needed. The expression of miR-21 has been found to be very high in pancreatic cancer cells and tissues. Importantly, the high expression of miR-21 was tightly associated with increased proliferation, invasion, chemoresistance, and liver metastasis [56, 77], suggesting that elevated expression of miR-21 could serve as a diagnostic and prognostic marker for pancreatic cancer. Indeed, studies have shown that high expression of miR-21 could be significantly correlated with worse prognosis in patients with pancreatic ductal adenocarcinoma ($P=0.045$). Furthermore, the downregulation of programmed cell death 4 (PDCD4) and tissue inhibitor of metalloproteinases-3 (TIMP3), two targets of miR-21, was significantly associated with the upregulation of miR-21 expression ($P<0.05$) and poor survival of the patients diagnosed with pancreatic ductal adenocarcinoma ($P=0.001$) [78], suggesting that miR-21 and its target PDCD4 and TIMP3 are biomarkers of prognosis in pancreatic cancer.

Significantly, high expression of miR-10b has been observed in pancreatic cancer cell lines and tissues compared to normal pancreatic ductal cells. Importantly, it has also been found that higher expression of miR-10b was associated with invasiveness of pancreatic cancer cells and shorter overall survival [72]. Similar to miR-21 and miR-10b, the expression of miR-155 and miR-196a-2 were also found to be overexpressed in pancreatic cancer cells, and the expression was correlated with tumor stage and poor survival [73, 76]. The high expression of miR-17–92 clusters has been commonly found in lymphomas and lung cancers. However, pancreatic cancer cells also showed higher levels of miR-17–5p compared to normal ductal cells, and that the high expression of miR-17–5p was associated with poor prognosis [74]. As discussed in the previous section, miR-200 family inhibits EMT and sensitizes pancreatic cancer cells to chemotherapy. It has also been found that pancreatic cancer patients with high expression of miR-200c have significantly better survival rates than those with low expression of miR-200c. The high expression of miR-200c in pancreatic cancer was also strongly correlated with high E-cadherin expression [75], suggesting that higher expression of miR-200c could inhibit EMT and could predict better survival. All of these findings suggest that specific miRNA expression could be useful as prognostic biomarkers for pancreatic cancers.

4 Regulation of miRNAs to Inhibit Pancreatic Cancer Aggressiveness

Emerging evidence clearly suggests that miRNAs play important roles in the development and progression of pancreatic cancer; therefore, targeting deregulated miRNAs in pancreatic cancer cells is an attractive novel strategy for the treatment of pancreatic cancer. The therapeutic strategies for targeting deregulated miRNAs in cancer include inhibition of oncogenic miRNAs, introduction of tumor suppressive miRNAs, and regulation of specific miRNAs that are known to control CSC, EMT, and chemoresistance. Oligonucleotide delivery or agent administration could be useful for systemic regulation of miRNAs; however, so far no promising strategy for the regulation of miRNA has been established.

4.1 Sense or Antisense Oligonucleotide Delivery Strategy for the Inhibition of Cancer

Sense or antisense oligonucleotide could be delivered by viral or nonviral vectors. However, the use of viral vectors has been limited because of a number of safety concerns including immunogenicity and risk of oncogenic integration and transformation. Nonviral vectors including lipid-based formulations such as liposomes, lipoplexes, etc. are attractive. Recently, investigators have utilized nanotechnology to create nanolipoparticles or other nanoparticles to facilitate oligonucleotide delivery [79, 80]. Experimental studies have shown that antisense oligonucleotides could bind to specific miRNAs and block the function of miRNAs [79, 81]. Studies have also shown that intravenous administration of chemically engineered antisense oligonucleotides against miR-16, miR-122, miR-192, and miR-194 into mice could significantly inhibit the expression of corresponding mRNA in several organs such as liver, lung, kidney, heart, intestine, fat, skin, bone marrow, muscle, ovaries, and adrenals [82]. These results demonstrated the feasibility for silencing of specific oncogenic miRNAs by antisense oligonucleotide delivery system, which could be a novel therapeutic strategy for cancer treatment in the future.

The introduction of tumor suppressive miRNAs by oligonucleotide delivery is another important strategy for pancreatic cancer treatment. Pramanik et al. reported a lipid-based nanoparticle for systemic delivery of miRNA expression vectors to pancreatic cancer cells [83]. This nanovector was so small that no apparent histopathologic or biochemical evidence of toxicity was observed after intravenous injection of the nanovector. They utilized this nanovector to deliver tumor suppressive miRNAs including miR-34a, miR-143, and miR-145 into pancreatic MiaPaCa-2 xenografts. They found that the intravenous delivery with either miR-34a or miR-143/145 nanovectors significantly inhibited the growth of MiaPaCa-2 subcutaneous tumors in xenograft model. The inhibitory effects were much stronger in the orthotopic tumor model. It was also found that tumor growth inhibition was due to increased apoptosis and decreased proliferation with significant upregulation of delivered miRNA and significant downregulation of miRNA targets such as SIRT1, CD44, aldehyde dehydrogenase, KRAS2, and RREB1 [83]. These results suggest that systemic delivery of tumor suppressive miRNA could be a novel therapeutic strategy for the treatment of pancreatic cancer.

Although experimental studies have focused on the knockdown or introduction of specific miRNAs by systemic oligonucleotide/nanoparticle delivery, which could induce drug sensitivity and inhibit cancer growth, invasion, and metastasis, the limitations still exist for the use of these oligonucleotide/vectors system because of several issues such as the poor stability, immune system stimulation, off-target effects, and delivery efficiency, and it is believed that overcoming such limitations would be useful for the implementation of these tools for the treatment of pancreatic cancer in the future.

4.2 Natural Agents for miRNA Regulation to Inhibit Pancreatic Cancer

To overcome the limitations of oligonucleotide/vectors system, natural agents including isoflavone, 3,3′-diindolylmethane (DIM), indole-3-carbinol (I3C), curcumin, (-)-epigallocatechin-3-gallate (EGCG), and others have been used to regulate the

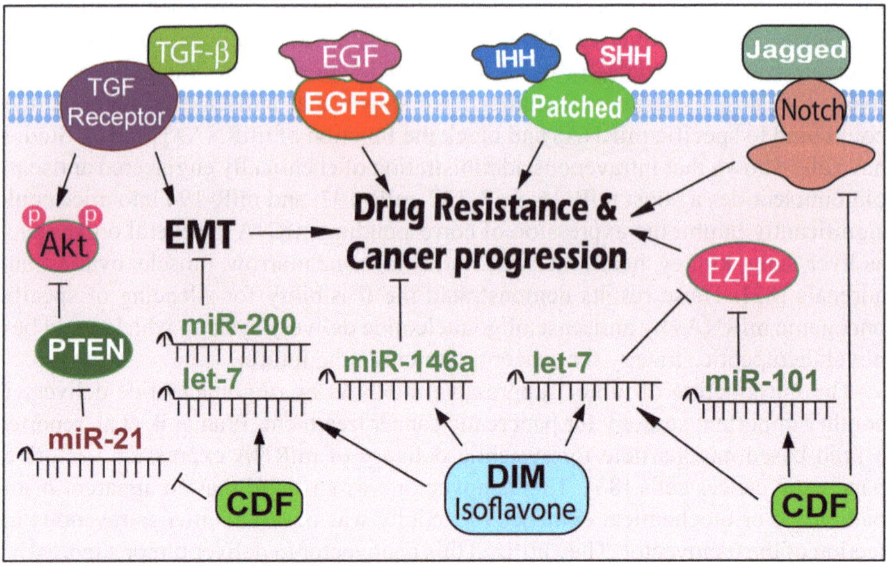

Fig. 10.3 The effects of natural agents on miRNA-mediated cellular signaling and biological behavior in the progression of pancreatic cancer

expression of specific miRNAs [84–87]. Through the regulation of specific miRNAs, the natural agents could increase the sensitivity of pancreatic cancer cells to conventional therapeutics and thereby inhibit invasion and metastasis of pancreatic cancer (Fig. 10.3). Due of the nontoxic features of natural agents, regulation of miRNAs by these natural agents combined with conventional chemotherapeutics could become a novel strategy for the treatment of pancreatic cancer with better treatment outcome, which requires in-depth investigation.

Pancreatic cancer shows many aggressive features, which is in part due to drug resistance characteristics. The existence of pancreatic CSCs and EMT-type cells in the tumor with deregulated expression of miRNAs could be responsible for drug resistance. We investigated the effects of several natural agents on miRNAs in pancreatic cancer cells especially in gemcitabine-resistance cells, which showed EMT phenotype. We found that isoflavone or DIM treatment could increase the expression of miR-200, decreased the expression of ZEB1, slug, and vimentin, and reversed EMT phenotype to epithelial characteristics (Fig. 10.3) [11]. Moreover, isoflavone or DIM treatment could also up-regulate the expression of let-7 [11], suggesting that these natural agents could be useful in the deregulated expression of miRNAs. Importantly, isoflavone or DIM treatment led to enhanced sensitivity of gemcitabine-resistant pancreatic cells to gemcitabine [11]. In addition, we also found that isoflavone or DIM could increase the expression of miR-146a and inhibit the invasion of pancreatic cancer through the regulation of EGFR and NF-κB signaling [88]. These results demonstrate that natural agents such as isoflavone and DIM could partially increase drug sensitivity and inhibit progression of pancreatic cancer through the regulation of miR-200, let-7, and miR-146a. Therefore, conventional chemotherapeutics combined with isoflavone or DIM could serve as novel strategies for better treatment of pancreatic cancer in the future.

Studies have also shown that curcumin and its synthetic analog could regulate the expression of miRNAs in pancreatic cancer. The miRNA expression profile altered by curcumin has been documented in pancreatic cancer cells [89]. It has been found that curcumin treatment could upregulate the expression of miR-22 and thereby could inhibit the expression of SP1 transcription factor and estrogen receptor 1, which are targets of miR-22 [89]. We have found that curcumin-analog diflourinated-curcumin (CDF) could inhibit the expression of miR-21 and could upregulate the expression of several miRNAs including miR-26a, miR-101, miR-146a, miR-200 family, and let-7 family whose expressions are typically lost in pancreatic cancer. This was found to be associated with the induction of drug sensitivity and the inhibition of tumor growth and aggressiveness in pancreatic cancer mediated in part through the regulation of specific miRNA targets such as EZH2 and Akt signaling (Fig. 10.3) [69, 90,–92].

5 Conclusions and Perspectives

In conclusion, emerging evidence clearly suggest that miRNAs play important roles in the development of normal pancreas and play critical roles in the development and progression of pancreatic cancer. During pancreas development, several miRNAs including miR-375, miR-7, miR-124, etc. have been shown to regulate exocrine and endocrine cell differentiation. These regulations could be mediated through the miRNA-specific target modulation of TGF-β, Notch, and Hedgehog signaling, which are the signal transduction pathways that are critically involved in organogenesis. Therefore, targeting these specific miRNAs could be a novel strategy for the regeneration of pancreatic endocrine cells toward the treatment of diabetes and other pancreatic endocrine diseases. In pancreatic cancer, the aberrant expression of miRNAs with deregulated cellular signaling could lead to the development and progression of pancreatic cancer. Moreover, several miRNAs are also known to regulate pancreatic CSCs and the acquisition and maintenance of EMT-phenotypic cells, which are typically responsible for de novo and acquired drug resistance in pancreatic cancer. Therefore, targeting specific miRNAs by oligonucleotides/nanoparticle vector delivery or regulation of miRNA by natural agents could become novel strategies for the treatment of pancreatic cancer in combination with conventional chemotherapeutics for achieving better treatment outcomes. Interestingly, because of the poor stability, immune system stimulation, and off-target effects of oligonucleotides, the use of nontoxic natural agents are very attractive for the regulation of miRNAs, and it is our expectation that natural agents could serve as a better strategy for targeting miRNAs for the combination treatment of pancreatic cancer. In addition, designing optimal nanoparticle vectors to deliver oligonucleotides efficiently could also improve the efficacy of treatment. However, more in vitro and in vivo experiments and clinical trials are needed to realize the value of targeting miRNAs for the treatment of pancreatic cancer. Therefore, newer treatment is urgently needed because such treatment will lead not only to a decrease in therapeutic resistance but also to the improvement of overall survival of patients diagnosed with pancreatic cancer.

Acknowledgments The authors' work cited in this review article was funded by grants from the National Cancer Institute, NIH (5R01CA083695, 5R01CA108535, 5R01CA131151, 5R01CA132794, 5R01CA154321, and 1R01CA164318 awarded to FHS). We also thank Puschelberg and Guido foundations for their generous contribution.

Reference

1. Lee RC, Feinbaum RL, Ambros V. The C. elegans heterochronic gene lin-4 encodes small RNAs with antisense complementarity to lin-14. Cell. 1993;75(5):843–54.
2. Reinhart BJ, Slack FJ, Basson M, et al. The 21-nucleotide let-7 RNA regulates developmental timing in Caenorhabditis elegans. Nature. 2000;403(6772):901–6.
3. Vandenboom Ii TG, Li, Philip PA, et al. MicroRNA and Cancer: tiny molecules with major implications. Curr Genomics. 2008;9:97–109.
4. Kasinski AL, Slack FJ. Epigenetics and genetics. MicroRNAs en route to the clinic: progress in validating and targeting microRNAs for cancer therapy. Nat Rev Cancer. 2011;11(12):849–64.
5. Pasquinelli AE. MicroRNAs and their targets: recognition, regulation and an emerging reciprocal relationship. Nat Rev Genet. 2012;13(4):271–82.
6. Pellettieri J, Sanchez AA. Cell turnover and adult tissue homeostasis: from humans to planarians. Annu Rev Genet. 2007;41:83–105.
7. Rando TA. Stem cells, ageing and the quest for immortality. Nature. 2006;441(7097):1080–86.
8. Peter ME. Regulating cancer stem cells the miR way. Cell Stem Cell. 2010;6(1):4–6.
9. Shimono Y, Zabala M, Cho RW, et al. Downregulation of miRNA-200c links breast cancer stem cells with normal stem cells. Cell. 2009;138(3):592–603.
10. Zheng T, Wang J, Chen X, et al. Role of microRNA in anticancer drug resistance. Int J Cancer. 2010;126(1):2–10.
11. Li Y, Vandenboom TG, Kong D, et al. Up-regulation of miR-200 and let-7 by natural agents leads to the reversal of epithelial-to-mesenchymal transition in gemcitabine-resistant pancreatic cancer cells. Cancer Res. 2009;69(16):6704–12.
12. Lynn FC, Skewes-Cox P, Kosaka Y, et al. MicroRNA expression is required for pancreatic islet cell genesis in the mouse. Diabetes. 2007;56(12):2938–45.
13. Massague J, Chen YG. Controlling TGF-beta signaling. Genes Dev. 2000;14(6):627–44.
14. Schonrock N, Humphreys DT, Preiss T, et al. Target gene repression mediated by miRNAs miR-181c and miR-9 both of which are down-regulated by amyloid-beta. J Mol Neurosci. 2012;46(2):324–35.
15. Altmae S, Martinez-Conejero JA, Esteban FJ, et al. MicroRNAs miR-30b, miR-30d, and miR-494 Regulate Human Endometrial Receptivity. Reprod Sci. 2013;20(3):308–17.
16. Gaedcke J, Grade M, Camps J, et al. The rectal cancer microRNAome—microRNA Expression in rectal cancer and matched normal mucosa. Clin Cancer Res. 2012;18(18):4919–30.
17. Gradwohl G, Dierich A, LeMeur M, et al. Neurogenin3 is required for the development of the four endocrine cell lineages of the pancreas. Proc Natl Acad Sci U S A. 2000;97(4):1607–11.
18. Apelqvist A, Li H, Sommer L, et al. Notch signalling controls pancreatic cell differentiation. Nature. 1999;400(6747):877–81.

19. Jensen J, Pedersen EE, Galante P, et al. Control of endodermal endocrine development by Hes-1. Nat Genet. 2000;24(1):36–44.
20. Schwitzgebel VM, Scheel DW, Conners JR, et al. Expression of neurogenin3 reveals an islet cell precursor population in the pancreas. Development. 2000;127(16):3533–42.
21. Knuckles P, Vogt MA, Lugert S, et al. Drosha regulates neurogenesis by controlling neurogenin 2 expression independent of microRNAs. Nat Neurosci. 2012;15(7):962–9.
22. Jee MK, Jung JS, Im YB, et al. Silencing of miR20a is crucial for Ngn1-mediated neuroprotection in injured spinal cord. Hum Gene Ther. 2012;23(5):508–20.
23. Hebrok M, Kim SK, St JB, et al. Regulation of pancreas development by hedgehog signaling. Development. 2000;127(22):4905–13.
24. Chen JS, Pedro MS, Zeller RW. miR-124 function during Ciona intestinalis neuronal development includes extensive interaction with the Notch signaling pathway. Development. 2011;138(22):4943–53.
25. Li Y, Zhang D, Chen C, et al. MicroRNA-212 displays tumor-promoting properties in non-small cell lung cancer cells and targets the hedgehog pathway receptor PTCH1. Mol Biol Cell. 2012;23(8):1423–34.
26. Northcott PA, Fernandez L, Hagan JP, et al. The miR-17/92 polycistron is up-regulated in sonic hedgehog-driven medulloblastomas and induced by N-myc in sonic hedgehog-treated cerebellar neural precursors. Cancer Res. 2009;69(8):3249–55.
27. Hornstein E, Mansfield JH, Yekta S, et al. The microRNA miR-196 acts upstream of Hoxb8 and Shh in limb development. Nature. 2005;438(7068):671–74.
28. Baroukh N, Ravier MA, Loder MK, et al. MicroRNA-124a regulates Foxa2 expression and intracellular signaling in pancreatic beta-cell lines. J Biol Chem. 2007;282(27):19575–88.
29. Joglekar MV, Parekh VS, Mehta S, et al. MicroRNA profiling of developing and regenerating pancreas reveal post-transcriptional regulation of neurogenin3. Dev Biol. 2007;311(2):603–12.
30. Simion A, Laudadio I, Prevot PP, et al. MiR-495 and miR-218 regulate the expression of the Onecut transcription factors HNF-6 and OC-2. Biochem Biophys Res Commun. 2010;391(1):293–98.
31. Poy MN, Eliasson L, Krutzfeldt J, et al. A pancreatic islet-specific microRNA regulates insulin secretion. Nature. 2004;432(7014):226–30.
32. El Ouaamari A, Baroukh N, Martens GA, et al. miR-375 targets 3†²-phosphoinositide-dependent protein kinase-1 and regulates glucose-induced biological responses in pancreatic beta-cells. Diabetes. 2008;57(10):2708–17.
33. Joglekar MV, Joglekar VM, Hardikar AA. Expression of islet-specific microRNAs during human pancreatic development. Gene Expr Patterns. 2009;9(2):109–13.
34. Dumortier O, Van Obberghen E. MicroRNAs in pancreas development. Diabetes Obes Metab. 2012;14 Suppl 3:22–8.
35. Wang C, Yao N, Lu CL, et al. Mouse microRNA-124 regulates the expression of Hes1 in P19 cells. Front Biosci (Elite Ed). 2010;2:127–32.
36. Plaisance V, Abderrahmani A, Perret-Menoud V, et al. MicroRNA-9 controls the expression of Granuphilin/Slp4 and the secretory response of insulin-producing cells. J Biol Chem. 2006;281(37):26932–42.
37. Tang X, Muniappan L, Tang G, et al. Identification of glucose-regulated miRNAs from pancreatic {beta} cells reveals a role for miR-30d in insulin transcription. RNA. 2009;15(2):287–93.
38. Lewis BP, Burge CB, Bartel DP. Conserved seed pairing, often flanked by adenosines, indicates that thousands of human genes are microRNA targets. Cell. 2005;120(1):15–20.
39. Jamieson NB, Morran DC, Morton JP, et al. MicroRNA molecular profiles associated with diagnosis, clinicopathologic criteria, and overall survival in patients with resectable pancreatic ductal adenocarcinoma. Clin Cancer Res. 2012;18(2):534–45.
40. Dillhoff M, Liu J, Frankel W, et al. MicroRNA-21 is overexpressed in pancreatic cancer and a potential predictor of survival. J Gastrointest Surg. 2008;12(12):2171–76.

41. Gregory PA, Bert AG, Paterson EL, et al. The miR-200 family and miR-205 regulate epithelial to mesenchymal transition by targeting ZEB1 and SIP1. Nat Cell Biol. 2008;10(5):593–601.
42. Ji Q, Hao X, Zhang M, et al. MicroRNA miR-34 inhibits human pancreatic cancer tumor-initiating cells. PLoS One. 2009;4(8):e6816.
43. Nalls D, Tang SN, Rodova M, et al. Targeting epigenetic regulation of miR-34a for treatment of pancreatic cancer by inhibition of pancreatic cancer stem cells. PLoS One. 2011;6(8):e24099.
44. Yu J, Li A, Hong SM, et al. MicroRNA alterations of pancreatic intraepithelial neoplasias. Clin Cancer Res. 2012;18(4):981–92.
45. Wang J, Sen S. MicroRNA functional network in pancreatic cancer: from biology to biomarkers of disease. J Biosci. 2011;36(3):481–91.
46. Szafranska AE, Davison TS, John J, et al. MicroRNA expression alterations are linked to tumorigenesis and non-neoplastic processes in pancreatic ductal adenocarcinoma. Oncogene. 2007;26(30):4442–52.
47. Chang TC, Wentzel EA, Kent OA, et al. Transactivation of miR-34a by p53 broadly influences gene expression and promotes apoptosis. Mol Cell. 2007;26(5):745–52.
48. Vogt M, Munding J, Gruner M, et al. Frequent concomitant inactivation of miR-34a and miR-34b/c by CpG methylation in colorectal, pancreatic, mammary, ovarian, urothelial, and renal cell carcinomas and soft tissue sarcomas. Virchows Arch. 2011;458(3):313–22.
49. Lodygin D, Tarasov V, Epanchintsev A, et al. Inactivation of miR-34a by aberrant CpG methylation in multiple types of cancer. Cell Cycle. 2008;7(16):2591–600.
50. Liu C, Kelnar K, Liu B, et al. The microRNA miR-34a inhibits prostate cancer stem cells and metastasis by directly repressing CD44. Nat Med. 2011;17(2):211–15.
51. Srivastava SK, Bhardwaj A, Singh S, et al. MicroRNA-150 directly targets MUC4 and suppresses growth and malignant behavior of pancreatic cancer cells. Carcinogenesis. 2011;32(12):1832–39.
52. Ribas J, Ni X, Haffner M, et al. miR-21: an androgen receptor-regulated microRNA that promotes hormone-dependent and hormone-independent prostate cancer growth. Cancer Res. 2009;69(18):7165–69.
53. Seike M, Goto A, Okano T, et al. MiR-21 is an EGFR-regulated anti-apoptotic factor in lung cancer in never-smokers. Proc Natl Acad Sci U S A. 2009;106(29):12085–90.
54. Si ML, Zhu S, Wu H, et al. miR-21-mediated tumor growth. Oncogene. 2007;26(19):2799–803.
55. Habbe N, Koorstra JB, Mendell JT, et al. MicroRNA miR-155 is a biomarker of early pancreatic neoplasia. Cancer Biol Ther. 2009;8(4):340–6.
56. Moriyama T, Ohuchida K, Mizumoto K, et al. MicroRNA-21 modulates biological functions of pancreatic cancer cells including their proliferation, invasion, and chemoresistance. Mol Cancer Ther. 2009;8(5):1067–74.
57. Park JK, Lee EJ, Esau C, et al. Antisense inhibition of microRNA-21 or -221 arrests cell cycle, induces apoptosis, and sensitizes the effects of gemcitabine in pancreatic adenocarcinoma. Pancreas. 2009;38(7):e190–9.
58. Basu A, Alder H, Khiyami A, et al. MicroRNA-375 and microRNA-221: potential noncoding RNAs associated with antiproliferative activity of benzyl isothiocyanate in pancreatic cancer. Genes Cancer. 2011;2(2):108–19.
59. Lee EJ, Gusev Y, Jiang J, et al. Expression profiling identifies microRNA signature in pancreatic cancer. Int J Cancer. 2007;120(5):1046–54.
60. Greither T, Grochola LF, Udelnow A, et al. Elevated expression of microRNAs 155, 203, 210 and 222 in pancreatic tumors is associated with poorer survival. Int J Cancer. 2010;126(1):73–80.
61. Gironella M, Seux M, Xie MJ, et al. Tumor protein 53-induced nuclear protein 1 expression is repressed by miR-155, and its restoration inhibits pancreatic tumor development. Proc Natl Acad Sci U S A. 2007;104(41):16170–75.

62. Gibbons DL, Lin W, Creighton CJ, et al. Contextual extracellular cues promote tumor cell EMT and metastasis by regulating miR-200 family expression. Genes Dev. 2009;23(18):2140–51.
63. Ji J, Yamashita T, Budhu A, et al. Identification of microRNA-181 by genome-wide screening as a critical player in EpCAM-positive hepatic cancer stem cells. Hepatology. 2009;50(2):472–80.
64. Jung DE, Wen J, Oh T, et al. Differentially expressed microRNAs in pancreatic cancer stem cells. Pancreas. 2011;40(8):1180–87.
65. Ji Q, Hao X, Zhang M, et al. MicroRNA miR-34 inhibits human pancreatic cancer tumor-initiating cells. PLoS One. 2009;4(8):e6816.
66. Hamada S, Satoh K, Masamune A, et al. Regulators of epithelial mesenchymal transition in pancreatic cancer. Front Physiol. 2012;3:254.
67. Bao B, Wang Z, Ali S, et al. Over-expression of FoxM1 leads to epithelial-mesenchymal transition and cancer stem cell phenotype in pancreatic cancer cells. J Cell Biochem. 2011;112(9):2296–306.
68. Bao B, Wang Z, Ali S, et al. Notch-1 induces epithelial-mesenchymal transition consistent with cancer stem cell phenotype in pancreatic cancer cells. Cancer Lett. 2011;307(1):26–36.
69. Bao B, Ali S, Banerjee S, et al. Curcumin analogue CDF inhibits pancreatic tumor growth by switching on suppressor microRNAs and attenuating EZH2 expression. Cancer Res. 2012;72(1):335–45.
70. Raponi M, Dossey L, Jatkoe T, et al. MicroRNA classifiers for predicting prognosis of squamous cell lung cancer. Cancer Res. 2009;69(14):5776–783.
71. Wang QZ, Xu W, Habib N, et al. Potential uses of microRNA in lung cancer diagnosis, prognosis, and therapy. Curr Cancer Drug Targets. 2009;9(4):572–94.
72. Nakata K, Ohuchida K, Mizumoto K, et al. MicroRNA-10b is overexpressed in pancreatic cancer, promotes its invasiveness, and correlates with a poor prognosis. Surgery. 2011;150(5):916–22.
73. Papaconstantinou IG, Manta A, Gazouli M, et al. Expression of microRNAs in patients with pancreatic cancer and its prognostic significance. Pancreas. 2012;42(1):67–71.
74. Yu J, Ohuchida K, Mizumoto K, et al. MicroRNA miR-17–5p is overexpressed in pancreatic cancer, associated with a poor prognosis, and involved in cancer cell proliferation and invasion. Cancer Biol Ther. 2010;10(8):748–57.
75. Yu J, Ohuchida K, Mizumoto K, et al. MicroRNA, hsa-miR-200c, is an independent prognostic factor in pancreatic cancer and its upregulation inhibits pancreatic cancer invasion but increases cell proliferation. Mol Cancer. 2010;9:169.
76. Bloomston M, Frankel WL, Petrocca F, et al. MicroRNA expression patterns to differentiate pancreatic adenocarcinoma from normal pancreas and chronic pancreatitis. JAMA. 2007;297(17):1901–08.
77. Roldo C, Missiaglia E, Hagan JP, et al. MicroRNA expression abnormalities in pancreatic endocrine and acinar tumors are associated with distinctive pathologic features and clinical behavior. J Clin Oncol. 2006;24(29):4677–84.
78. Nagao Y, Hisaoka M, Matsuyama A, et al. Association of microRNA-21 expression with its targets, PDCD4 and TIMP3, in pancreatic ductal adenocarcinoma. Mod Pathol. 2012;25(1):112–21.
79. Babar IA, Cheng CJ, Booth CJ, et al. Nanoparticle-based therapy in an in vivo microRNA-155 (miR-155)-dependent mouse model of lymphoma. Proc Natl Acad Sci U S A. 2012;109(26):E1695–704.
80. Sureban SM, May R, Mondalek FG, et al. Nanoparticle-based delivery of siDCAMKL-1 increases microRNA-144 and inhibits colorectal cancer tumor growth via a Notch-1 dependent mechanism. J Nanobiotechnology. 2011;9:40.
81. Fonsato V, Collino F, Herrera MB, et al. Human liver stem cell-derived microvesicles inhibit hepatoma growth in SCID mice by delivering antitumor microRNAs. Stem Cells. 2012;30(9):1985–98.
82. Krutzfeldt J, Rajewsky N, Braich R, et al. Silencing of microRNAs in vivo with 'antagomirs'. Nature. 2005;438(7068):685–9.

83. Pramanik D, Campbell NR, Karikari C, et al. Restitution of tumor suppressor microRNAs using a systemic nanovector inhibits pancreatic cancer growth in mice. Mol Cancer Ther. 2011;10(8):1470–80.
84. Li Y, VandenBoom TG, Kong D, et al. Up-regulation of miR-200 and let-7 by natural agents leads to the reversal of epithelial-to-mesenchymal transition in gemcitabine-resistant pancreatic cancer cells. Cancer Res. 2009;69(16):6704–12.
85. Sun M, Estrov Z, Ji Y, et al. Curcumin (diferuloylmethane) alters the expression profiles of microRNAs in human pancreatic cancer cells. Mol Cancer Ther. 2008;7(3):464–73.
86. Melkamu T, Zhang X, Tan J, et al. Alteration of microRNA expression in vinyl-carbamate-induced mouse lung tumors and modulation by the chemopreventive agent indole-3-carbinol. Carcinogenesis. 2010;31(2):252–8.
87. Tsang WP, Kwok TT. Epigallocatechin gallate up-regulation of miR-16 and induction of apoptosis in human cancer cells. J Nutr Biochem. 2010;21(2):140–6.
88. Li Y, Vandenboom TG, Wang Z, et al. miR-146a suppresses invasion of pancreatic cancer cells. Cancer Res. 2010;70(4):1486–95.
89. Sun M, Estrov Z, Ji Y, et al. Curcumin (diferuloylmethane) alters the expression profiles of microRNAs in human pancreatic cancer cells. Mol Cancer Ther. 2008;7(3):464–73.
90. Bao B, Ali S, Kong D, et al. Anti-tumor activity of a novel compound-CDF is mediated by regulating miR-21, miR-200, and PTEN in pancreatic cancer. PLoS One. 2011;6(3):e17850.
91. Ali S, Ahmad A, Banerjee S, et al. Gemcitabine sensitivity can be induced in pancreatic cancer cells through modulation of miR-200 and miR-21 expression by curcumin or its analogue CDF. Cancer Res. 2010;70(9):3606–17.
92. Soubani O, Ali AS, Logna F, et al. Re-expression of miR-200 by novel approaches regulates the expression of PTEN and MT1-MMP in pancreatic cancer. Carcinogenesis. 2012;33(8):1563–71.

Chapter 11
Epigenetic Regulation of microRNA Genes in Colorectal Cancer

Hiromu Suzuki, Eiichiro Yamamoto and Reo Maruyama

Abstract DNA methylation and histone modification are epigenetic changes that play key roles in the dysregulation of tumor-related genes, thereby affecting numerous cellular processes, including cell proliferation, cell adhesion, apoptosis, and metastasis. In recent years, studies have shown that microRNAs (miRNAs) play important roles in the development of colorectal cancer (CRC), and that epigenetic mechanisms are deeply involved in their dysregulation. Specifically, technological advances that enable comprehensive analysis of miRNA expression profiles and the epigenome in CRC cells have led to the identification of a large number of epigenetically regulated miRNAs. As with protein-coding genes, it appears that miRNA genes involved in regulating cancer-related pathways are silenced in association with CpG island hypermethylation and altered histone modification. Aberrant DNA methylation of miRNA genes is a potentially useful biomarker for detecting CRC or predicting its outcome. Moreover, re-expression of the miRNAs could be an effective approach to cancer therapy, and unraveling the relationship between epigenetic alteration and miRNA dysregulation may lead to the discovery of new therapeutic targets.

Keywords Colorectal cancer (CRC) · Epigenetic silencing · CpG island (CGI) · Methylation · Histone · Chromatin signature · Promoter · Biomarker

1 Introduction

Epigenetics are inherited factors that influence gene activity but do not alter primary DNA sequences; among them, DNA methylation and histone modification are key events that silence gene expression [1, 2]. Colorectal cancer (CRC) is thought to arise through the accumulation of multiple genetic alterations that lead to activation of oncogenes and loss of function of tumor suppressor genes. However, a growing body of evidence now suggests that, in addition to genetic alterations, epigenetic changes such as DNA methylation and histone modification also play crucial roles in the development and progression of CRC [3, 4].

H. Suzuki (✉) · E. Yamamoto · R. Maruyama
Department of Molecular Biology, Sapporo Medical University, S1 W17, Chuo-Ku, Sapporo 060-8556, Japan
e-mail: hsuzuki@sapmed.ac.jp

S. R. Singh, P. Rameshwar (eds.), *MicroRNA in Development and in the Progression of Cancer,* DOI 10.1007/978-1-4899-8065-6_11,

The first-identified cancer-related change in DNA methylation was genome-wide hypomethylation [5]. Subsequently, it became apparent that hypermethylation of 5′ CpG islands (CGIs) is crucial for silencing tumor suppressor genes [5]. Although the classical two-hit theory posits that tumor suppressor genes are inactivated by gene mutation or deletion, it is now recognized that DNA hypermethylation is a third mechanism by which inactivation of tumor suppressor genes occurs, and that it plays a significant role in tumorigenesis. In addition to DNA methylation, acetylation and methylation of lysine residues in histones H3 and H4 also play key roles in gene regulation [1, 2]. In general, increase in histone acetylation is associated with increase in transcriptional activity, while decrease in acetylation is associated with gene repression. Methylation of histones H3 and H4 is also deeply involved in gene transcription, and it is well documented that methylation of H3 lysine 4 is associated with active transcription, while methylation of H3 lysines 9 and 27 is associated with gene repression.

In recent years, microRNAs (miRNAs) have been attracting the interest of many researchers, and the accumulated evidence suggests that they are strongly involved in cancer [6, 7]. miRNAs are a class of small noncoding RNAs that regulate gene expression by inducing translational inhibition or direct degradation of target mRNAs through base pairing to partially complementary sites [6]. miRNA genes are transcribed as large precursors, called pri-miRNAs, which may encode multiple miRNAs in a polycistronic arrangement. Pri-miRNAs are then processed by the RNase III enzyme Drosha and its cofactor Pasha to produce ~70 nucleotide (nt) hairpin-structured second precursors, called pre-miRNAs. The pre-miRNAs are transported to the cytoplasm and processed by another RNase III enzyme, Dicer, to generate mature miRNA products. miRNAs are highly conserved among species and play critical roles in a variety of biological processes, including cell proliferation, development, differentiation, and apoptosis. In addition, subsets of miRNAs are thought to act as tumor suppressor genes or oncogenes, and their dysregulation is a common feature of human cancers. Expression of miRNAs is generally downregulated in tumor tissues, as compared to corresponding healthy tissues [8], which suggests that some miRNAs may behave as tumor suppressors in some tumors. Although the mechanism underlying the alteration of miRNA expression in cancer is still not fully understood, recent studies have shown that epigenetic alterations are important mechanisms by which the normal patterns of miRNA expression are disrupted in cancer [9, 10]. In this chapter, we highlight the contribution made by epigenetic alteration to the dysregulation of miRNA in CRC and discuss their clinical application as biomarkers and therapeutic targets.

2 Screening for Epigenetically Silenced miRNA Genes in CRC

Genome-wide gene expression analysis is a powerful tool for identifying epigenetically silenced genes in cancer [11]. Expression of epigenetically silenced genes can be restored by inhibiting DNA methyltransferases (DNMTs) or histone deacetylases (HDACs), and the first evidence of an epigenetic mechanism involved in silencing miRNAs in cancer came from just such a pharmacological unmasking experiment. Microarray-based screening of miRNAs in T24 human bladder cancer cells and normal fibroblasts (LD419) followed by treatment with 5-aza-2′-deoxycytidine (5-aza-dC), a DNMT inhibitor, and 4-phenylbutyric acid (PBA), an HDAC inhibitor, revealed cancer-specific upregulation of miR-127 by the drugs [12]. The miR-127 gene is embedded within a CGI, and its upregulation was associated not only with DNA demethylation but also with acetylation of histone H3 and trimethylation of histone H3 lysine 4 (H3K4me3), which are marks of active transcription.

Lujambio et al. carried out similar genetic unmasking experiments in CRCusing the HCT116 cell line plus the same cell line with genetic disruption of both DNMT1 and DNMT3B (DNMTs KO) [13]. Microarray analysis revealed that 18 of 320 miRNAs were upregulated in the DNMTs knockout (KO) cells. In addition, CGI hypermethylation was found for miR-124, miR-517c, and miR-373 in HCT116 cells. Among these, miR-517c and miR-373 are also densely methylated in normal colonic tissue, suggesting tissue-specific methylation. By contrast, all three miR-124 family genes (miR-124-1, miR-124-2, and miR-124-3) exhibited tumor-specific methylation, and are frequently methylated in primary CRC tissues (42 of 56; 75 %). miR-124 is thought to exert tumor suppressor effects by targeting cyclin-dependent kinase 6 (CDK6), and epigenetic silencing of miR-124 leads to CDK6 activation and Rb phosphorylation [13, 14]. We carried out similar screening by using quantitative reverse transcriptase polymerase chain reaction (qRT-PCR) to analyze 157 miRNAs in three CRC cell lines (HCT116, DLD1, and RKO) treated with or without 5-aza-dC, as well as in DNMTs KO cells [15]. We identified 37 miRNAs that were significantly upregulated upon inhibition of DNMTs in CRC cells. These included miRNAs known to be silenced in CRC (e.g., miR-127 and miR-124) as well as novel targets not previously reported to be epigenetically silenced in CRC (e.g., miR-34b and miR-34c; see below).

Epigenetically silenced miRNA genes have also been identified through genome-wide DNA methylation analysis. Yang et al. performed deep sequencing of methylated DNA binding domain (MBD)-isolated DNA in HCT116 cells [16]. They found that 64 miRNA genes were located within 500 base pairs (bp) of robustly methylated regions. Among them, 5 were located in imprinting loci and 13 were previously reported to be methylated. In addition, 46 miRNA genes showed potentially novel DNA methylation. Among these, the expression of 18 were consistent with the DNA methylation status, and 8 were confirmed to be novel miRNAs regulated by DNA methylation. Subsequent functional assays revealed that ectopic expression of miR-941 and miR-1247 suppressed the growth and migration of CRC cells.

3 DNA Methylation of miRNA Genes in CRC

As mentioned, recent advances in microarray and sequencing technologies have enabled comprehensive analysis of the epigenome and miRNA expression in cancer cells. As a result, the list of miRNA genes silenced by methylation in cancer is

Table 11.1 miRNA genes silenced in association with CGI methylation in CRC

Name	Target genes	Reference
miR-1-1	FOXP1, MET, HDAC4, ANXA2, BDNF	[41, 55]
miR-9 family	FGFR1, CDK6, CDX2, MMP14	[23, 26, 56, 57, 58]
miR-34 family	MET, CDK4, CCNE2, CDK6, C-MYC, E2F3	[15,19, 23]
miR-124 family	CDK6	[13]
miR-127	BCL6	[12]
miR-129-2	SOX4	[26, 59]
miR-137	CDK6, LSD1	[40]
miR-148a	TGIF2	[19]
miR-200 family	ZEB1, ZEB2	[32]
miR-345	BAG3	[60]
miR-373		[61]
miR-941		[16]
miR-1237		[16]
miR-1247		[16]

rapidly growing [9, 10]. In this section, we describe representative miRNA genes silenced in association with CGI methylation in CRC (see also Table 11.1).

3.1 miR-34 Family

Members of the miR-34 gene family (miR-34a, miR-34b, and miR-34c) are direct targets of p53, and their ectopic expression in cancer cells induces cell cycle arrest and apoptosis [17, 18]. Within the human genome, miR-34a is located on chromosome 1p36, while miR-34b and miR-34c are co-transcribed from a single transcription unit on chromosome 11q23, and both are targets of CGI hypermethylation in various malignancies, including CRC [15, 19–22]. As mentioned above, our screening experiment identified miR-34b/c as a novel epigenetically silenced miRNA gene in CRC (Fig. 11.1) [15]. The CGI located in the proximal upstream of miR-34b/c is densely methylated in CRC cells, and restoration of its expression is associated with reduced DNA methylation and increased H3K4me3 in this region. Interestingly, this CGI region functions as a bidirectional promoter, and its hypermethylation also leads to epigenetic silencing of another tumor suppressor candidate, BTG4, which is transcribed in the opposite direction of miR-34b/c. Introduction of miR-34b/c into cancer cells lead to the downregulation of candidate target genes, including MET, cyclin-dependent kinase 4 (CDK4), cyclin E2 (CCNE2), and MYC [15, 23]. Likewise, restoration of endogenous miRNA expression through demethylation also downregulates target genes, suggesting miRNAs could be important targets for epigenetic cancer therapy [15]. These findings, as well as its contribution to the p53 network, strongly suggest that miR-34 family members act as tumor suppressors in cancer.

The miR-34b/c CGI is methylated in approximately 90 % of primary CRCs [15]. The high frequency of its methylation in CRC implies not only that miR-34b/c may function as an important tumor suppressor, but also that it could be an useful

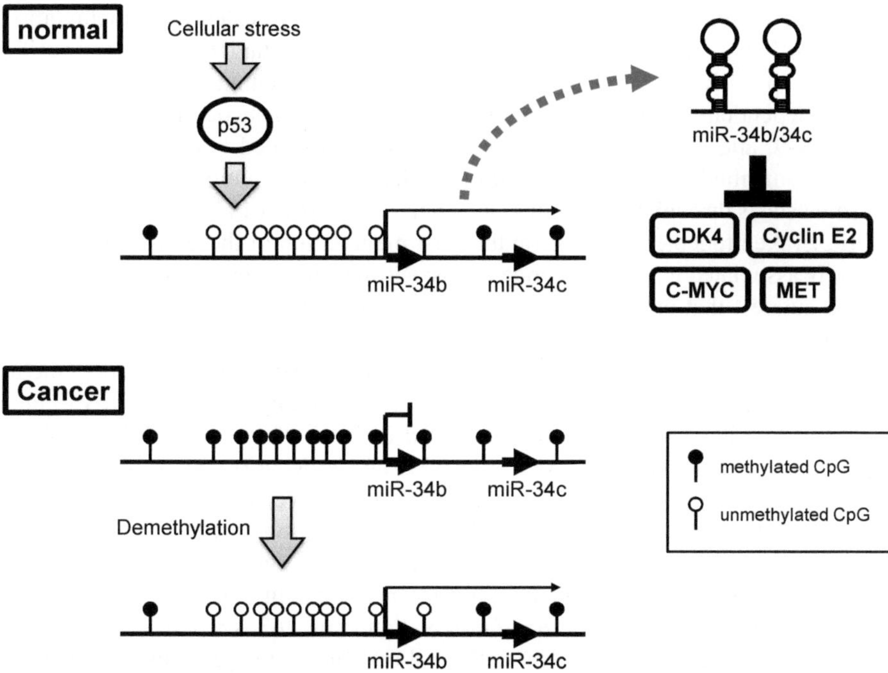

Fig. 11.1 Epigenetic silencing of miR-34b/c in CRC. In normal cells, the miR-34b/c CGI is unmethylated and miR-34b/c is transcribed upon p53 activation. In CRC cells, the CGI is densely methylated and miR-34b/c is transcriptionally inactivated

biomarker for detection of CRC. For instance, Kalimutho et al. reported that miR-34b/c is methylated in 97.5 % (79 out of 82) of primary CRCs, and its methylation can be detected in 75 % (21 out of 28) of fecal specimens from CRC patients, which suggests that miR-34b/c methylation could be an ideal biomarker for CRC screening [24]. In addition, we showed the usefulness of DNA methylation detected in the mucosal wash fluid obtained during colonoscopy [25]. Cytologic and KRAS mutation analysis suggest that mucosal wash fluid from invasive tumors contain greater number of tumor cells than wash fluid from noninvasive tumors. Taking advantage of this finding, we showed that DNA methylation of a panel of four genes, including miR-34b/c, in mucosal wash fluid could be a predictive marker of the invasiveness of CRC.

3.2 miRNA Gene Methylation Associated with Metastasis

Methylation of several miRNA genes has been implicated in cancer metastasis. Lujambio et al. searched for epigenetically silenced miRNAs in cancer cell lines established from lymph node metastases from colon cancer (SW620), melanoma (IGR37), and head and neck cancer (SIHN001B) [23]. Of the 389 miRNAs analyzed in a microarray, 57 were upregulated by 5-aza-dC treatment, among which 27 were embedded within CGIs. Bisulfite sequencing analysis revealed hypermethylation of 16 miRNA genes, among which they identified 3 as targets of tumor-specific methylation (miR-148a, miR-34b/c, and miR-9 family). Within the human genome, three independent loci (miR-9-1, miR-9-2, and miR-9-3) encode the identical mature miR-9, and all three of these loci are hypermethylated in cancer. Functional studies confirmed that miR-148a and miR-34b/c suppress tumor growth, invasion, and metastasis through the targeting oncogenes, such as C-MYC, E2F3, CDK6, and TGIF2. Taken together with the finding that methylation of these miRNA genes is strongly associated with lymph node metastasis in primary tumors, the summarized results suggest that epigenetic silencing contributes to the development of cancer metastasis.

Several independent studies also confirmed an association between miR-9 methylation and metastasis. Bandres et al. used a sequential approach to identify tumor-suppressive miRNAs epigenetically silenced in CRC. They selected five miRNA genes (miR-9, miR-124, miR-129, miR-137, and miR-149) that were frequently downregulated in CRC and were located within 1,000 bp of CGIs. Of those, three genes (miR-9-1, miR-129-2, and miR-137) exhibited tumor-specific methylation in primary CRC [26]. The methylation of miR-9-1 was frequently seen in advanced CRCs and was associated with lymph node involvement, vascular invasion, and metastasis. Moreover, methylation of miR-9-1 and miR-9-3 is reportedly associated with metastatic recurrence of clear cell-renal cell carcinoma [27], and methylation of miR-9-3 is associated with lymphovascular involvement and reduced survival rate in gastric cancer [28]. Thus, methylation of these miRNA genes could be a useful maker for predicting tumor prognosis and metastatic behavior. In addition, restoration of the silenced miRNAs could be an effective strategy for treating cancer patients with metastatic disease.

3.3 miR-200 Family

The miR-200 gene family (miR-200a, miR-200b, miR-200c, miR-141, and miR-429) encode key regulators of epithelial–mesenchymal transition (EMT). They act by directly targeting zinc finger E-box binding homeobox 1 (ZEB1) and ZEB2, which are transcriptional repressors that downregulate CDH1 [29–31]. Within the human genome, the miR-200 family genes are grouped into two polycistronic units, miR-200b/200a/429 and miR-200c/141, located on chromosomes 1 and 12, respectively [32]. Several studies have shown that expression of the miR-200 family is regulated by epigenetic mechanisms in both normal and cancer cells [33, 34], and that methylation of miR-200 family genes is associated with an invasive phenotype in breast, bladder, and non-small cell lung cancer [35–37].

Davalos et al. demonstrated that the CGIs upstream of both miR200 polycistronic units (miR-200b/200a/429 and miR-200c/141) are unmethylated in cancer cells with epithelial features, but are methylated and silenced in transformed cells with mesenchymal characteristics [32]. They also found that in primary normal colonic mucosa, the CGIs of miR-200 family genes are unmethylated within crypts (epithelial component), but are hypermethylated in stromal cells (mesenchymal component). In primary CRC tumors, the stromal component again shows hypermethylation of both CGIs, while the tumor epithelium exhibited hypermethylation of the miR-200c CGI and reduced expression of E-cadherin. These results indicate that the epigenetic plasticity of the miR-200 family genes plays key roles in the evolution of malignant tumors.

3.4 miR-137

miR-137 is reportedly methylated in various human malignancies, including oral cancer, head and neck squamous carcinoma, gastric cancer, and CRC [20, 26, 38, 39]. One recent study by Balaguer et al. showed that miR-137 is frequently methylated in CRC cell lines (100 %), primary CRC (81.4 %), and adenoma (82.3 %). By contrast, it is seen less frequently in normal colonic mucosae from CRC patients (14.4 %) and from healthy individuals (4.7 %), which suggests miR-137 methylation is an early event during colorectal tumorigenesis [40]. Ectopic expression of miR-137 in CRC cells suppresses cell proliferation, and gene expression microarray analysis revealed that a number of candidate target genes, including LDS1, AURKA, and CDK6 are downregulated by miR-137. These results suggest the potential therapeutic value of miR-137 in CRC patients.

4 Chromatin Signatures miRNA Genes in CRC

4.1 Identification of miRNA Gene Promoters

Most of the miRNA genes reportedly hypermethylated in cancer are embedded within CGIs or are located proximal to them. However, miRNA expression profiles in cancer cells suggest that a larger number of miRNAs are potential targets of epigenetic silencing. For instance, by comparing miRNA expression signatures in CRC cell lines and normal colonic mucosa, we found that a majority of miRNAs are downregulated in CRC [41]. We also found that more than half of the downregulated miRNAs were upregulated by DNA demethylation. However, because of the poor annotation of primary miRNA genes, the precise locations of the promoters and transcription start sites (TSSs) are not fully elucidated, which in turn hampers our understanding of the epigenetic dysregulation of miRNAs.

In previous studies, miRNA gene promoters were identified based on specific genomic features such as RNA pol II-binding patterns [23, 24], evolutionally conserved regions [25], and computational prediction [42, 43]. It is also well documented that active promoters are marked by H3K4me3, and such histone modifications have been used successfully to identify miRNA gene promoters or TSSs in human embryonic stem (ES) cells and cancer cells [44, 45]. Recently, we carried

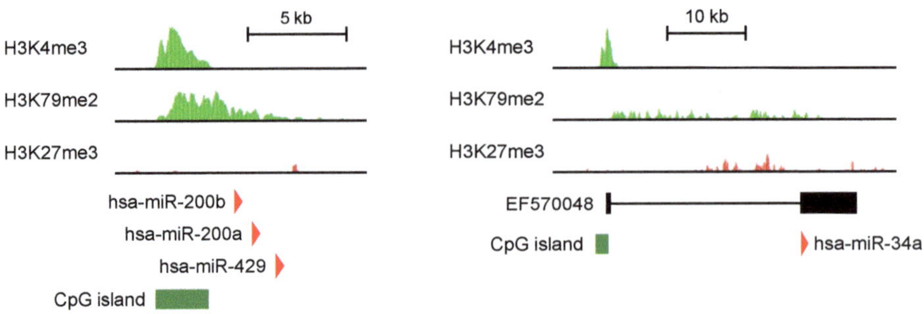

Fig. 11.2 Chromatin signatures of transcriptionally active miRNA genes in CRC cells. ChIP-seq results for H3K4me3, H3K79me2, and H3K27me3 in transcriptionally active miRNA genes in HCT116 cells are shown

out genome-wide profiling of chromatin signatures in CRC cells and identified the active promoter regions of miRNA genes [41]. Using HCT116 and isogenic DNMTs KO cells, we performed chromatin immunoprecipitation (ChIP)-seq analysis with antibodies against H3K4me3, which marks active promoters; dimethylated histone H3 lysine 79 (H3K79me2), which is associated with transcriptional elongation; and trimethylated histone H3 lysine 27 (H3K27me3), which is a repressive mark. Representative chromatin signatures of the actively transcribed miRNA genes are shown in Fig. 11.2. Both the miR-200b cluster (miR-200b, miR-200a, and miR-429) and miR-34a are abundantly expressed in HCT116 cells, and the H3K4me3 mark is enriched in the upstream CGI region. Gene bodies are marked by H3K79me2, whereas both genes almost completely lack the repressive H3K27me3 mark. Histone marks are also strong indicators of promoter regions located far upstream of the miRNA coding region. For instance, although the TSS of the host gene of miR-34a is located more than 20 kb upstream of the pre-miR-34a coding region, the H3K4me3 mark is clearly enriched in the TSS region (Fig. 11.2).

We used histone marks to identify active promoter regions in HCT116 and DNMTs KO cells [41]. More than half of miRNA genes are located within the introns of protein-coding or long noncoding RNA genes, and it is generally believed that such intragenic miRNAs share common promoters with their host genes [46]. Consistent with that idea, we identified the putative promoter regions of 166 intragenic miRNAs, and most were at the TSS of the host genes. For the intergenic miRNAs, we searched 10 kb upstream for H3K4me3 marks and identified the putative promoters for 67 of the genes. In total, we identified the putative promoters of 174 transcript units, encoding 233 distinct pre-miRNAs. On the other hand, the promoters of 135 miRNAs, most of which were located in intergenic regions, remain unidentified despite their positive expression in CRC cells.

4.2 Identification of Epigenetically Silenced miRNA Genes

We next identified epigenetically silenced miRNA genes using the following criteria: (a) miRNA showing little or no expression and an H3K4me3 mark in HCT116 cells or (b) miRNA showing restored expression and an H3K4me3 mark in DNMTs

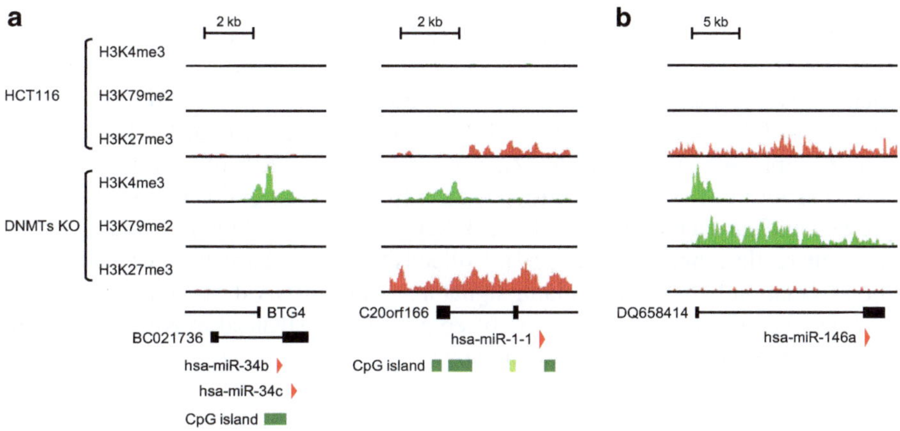

Fig. 11.3 Chromatin signatures of epigenetically silenced miRNA genes in CRC cells. **a** ChIP-seq results for CGI-methylated miRNA genes in HCT116 cells. Note that the CGI is hypermethylated in HCT116 cells but is unmethylated in DNMTs KO cells. **b** ChIP-seq results for a miRNA gene without a promoter CGI

KO cells. We identified 47 pre-miRNA genes encoded in 37 pri-miRNAs as potential targets of epigenetic silencing in HCT116 cells. Among them, the promoters of 22 pri-miRNAs were associated with CGIs, and all of them were hypermethylated in HCT116 cells. In most cases, DNA demethylation led to increase in H3K4me3 and H3K27me3 marks in the methylated CGIs of the miRNA genes, and H3K79me2 marks were not restored by demethylation (Fig. 11.3a). This situation is similar to that of protein-coding genes hypermethylated in cancer. Studies have shown that tumor suppressor genes with CGI methylation retain repressive histone modifications (H3K9me3 and H3K27me3), even after DNA demethylation [47, 48]. Moreover, it was further demonstrated that DNA demethylation never results in restoration of the H3K79me2 mark in tumor suppressor genes with CGI hypermethylation [49]. These observations suggest that once a CGI is hypermethylated in cancer, DNA demethylation alone is not sufficient to return to a full euchromatin state.

The CGI-methylated miRNA genes include a number of previously reported cancer-related miRNAs, including the miR-124 family, miR-9 family, miR-34b/c, and miR-129-2. In addition, we identified miR-1-1 as a novel target of epigenetic silencing in CRC (Fig. 11.3a). The CGI upstream of miR-1-1 is hypermethylated in 67 % of CRC cell lines (6 out of 9), 78 % of primary CRCs (70 out of 90), and 69 % of colorectal adenoma (54 out of 78), suggesting that methylation of miR-1-1 is an early and frequent event in colorectal tumorigenesis. Functional analysis revealed that ectopic expression of miR-1 in CRC cells suppresses its putative target genes, including annexin A2 (ANXA2) and brain-derived neurotrophic factor (BDNF), and inhibits cell growth, migration, and invasion. Reduced expression of miR-1 is also reported in primary CRCs [50], and one recent study showed that miR-1 exerts tumor-suppressive effects by targeting MET in CRC cells [51].

Our analysis revealed that a number of miRNAs without promoter CGIs are also potential targets of epigenetic silencing in CRC. These miRNAs were identified through restoration of both their expression and H3K4me3 mark upon DNA demethylation, but the H3K79me2 and H3K27me3 signatures varied among genes. Interestingly, miR-146a is characterized by lack of active histone marks and enrichment of H3K27me3 in HCT116 cells, but both H3K4me3 and H3K79me2 were restored in DNMTs KO cells (Fig. 11.3b). Earlier studies provided evidence of its tumor suppressor roles in various cancers. For instance, loss of miR-146a was reported in hormone-refractory prostate cancer [52], and expression of miR-146a suppressed nuclear factor-kappa B (NF-κB) activity and metastatic potential in breast and pancreatic cancer cells [53, 54]. Collectively, these findings offer insight into the association between chromatin signatures and miRNA dysregulation in CRC, and they suggest that re-expression of epigenetically silenced miRNAs may contribute to the effects of epigenetic therapy.

5 Concluding Remarks

In this chapter, we highlighted the roles of epigenetic alteration in the dysregulation of miRNAs in CRC. Aberrant DNA methylation and histone modification are major mechanisms underlying miRNA dysregulation in CRC, and methylation of a subset of miRNA genes may be useful biomarkers for detecting cancer and/or predicting clinical outcome. Furthermore, restoration of silenced tumor-suppressive miRNAs in cancer cells could be a promising strategy for cancer treatment. We anticipate that further studies of the cancer epigenome and miRNAs will lead to the discovery of a variety of novel biomarkers and potential therapeutic targets.

Acknowledgments We thank Dr. William Goldman for editing the manuscript.

References

1. Baylin SB, Jones PA. A decade of exploring the cancer epigenome—biological and translational implications. Nat Rev Cancer. 2011;11(10):726–34.
2. Dawson MA, Kouzarides T. Cancer epigenetics: from mechanism to therapy. Cell. 2012;150(1):12–27.
3. Lao VV, Grady WM. Epigenetics and colorectal cancer. Nat Rev Gastroenterol Hepatol. 2011;8(12):686–700.
4. van Engeland M, Derks S, Smits KM, Meijer GA, Herman JG. Colorectal cancer epigenetics: complex simplicity. J Clin Oncol. 2011;29(10):1382–91.
5. Feinberg AP, Tycko B. The history of cancer epigenetics. Nat Rev Cancer. 2004;4(2):143–53.
6. Esquela-Kerscher A, Slack FJ. Oncomirs—microRNAs with a role in cancer. Nat Rev Cancer. 2006;6(4):259–69.
7. Croce CM. Causes and consequences of microRNA dysregulation in cancer. Nat Rev Genet. 2009;10(10):704–14.
8. Lu J, Getz G, Miska EA, Alvarez-Saavedra E, Lamb J, et al. MicroRNA expression profiles classify human cancers. Nature. 2005;435(7043):834–8.
9. Suzuki H, Maruyama R, Yamamoto E, Kai M. DNA methylation and microRNA dysregulation in cancer. Mol Oncol. 2012;6(6):567–78.

10. Lopez-Serra P, Esteller M. DNA methylation-associated silencing of tumor-suppressor microRNAs in cancer. Oncogene. 2012;31(13):1609–22.
11. Suzuki H, Gabrielson E, Chen W, Anbazhagan R, van Engeland M, et al. A genomic screen for genes upregulated by demethylation and histone deacetylase inhibition in human colorectal cancer. Nat Genet. 2002;31(2):141–9.
12. Saito Y, Liang G, Egger G, Friedman JM, Chuang JC, et al. Specific activation of microRNA-127 with downregulation of the proto-oncogene BCL6 by chromatin-modifying drugs in human cancer cells. Cancer Cell. 2006;9(6):435–43.
13. Lujambio A, Ropero S, Ballestar E, Fraga MF, Cerrato C, et al. Genetic unmasking of an epigenetically silenced microRNA in human cancer cells. Cancer Res. 2007;67(4):1424–9.
14. Agirre X, Vilas-Zornoza A, Jimenez-Velasco A, Martin-Subero JI, Cordeu L, et al. Epigenetic silencing of the tumor suppressor microRNA Hsa-miR-124a regulates CDK6 expression and confers a poor prognosis in acute lymphoblastic leukemia. Cancer Res. 2009;69(10):4443–53.
15. Toyota M, Suzuki H, Sasaki Y, Maruyama R, Imai K, et al. Epigenetic silencing of microRNA-34b/c and B-cell translocation gene 4 is associated with CpG island methylation in colorectal cancer. Cancer Res. 2008;68(11):4123–32.
16. Yan H, Choi AJ, Lee BH, Ting AH. Identification and functional analysis of epigenetically silenced microRNAs in colorectal cancer cells. PLoS One. 2011;6(6):e20628.
17. He L, He X, Lim LP, de Stanchina E, Xuan Z, et al. A microRNA component of the p53 tumour suppressor network. Nature. 2007;447(7148):1130–4.
18. Bommer GT, Gerin I, Feng Y, Kaczorowski AJ, Kuick R, et al. p53-mediated activation of miRNA34 candidate tumor-suppressor genes. Curr Biol. 2007;17(15):1298–307.
19. Lodygin D, Tarasov V, Epanchintsev A, Berking C, Knyazeva T, et al. Inactivation of miR-34a by aberrant CpG methylation in multiple types of cancer. Cell Cycle. 2008;7(16):2591–600.
20. Kozaki K, Imoto I, Mogi S, Omura K, Inazawa J. Exploration of tumor-suppressive microRNAs silenced by DNA hypermethylation in oral cancer. Cancer Res. 2008;68(7):2094–105.
21. Suzuki H, Yamamoto E, Nojima M, Kai M, Yamano HO, et al. Methylation-associated silencing of microRNA-34b/c in gastric cancer and its involvement in an epigenetic field defect. Carcinogenesis. 2010;31(12):2066–73.
22. Corney DC, Hwang CI, Matoso A, Vogt M, Flesken-Nikitin A, et al. Frequent downregulation of miR-34 family in human ovarian cancers. Clin Cancer Res. 2010;16(4):1119–28.
23. Lujambio A, Calin GA, Villanueva A, Ropero S, Sanchez-Cespedes M, et al. A microRNA DNA methylation signature for human cancer metastasis. Proc Natl Acad Sci U S A. 2008;105(36):13556–61.
24. Kalimutho M, Di Cecilia S, Del Vecchio Blanco G, Roviello F, Sileri P, et al. Epigenetically silenced miR-34b/c as a novel faecal-based screening marker for colorectal cancer. Br J Cancer. 2011;104(11):1770–8.
25. Kamimae S, Yamamoto E, Yamano HO, Nojima M, Suzuki H, et al. Epigenetic alteration of DNA in mucosal wash fluid predicts invasiveness of colorectal tumors. Cancer Prev Res (Phila). 2011;4(5):674–83.
26. Bandres E, Agirre X, Bitarte N, Ramirez N, Zarate R, et al. Epigenetic regulation of microRNA expression in colorectal cancer. Int J Cancer. 2009;125(11):2737–43.
27. Hildebrandt MA, Gu J, Lin J, Ye Y, Tan W, et al. Hsa-miR-9 methylation status is associated with cancer development and metastatic recurrence in patients with clear cell renal cell carcinoma. Oncogene. 2010;29(42):5724–8.
28. Tsai KW, Liao YL, Wu CW, Hu LY, Li SC, et al. Aberrant hypermethylation of miR-9 genes in gastric cancer. Epigenetics. 2011;6(10):1189–97.
29. Gregory PA, Bert AG, Paterson EL, Barry SC, Tsykin A, et al. The miR-200 family and miR-205 regulate epithelial to mesenchymal transition by targeting ZEB1 and SIP1. Nat Cell Biol. 2008;10(5):593–601.
30. Korpal M, Lee ES, Hu G, Kang Y. The miR-200 family inhibits epithelial-mesenchymal transition and cancer cell migration by direct targeting of E-cadherin transcriptional repressors ZEB1 and ZEB2. J Biol Chem. 2008;283(22):14910–14.

31. Park SM, Gaur AB, Lengyel E, Peter ME. The miR-200 family determines the epithelial phenotype of cancer cells by targeting the E-cadherin repressors ZEB1 and ZEB2. Genes Dev. 2008;22(7):894–907.
32. Davalos V, Moutinho C, Villanueva A, Boque R, Silva P, et al. Dynamic epigenetic regulation of the microRNA-200 family mediates epithelial and mesenchymal transitions in human tumorigenesis. Oncogene. 2012;31(16):2062–74.
33. Vrba L, Jensen TJ, Garbe JC, Heimark RL, Cress AE, et al. Role for DNA methylation in the regulation of miR-200c and miR-141 expression in normal and cancer cells. PLoS One. 2010;5(1):e8697.
34. Vrba L, Garbe JC, Stampfer MR, Futscher BW. Epigenetic regulation of normal human mammary cell type-specific miRNAs. Genome Res. 2011;21(12):2026–37.
35. Neves R, Scheel C, Weinhold S, Honisch E, Iwaniuk KM, et al. Role of DNA methylation in miR-200c/141 cluster silencing in invasive breast cancer cells. BMC Res Notes. 2010;3:219.
36. Wiklund ED, Bramsen JB, Hulf T, Dyrskjot L, Ramanathan R, et al. Coordinated epigenetic repression of the miR-200 family and miR-205 in invasive bladder cancer. Int J Cancer. 2011;128(6):1327–34.
37. Ceppi P, Mudduluru G, Kumarswamy R, Rapa I, Scagliotti GV, et al. Loss of miR-200c expression induces an aggressive, invasive, and chemoresistant phenotype in non-small cell lung cancer. Mol Cancer Res. 2010;8(9):1207–16.
38. Langevin SM, Stone RA, Bunker CH, Grandis JR, Sobol RW, et al. MicroRNA-137 promoter methylation in oral rinses from patients with squamous cell carcinoma of the head and neck is associated with gender and body mass index. Carcinogenesis. 2010;31(5):864–70.
39. Chen Q, Chen X, Zhang M, Fan Q, Luo S, et al. miR-137 is frequently down-regulated in gastric cancer and is a negative regulator of Cdc42. Dig Dis Sci. 2011;56(7):2009–16.
40. Balaguer F, Link A, Lozano JJ, Cuatrecasas M, Nagasaka T, et al. Epigenetic silencing of miR-137 is an early event in colorectal carcinogenesis. Cancer Res. 2010;70(16):6609–18.
41. Suzuki H, Takatsuka S, Akashi H, Yamamoto E, Nojima M, et al. Genome-wide profiling of chromatin signatures reveals epigenetic regulation of MicroRNA genes in colorectal cancer. Cancer Res. 2011;71(17):5646–58.
42. Zhou X, Ruan J, Wang G, Zhang W. Characterization and identification of microRNA core promoters in four model species. PLoS Comput Biol. 2007;3(3):e37.
43. Long YS, Deng GF, Sun XS, Yi YH, Su T, et al. Identification of the transcriptional promoters in the proximal regions of human microRNA genes. Mol Biol Rep. 2011;38(6):4153–7.
44. Marson A, Levine SS, Cole MF, Frampton GM, Brambrink T, et al. Connecting microRNA genes to the core transcriptional regulatory circuitry of embryonic stem cells. Cell. 2008;134(3):521–33.
45. Ozsolak F, Poling LL, Wang Z, Liu H, Liu XS, et al. Chromatin structure analyses identify miRNA promoters. Genes Dev. 2008;22(22):3172–83.
46. Rodriguez A, Griffiths-Jones S, Ashurst JL, Bradley A. Identification of mammalian microRNA host genes and transcription units. Genome Res. 2004;14(10A):1902–10.
47. McGarvey KM, Fahrner JA, Greene E, Martens J, Jenuwein T, et al. Silenced tumor suppressor genes reactivated by DNA demethylation do not return to a fully euchromatic chromatin state. Cancer Res. 2006;66(7):3541–9.
48. McGarvey KM, Van Neste L, Cope L, Ohm JE, Herman JG, et al. Defining a chromatin pattern that characterizes DNA-hypermethylated genes in colon cancer cells. Cancer Res. 2008;68(14):5753–9.
49. Jacinto FV, Ballestar E, Esteller M. Impaired recruitment of the histone methyltransferase DOT1 L contributes to the incomplete reactivation of tumor suppressor genes upon DNA demethylation. Oncogene. 2009;28(47):4212–24.
50. Sarver AL, French AJ, Borralho PM, Thayanithy V, Oberg AL, et al. Human colon cancer profiles show differential microRNA expression depending on mismatch repair status and are characteristic of undifferentiated proliferative states. BMC Cancer. 2009;9:401.
51. Reid JF, Sokolova V, Zoni E, Lampis A, Pizzamiglio S, et al. miRNA profiling in colorectal cancer highlights miR-1 involvement in MET-dependent proliferation. Mol Cancer Res. 2012;10(4):504–15.

52. Lin SL, Chiang A, Chang D, Ying SY. Loss of mir-146a function in hormone-refractory prostate cancer. RNA. 2008;14(3):417–24.
53. Bhaumik D, Scott GK, Schokrpur S, Patil CK, Campisi J, et al. Expression of microRNA-146 suppresses NF-kappaB activity with reduction of metastatic potential in breast cancer cells. Oncogene. 2008;27(42):5643–7.
54. Li Y, Vandenboom TG 2nd, Wang Z, Kong D, Ali S, et al. miR-146a suppresses invasion of pancreatic cancer cells. Cancer Res. 2010;70(4):1486–95.
55. Datta J, Kutay H, Nasser MW, Nuovo GJ, Wang B, et al. Methylation mediated silencing of MicroRNA-1 gene and its role in hepatocellular carcinogenesis. Cancer Res. 2008; 68(13):5049–58.
56. Rodriguez-Otero P, Roman-Gomez J, Vilas-Zornoza A, Jose-Eneriz ES, Martin-Palanco V, et al. Deregulation of FGFR1 and CDK6 oncogenic pathways in acute lymphoblastic leukaemia harbouring epigenetic modifications of the MIR9 family. Br J Haematol. 2011;155(1):73–83.
57. Rotkrua P, Akiyama Y, Hashimoto Y, Otsubo T, Yuasa Y. MiR-9 downregulates CDX2 expression in gastric cancer cells. Int J Cancer. 2011;129(11):2611–20.
58. Zhang H, Qi M, Li S, Qi T, Mei H, et al. microRNA-9 targets matrix metalloproteinase 14 to inhibit invasion, metastasis, and angiogenesis of neuroblastoma cells. Mol Cancer Ther. 2012;11(7):1454–66.
59. Huang YW, Liu JC, Deatherage DE, Luo J, Mutch DG, et al. Epigenetic repression of microRNA-129–2 leads to overexpression of SOX4 oncogene in endometrial cancer. Cancer Res. 2009;69(23):9038–46.
60. Tang JT, Wang JL, Du W, Hong J, Zhao SL, et al. MicroRNA 345, a methylation-sensitive microRNA is involved in cell proliferation and invasion in human colorectal cancer. Carcinogenesis. 2011;32(8):1207–15.
61. Tanaka T, Arai M, Wu S, Kanda T, Miyauchi H, et al. Epigenetic silencing of microRNA-373 plays an important role in regulating cell proliferation in colon cancer. Oncol Rep. 2011;26(5):1329–35.

Chapter 12
MicroRNAs in the Development and Progression of Kidney Cancer

Sharanjot Saini, Sumit Arora, Shahana Majid, Hiroshi Hirata and Rajvir Dahiya

Abstract Renal cell carcinoma (RCC), a genetically and histopathologically heterogeneous disorder, is the most lethal of all genitourinary cancers and is generally refractory to current treatment regimens, including chemotherapy and radiation therapy. Targeted therapies against critical signaling pathways associated with RCC pathogenesis, such as von Hippel–Lindau (VHL) tumor suppressor, vascular endothelial growth factor (VEGF), and mammalian target of rapamycin (mTOR), have shown limited efficacy so far. Therefore, there has been much interest in identifying novel biomarkers for early diagnosis, risk assessment, and the design of novel therapeutic interventions for the disease. MicroRNAs (miRNAs) have been shown to be differentially expressed in RCC and play an important role in RCC pathogenesis. Studies have analyzed global miRNA expression profiles and the functional role of specific miRNAs in RCC. Here, we review our current understanding about the role of miRNAs in RCC by summarizing findings from various studies. Several miRNA-profiling studies have been conducted to identify specific miRNA signatures capable of distinguishing tumor from normal tissue, identifying RCC subtypes and the potential use of miRNAs in prognosis. Specific miRNAs have been found to be associated with key signaling pathways implicated in RCC pathogenesis (including pVHL-HIF, VEGF, mTOR signaling). Although current knowledge of the role of miRNAs in RCC pathogenesis is far from complete, key future challenges await in the use of miRNAs as novel biomarkers for improved diagnosis, prognosis, and the development of novel therapies for improved clinical management of the disease.

Keywords MicroRNAs · Renal cell carcinomas · Pathogenesis · VHL · Apoptosis · Proliferation · Metastasis

R. Dahiya (✉)
Department of Urology, Veterans Affairs Medical Center, University of California,
4150 Clement Street, San Francisco, CA 94121, USA
e-mail: rdahiya@urology.ucsf.edu

S. Saini · S. Arora · S. Majid · H. Hirata
Department of Urology, Veterans Affairs Medical Center, University of California,
San Francisco, CA, USA

S. R. Singh, P. Rameshwar (eds.), *MicroRNA in Development and in the Progression of Cancer,* DOI 10.1007/978-1-4899-8065-6_12,

1 Introduction

Kidney cancer accounts for about 2% of all cancers and worldwide >250,000 new cases of kidney cancer are diagnosed each year. The incidence and mortality rates for kidney cancer have increased over the past two decades [1]. Renal cell carcinoma (RCC) is the most common form of adult kidney cancer, and the most lethal genitourinary malignancy with more than 40% mortality [2]. RCC is not a single disease; rather, it is comprised of a number of different histologically and genetically distinct types of cancer, each with a different clinical course and a different response to therapy [3–5]. These tumors typically originate in the lining of the proximal renal tubules [6]. Based on cytomorphological characteristics and presumed cellular origin, the major subtypes of RCC include clear cell RCC (ccRCC), papillary RCC (PRCC), chromophobe RCC (chRCC), and collecting duct carcinoma [7–9]. ccRCC is the major subtype, accounting for nearly 75–80% of kidney tumors, and has the highest mortality rate (at more than 40%) [10, 11]. This renal cortical tumor is typically characterized by malignant epithelial cells with clear cytoplasm filled with lipids and glycogen and a compact-alveolar (nested) or acinar growth pattern interspersed with intricate, arborizing vasculature. Frequent deletions of the chromosome 3p region harboring the von Hippel–Lindau (VHL) tumor suppressor gene(*VHL*) has been associated with ccRCC [12]. PRCC is the second most common RCC subtype (10–15% of cases) that also originates from the proximal convoluted tubule. However, PRCC cases have significantly low cancer-related deaths and tumor recurrence [13]. PRCC is traditionally subclassified by histological morphology as type 1 and type 2. Type 1 PRCC is characterized by tightly formed, cribiform papillae covered by small cuboidal cells with a single line of uniform nuclei and small nucleoli [14, 15]. This subtype is much more common than type 2 PRCC and tends to have a significantly better prognosis. Type 1 papillary kidney cancer is often multifocal, bilateral, with a slow growth rate. Type 2 papillary kidney cancer is characterized by large, less well-organized papillae with pleomorphic nuclei, prominent nucleoli, and nuclear pseudostratification [14, 15]. Type 2 papillary kidney cancer tends to be solitary and has a tendency to metastasize early. chRCC originates from the distal tubule/collecting tubule, accounts for approximately 5% of RCC [16], and has a significantly better prognosis than ccRCC [13]. Cells have a characteristic pale or eosinophilic granular cytoplasm, corresponding to the variable number of cytoplasmic microvesicles as observed by electron microscopy, and usually grow in large solid sheets [8]. Oncocytomas (5%) are benign kidney tumors that do not metastasize [3, 17].

Initial treatment for RCC is commonly a radical or partial nephrectomy and remains the mainstay for curative treatment [18]. When the tumor is confined to the renal parenchyma, the 5-year survival rate is 60–70%, but the rate is considerably lower when metastases occur. Approximately 30% of localized RCC cases develop metastatic recurrence [19] with poor prognosis due to the refractory nature of RCC to current treatment regimens. It is relatively resistant to radiation therapy and chemotherapy, although some cases respond to immunotherapy. Antitumor agents targeting angiogenesis and mammalian target of rapamycin (mTOR) signaling pathways are

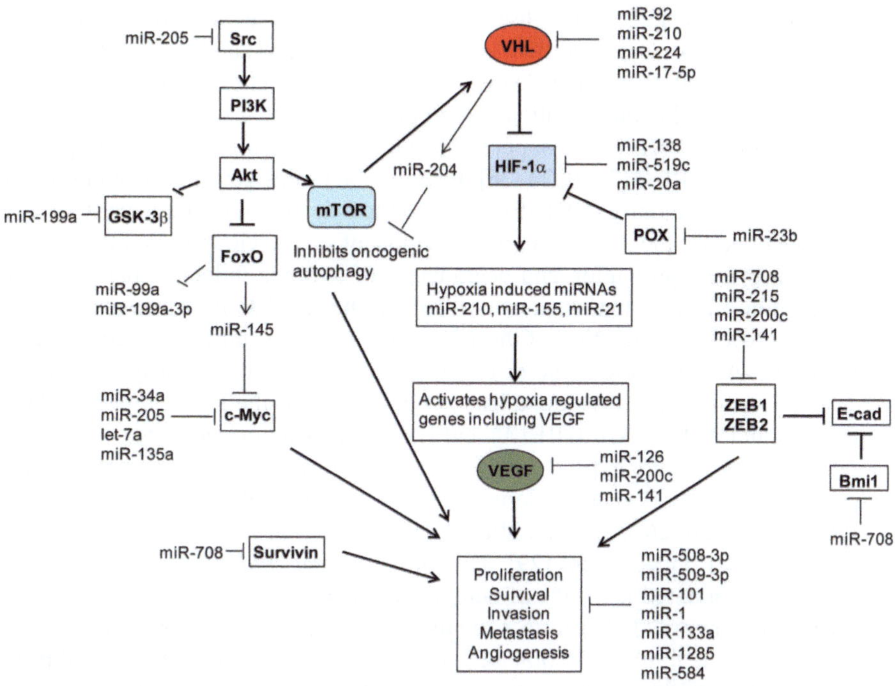

Fig. 12.1 MicroRNAs in the development and progression of RCC. Schematic illustration showing validated interactions of miRNAs with known dysregulated signaling pathways in RCC. Specific miRNAs have been found to be associated with key signaling pathways (including pVHL-HIF, VEGF, mTOR signaling) that lead to development and progression of RCC

currently used for treating metastatic RCC [20]. Targeted cancer therapies such as the anti-vascular endothelial growth factor (VEGF) monoclonal antibody (bevacizumab), receptor tyrosine kinase inhibitors (sunitinib, sorafenib, and pazopanib) as well as the mTOR inhibitors (temsirolimus and everolimus) have improved the outlook for RCC (progression-free survival), although they have not yet demonstrated improved survival and their efficiency is limited. Therefore, there has been much interest in identifying novel molecular biomarkers for early diagnosis and better risk assessment and to design novel therapeutic interventions for the disease.

2 Genetics of RCC

The majority of RCC occurs sporadically. However, several inherited RCC syndromes and familial RCC cases have been reported [21] such as VHL (inherited ccRCC), hereditary type 1 PRCC, Birt–Hogg–Dubé (BHD) syndrome (inherited-chromophobe, oncocytic hybrid RCC), and hereditary leiomyomatosis RCC (HLRCC) [3, 22]. Studies of the inherited forms of kidney cancer have led to the identification of a number of genes related to this disease and major advances in understanding the molecular basis of RCC. In addition, large-scale genomic studies of sporadic kidney tumors are beginning to provide clearer pictures of the genetics of RCC and the key pathways implicated in the initiation and progression of the disease [23].

The role of the *VHL* tumor suppressor gene, located on chromosome 3p25, is well established in RCC [3, 16, 24–26]. Inactivation of this gene plays an important role in 100 % of hereditary (VHL disease) and the majority of sporadic ccRCC. The protein encoded by the *VHL* gene is a component of the elongin complex and is involved in the ubiquitination and degradation of hypoxia-inducible factor (HIF), a transcription factor that plays a central role in the regulation of gene expression by oxygen. Allelic inactivation of the *VHL* gene has been shown to occur through mutation, DNA methylation, and/or chromosomal loss in the majority of ccRCCs [26]. Inactivation of the *VHL* gene leads to activation of the hypoxia pathway via hypoxia-inducible factor-1α (HIF1A) and HIF2α. This in turn activates expression of genes involved in the hypoxia response, angiogenesis, and other signaling pathways involving VEGF, platelet-derived growth factor (PDGF), and transforming growth factor α [24, 26–28]. In addition, VHL has number of other important functions that are HIF-independent [29]. The *Mesenchymal epithelial transition factor (MET)* gene is mutated in 100 % of hereditary PRCC families and is also mutated in a subset of tumors (13 %) from patients with sporadic, type 1 papillary kidney cancer [3, 30]. The fumarate hydratase gene (*FH*) is the gene for the inherited form of type 2 papillary kidney cancer associated with HLRCC [31]. Sporadic papillary type 2 RCC is composed of a mixture of kidney cancers including collecting duct RCC and medullary RCC. The gene(s) for sporadic type 2 papillary RCC, collecting duct RCC, and medullary RCC is not known. The folliculin gene (*FLCN*) is mutated in 96 % of families affected with the inherited form of chRCC and oncocytoma associated with BHD syndrome [3, 32–34]. The genes for the sporadic forms of chRCC and oncocytoma are also not known. However, identification of the genes that cause kidney cancer has improved the management of patients and provided a foundation for the development of therapeutic approaches for those patients with advanced forms of this disease [3].

3 MicroRNAs: New Players in RCC Pathogenesis

Recently, microRNAs (miRNAs) have been shown to be differentially expressed in RCC and their role in RCC pathogenesis is emerging. miRNAs are small noncoding RNAs that negatively regulate expression of multiple genes either by inducing translational silencing or by causing degradation of the messenger RNA (mRNA) of the targeted gene [35]. It has been firmly established that miRNAs control various key cellular processes such as proliferation, apoptosis, differentiation, and development [36], and are implicated in human diseases, including cancer [37]. Examination of tumor-specific miRNA expression profiles has revealed widespread dysregulation of these molecules in diverse cancers [38, 39], including RCC [40–43]. In fact, miRNA expression profiles have been found to be more informative than mRNA profiling [39]. miRNAs have been identified that function as classical oncogenes or tumor suppressor genes [37]. Genomic deletion or epigenetic silencing of an miRNA that normally represses expression of one or more oncogenes might lead to increased oncogenic expression. Alternatively, amplification, overexpression, or loss of epigenetic silencing of a gene encoding an miRNA that targets one or more tumor suppressor genes could inhibit the activity of an anti-oncogenic pathway [37]. Due to their tissue-specific and disease-specific expression patterns and tremendous regulatory potential, miRNAs may be important biomarkers for diagnosis and prognosis of human malignancies [44]. Due to their secondary structure, miRNAs are well preserved in formalin-fixed, paraffin-embedded (FFPE) samples compared to mRNA and can be extracted in intact form allowing accurate identification in archived clinical materials [45]. Their tissue specificity and tight regulation make them highly effective as biomarkers for tracing the tissue of origin of cancers of unknown primary origin [46].

The following sections provide a comprehensive review of the current knowledge of the role of miRNAs in the development and progression of RCC. Future challenges relate to the utilization of the knowledge gained from these miRNA studies and their development as novel biomarkers for diagnosis, prognosis, and novel targeted therapies for RCC.

4 MicroRNA Profiling Studies in RCC

Considering that several molecular pathways are deregulated in tumor cells, high-throughput analysis of miRNA expression is more likely to identify molecular mechanisms of carcinogenesis than the study of a single miRNA [47, 48]. Several miRNA-profiling studies have been conducted in RCC to identify specific miRNA signatures capable of the following: (1) distinguishing tumor from normal tissue, (2) identifying different RCC subtypes, and (3) searching for miRNAs with prognostic information [40–43]. An overview of these studies is provided below and Table 12.1 summarizes the key findings of these miRNA-profiling studies.

4.1 Profiling Studies to Identify RCC-Specific miRNA Signatures

There have been several miRNA-profiling reports in RCC in this direction, though still a consensus has not been reached on RCC-specific miRNA signatures.

4.1.1 Profiling Studies to Identify RCC-Specific miRNA Signatures in RCC Tissues

Gottardo et al. [40] analyzed the expression profile of 245 miRNAs in kidney primary tumors and reported that a set of four human miRNAs (miR-28, miR-185,

Table 12.1 Summary of miRNA-profiling studies in RCC

Study	Summary of results	Reference
1. Profiling studies to identify RCC-specific miRNA signature		
Profiling studies to identify RCC-specific miRNA signatures in RCC tissues		
Gottardo et al.	miR-28, miR-185, miR-27, let-7f-2 significantly upregulated in cancer vs. normal	[40]
Nakada et al.	miR-141 and miR-200c significantly downregulated in RCC vs. normal kidney	[49]
Juan et al.	Identified a panel of 10 miRNAs (miR-200c, miR-185, miR-34a, miR-142-3p, miR-21, miR-155, miR-224, miR-210, and miR-592) that could distinguish ccRCC from normal kidney	[42]
Yi et al.	48 miRNAs significantly downregulated (including miR-141 and miR-200c) and 38 upregulated in RCC vs. normal kidney	[50]
Jung et al.	13 overexpressed (including miR-16, -452*, -224, -155, and -210) and 20 downregulated miRNAs (miR-200b, -363, -429, -200c, -514 and -141) in ccRCC vs. normal. Combination of miR-141 and miR-155 expression as classifier for ccRCC	[43]
White et al.	Significant upregulation of miR-122, miR-155, and miR-210 and downregulation of miR-200c, miR-335, and miR-218	[51]
Chow et al.	Significant upregulation of miR-122 and downregulation of miR-199* and miR-200c	[52]
Weng et al.	Confirmed significant downregulation of miR-141 and miR-200 in RCC	[54]
Osanto et al.	100-miRNA signature distinguishes ccRCC from normal kidney	[55]
Profiling studies to identify serum biomarkers for RCC		
Wulfken et al.	Identified high serum levels of miR-1233 as a potential biomarker for RCC	[58]
Redova et al.	Increased miR-378 and decreased miR-451 serum levels in RCC patients	[59]
2. Profiling studies to identify RCC subtypes		
Nakada et al.	ccRCC and chromophobe RCC have significantly different miRNA expression patterns	[49]
Petillo et al.	Unique miRNA signature for each of the four major subtypes of RCC	[61]
Fridman et al.	Accurate classification of each RCC subtype using a decision tree classifier based on six miRNAs	[63]
Youssef et al.	Developed a classification system to distinguish the different RCC subtypes using unique miRNA signatures	[64]
Powers et al.	Three miRNAs could differentiate papillary RCC from ccRCC and four miRNAs could differentiate oncocytomas from chRCCs	[66]
3. Profiling studies to identify RCC specific miRNAs with prognostic potential Metastasis associated miRNAs		
Heinzelmann et al.	33 downregulated miRNAs distinguish metastatic and non metastatic ccRCC (including miR-451, miR-221, miR-30a, miR-10b, and miR-29a)	[69]
White et al.	Identified miR-10b, miR-196a and miR-27b as downregulated and miR-638, miR-1915 and miR-149* as significantly upregulated in metastasis vs primary RCC	[70]
Khella et al.	Downregulation of miRNA-10b, miR-126, miR-196a, miR-204, miR-215, miR-192, and miR-194 in metastatic ccRCC	[71]
Wu et al.	Four-miRNA signature for ccRCC metastasis and progression (downregulated miR- 10b, miR-139-5p and upregulated miR-130b, miR-199b-5p)	[72]

Table 12.1 (continued)

Study	Summary of results	Reference
Wotschofsky et al.	Downregulation of 23 miRNAs (miR-10b/-19a/-19b/-20a/-29a/-29b/-29c/-100/-101/-126/-127/-130/-141/-143/-145/-148a/-192/-194/-200c/-210/-215/-370/-514) in metastatic RCC	[73]
miRNAs associated with early relapse		
Slaby et al.	64 miRNAs associated with relapse after nephrectomy (including downregulation of miR- 143, miR-26a, miR-145, miR-10b, miR-195, and miR-126 expression)	[75]

This table summarizes the key findings of miRNA-profiling studies in RCC. Several of these studies have identified (1) RCC-specific miRNA signatures capable of distinguishing tumor from normal tissue, (2) miRNA signatures capable of distinguishing between different subtypes of RCC, and (3) miRNAs for prognosis

miR-27, and let-7f-2) were significantly upregulated in RCC compared to normal kidney. Nakada et al. [49] performed a systematic miRNA expression analysis in ccRCC and chRCC by using an miRNA microarray platform which covered a total of 470 human miRNAs (Sanger miRBase release 9.1) and reported numerous miRNAs whose expression was altered. They reported that miR-141 and miR-200c were the most significantly downregulated miRNAs in RCC that target ZEB2, a transcriptional repressor of E-cadherin [49]. Juan et al. [42] identified a panel of 10 miRNAs (miR-200c, miR-185, miR-34a, miR-142-3p, miR-21, miR-155, miR-224, miR-210, and miR-592) that could distinguish ccRCC from normal kidney. Four of the upregulated miRNAs identified in this study (miR-21, miR-155, miR-34a, miR-210) have strong correlation with other tumorigenic states and are commonly dysregulated in other solid tumors. Yi et al. [50] profiled miRNA expression in 30 pairs of RCC and adjacent nontumorous tissues using an miRNA microarray platform which had a total of 847 human miRNAs (Sanger miRBase release 11.0). They reported that 86 miRNAs were differentially expressed between RCC and normal kidney tissue of which 48 were significantly downregulated in RCC (including miR-141 and miR-200c) and 38 were upregulated. miRNA profiling of ccRCC samples by Jung et al. identified 13 overexpressed and 20 downregulated miRNAs including increased expression of miR-16, miR-452*, miR-224, miR-155, and miR-210, and decreased expression of miR-200b, miR-363, miR-429, miR-200c, miR-514, and miR-141 [43]. No significant associations between these differentially expressed miRNAs and clinicopathological factors (tumor stage, tumor grade, and survival rate) were observed, though miRNA profiles could clearly differentiate between malignant and nonmalignant tissues. In particular, a combination of miR-141 and miR-155 expression profiles resulted in 97% of overall correct classification of ccRCC tumors [43]. In another profiling study, White et al. reported 166 miRNAs that were significantly dysregulated in ccRCC, including miR-122, miR-155, and miR-210, which had the highest overexpression, and miR-200c, miR-335, and miR-218, which were the most downregulated [51]. Their analysis showed that miR-155 expression correlated with ccRCC size [51]. Profiling of ccRCC samples by Chow et al. identified upregulated expression of miR-122 and downregulated levels of miR-199* and miR-200c among other miRNAs [52]. Liu et al. identified a subset of robust, tissue-specific, direct functional mRNA targets of dysregulated miRNA in ccRCC in patient-matched tumor/normal samples [53]. This study revealed many new regulatory pathways in ccRCC. For instance, loss of miR-149, miR-200c, and miR-141 caused gain of function of oncogenes (KCNMA1, LOX), VEGF-A, and SEMA6A, respectively, and increased levels of miR-142-3p, miR-185, miR-34a, miR-224, and miR-21 caused loss of function of tumor suppressors LRRC2, PTPN13, SFRP1, ERBB4, and (SLC12A1, TCF21), respectively. They also found strong anticorrelation between VEGF-A and the miR-200 family (miR-200a*, miR-200b, miR-200c, and miR-141) [53].

Recently, a newly advanced next-generation sequencing (NGS) technology has been replacing conventional microarray-based platforms at least at the discovery level for whole-genome miRNA expression in RCC. This technology enables genome-wide expression profiling of known miRNAs and discovery of novel miRNAs with high quantitative and qualitative accuracy. Weng et al. profiled both frozen and paraffin-embedded RCC samples using microarray, deep-sequencing, and reverse transcriptase polymerase chain reaction (RT-PCR) methodologies and found that there was a high correlation between the three technologies. Their study confirmed the importance of miR-141 and miR-200 downregulation in RCC [54]. This was the first study to demonstrate that FFPE specimens can be used reliably for miRNA deep-sequencing analysis, making future large-scale clinical cohort/trial-based studies possible [54]. In another recent study, NGS allowed the identification of a 100-miRNA signature distinguishing ccRCC from normal kidney [55]. In summary, the expression profiling results of the various array studies are not consistent, though these studies show some overlapping results. The discrepancies between results may arise from differences in samples, sample preparation, and the different microarray platforms used. Nonetheless, the data indicate that dysregulated miRNAs may play a pivotal role in the pathogenesis of ccRCC.

4.1.2 Profiling Studies to Identify Serum Biomarkers for RCC

miRNAs are highly stable and are resistant to RNAse, extreme pH, and temperature in body fluids [56, 57] and consequently are highly abundant in plasma, serum, and other body fluids. Circulating miRNAs in biological fluids have shown great potential to serve as novel biomarkers for noninvasive diagnosis and prognosis of a wide range of solid cancers. The use of miRNAs as serum and urine biomarkers is particularly attractive for RCC patients as the kidney is in direct contact with blood and urine. In this direction, Wulfken et al. 2011. identified high circulating levels of miR-1233 in serum as a potential biomarker for RCC patients [58]. A recent study also suggests that levels of miR-378 are increased, while those of miR-451 are decreased in the serum of RCC patients and that a combination of miR-378 and miR-451 levels enable identification of RCC with high specificity and sensitivity [59]. Another recent study suggests that miR-378 is unlikely to provide helpful diagnostic/prognostic information in RCC as the serum levels of this miRNA were not correlated with clinicopathological characteristics such as pT stage, Fuhrman grade, and lymph node/distant metastasis [60].

4.2 Profiling Studies to Identify RCC Subtypes

As summarized in Table 12.1, studies suggest that miRNA profiles can be used to pathologically differentiate the various histological subtypes of RCC. Nakada et al. [49] reported that ccRCC and chRCC had significantly different miRNA expression patterns. Petillo et al. screened the expression levels of miRNAs in the four histological subtypes of RCC (clear cell, papillary, and chromophobe and benign renal oncocytomas) and found a unique miRNA signature for each subtype of renal tumor [61]. Furthermore, they identified unique patterns of miRNA expression distinguishing ccRCC cases with favorable vs. unfavorable outcome. Specifically, they documented the overexpression of miR-424 and miR-203 in ccRCC relative to PRCC, as well as the inversion of expression of miR-203 in benign oncocytomas (where it is underexpressed relative to normal kidney) as compared to the malignant chRCC (where it is overexpressed relative to normal kidney) [61]. Five miRNAs (including miR-200b) could separate histologically similar tumors, such as oncocytomas and chRCCs [62]. In another study of paraffin-embedded tumor samples from different histological RCC subtypes, similarity in miRNA expression was found between oncocytoma and chromophobe subtypes as well as between clear cell and papillary tumors [63]. Accurate classification of each RCC subtype was demonstrated using a decision tree classifier based on six miRNAs, and this approach was validated in an independent population, achieving a correct diagnosis in 93 % of cases. This study defined a two-step decision tree classifier that uses expression levels of six miRNAs: The first step uses expression levels of hsa-miR-210 and hsa-miR-221 to distinguish between the two pairs of subtypes; the second step uses either hsa-miR-200c with hsa-miR-139-5p to distinguish oncocytoma from chromophobe or hsa-miR-31 with hsa-miR-126 to distinguish conventional from papillary tumors. This classifier was validated on an independent patient cohort [63]. Youseff et al. confirmed the concept of miRNAs as molecular markers to identify different RCC subtypes and developed a classification system that could distinguish the different RCC subtypes using unique miRNA signatures in a maximum of four steps [47, 64]. Their analysis showed that nearly 100 miRNAs were differentially expressed among RCC subtypes. The clear cell subtype was more related to papillary subtype and both these subtypes were completely different from oncocytoma and chromophobe tumors. Using a stepwise decision tree, the authors demonstrated that the system has a sensitivity of 97 % in distinguishing normal from RCC, 100 % for ccRCC subtype, 97 % for PRCC subtype, and 100 % accuracy in distinguishing oncocytoma from chRCC subtype. This system was cross-validated and it showed an accuracy of about 90 % [47, 64]. In a recent study, Powers et al. identified a set of 18 miRNAs that were significantly different among the four tumor RCC subtypes. Four miRNAs could differentiate oncocytomas from chRCCs and a set of three miRNAs could differentiate PRCC from ccRCC, including miR-126, a known vasculogenic miRNA [65, 66].

4.3 Profiling Studies to Identify RCC Specific miRNAs with Prognostic Potential

miRNAs possess significant potential as prognostic biomarkers and have been shown to play important roles in tumor progression and metastasis [67, 68]. In RCC, several profiling studies have been done to identify miRNAs with prognostic potential as summarized in the following sections (Table 12.1).

4.3.1 Metastasis-associated miRNAs

Studies have examined miRNA expression profiles in metastatic ccRCC tissues compared to the primary tumors. Heinzelmann et al. reported a signature of 33 downregulated miRNAs that distinguish metastatic and nonmetastatic ccRCC, including miR-451, miR-221, miR-30a, miR-10b, and miR-29a [69]. This study also found correlations between expression levels of specific miRNAs with progression-free survival and overall survival [69]. White et al. identified 65 miRNAs that were significantly altered in metastatic compared with primary RCCs. In all, nine (14 %) had increased expression, whereas 56 (86 %) had decreased expression. miR-10b, miR-196a, and miR-27b were the most downregulated, whereas miR-638, miR-1915, and miR-149* were the most upregulated in metastasis when compared with primary RCCs [70]. The downregulation of miRNA-10b, miR-126, miR-196a, miR-204, miR-215, miR-192, and miR-194 in metastatic ccRCC was validated in another study [71]. Target prediction analysis showed that miRNAs that are differentially expressed in metastasis can play a role in metastatic ccRCC pathogenesis by targeting key molecules such as VEGF, HIF1α subunit, platelet-derived growth factor B (PDGFB), and platelet-derived growth factor C (PDGFC) [71]. Wu et al. identified a four-miRNA signature for ccRCC metastasis and progression where the expression levels of miR-10b, miR-139-5p, miR-130b, and miR-199b-5p were used to determine the status of ccRCC metastasis [72]. miR-199b-5p and miR-130b were overexpressed in metastatic tumors, while miR-10b and miR-139-5p were downregulated [72]. A recent study reported a total of 23 miRNAs (miR-10b/-19a/-19b/-20a/-29a/-29b/-29c/-100/-101/-126/-127/-130/-141/-143/-145/-148a/-192/-194/-200c/-210/-215/-370/-514) that were downregulated in metastatic tissue samples compared with normal tissues primarily due to epigenetic modifications [73]. Another study investigated the molecular signatures and underlying genomic aberrations associated with RCC metastasis and found that increased metastatic activity is associated with acquisition of a myofibroblast-like signature in both tumor cell lines and in metastatic tumor biopsies [74]. miRNA-induced expression and epigenetic silencing accounted for the change in expression of a significant number of genes, including myofibroblastic marker S100A4 [74].

4.3.2 miRNAs Associated with Early Relapse

One of the clinical challenges in RCC is the lack of prognostic biomarkers enabling identification of patients at high risk of relapse after nephrectomy, and individualized therapy and follow-up for these patients. Slaby et al. described a tumor relapse-miRNA signature based on the expression of 64 miRNAs differentially expressed between relapse-free RCC patients and those who developed relapse (20 miRNAs were increased, 44 miRNAs were decreased) [75]. In the validation phase of the study, the authors confirmed that expression levels of miR-143, miR-26a, miR-145, miR-10b, miR-195, and miR-126 are lower in the tumors of RCC patients who developed tumor relapse and the lowest levels of these miRNAs were observed in primary metastatic tumors. By using Kaplan–Meier analysis, this study identified that miR-127-3p, miR-145, and miR-126 are significantly correlated with relapse-free survival of nonmetastatic RCC patients [75]. Petillo et al. reported that overexpression of miR-32 is associated with poor outcome [61].

5 MicroRNAs in RCC Pathogenesis

miRNAs impact key cellular processes in RCC including cell proliferation, apoptosis, invasion, migration, and metastasis. Specific miRNAs have been found to be associated with key signaling pathways implicated in RCC pathogenesis (including pVHL-HIF, VEGF, mTOR signaling). The following section and Fig. 12.1 summarize the studies examining the functional role of specific miRNAs in RCC pathogenesis.

5.1 miRNAs Related to the pVHL-HIF Pathway

The VHL protein (pVHL), which is defective in the majority of patients with ccRCC, plays a well-characterized role in the pathogenesis of kidney cancer through both HIF-dependent and HIF-independent pathways [24, 26, 27, 29]. pVHL is mainly involved in the ubiquitination and degradation of HIF1α, a transcription factor that plays a central role in the regulation of gene expression by oxygen. In the presence of oxygen, hydroxylation of two proline residues and acetylation of a HIF1α lysine residue occurs and pVHL binds and degrades HIF1α through the ubiquitin-26S proteasome system. Hence, under normoxic and VHL-activated conditions, HIF1α is inactivated. However, under hypoxic and VHL-inactivated conditions, pVHL cannot bind to HIF1α, which remains stable and accumulates and binds to HIF1β to form the HIF1α–HIF1β heterodimer. This leads to activation of the hypoxic pattern of gene expression including increased expression of genes involved in angiogenesis and other signaling pathways such as VEGF, PDGF, etc. [28, 76]. Cancer states are typically characterized by hypoxia, enhanced HIF1α levels, and increased expression of hypoxia-regulated genes, which correlate with tumor progression and patient outcome [77, 78]. In agreement with this, ccRCC tumor growth depends on abundant angiogenesis and activation of glycolytic metabolic pathways that result from VHL loss and induction of the HIF gene and its targets. Tumor hypoxia has been shown to be a prognostic factor in solid tumors [79–82] and miRNAs have also been shown to be altered in response to hypoxia [83]. In particular, the hypoxia-responsive miR-210 has been described to be ubiquitously upregulated by HIF1α in hypoxic tumors [84–86]. Camps et al. reported that under hypoxic conditions, miR-210 exhibits a significant change in breast cancer, mediated through the HIF1α/VHL system [87]. Increasing evidence suggests that miRNAs are involved in the regulation of VHL expression in RCC, though regulation of VHL pathway components by miRNAs remain underexplored in RCC [88]. Although interactions between miRNAs and the pVHL/HIF1α signaling pathway have been hypothesized, direct interactions between miRNAs and the components of this pathway have only recently been described [88, 89]. An overview of the miRNAs related to VHL signaling pathway is schematically summarized in Fig. 12.1. Fendler et al. [88] performed an analysis of miRNAs predicted to regulate the VHL pathway based on target prediction algorithm TargetScan 5.1 [90] and reported that a number of miRNAs may target six genes that are components of this signaling pathway. The miR-17-5p cluster targets VHL, HIF1A, egl nine homolog 1 (Caenorhabditis elegans) (EGLN1), egl nine homolog 3 (C. elegans; EGLN3), E1A binding protein p300 (EP300); and CREB binding protein (CREBBP). miR-200b, miR-200c, and miR-429 target VHL, transcription elongation factor B (SIII), polypeptide 1 (15 kDa, elongin C;TCEB1), EGLN1/3; EP300; and CREBBP [88].

Valera et al. reported that ccRCC, in either sporadic or hereditary forms, showed hypoxic activation of miR-210 when compared to tumors of non-clear cell histology. Also, they reported an inverse correlation between miR-92 and VHL mRNA levels suggesting a regulatory effect of miR-92 on the VHL gene [91]. Gleadle and colleagues showed that miRNA expression can be VHL-dependent in RCC [92, 93]. To examine the VHL-dependent regulation of miRNAs, they performed microarray analysis of renal cell line RCC4 with mutated VHL (RCC4-VHL) and reintroduced wild-type VHL (RCC4 + VHL). This identified a number of significantly dysregulated VHL-dependent miRNAs. Interestingly, some of these miRNAs had altered expression as a consequence of HIF upregulation as previously documented in the literature, other miRNAs were dysregulated in an HIF-independent manner [92, 93]. This pattern of regulation was also seen in ccRCC tissue for several miRNAs (miR-210, miR-155, let-7i, and members of the miR-17-92 cluster) when compared with normal tissue. A significant increase in miR-210, miR-155, and miR-21 expression was observed in the tumor tissue. The level of HIF- and VHL-regulated miR-210 showed marked increases in expression in renal cancer tissues. miR-210 levels showed correlation with VHL mutation or promoter methylation and also a

HIF-regulated mRNA, carbonic anhydrase IX (CAIX). An inverse correlation was observed between miR-210 expression and patient survival, and a putative target of miR-210, iron–sulfur cluster assembly protein (ISCU1/2), had reciprocal levels of mRNA expression in tumors [92]. A recent study suggests that silencing of miR-210 expression in RCC cell lines led to decreased viability, G2 phase cell cycle arrest, and reduced migratory and invasive potential through downregulation of HIF1α and other target mRNAs [94]. Lichner et al. demonstrated that miR-17-5p and miR-224 directly target pVHL and HIF1α supporting the concept that miRNAs represent an alternative mechanism for the inactivation of pVHL in ccRCC [95]. Recently, it has been reported that miR-204 is a VHL-regulated tumor suppressor miRNA in ccRCC that inhibits macroautophagy, with MAP1LC3B (LC3B) as a direct and functional target [96]. Using human ccRCC specimens, VHL-deficient cells, and xenograft models, the authors show that miR-204 is a VHL-regulated tumor suppressor that inhibits LC3B-mediated autophagic program that is necessary for tumor growth. Of note, higher tumor grade of human ccRCC was correlated with a concomitant decrease in miR-204 and increase in LC3B levels, indicating that LC3B-mediated macroautophagy is necessary for RCC progression. The authors show that loss of VHL promotes in an HIF-independent manner, access to nutrients from intracellular sources through activation of LC3B-mediated autophagy. Dependence of cancer cells on oncogenic autophagy has been demonstrated previously [97, 98] whereby cells eliminate defective organelles and molecules to recycle nutrients for survival under deprived conditions. This autophagic program is necessary for tumor growth. VHL, in addition to inducing endogenous miR-204, triggered the expression of LC3C, an HIF-regulated LC3B paralog, that suppressed tumor growth. These data revealed the miRNA-mediated role of VHL as a tumor-suppressing regulator of autophagic programs. The dependence of cancer cells on oncogenic autophagy creates the possibility that miR-204 could be used in ccRCC therapy with minimal effects on normal renal cells [96]. Liu et al. reported that miR-23b, which is upregulated in RCC, can directly target proline oxidase, a novel mitochondrial tumor suppressor that induces apoptosis through the generation of reactive oxygen species (ROS) and decreased HIF signaling [99]. In addition, miR-21 is a hypoxia-regulated miRNA that is overexpressed in several solid tumors including renal cancers and plays a key role in regulating cell apoptosis by targeting multiple genes in RCC including tissue inhibitor of metalloproteinase 3 (TIMP3) [100, 101]. HIF1α is a key downstream target of miR-21 in regulating tumor angiogenesis [102]. In prostate cancer cell line Du145, miR-21 induces tumor angiogenesis through targeting phosphatase and tensin homolog (PTEN), leading to activation of AKT and ERK1/2 signaling pathways, and thereby enhancing HIF1α and VEGF expression [102]. Song et al. reported that miR-138 directly targets HIF1α in a ccRCC cell line (786-O) and thereby regulates apoptosis and migration of ccRCC cells [103]. Based on target prediction algorithms, miR-20a, miR-130a, miR-519a, miR-519c, and miR-338 potentially target the HIF1α 3′-untranslated region (UTR) [64]. Out of these, miR-519c and miR-20a suppressed luciferase activity of the HIF1α 3′-UTR in 293T human embryonic kidney cells. This study showed miR-519c as a hypoxia-independent regulator of HIF1α, acting through direct binding to the HIF1α 3′ untranslated region and leading to reduced tumor angiogenesis [104]. However, the role of this miRNA has not been studied in RCC. Similarly, miR-22 and miR-20b have been shown to regulate HIF1α and VEGF in other cancers [105, 106] and remain unexplored in RCC.

5.2 miRNAs and the VEGF Pathway

The VEGF signaling pathway, a relevant target for molecular therapy in RCC, is transcriptionally regulated by the HIF transcription complex during hypoxia or due to loss of VHL in renal cell cancer. VEGF is a potent proangiogenic protein which plays a key role in tumor angiogenesis [107] and exerts its effects by binding to the VEGF receptor (VEGFR) on endothelial cells [108]. The VEGF signaling pathway is an attractive target for existing and novel RCC therapies [109]. Several drugs inhibiting VEGF signaling have been approved for treatment of mRCC including bevacizumab (a monoclonal antibody that binds circulating VEGF) and tyrosine kinase inhibitors (TKIs) (sunitinib, sorafenib, and pazopanib). miRNAs are promising new molecules that regulate VEGF signaling and are potential novel targets for therapy of RCC [88]. Target prediction algorithms suggest that the VEGF signal transduction pathway might be targeted by well-known dysregulated miRNAs in renal cancer, including miR-200b, miR-200c, miR-141, or the miR-17-5p-92 cluster [88]. In addition, miR-361-5p, miR-29c, miR-29b, miR-29a, and miR-126 are predicted to target VEGF-A [71]. Interactions between miRNAs and the VEGF signaling pathway have not been well studied in RCC [88] though there are a few studies in this direction. A recent report suggests a negative correlation of expression of miR-126 and VEGF-A [71]. A strong inverse correlation between the miR-200 family (miR-200a-c, miR-141) and VEGF-A has been reported suggesting that, VEGF-A is a direct target of these miRNAs [53]. Sinha et al. reported that miR-29b indirectly regulates VEGF expression in renal cancer cells [110].

5.3 miRNAs and the mTOR Pathway

mTOR is a kinase that plays a central role in integrating input from upstream pathways, including insulin-like growth factors (such as IGF-1, IGF-2), amino acids, cellular nutrients, energy levels, and redox status to regulate cell growth, metabolism, and survival [111]. mTOR has emerged as a major effector of cell growth and proliferation via the regulation of protein synthesis through a number of direct or indirect downstream targets. The mTOR pathway is dysregulated in human cancers including RCC [112]. Activated mTOR contributes to cancer progression through its effect on the cell cycle such as acceleration of G1 to S phase transition, or by promoting the translation of growth-associated mRNAs. HIF1α expression and function is induced by mTOR [113]. In RCC, activated mTOR leads to HIF1α accumulation leading to increased expression of downstream target genes such as

VEGF. mTOR is clearly an important therapeutic target for advanced RCC and mTOR inhibitors temsirolimus and everolimus are currently being used for treatment of metastatic RCC patients [20]. Given the clinical relevance of the mTOR pathway in RCC, an understanding of the role of miRNAs in relation to mTOR complex signaling may inform their role in RCC pathogenesis and guide further drug development strategies in this direction.

The mTOR complex 1 (mTORC1), composed of mTOR, is hyperactivated in the majority of human RCC samples [114, 115]. Increased knowledge of mTORC1 signaling has demonstrated that mTORC1 acts both downstream and upstream of PI3K-AKT signaling. The PI3K–AKT axis is activated in virtually all human cancers and the wide range of tumorigenic phenotypes mediated by PI3K–AKT signaling is consistent with the existence of diverse downstream effectors including Forkhead box O-class transcription factors (FoxOs), glycogen synthase kinase 3 (GSK3), and MDM2 (Fig. 12.1) [116, 117]. The mammalian FoxO transcription factors (FoxO1, FoxO3, FoxO4) function in the nucleus to direct transcription of specific gene targets governing cellular survival, proliferation, metabolism, differentiation, and oxidative defense. Activation of PI3K by extracellular growth factors leads to AKT-mediated phosphorylation of FoxO1, FoxO3, and FoxO4, resulting in their sequestration in the cytoplasm such that they are unable to regulate their gene targets. Gan et al. showed that expression of FoxO transcription factors is downregulated in most human ccRCC and papillary carcinomas. Their study demonstrated that FoxOs enforce a critical checkpoint that functions to constrain mTORC1-mediated renal tumorigenesis by inhibiting Myc via transcriptional regulation of miR-145. miR-145, which is known to target Myc [118], harbors two putative FoxO binding elements within 2 kb upstream of the miR-145 transcriptional start site. Thus, FoxOs suppress Myc expression through miR-145 regulation in renal cancer [117] miR-27a also negatively regulates the tumor suppressor FOXO1. Upregulation of FOXO1 by miR-27a inhibitor led to an antiproliferative effect in RCC cell lines [119]. Recently Cui et al. reported that miR-99a is downregulated in RCC and low expression levels of miR-99a is correlated with poor survival of RCC patients. Functional studies suggested that miR-99a restoration suppressed RCC growth, migration, and invasion as well as induced G1-phase cell cycle arrest in vitro and inhibited tumor growth in vivo partially via direct targeting of mTOR [120]. Another study showed that miR-199a-3p targets mTOR and c-MET and may have therapeutic benefit in highly lethal cancers including RCC (ccRCC and PRCC) [121].

5.4 MicroRNAs Related to Cell Survival/Apoptosis, Cell Proliferation, Migration, and Invasion

Cell proliferation, survival, migration, and invasion are among the common elements required by tumor cells for growth and progression in target microenvironments. miRNAs influence all these fundamental cellular processes involved in carcinogenesis (Fig. 12.1). The reports on miRNAs involved in RCC pathogenesis that influence these cellular processes are summarized in the following sections.

5.4.1 MicroRNAs Related to Apoptosis and Cell Proliferation

Apoptosis is a well-orchestrated cellular mechanism that balances cell proliferation and cell death and the ability to evade apoptosis is a hallmark of tumorigenesis. Several miRNAs have been implicated to influence this cellular process and also cellular proliferation as outlined below.

Tumor Suppressive miRNAs Related to Apoptosis and Cell Proliferation

Our laboratory identified miR-708 as an important pro-apoptotic microRNA. miR-708 expression was widely attenuated in human RCC specimens. Restoration of miR-708 expression in RCC cell lines decreased cell growth, clonability, invasion, and migration and elicited a dramatic increase in apoptosis through the cleavage of caspase-7 and caspase-3 in RCC cell lines. Moreover, intratumoral delivery of miR-708 was sufficient to trigger in vivo regression of established tumors in a murine xenograft model of human RCC [122]. We also showed that miR-205 is an important tumor suppressor miRNA that inhibits Src-mediated oncogenic pathways in RCC. Transient and stable overexpression of miR-205 in A498 cells resulted in induction of G(0)/G(1) cell cycle arrest and apoptosis, as indicated by decreased levels of cyclin D1 and c-Myc, suppressed cell proliferation, colony formation, migration, and invasion in RCC cells in vitro, and also inhibited tumor growth in vivo [123]. miR-508-3p and miR-509-3p have been reported to be tumor suppressor miRNAs whose overexpression in RCC cell line led to induction of apoptosis and inhibition of cell proliferation and migration [124]. Similarly, restoration of tumor-suppressive miR-135a inhibited cancer cell proliferation and induced G(0)/G(1) arrest in the RCC cell lines by targeting the c-Myc oncogene [125]. Another tumor suppressor miRNA, miR-199a expression is downregulated in RCC and lower expression correlates with higher tumor stage. miR-199a acts as a negative regulator of glycogen synthase kinase-3 β (GSK-3β), thereby regulating renal cancer cell proliferation and survival [126]. miR-101 is downregulated in RCC and acts as a negative regulator of EZH2, a histone methyltransferase, and renal cancer cell proliferation [127]. Let -7a acts as a tumor suppressor in RCC cell lines by down-regulating c-myc and c-myc target genes such as proliferating cell nuclear antigen (PCNA), cyclin D1 (CCND1), and the miR17–92 cluster, which is accompanied by inhibition of proliferation and cell cycle arrest [128]. Functional screening identified that 14 miRNAs (miR-1285, miR-206, miR-1, miR-135a, miR-429, miR-200c, miR-1291, miR-133b, miR-508-3p, miR-360-3p, miR-509-5p, miR-218, miR-335, miR-1255b) markedly inhibit RCC cell proliferation, suggesting that these miRNAs are candidate RCC tumor suppressor miRNAs [129]. Tumor suppressor miR-1285 inhibits cancer cell proliferation, invasion, and migration by directly targeting the oncogene transglutaminase 2 (TGM2) [129]. Similarly, miR-1 and miR-133a are significantly downregulated in RCC and ectopic restoration of their expression led to significant inhibition of cell proliferation, invasion, and induction of apoptosis and cell cycle arrest. The oncogene transgelin-2 (TAGLN2) was found to be directly regulated by both miR-1 and miR-133a [130]. Insulin-like growth factor 1 (IGF1) and VEGF-A, positive regulators of cellular proliferation, are targeted by several miRNAs and miR-126, respectively. Dysregulation of miRNA networks in RCC thereby influences cellular proliferation [71, 131].

Oncogenic miRNAs Related to Apoptosis and Cell Proliferation

miR-21, a known oncogenic miRNA in various tumors, is upregulated in RCC [100, 101, 132]. High expression of miR-21 in RCC is correlated with low disease-free and overall survival rates and with increased pathological stage and grade [101, 132]. miR-21 expression showed the highest expression in ccRCC and PRCC and could distinguish these two histological subtypes from chRCC and oncocytomas [132]. Functional studies after inhibiting miR-21 in RCC cell lines showed cell cycle arrest, induction of apoptosis, and reduced invasion and migration with an increase in p21 and p38 MAP kinase expression and a reduction in cyclin E2 [101]. Another study showed that miR-21 plays a key role in regulating cell apoptosis by targeting multiple genes in RCC including tumor necrosis factor (TNF) receptor superfamily member 6 (FAS) ligand and TIMP3 [100]. miR-15a has been reported to be upregulated in malignant RCC and downregulated in oncocytoma. The expression of miR-15a has been shown to be a potential marker to differentiate between benign and malignant renal tumors in biopsy and urine samples. The levels of miR-15a are inversely correlated to protein kinase C (PKC) α, a component of the transcription complex that regulates NF-κB signal transduction in tumors [133]. Li et al. reported that miR-155 may function as an oncogene in RCC by targeting transcription regulator BACH1 [134]. The expression of miR-155 was upregulated in ccRCC tissue and renal cancer cell lines. The suppression of miR-155 inhibited cell proliferation and migration and induced apoptosis in renal cancer cells [134].

Tumor suppressor gene p53 has been shown to induce the expression of miRNAs including miR-34a and miR-34b/c, which are direct p53 target genes that are inactivated in several tumors. Ectopic miR-34 expression induces apoptosis, cell cycle arrest, or senescence [135, 136]. The gene promoters of miR-34a and miR-34b/c are inactivated by CpG methylation in various tumor types including RCC [135, 137, 138] though a few studies suggest that miR-34a is upregulated in RCC [53, 139]. Wirsing et al. showed that miR-34a and miR-21 cooperate in downregulation of hepatocyte nuclear factor 4α (HNF4α) that has been linked to RCC [139]. Another study suggests that upregulation of miR-34 in RCC correlates with decreased expression of tumor suppressor secreted frizzled-related protein 1 (SFRP1) [53].

5.4.2 miRNAs Related to EMT, Invasion, and Metastasis

Epithelial–mesenchymal transition (EMT) is the initial and critical step in invasion and metastasis that is characterized by changes in gene expression, including decrease in epithelial genes, such as E-cadherin, and increase in mesenchymal genes.

This is driven by transcription factors (Snail, Twist, and ZEB families) and is associated with altered cell morphology (loss of apical–basal polarity) and a concomitant initiation of a migratory state [140]. A role for miRNAs that affects the EMT program and progression and metastasis of human cancers has begun to emerge [141–143]. It has been reported that the miR-200 family and miR-205 regulate EMT by direct targeting of ZEB1 and ZEB2 [141, 144, 145]. We reported that miR-708 targets transcription factor ZEB2 and polycomb repressor, BMI1 in RCC [122]. ZEB2, a member of the ZEB family of transcription factors, is a transcriptional repressor that regulates the expression of E-cadherin and EMT [146, 147]. In keeping with this novel finding, E-cadherin levels in miR-708-overexpressing cells showed that miR-708 reexpression increases the levels of this epithelial marker and represses fibronectin. BMI1, a member of the polycomb-repressive complex 1 [148, 149] is frequently overexpressed in cancers [150–152]. Its overexpression in carcinoma cell lines resulted in the acquisition of EMT characteristics, induction of stem-cell markers, and enhancement of tumor-initiating capability [153]. miRNA-34a was identified as being downregulated in hypoxic renal tubular epithelial cells. Inhibition of miR-34a expression in HK-2 cells, which highly express endogenous miR-34a, promoted EMT in renal tubular epithelial cells by directly targeting Notch1 and Jagged1 and Notch downstream signaling. Conversely, miR-34a mimics effectively prevented hypoxia-induced EMT [154]. miR-215 overexpression induces cell migration and invasion in vivo through direct targeting of ZEB2, thereby regulating EMT and the metastatic potential of RCC cells [70, 131]. miR-138a exerts tumor-suppressive effects in RCC by directly targeting vimentin, inhibiting cell migration and invasion [155].

A novel role of miR-34a in the regulation of transcription by c-Myc was recently identified whereby miR-34a downregulated expression of multiple oncogenes including c-Myc by targeting its 3′ untranslated region. In addition, miR-34a repressed Ras homolog gene family, member A (RhoA) expression by suppressing the Myc-S-phase kinase-associated protein 2 (Skp2)–Myc-associated zinc-finger protein 1 (Miz1) transcriptional complex that normally activates RhoA. Overexpression of c-Myc reversed miR-34a suppression of RhoA expression and inhibition of cell invasion, suggesting that miR-34a inhibits invasion by suppressing RhoA through c-Myc. miR-34a was also found to repress the c-Myc–positive transcription elongation factor (P-TEFb) transcription elongation complex that extends transcription through RNA polymerase II, indicating one of the mechanisms by which miR-34a has profound effects on cellular functions [156]. miR-584 is a tumor suppressor miRNA in ccRCC that inhibits cell motility through downregulation of oncogene ROCK-1 [157]. Expression levels of miRNA-106b were found to be significantly lower in tumors of patients who developed metastasis and miR-106b is a potential predictive marker of early metastasis after nephrectomy in RCC patients [158]. Ectopic overexpression of miR-205 and miR-99a inhibits cellular migration and invasion in RCC cell lines [120, 123]. miR-1285 apart from its effects on cellular proliferation also inhibits migration and invasion by targeting oncogene TGM2, which is widely overexpressed in RCC tissues [71]. Oncogenic miR-21 promotes tumor invasion in RCC cell lines in vitro [101]. hsa-miR-9 has been shown to be involved in the development of ccRCC while also having a role in the development of metastatic recurrence. miR-9-1 and miR-9-3 were both significantly downregulated in RCC. The methylation levels of these miRNA genes were found to be higher in the primary tumors of patients who developed recurrence, compared to those without recurrence [159].

6 Clinical Applications: miRNAs in Diagnosis, Prognosis, and Therapy of RCC

The spectrum of potential clinical applications of miRNAs in RCC is broad and continues to grow. miRNAs can be used for diagnosis, tumor classification, prognosis, prediction of treatment efficiency, and therapeutic applications [160].

Diagnostic application has three aspects and can be used to distinguish normal from malignant tissues; to identify the tissue of origin in poorly differentiated tumors or tumors of unknown origin; and finally, to distinguish the different subtypes of the same tumor [160]. The superior ability of miRNAs in classifying tumors compared to mRNA profiling [39] and their marked stability in body fluids [56, 57] indicate that miRNAs have great potential as novel biomarkers in RCC. In this direction, miRNA profiling has been done in RCC clinical specimens and serum/blood/urine searching for miRNA signatures that distinguish tumor from normal tissues and for identifying different RCC subtypes (as summarized in Sect. 4.1). Detecting miRNAs is easy and relatively inexpensive using standard quantitative reverse transcriptase polymerase chain reaction (qRT-PCR) assays [131, 161]. The potential diagnostic application of miRNAs in RCC is elegantly demonstrated in a study by Wulfken et al. where miR-1233 serum levels distinguishes RCC patients from healthy controls [58].

The differential diagnosis of RCC subtypes is a frequent clinical challenge as the subtypes have different genetic backgrounds, prognoses, and responses to surgical and medical treatment. Novel, reliable techniques for the exact classification of renal tumors, especially in cases with morphological ambiguity, are important for diagnosis and therapy. Thus, miRNA expression profiling is an effective molecular bioassay for classification of renal tumors and offers a quantitative standardized complement to current methods of tumor classification. Reliable and specific characterization of renal tumor subtypes, such as ccRCC, PRCC, chRCC, and oncocytoma, has already been demonstrated by miRNA-profiling methods [61, 63–66] (summarized in Sect. 4.2).

Another major clinical challenge in RCC is the lack of reliable prognostic biomarkers that can identify those patients who are at high risk for progression, metastasis, and relapse after nephrectomy. Studies in RCC have shown that aberrant miRNA expression profiles are related to overall survival, disease progression, recurrence, and metastasis [131] suggesting that miRNAs can be useful biomarkers with prognostic and predictive potential. For instance, miR-708 could be a potential RCC prognostic marker as it targets survivin, an independent predictor of progression and death in RCC [122, 131]. Significant downregulation of miR-106b has been shown to be a predictive marker of early metastasis after nephrectomy [158]. High levels of miR-21 is inversely correlated with the overall survival of RCC patients [101, 131]. Prognostic miRNA expression profiling can potentially be used for personalized therapy and targeted treatments. Therefore, large independent external validation studies on miRNAs have been recommended [131].

Recently, miRNAs have gained considerable attention as potential promising targets for cancer therapeutics. miRNA-based cancer therapy offers the theoretical appeal of targeting multiple gene networks controlled by a single, aberrantly expressed miRNA [160]. miRNAs targeting key signaling pathways in RCC pathogenesis (including pVHL-HIF, mTOR, and VEGF) have been identified and might be useful targets for RCC therapeutics. Reconstitution of tumor suppressor miRNAs, or sequence-specific knockdown of oncogenic miRNAs has produced favorable antitumor outcomes in experimental animal models. We showed that intratumoral administration of miR-708 and miR-205 precursors induced regression of tumors in a murine model of RCC [122, 123]. Genistein, a chemopreventative agent, reduced the expression of oncogenic miR-21 in vitro and in vivo [101]. However, several important issues need to be resolved prior to the use of miRNA-based cancer therapy, such as the possibility for nonspecific immune activation and the lack of well-defined delivery systems. Also, there is an urgent need for definitive mRNA target validation and a comprehensive understanding of rate-limiting cellular components that may impact the efficiency of this gene-silencing phenomenon [160, 162].

7 Concluding Remarks

In conclusion, experimental evidence suggests an important role for miRNAs in the etiology of RCC. It is evident that miRNAs can target various signaling cascades that are altered in RCC pathogenesis. However, our current understanding in this area is still limited. Although we are in an early phase of miRNA research, it is anticipated that miRNAs will have a significant impact in improving RCC patient management. miRNAs represent a novel group of potential biomarkers that may improve diagnostic, prognostic, and predictive abilities and are potential cancer therapeutic agents. Studies characterizing the role of miRNAs in RCC have the potential to greatly advance cancer biomarker development and offer potential for development of target-specific personalized molecular therapies. However, there are significant challenges that need to be addressed. In particular, the transition of miRNA applications from the research setting to the clinical stage poses several challenges. For example, conclusions obtained from in vitro studies performed in cell lines and a limited number of clinical samples may be very different from in vivo preclinical studies and those from a large number of human samples. Thus, focused, functional, and clinical studies are required in order to elucidate the complex relationship of miRNAs in the development and progression of RCC. These types of studies will aid in the development of novel miRNA-based diagnostic, prognostic, and therapeutic strategies for effective clinical management of this disease.

Acknowledgments We thank Dr. Roger Erickson for his support and assistance with the preparation of the manuscript. The work in Dr. Dahiyas's lab is supported by Grants RO1CA130860, RO1CA111470, RO1CA138642 from the NIH, VA Program Project, and Merit Review grants.

References

1. Siegel R, Naishadham D, Jemal A. Cancer statistics. CA: a Cancer J Clin. 2012;62(1):10–29. Epub 13 Jan 2012.
2. Rini BI, Campbell SC, Escudier B. Renal cell carcinoma. Lancet. 2009;373(9669):1119–32. Epub 10 Mar 2009.
3. Linehan WM. Genetic basis of kidney cancer: Role of genomics for the development of disease-based therapeutics. Genome Res. 2012;22(11):2089–100. Epub 6 Oct 2012.
4. Linehan WM, Walther MM, Zbar B. The genetic basis of cancer of the kidney. J Urol. 2003;170(6 Pt 1):2163–72. Epub 25 Nov 2003.
5. Linehan WM, Zbar B. Focus on kidney cancer. Cancer Cell. 2004;6(3):223–8. Epub 24 Sept 2004.
6. Reuter VE, Presti JC Jr. Contemporary approach to the classification of renal epithelial tumors. Semin Oncol. 2000;27(2):124–37. Epub 18 Apr 2000.
7. Zambrano NR, Lubensky IA, Merino MJ, Linehan WM, Walther MM. Histopathology and molecular genetics of renal tumors toward unification of a classification system. J Urol. 1999;162(4):1246–58. Epub 24 Sept 1999.
8. Kovacs G, Akhtar M, Beckwith BJ, Bugert P, Cooper CS, Delahunt B, et al. The Heidelberg classification of renal cell tumours. J Pathol. 1997;183(2):131–3. Epub 9 Dec 1997.
9. Thoenes W, Storkel S, Rumpelt HJ. Histopathology and classification of renal cell tumors (adenomas, oncocytomas and carcinomas). The basic cytological and histopathological elements and their use for diagnostics. Pathol Res Pract. 1986;181(2):125–43. Epub 1 May 1986.
10. Motzer RJ, Bander NH, Nanus DM. Renal-cell carcinoma. New Engl J Med. 1996;335(12): 865–75. Epub 19 Sept 1996.
11. van Spronsen DJ, Mulders PF, De Mulder PH. Novel treatments for metastatic renal cell carcinoma. Crit Rev Oncol Hematol. 2005;55(3):177–91. Epub 28 June 2005.
12. Dondeti VR, Wubbenhorst B, Lal P, Gordan JD, D'Andrea K, Attiyeh EF, et al. Integrative genomic analyses of sporadic clear cell renal cell carcinoma define disease subtypes and potential new therapeutic targets. Cancer Res. 2012;72(1):112–21. Epub 19 Nov 2011.
13. Teloken PE, Thompson RH, Tickoo SK, Cronin A, Savage C, Reuter VE, et al. Prognostic impact of histological subtype on surgically treated localized renal cell carcinoma. J Urol. 2009;182(5):2132–6. Epub 18 Sept 2009.
14. Delahunt B, Bethwaite PB, Thornton A. Prognostic significance of microscopic vascularity for clear cell renal cell carcinoma. Brit J Urol. 1997;80(3):401–4. Epub 6 Oct 1997.
15. Leroy X, Zini L, Leteurtre E, Zerimech F, Porchet N, Aubert JP, et al. Morphologic subtyping of papillary renal cell carcinoma: correlation with prognosis and differential expression of MUC1 between the two subtypes. Mod Pathol. 2002;15(11):1126–30. Epub 14 Nov 2002.
16. Arai E, Kanai Y. Genetic and epigenetic alterations during renal carcinogenesis. Int J Clin Exp Pathol. 2010;4(1):58–73. Epub 14 Jan 2011.
17. Sukov WR, Lohse CM, Leibovich BC, Thompson RH, Cheville JC. Clinical and pathological features associated with prognosis in patients with papillary renal cell carcinoma. J Urol. 2012;187(1):54–9. Epub 18 Nov 2011.

18. Rini BI, Rathmell WK, Godley P. Renal cell carcinoma. Curr Opin Oncol. 2008;20(3):300–6. Epub 9 Apr 2008.
19. Pantuck AJ, Zisman A, Belldegrun AS. The changing natural history of renal cell carcinoma. J Urol. 2001;166(5):1611–23. Epub 5 Oct 2001.
20. Sonpavde G, Choueiri TK, Escudier B, Ficarra V, Hutson TE, Mulders PF, et al. Sequencing of agents for metastatic renal cell carcinoma: can we customize therapy? Eur Urol. 2012;61(2):307–16. Epub 8 Nov 2011.
21. Takahashi M, Kahnoski R, Gross D, Nicol D, Teh BT. Familial adult renal neoplasia. J Med Genet. 2002;39(1):1–5. Epub 5 Feb 2002.
22. Linehan WM, Pinto PA, Bratslavsky G, Pfaffenroth E, Merino M, Vocke CD, et al. Hereditary kidney cancer: unique opportunity for disease-based therapy. Cancer. 2009;115(10 Suppl):2252–61. Epub 30 Apr 2009.
23. Maher ER. Genomics and epigenomics of renal cell carcinoma. Semin Cancer Biol. 2013;23(1):10–7. Epub 4 July 2012.
24. Kaelin WG, Jr. The von Hippel-Lindau tumor suppressor protein and clear cell renal carcinoma. Clin Cancer Res. 2007;13(2 Pt 2):680s–4s. Epub 27 Jan 2007.
25. Kim WY, Kaelin WG. Role of VHL gene mutation in human cancer. J Clin Oncol. 2004;22(24):4991–5004. Epub 22 Dec 2004.
26. Linehan WM, Rubin JS, Bottaro DP. VHL loss of function and its impact on oncogenic signaling networks in clear cell renal cell carcinoma. Int J Biochem Cell Biol. 2009;41(4):753–6. Epub 28 Oct 2008.
27. Cairns P. Renal cell carcinoma. Cancer Biomarkers: Sect A of disease markers. 2010;9(1–6):461–73. Epub 25 Nov 2011.
28. Wiesener MS, Munchenhagen PM, Berger I, Morgan NV, Roigas J, Schwiertz A, et al. Constitutive activation of hypoxia-inducible genes related to overexpression of hypoxia-inducible factor-1alpha in clear cell renal carcinomas. Cancer Res. 2001;61(13):5215–22. Epub 30 June 2001.
29. Young AP, Schlisio S, Minamishima YA, Zhang Q, Li L, Grisanzio C, et al. VHL loss actuates a HIF-independent senescence programme mediated by Rb and p400. Nat Cell Biol. 2008;10(3):361–9. Epub 26 Feb 2008.
30. Schmidt L, Duh FM, Chen F, Kishida T, Glenn G, Choyke P, et al. Germline and somatic mutations in the tyrosine kinase domain of the MET proto-oncogene in papillary renal carcinomas. Nat Genet. 1997;16(1):68–73. Epub 1 May 1997.
31. Tomlinson IP, Alam NA, Rowan AJ, Barclay E, Jaeger EE, Kelsell D, et al. Germline mutations in FH predispose to dominantly inherited uterine fibroids, skin leiomyomata and papillary renal cell cancer. Nat Genet. 2002;30(4):406–10. Epub 28 Feb 2002.
32. Nickerson ML, Warren MB, Toro JR, Matrosova V, Glenn G, Turner ML, et al. Mutations in a novel gene lead to kidney tumors, lung wall defects, and benign tumors of the hair follicle in patients with the Birt-Hogg-Dube syndrome. Cancer Cell. 2002;2(2):157–64. Epub 3 Sept 2002.
33. Toro JR, Nickerson ML, Wei MH, Warren MB, Glenn GM, Turner ML, et al. Mutations in the fumarate hydratase gene cause hereditary leiomyomatosis and renal cell cancer in families in North America. Am J Hum Genet. 2003;73(1):95–106. Epub 29 May 2003.
34. Vocke CD, Yang Y, Pavlovich CP, Schmidt LS, Nickerson ML, Torres-Cabala CA, et al. High frequency of somatic frameshift BHD gene mutations in Birt-Hogg-Dube-associated renal tumors. J Natl Cancer Inst. 2005;97(12):931–5. Epub 16 June 2005.
35. Bartel DP. MicroRNAs: target recognition and regulatory functions. Cell. 2009;136(2):215–33. Epub 27 Jan 2009.
36. Schickel R, Boyerinas B, Park SM, Peter ME. MicroRNAs: key players in the immune system, differentiation, tumorigenesis and cell death. Oncogene. 2008;27(45):5959–74. Epub 7 Oct 2008.
37. Ventura A, Jacks T. MicroRNAs and cancer: short RNAs go a long way. Cell. 2009;136(4):586–91. Epub 26 Feb 2009.

38. Calin GA, Croce CM. MicroRNA signatures in human cancers. Nat Rev Cancer. 2006;6(11):857–66. Epub 25 Oct 2006.
39. Lu J, Getz G, Miska EA, Alvarez-Saavedra E, Lamb J, Peck D, et al. MicroRNA expression profiles classify human cancers. Nature. 2005;435(7043):834–8. Epub 10 June 2005.
40. Gottardo F, Liu CG, Ferracin M, Calin GA, Fassan M, Bassi P, et al. Micro-RNA profiling in kidney and bladder cancers. Urol Oncol. 2007;25(5):387–92. Epub 11 Sept 2007.
41. Huang Y, Dai Y, Yang J, Chen T, Yin Y, Tang M, et al. Microarray analysis of microRNA expression in renal clear cell carcinoma. Eur J Surg Oncol. 2009;35(10):1119–23. Epub 16 May 2009.
42. Juan D, Alexe G, Antes T, Liu H, Madabhushi A, Delisi C, et al. Identification of a microRNA panel for clear-cell kidney cancer. Urology. 2010;75(4):835–41. Epub 29 Dec 2009.
43. Jung M, Mollenkopf HJ, Grimm C, Wagner I, Albrecht M, Waller T, et al. MicroRNA profiling of clear cell renal cell cancer identifies a robust signature to define renal malignancy. J Cell Mol Med. 2009;13(9B):3918–28. Epub 21 Feb 2009.
44. Bartels CL, Tsongalis GJ. MicroRNAs: novel biomarkers for human cancer. Clin Chem. 2009;55(4):623–31. Epub 28 Feb 2009.
45. Verghese ET, Hanby AM, Speirs V, Hughes TA. Small is beautiful: microRNAs and breast cancer-where are we now? J Pathol. 2008;215(3):214–21. Epub 1 May 2008.
46. Rosenfeld N, Aharonov R, Meiri E, Rosenwald S, Spector Y, Zepeniuk M, et al. MicroRNAs accurately identify cancer tissue origin. Nat Biotechnol. 2008;26(4):462–9. Epub 26 March 2008.
47. Ribal MJ. Molecular profiling of renal cancer: the journey to clinical application. Eur Urol. 2011;59(5):731–3. Epub 8 Feb 2011.
48. Schaefer A, Jung M, Kristiansen G, Lein M, Schrader M, Miller K, et al. MicroRNAs and cancer: current state and future perspectives in urologic oncology. Urol Oncol. 2010;28(1):4–13. Epub 2 Jan 2009.
49. Nakada C, Matsuura K, Tsukamoto Y, Tanigawa M, Yoshimoto T, Narimatsu T, et al. Genome-wide microRNA expression profiling in renal cell carcinoma: significant down-regulation of miR-141 and miR-200c. J Pathol. 2008;216(4):418–27. Epub 18 Oct 2008.
50. Yi Z, Fu Y, Zhao S, Zhang X, Ma C. Differential expression of miRNA patterns in renal cell carcinoma and nontumorous tissues. J Cancer Res Clin Oncol. 2010;136(6):855–62. Epub 19 Nov 2009.
51. White NM, Bao TT, Grigull J, Youssef YM, Girgis A, Diamandis M, et al. miRNA profiling for clear cell renal cell carcinoma: biomarker discovery and identification of potential controls and consequences of miRNA dysregulation. J Urol. 2011;186(3):1077–83. Epub 26 July 2011.
52. Chow TF, Youssef YM, Lianidou E, Romaschin AD, Honey RJ, Stewart R, et al. Differential expression profiling of microRNAs and their potential involvement in renal cell carcinoma pathogenesis. Clin Biochem. 2010;43(1–2):150–8. Epub 04 Aug 2009.
53. Liu H, Brannon AR, Reddy AR, Alexe G, Seiler MW, Arreola A, et al. Identifying mRNA targets of microRNA dysregulated in cancer: with application to clear cell Renal Cell Carcinoma. BMC Syst Biol. 2010;4:51. Epub 28 Apr 2010.
54. Weng L, Wu X, Gao H, Mu B, Li X, Wang JH, et al. MicroRNA profiling of clear cell renal cell carcinoma by whole-genome small RNA deep sequencing of paired frozen and formalin-fixed, paraffin-embedded tissue specimens. J Pathol. 2010;222(1):41–51. Epub 2 July 2010.
55. Osanto S, Qin Y, Buermans HP, Berkers J, Lerut E, Goeman JJ, et al. Genome-wide microRNA expression analysis of clear cell renal cell carcinoma by next generation deep sequencing. PloS One. 2012;7(6):e38298. Epub 30 June 2012.
56. Chen X, Ba Y, Ma L, Cai X, Yin Y, Wang K, et al. Characterization of microRNAs in serum: a novel class of biomarkers for diagnosis of cancer and other diseases. Cell Res. 2008;18(10):997–1006. Epub 4 Sept 2008.
57. Hanke M, Hoefig K, Merz H, Feller AC, Kausch I, Jocham D, et al. A robust methodology to study urine microRNA as tumor marker: microRNA-126 and microRNA-182 are related to urinary bladder cancer. Urol Oncol. 2010;28(6):655–61. Epub 21 Apr 2009.

58. Wulfken LM, Moritz R, Ohlmann C, Holdenrieder S, Jung V, Becker F, et al. MicroRNAs in renal cell carcinoma: diagnostic implications of serum miR-1233 levels. PloS One. 2011;6(9):e25787. Epub 11 Oct 2011.
59. Redova M, Poprach A, Nekvindova J, Iliev R, Radova L, Lakomy R, et al. Circulating miR-378 and miR-451 in serum are potential biomarkers for renal cell carcinoma. J Transl Med. 2012;10:55. Epub 24 Mar 2012.
60. Hauser S, Wulfken LM, Holdenrieder S, Moritz R, Ohlmann CH, Jung V, et al. Analysis of serum microRNAs (miR-26a-2*, miR-191, miR-337–3p and miR-378) as potential biomarkers in renal cell carcinoma. Cancer Epidemiol. 2012;36(4):391–4. Epub 1 May 2012.
61. Petillo D, Kort EJ, Anema J, Furge KA, Yang XJ, Teh BT. MicroRNA profiling of human kidney cancer subtypes. Int J Oncol. 2009;35(1):109–14. Epub 11 June 2009.
62. Catto JW, Alcaraz A, Bjartell AS, De Vere White R, Evans CP, Fussel S, et al. MicroRNA in prostate, bladder, and kidney cancer: a systematic review. Eur Urol. 2011;59(5):671–81. Epub 8 Feb 2011.
63. Fridman E, Dotan Z, Barshack I, David MB, Dov A, Tabak S, et al. Accurate molecular classification of renal tumors using microRNA expression. J Mol Diagn. 2010;12(5):687–96. Epub 3 July 2010.
64. Youssef YM, White NM, Grigull J, Krizova A, Samy C, Mejia-Guerrero S, et al. Accurate molecular classification of kidney cancer subtypes using microRNA signature. Euro Urol. 2011;59(5):721–30. Epub 29 Jan 2011.
65. Maj M, Wagner L. Molecular approaches to classify adult renal epithelial neoplasms. Expert Rev Mol Diagn. 2011;11(8):803–6. Epub 26 Oct 2011.
66. Powers MP, Alvarez K, Kim HJ, Monzon FA. Molecular classification of adult renal epithelial neoplasms using microRNA expression and virtual karyotyping. Diagn Mol pathol: Am J Surg Pathol, Part B. 2011;20(2):63–70. Epub 3 May 2011.
67. White NM, Fatoohi E, Metias M, Jung K, Stephan C, Yousef GM. Metastamirs: a stepping stone towards improved cancer management. Nat Rev Clin Oncol. 2011;8(2):75–84. Epub 4 Nov 2010.
68. Zhang J, Ma L. MicroRNA control of epithelial-mesenchymal transition and metastasis. Cancer Metast Rev. 2012;31(3–4):653–62. Epub 12 June 2012.
69. Heinzelmann J, Henning B, Sanjmyatav J, Posorski N, Steiner T, Wunderlich H, et al. Specific miRNA signatures are associated with metastasis and poor prognosis in clear cell renal cell carcinoma. World J Urol. 2011;29(3):367–73. Epub 14 Jan 2011.
70. White NM, Khella HW, Grigull J, Adzovic S, Youssef YM, Honey RJ, et al. miRNA profiling in metastatic renal cell carcinoma reveals a tumour-suppressor effect for miR-215. Br J Cancer. 2011;105(11):1741–9. Epub 29 Oct 2011.
71. Khella HW, White NM, Faragalla H, Gabril M, Boazak M, Dorian D, et al. Exploring the role of miRNAs in renal cell carcinoma progression and metastasis through bioinformatic and experimental analyses. Tumour Biol: J Int Soc Oncodev Biol Med. 2012;33(1):131–40. Epub 17 Nov 2011.
72. Wu X, Weng L, Li X, Guo C, Pal SK, Jin JM, et al. Identification of a 4-microRNA signature for clear cell renal cell carcinoma metastasis and prognosis. PloS One. 2012;7(5):e35661. Epub 25 May 2012.
73. Wotschofsky Z, Liep J, Meyer HA, Jung M, Wagner I, Disch AC, et al. Identification of Metastamirs as Metastasis-associated MicroRNAs in Clear Cell Renal Cell Carcinomas. Int J Biol Sci. 2012;8(10):1363–74. Epub 10 Nov 2012.
74. Lopez-Lago MA, Thodima VJ, Guttapalli A, Chan T, Heguy A, Molina AM, et al. Genomic deregulation during metastasis of renal cell carcinoma implements a myofibroblast-like program of gene expression. Cancer Res. 2010;70(23):9682–92. Epub 19 Oct 2010.
75. Slaby O, Redova M, Poprach A, Nekvindova J, Iliev R, Radova L, et al. Identification of MicroRNAs associated with early relapse after nephrectomy in renal cell carcinoma patients. Genes Chromosome Canc. 2012;51(7):707–16. Epub 12 Apr 2012.

76. Turner KJ, Moore JW, Jones A, Taylor CF, Cuthbert-Heavens D, Han C, et al. Expression of hypoxia-inducible factors in human renal cancer: relationship to angiogenesis and to the von Hippel-Lindau gene mutation. Cancer Res. 2002;62(10):2957–61. Epub 23 May 2002.
77. Beasley NJ, Leek R, Alam M, Turley H, Cox GJ, Gatter K, et al. Hypoxia-inducible factors HIF-1alpha and HIF-2alpha in head and neck cancer: relationship to tumor biology and treatment outcome in surgically resected patients. Cancer Res. 2002;62(9):2493–7. Epub 1 May 2002.
78. Hutchison GJ, Valentine HR, Loncaster JA, Davidson SE, Hunter RD, Roberts SA, et al. Hypoxia-inducible factor 1alpha expression as an intrinsic marker of hypoxia: correlation with tumor oxygen, pimonidazole measurements, and outcome in locally advanced carcinoma of the cervix. Clin Cancer Res. 2004;10(24):8405–12. Epub 30 Dec 2004.
79. Fischer U, Radermacher J, Mayer J, Mehraein Y, Meese E. Tumor hypoxia: Impact on gene amplification in glioblastoma. Int J Oncol. 2008;33(3):509–15. Epub 13 Aug 2008.
80. Kirkpatrick JP. Tumor hypoxia and prognosis in human gliomas. Cancer J. 2006;12(6): 451–4. Epub 9 Jan 2007.
81. Vaupel P. The role of hypoxia-induced factors in tumor progression. Oncology. 2004;9 (Suppl 5):10–7. Epub 14 Dec 2004.
82. Vaupel P, Mayer A, Hockel M. Tumor hypoxia and malignant progression. Meth Enzymol. 2004;381:335–54. Epub 6 Apr 2004.
83. Kulshreshtha R, Ferracin M, Wojcik SE, Garzon R, Alder H, Agosto-Perez FJ, et al. A microRNA signature of hypoxia. Mol Cell Biol. 2007;27(5):1859–67. Epub 30 Dec 2006.
84. Gee HE, Camps C, Buffa FM, Patiar S, Winter SC, Betts G, et al. hsa-mir-210 is a marker of tumor hypoxia and a prognostic factor in head and neck cancer. Cancer. 2010;116(9): 2148–58. Epub 27 Feb 2010.
85. Ho AS, Huang X, Cao H, Christman-Skieller C, Bennewith K, Le QT, et al. Circulating miR-210 as a Novel Hypoxia Marker in Pancreatic Cancer. Transl Oncol. 2010;3(2):109–13. Epub 3 Apr 2010.
86. McCormick R, Buffa FM, Ragoussis J, Harris AL. The role of hypoxia regulated microRNAs in cancer. Curr Top Microbiol Immunol. 2010;345:47–70. Epub 16 June 2010.
87. Camps C, Buffa FM, Colella S, Moore J, Sotiriou C, Sheldon H, et al. hsa-miR-210 Is induced by hypoxia and is an independent prognostic factor in breast cancer. Clin Cancer Res. 2008;14(5):1340–8. Epub 5 March 2008.
88. Fendler A, Stephan C, Yousef GM, Jung K. MicroRNAs as regulators of signal transduction in urological tumors. Clin Chem. 2011;57(7):954–68. Epub 3 June 2011.
89. Chow TF, Mankaruos M, Scorilas A, Youssef Y, Girgis A, Mossad S, et al. The miR-17–92 cluster is over expressed in and has an oncogenic effect on renal cell carcinoma. J Urol. 2010;183(2):743–51. Epub 22 Dec 2009.
90. Lewis BP, Burge CB, Bartel DP. Conserved seed pairing, often flanked by adenosines, indicates that thousands of human genes are microRNA targets. Cell. 2005;120(1):15–20. Epub 18 Jan 2005.
91. Valera VA, Walter BA, Linehan WM, Merino MJ. Regulatory effects of microRNA-92 (miR-92) on VHL gene expression and the hypoxic activation of miR-210 in clear cell renal cell carcinoma. J Cancer. 2011;2:515–26. Epub 2 Nov 2011.
92. Neal CS, Michael MZ, Rawlings LH, Van der Hoek MB, Gleadle JM. The VHL-dependent regulation of microRNAs in renal cancer. BMC Med. 2010;8:64. Epub 23 Oct 2010.
93. White NM, Yousef GM. MicroRNAs: exploring a new dimension in the pathogenesis of kidney cancer. BMC Med. 2010;8:65. Epub 23 Oct 2010.
94. Redova M, Poprach A, Besse A, Iliev R, Nekvindova J, Lakomy R, et al. MiR-210 expression in tumor tissue and in vitro effects of its silencing in renal cell carcinoma. Tumour Biol. 2013;34(1):481–91. Epub 15 Nov 2012.
95. Lichner Z, Mejia-Guerrero S, Ignacak M, Krizova A, Bao TT, Girgis AH, et al. Pleiotropic action of renal cell carcinoma-dysregulated miRNAs on hypoxia-related signaling pathways. Am J Pathol. 2012;180(4):1675–87. Epub 14 Feb 2012.

96. Mikhaylova O, Stratton Y, Hall D, Kellner E, Ehmer B, Drew AF, et al. VHL-regulated MiR-204 suppresses tumor growth through inhibition of LC3B-mediated autophagy in renal clear cell carcinoma. Cancer Cell. 2012;21(4):532–46. Epub 21 Apr 2012.
97. Degenhardt K, Mathew R, Beaudoin B, Bray K, Anderson D, Chen G, et al. Autophagy promotes tumor cell survival and restricts necrosis, inflammation, and tumorigenesis. Cancer cell. 2006;10(1):51–64. Epub 18 July 2006.
98. Jin S, White E. Role of autophagy in cancer: management of metabolic stress. Autophagy. 2007;3(1):28–31. Epub 14 Sept 2006.
99. Liu W, Zabirnyk O, Wang H, Shiao YH, Nickerson ML, Khalil S, et al. miR-23b targets proline oxidase, a novel tumor suppressor protein in renal cancer. Oncogene. 2010;29(35):4914–24. Epub 22 June 2010.
100. Zhang A, Liu Y, Shen Y, Xu Y, Li X. miR-21 modulates cell apoptosis by targeting multiple genes in renal cell carcinoma. Urology. 2011;78(2):474 e13–9. Epub 9 Aug 2011.
101. Zaman MS, Shahryari V, Deng G, Thamminana S, Saini S, Majid S, et al. Up-regulation of microRNA-21 correlates with lower kidney cancer survival. PloS One. 2012;7(2):e31060. Epub 22 Feb 2012.
102. Liu LZ, Li C, Chen Q, Jing Y, Carpenter R, Jiang Y, et al. MiR-21 induced angiogenesis through AKT and ERK activation and HIF-1alpha expression. PloS One. 2011;6(4):e19139. Epub 6 May 2011.
103. Song T, Zhang X, Wang C, Wu Y, Cai W, Gao J, et al. MiR-138 suppresses expression of hypoxia-inducible factor 1alpha (HIF-1alpha) in clear cell renal cell carcinoma 786-O cells. Asian Pac J Cancer Prev. 2011;12(5):1307–11. Epub 31 Aug 2011.
104. Cha ST, Chen PS, Johansson G, Chu CY, Wang MY, Jeng YM, et al. MicroRNA-519c suppresses hypoxia-inducible factor-1alpha expression and tumor angiogenesis. Cancer Res. 2010;70(7):2675–85. Epub 18 March 2010.
105. Lei Z, Li B, Yang Z, Fang H, Zhang GM, Feng ZH, et al. Regulation of HIF-1alpha and VEGF by miR-20b tunes tumor cells to adapt to the alteration of oxygen concentration. PloS One. 2009;4(10):e7629. Epub 7 Nov 2009.
106. Yamakuchi M, Yagi S, Ito T, Lowenstein CJ. MicroRNA-22 regulates hypoxia signaling in colon cancer cells. PloS One. 2011;6(5):e20291. Epub 2 June 2011.
107. Kerbel RS. Tumor angiogenesis. N Eng J Med. 2008;358(19):2039–49. Epub 9 May 2008.
108. Ferrara N, Gerber HP, LeCouter J. The biology of VEGF and its receptors. Nat Med. 2003;9(6):669–76. Epub 5 June 2003.
109. van der Veldt AA, Haanen JB, van den Eertwegh AJ, Boven E. Targeted therapy for renal cell cancer: current perspectives. Disc Med. 2010;10(54):394–405. Epub 3 Dec 2010.
110. Sinha S, Dutta S, Datta K, Ghosh AK, Mukhopadhyay D. Von Hippel-Lindau gene product modulates TIS11B expression in renal cell carcinoma: impact on vascular endothelial growth factor expression in hypoxia. J Biol Chem. 2009;284(47):32610–8. Epub 6 Oct 2009.
111. Hay N, Sonenberg N. Upstream and downstream of mTOR. Genes Dev. 2004;18(16):1926–45. Epub 18 Aug 2004.
112. Hudes GR. Targeting mTOR in renal cell carcinoma. Cancer. 2009;115(10 Suppl):2313–20. Epub 30 Apr 2009.
113. Hudson CC, Liu M, Chiang GG, Otterness DM, Loomis DC, Kaper F, et al. Regulation of hypoxia-inducible factor 1alpha expression and function by the mammalian target of rapamycin. Mol Cell Biol. 2002;22(20):7004–14. Epub 21 Sept 2002.
114. Pantuck AJ, Seligson DB, Klatte T, Yu H, Leppert JT, Moore L, et al. Prognostic relevance of the mTOR pathway in renal cell carcinoma: implications for molecular patient selection for targeted therapy. Cancer. 2007;109(11):2257–67. Epub 19 Apr 2007.
115. Robb VA, Karbowniczek M, Klein-Szanto AJ, Henske EP. Activation of the mTOR signaling pathway in renal clear cell carcinoma. J Urol. 2007;177(1):346–52. Epub 13 Dec 2006.
116. Cully M, You H, Levine AJ, Mak TW. Beyond PTEN mutations: the PI3K pathway as an integrator of multiple inputs during tumorigenesis. Nat Rev Cancer. 2006;6(3):184–92. Epub 3 Feb 2006.

117. Gan B, Lim C, Chu G, Hua S, Ding Z, Collins M, et al. FoxOs enforce a progression checkpoint to constrain mTORC1-activated renal tumorigenesis. Cancer Cell. 2010;18(5):472–84. Epub 16 Nov 2010.
118. Sachdeva M, Zhu S, Wu F, Wu H, Walia V, Kumar S, et al. p53 represses c-Myc through induction of the tumor suppressor miR-145. Proc Natl Acad Sci U S A. 2009;106(9):3207–12. Epub 10 Feb 2009.
119. Zhou L, Yin B, Liu Y, Hong Y, Zhang C, Fan J. Mechanism and function of decreased FOXO1 in renal cell carcinoma. J Surg Oncol. 2012;105(8):841–7. Epub 4 Jan 2012.
120. Cui L, Zhou H, Zhao H, Zhou Y, Xu R, Xu X, et al. MicroRNA-99a induces G1-phase cell cycle arrest and suppresses tumorigenicity in renal cell carcinoma. BMC Cancer. 2012;12:546. Epub 24 Nov 2012.
121. Fornari F, Milazzo M, Chieco P, Negrini M, Calin GA, Grazi GL, et al. MiR-199a-3p regulates mTOR and c-Met to influence the doxorubicin sensitivity of human hepatocarcinoma cells. Cancer Res. 2010;70(12):5184–93. Epub 27 May 2010.
122. Saini S, Yamamura S, Majid S, Shahryari V, Hirata H, Tanaka Y, et al. MicroRNA-708 induces apoptosis and suppresses tumorigenicity in renal cancer cells. Cancer Res. 2011;71(19):6208–19. Epub 20 Aug 2011.
123. Majid S, Saini S, Dar AA, Hirata H, Shahryari V, Tanaka Y, et al. MicroRNA-205 inhibits Src-mediated oncogenic pathways in renal cancer. Cancer Res. 2011;71(7):2611–21. Epub 19 Feb 2011.
124. Zhai Q, Zhou L, Zhao C, Wan J, Yu Z, Guo X, et al. Identification of miR-508–3p and miR-509–3p that are associated with cell invasion and migration and involved in the apoptosis of renal cell carcinoma. Biochem Biophys Res Commun. 2012;419(4):621–6. Epub 01 March 2012.
125. Yamada Y, Hidaka H, Seki N, Yoshino H, Yamasaki T, Itesako T, et al. Tumor-suppressive microRNA-135a inhibits cancer cell proliferation by targeting the c-MYC oncogene in renal cell carcinoma. Cancer Sci. 2013;104(3):304–12. Epub 28 Nov 2012.
126. Pichler M, Hutterer GC, Chromecki TF, Jesche J, Groselj-Strele A, Kampel-Kettner K, et al. Prognostic value of the Leibovich prognosis score supplemented by vascular invasion for clear cell renal cell carcinoma. J Urol. 2012;187(3):834–9. Epub 17 Jan 2012.
127. Sakurai T, Bilim VN, Ugolkov AV, Yuuki K, Tsukigi M, Motoyama T, et al. The enhancer of zeste homolog 2 (EZH2), a potential therapeutic target, is regulated by miR-101 in renal cancer cells. Biochem Biophys Res Commun. 2012;422(4):607–14. Epub 23 May 2012.
128. Liu Y, Yin B, Zhang C, Zhou L, Fan J. Hsa-let-7a functions as a tumor suppressor in renal cell carcinoma cell lines by targeting c-myc. Biochem Biophys Res Commun. 2012;417(1):371–5. Epub 14 Dec 2011.
129. Hidaka H, Seki N, Yoshino H, Yamasaki T, Yamada Y, Nohata N, et al. Tumor suppressive microRNA-1285 regulates novel molecular targets: aberrant expression and functional significance in renal cell carcinoma. Oncotarget. 2012;3(1):44–57. Epub 2 Feb 2012.
130. Kawakami K, Enokida H, Chiyomaru T, Tatarano S, Yoshino H, Kagara I, et al. The functional significance of miR-1 and miR-133a in renal cell carcinoma. Eur J Cancer. 2012;48(6):827–36. Epub 13 July 2011.
131. Al-Ali BM, Ress AL, Gerger A, Pichler M. MicroRNAs in renal cell carcinoma: implications for pathogenesis, diagnosis, prognosis and therapy. Anticancer Res. 2012;32(9):3727–32. Epub 21 Sept 2012.
132. Faragalla H, Youssef YM, Scorilas A, Khalil B, White NM, Mejia-Guerrero S, et al. The clinical utility of miR-21 as a diagnostic and prognostic marker for renal cell carcinoma. J Mol Diagn. 2012;14(4):385–92. Epub 15 May 2012.
133. von Brandenstein M, Pandarakalam, Braun G, et al. MicroRNA 15a, inversely correlated to PKCalpha, is a potential marker to differentiate between benign and malignant renal tumors in biopsy and urine samples. Am J Pathol. 2012;180(5):1787–97. Epub 21 March 2012.
134. Li S, Chen T, Zhong Z, Wang Y, Li Y, Zhao X. microRNA-155 silencing inhibits proliferation and migration and induces apoptosis by upregulating BACH1 in renal cancer cells. Mol Med Rep. 2012;5(4):949–54. Epub 7 Feb 2012.

135. Hermeking H. The miR-34 family in cancer and apoptosis. Cell Death Differ. 2010;17(2):193–9. Epub 23 May 2009.
136. Medina PP, Slack FJ. microRNAs and cancer: an overview. Cell Cycle. 2008;7(16):2485–92. Epub 23 Aug 2008.
137. Lodygin D, Tarasov V, Epanchintsev A, Berking C, Knyazeva T, Korner H, et al. Inactivation of miR-34a by aberrant CpG methylation in multiple types of cancer. Cell Cycle. 2008;7(16):2591–600. Epub 23 Aug 2008.
138. Vogt M, Munding J, Gruner M, Liffers ST, Verdoodt B, Hauk J, et al. Frequent concomitant inactivation of miR-34a and miR-34b/c by CpG methylation in colorectal, pancreatic, mammary, ovarian, urothelial, and renal cell carcinomas and soft tissue sarcomas. Virchows Archiv. 2011;458(3):313–22. Epub 13 Jan 2011.
139. Wirsing A, Senkel S, Klein-Hitpass L, Ryffel GU. A systematic analysis of the 3'UTR of HNF4A mRNA reveals an interplay of regulatory elements including miRNA target sites. PloS One. 2011;6(11):e27438. Epub 6 Dec 2011.
140. Gravdal K, Halvorsen OJ, Haukaas SA, Akslen LA. A switch from E-cadherin to N-cadherin expression indicates epithelial to mesenchymal transition and is of strong and independent importance for the progress of prostate cancer. Clin Cancer Res. 2007;13(23):7003–11. Epub 7 Dec 2007.
141. Bracken CP, Gregory PA, Khew-Goodall Y, Goodall GJ. The role of microRNAs in metastasis and epithelial-mesenchymal transition. Cell Mol Life Sci. 2009;66(10):1682–99. Epub 21 Jan 2009.
142. Kang Y, Massague J. Epithelial-mesenchymal transitions: twist in development and metastasis. Cell. 2004;118(3):277–9. Epub 6 Aug 2004.
143. Zhang J, Ma L. MicroRNA control of epithelial-mesenchymal transition and metastasis. Cancer Metastasis Rev. 2012;31(3-4):653–62. Epub 12 June 2012.
144. Gregory PA, Bracken CP, Bert AG, Goodall GJ. MicroRNAs as regulators of epithelial-mesenchymal transition. Cell Cycle. 2008;7(20):3112–8. Epub 18 Oct 2008.
145. Gregory PA, Bert AG, Paterson EL, Barry SC, Tsykin A, Farshid G, et al. The miR-200 family and miR-205 regulate epithelial to mesenchymal transition by targeting ZEB1 and SIP1. Nat Cell Biol. 2008;10(5):593–601. Epub 1 Apr 2008.
146. Vandewalle C, Comijn J, De Craene B, Vermassen P, Bruyneel E, Andersen H, et al. SIP1/ZEB2 induces EMT by repressing genes of different epithelial cell-cell junctions. Nucleic Acids Res. 2005;33(20):6566–78. Epub 30 Nov 2005.
147. Comijn J, Berx G, Vermassen P, Verschueren K, van Grunsven L, Bruyneel E, et al. The two-handed E box binding zinc finger protein SIP1 downregulates E-cadherin and induces invasion. Mol Cell. 2001;7(6):1267–78. Epub 30 June 2001.
148. Valk-Lingbeek ME, Bruggeman SW, van Lohuizen M. Stem cells and cancer; the polycomb connection. Cell. 2004;118(4):409–18. Epub 19 Aug 2004.
149. Widschwendter M, Fiegl H, Egle D, Mueller-Holzner E, Spizzo G, Marth C, et al. Epigenetic stem cell signature in cancer. Nat Genet. 2007;39(2):157–8. Epub 4 Jan 2007.
150. Chiba T, Miyagi S, Saraya A, Aoki R, Seki A, Morita Y, et al. The polycomb gene product BMI1 contributes to the maintenance of tumor-initiating side population cells in hepatocellular carcinoma. Cancer Res. 2008;68(19):7742–9. Epub 3 Oct 2008.
151. Leung C, Lingbeek M, Shakhova O, Liu J, Tanger E, Saremaslani P, et al. Bmi1 is essential for cerebellar development and is overexpressed in human medulloblastomas. Nature. 2004;428(6980):337–41. Epub 19 March 2004.
152. Prince ME, Sivanandan R, Kaczorowski A, Wolf GT, Kaplan MJ, Dalerba P, et al. Identification of a subpopulation of cells with cancer stem cell properties in head and neck squamous cell carcinoma. Proc Natl Acad Sci U S A. 2007;104(3):973–8. Epub 11 Jan 2007.
153. Yang MH, Hsu DS, Wang HW, Wang HJ, Lan HY, Yang WH, et al. Bmi1 is essential in Twist1-induced epithelial-mesenchymal transition. Nat Cell Biol.12(10):982–92. Epub 8 Sept 2010.

154. Du R, Sun W, Xia L, Zhao A, Yu Y, Zhao L, et al. Hypoxia-induced down-regulation of microRNA-34a promotes EMT by targeting the Notch signaling pathway in tubular epithelial cells. PloS One. 2012;7(2):e30771. Epub 1 March 2012.
155. Yamasaki T, Seki N, Yamada Y, Yoshino H, Hidaka H, Chiyomaru T, et al. Tumor suppressive microRNA138 contributes to cell migration and invasion through its targeting of vimentin in renal cell carcinoma. Int J Oncol. 2012;41(3):805–17. Epub 7 July 2012.
156. Yamamura S, Saini S, Majid S, Hirata H, Ueno K, Chang I, et al. MicroRNA-34a suppresses malignant transformation by targeting c-Myc transcriptional complexes in human renal cell carcinoma. Carcinogenesis. 2012;33(2):294–300. Epub 14 Dec 2011.
157. Ueno K, Hirata H, Shahryari V, Chen Y, Zaman MS, Singh K, et al. Tumour suppressor microRNA-584 directly targets oncogene Rock-1 and decreases invasion ability in human clear cell renal cell carcinoma. Br J Cancer. 2011;104(2):308–15. Epub 2 Dec 2010.
158. Slaby O, Jancovicova J, Lakomy R, Svoboda M, Poprach A, Fabian P, et al. Expression of miRNA-106b in conventional renal cell carcinoma is a potential marker for prediction of early metastasis after nephrectomy. J Exp Clin Cancer Res. 2010;29:90. Epub 9 July 2010.
159. Hildebrandt MA, Gu J, Lin J, Ye Y, Tan W, Tamboli P, et al. Hsa-miR-9 methylation status is associated with cancer development and metastatic recurrence in patients with clear cell renal cell carcinoma. Oncogene. 2010;29(42):5724–8. Epub 3 Aug 2010.
160. Metias SM, Lianidou E, Yousef GM. MicroRNAs in clinical oncology: at the crossroads between promises and problems. J Clin Pathol. 2009;62(9):771–6. Epub 8 Sept 2009.
161. Cortez MA, Bueso-Ramos C, Ferdin J, Lopez-Berestein G, Sood AK, Calin GA. MicroRNAs in body fluids–the mix of hormones and biomarkers. Nat Rev Clin Oncol. 2011;8(8):467–77. Epub 8 June 2011.
162. Tong AW, Nemunaitis J. Modulation of miRNA activity in human cancer: a new paradigm for cancer gene therapy? Cancer Gene Ther. 2008;15(6):341–55. Epub 29 March 2008.

Chapter 13
MicroRNAs in the Development and Progression of Skin Cancer

Vivek Choudhary, Meg Gullotto, Lisa Sato and Wendy B. Bollag

Abstract MicroRNAs are known to regulate the messenger RNA (mRNA) levels and protein expression of multiple genes and their products by binding to the 3′ untranslated region (3′-UTR) of target mRNAs and causing their degradation or inhibiting their translation. Numerous reports have demonstrated the importance of microRNAs in skin biology, with microRNAs regulating multiple processes in the skin in vivo and in skin cells in vitro. Keratinocytes are the predominant cells of the outer layer of skin, the epidermis, and microRNAs have been shown to regulate proliferation, differentiation, apoptosis, senescence, migration/invasion, and morphogenesis of these cells. In addition, microRNAs modulate proteins involved in angiogenesis and genome stability, as well as immune function. Since dysregulation of all of these cellular processes can contribute to the development, progression, and metastasis of cancer, it is perhaps not surprising that recent studies are beginning to show alterations in microRNA profiles in different skin cancers. Thus, differences in various microRNAs have been detected in melanoma and in the nonmelanoma skin cancers, basal and squamous cell carcinoma. In this chapter, we discuss the literature indicating a role of microRNAs in regulating skin structure and function as well as in the development, progression, and metastasis of skin cancer, focusing primarily on the nonmelanoma skin cancers.

Keywords Basal cell carcinoma · Epidermis · Keratinocyte · Melanoma · MicroRNA · Nonmelanoma skin cancer · p63 · Skin cancer · (Cutaneous) squamous cell carcinoma

W. B. Bollag (✉) · V. Choudhary
Charlie Norwood VA Medical Center, Augusta, GA 30904, USA
e-mail: WB@gru.ed

V. Choudhary · W. B. Bollag
Department of Physiology, Medical College of Georgia,
Georgia Regents University, 1120 15th Street, Augusta, GA 30912, USA

M. Gullotto · L. Sato
Department of Medical Illustration, Georgia Regents University, Augusta, GA 30912, USA

W. B. Bollag
Departments of Cell Biology and Anatomy, Medicine, Oral Biology and Orthopaedic Surgery,
Medical College of Georgia, Georgia Regents University, Augusta, GA 30912, USA

S. R. Singh, P. Rameshwar (eds.), *MicroRNA in Development and in the Progression of Cancer,* DOI 10.1007/978-1-4899-8065-6_13,

List of Abbreviations

DMBA	7,12-dimethylbenz(a)anthracene
GRHL-3	Grainyhead-like 3
miRNA	MicroRNA
miR	MicroRNA
RISC	RNA-induced silencing complex
TPA	12-O-tetradecanoylphorbol 13-acetate
VEGF	Vascular endothelial cell growth factor

MicroRNAs (miRNAs or miRs) are emerging as important regulators of many physiological and pathophysiological processes including cancer. In fact, the expression of approximately 30 % of all genes is thought to be modulated by miRNAs [1–3]. These small RNAs are generated from long RNA polymerase II transcripts called primary microRNAs (pri-microRNAs) by the sequential action of two enzymes, Drosha, a nuclear RNase III that converts pri-microRNAs to shorter pre-microRNAs, and Dicer, a cytoplasmic RNase III that processes pre-microRNAs to microRNAs. Drosha forms a complex with the double-stranded RNA-binding protein DGCR8, the pri-microRNA, and other components in the nucleus to generate the pre-microRNAs. The pre-microRNA is then transported out of the nucleus where it interacts with cytoplasmic Dicer to be trimmed, and one strand of the resulting short duplex molecule (the mature miR) is loaded into the RNA-induced silencing complex, or RISC (the other strand is termed the miR*) [2]. The mature miRs are approximately 19–22 nucleotides in length and downregulate protein expression by binding to complementary mRNAs in the 3′ untranslated region (3′-UTR) and either eliciting their degradation or inhibiting their translation. Recent studies examining the role of miRNAs in skin have shown a key role of miRNAs in the development of this largest organ of the body. Initial research is beginning to discover an involvement of miRNAs in the skin cancer melanoma, and evidence is emerging as well concerning the contribution of these small RNAs in the nonmelanoma skin cancers. In this chapter, we first provide an introduction to skin and skin cancers, briefly describe the data implicating miRNAs in melanoma, and finally discuss the research suggesting a role of miRNAs in regulating skin development as well as their potential participation in the development and progression of nonmelanoma skin cancers.

1 Overview of the Skin

1.1 Structure of the Skin

Skin is composed of two main layers, the outer epidermis and the inner dermis. Sometimes, the subcutaneous layer, also called the hypodermis or fat layer, is also regarded as a layer of the skin. Each layer has different functions and characteristics as indicated in Table 13.1.

Table 13.1 Skin structure and function

Layer	Cells present	Structure	Function(s)
Epidermis	Keratinocytes, melanocytes, Langerhans cells, Merkel cells	Stratified epithelium with different epidermal layers	Provides a mechanical and waterproof barrier and creates skin tone
Dermis	Fibroblasts, monocytes, macrophages, dermal dendrocytes	Connective tissue with embedded hair follicles, sweat and sebaceous glands, hair, blood and lymph vessels, nerves	Helps to maintain skin structure through production of extracellular matrix support fibers
Hypodermis	Adipocytes, fibroblasts, macrophages	Adipose tissue with blood and lymph vessels, nerves, adipocytes	Insulates to prevent excessive heat loss

1.1.1 Epidermis: The Outer Layer of the Skin

The epidermis is a stratified squamous epithelium that continually renews itself through proliferation and progressive differentiation. The epidermis is relatively thin but tough. It is composed of many different types of cells, such as keratinocytes, melanocytes, Langerhans cells, and Merkel cells.

Keratinocytes

The majority of the cells present in the epidermis are the keratinocytes (90–95 %). The term "keratinocyte" is derived from the fact that these cells synthesize a family of structural proteins called keratins. Keratins assemble into bundles to produce intermediate filaments, which help to form a tough insoluble but unmineralized tissue contributing to the mechanical barrier function of the skin. Keratinocytes originate from epidermal stem cells that reside in the layer at the dermal/epidermal junction, the basal layer or stratum basale [4]. These keratinocytes undergo cell division (mitosis) and migrate to the surface of the skin to replace the dead epidermal cells that have been shed or lost to the environment. After dividing, keratinocytes undergo a distinct pattern of differentiation to form different layers of the epidermis; this process is essential for the function of the skin as a protective barrier. The epidermal layers from lower (deepest) to upper (most superficial) layers are as follows: the strata basale, spinosum, granulosum, and corneum. Figure 13.1 shows a photomicrograph of human skin and Fig. 13.2 illustrates these layers schematically.

Stratum Basale This layer consists primarily of mitotically active keratinocytes and other cell types (melanocytes and Langerhans and Merkel cells). The basal keratinocytes are attached to one another through desmosomes and are structurally and functionally associated with the basement membrane at the dermal–epidermal junction via hemidesmosomes [5]. These proliferating keratinocytes express keratins 5 and 14, immature keratins regarded as markers of proliferation. After cell division, the daughter keratinocytes differentiate as they migrate up through the epidermis.

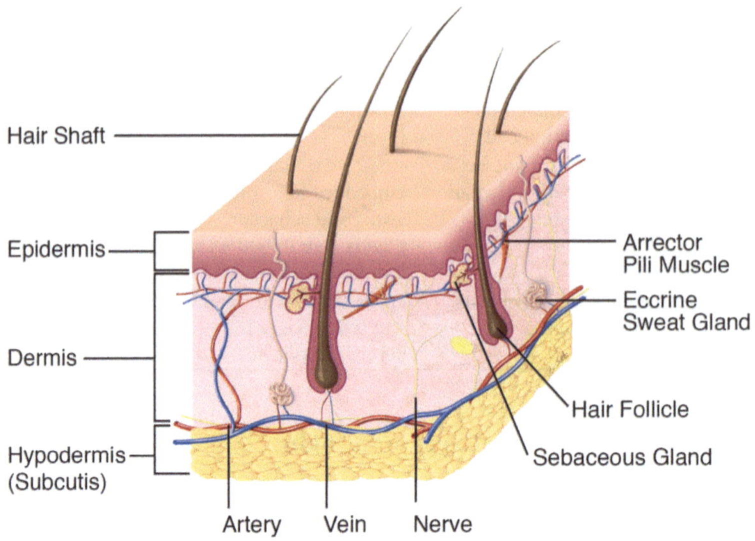

Fig. 13.1 Skin structure: Shown is a schematic illustrating the structure of skin, including the hypodermis (subcutis), dermis, and epidermis as well as various skin appendages (hair follicles and shafts, sebaceous glands, and eccrine sweat glands), the vasculature, nerves, and muscles

Stratum Spinosum Keratinocytes proliferate in the basal layer and eventually migrate towards the outer layer of the skin into the stratum spinosum. This suprabasal layer is also referred to as the prickle-cell layer, because of its appearance in histology. The spininess of the cells in this layer is due to abundant desmosomal connections between adjacent keratinocytes. This layer is marked by growth arrest and the onset of keratinization, a process in which cells begin synthesizing the mature keratins 1 and 10, markers of early keratinocyte differentiation.

Stratum Granulosum In this last living layer, the keratinocytes continue to traverse towards the surface, flattening and becoming granular due to the basophilic keratohyalin granules that accumulate in the cytoplasm. These granules are composed primarily of profilaggrin and promote hydration and cross-linking of keratin. The expression of intermediate (involucrin and transglutaminase) and late (filaggrin and loricrin) differentiation markers is evident in this layer. At the transition between this layer and the stratum corneum, cells secrete lamellar bodies, which are lipid-filled granules containing glycoproteins, glycolipids, phospholipids, free sterols, and hydrolases, into the extracellular space. Cells then lose their nuclei and organelles to become corneocytes in the stratum corneum.

Stratum Corneum The outermost layer of the epidermis is composed of multiple layers of nonviable, terminally differentiated corneocytes. These cells are surrounded by stacked layers of lipid bilayers in the extracellular space (secreted as lamellar bodies and processed by the hydrolases also released) and together provide the natural physical and water-retaining barrier of the skin. Changes in the structure and function of corneocytes occur as they move towards the outer skin surface.

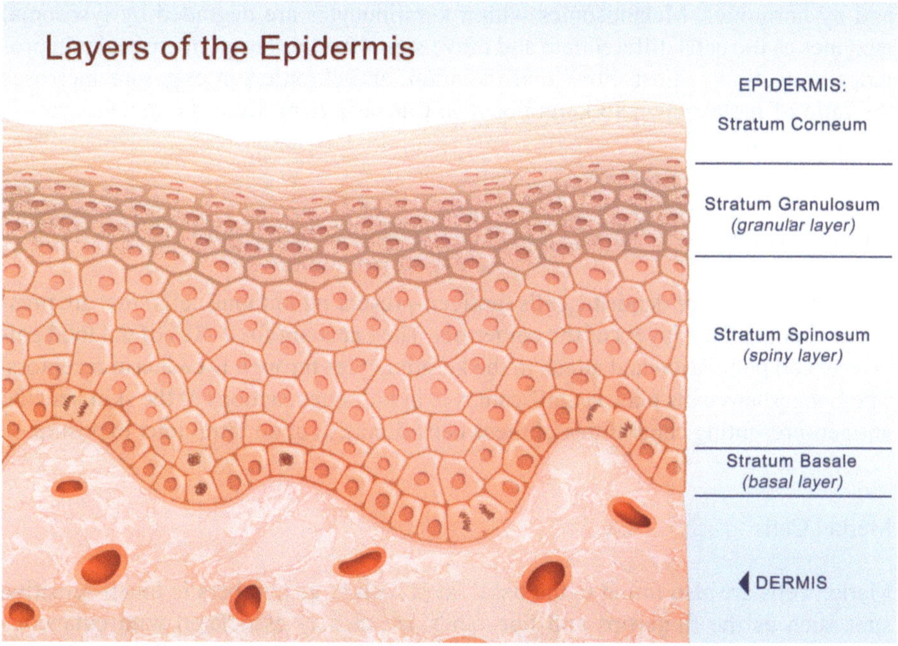

Fig. 13.2 Schematic of the epidermal layers. The epidermal layers, the stata basale (basal layer), spinosum (spinous layer), granulosum (granular layer), and corneum (cornified layer), are shown schematically. Note that some keratinocytes in the basal layer are undergoing mitosis in order to replace cells that are lost to the environment at the skin surface

The cells in the outermost stratum corneum undergo proteolytic degradation of desmosomes that helps in shedding, or desquamation, of the outermost layer. In adult humans, the turnover from stem cells to desquamation takes approximately 30–45 days ([6] and http://webtools.delmarlearning.com/sample_chapters/1401815553_CHAP1.pdf).

The balance between keratinocyte proliferation and differentiation is very important for normal functioning of the skin. Alterations in these processes may lead to various skin diseases such as cancer. Nonmelanoma skin cancers, including basal cell carcinoma and squamous cell carcinoma, arise as cancers of the keratinocytes (see below).

Melanocytes

Melanocytes are the pigment (melanin)-producing cells and developmentally arise from the neural crest. These cells migrate and establish themselves in the stratum basale of the epidermis. They possess dendritic processes that stretch and communicate with a large number of neighboring keratinocytes, allowing the transfer of the pigment synthesized within the melanocytes to the keratinocytes, where it remains as granules (melanosomes). Melanosome transfer is stimulated by ultraviolet light and by hormones. Melanosomes within keratinocytes are degraded by lysosomal enzymes as the cells differentiate and move suprabasally. The pigment melanin provides protection against ultraviolet radiation, and chronic sun exposure increases the ratio of melanocytes to keratinocytes. Cancer arising from the melanocytes is called melanoma.

Langerhans Cells

Langerhans cells are dendritic cells (antigen-presenting immune cells) that are present in almost all the layers of the epidermis. They account for 2–8 % of the total epidermal cell population and attach to the keratinocytes through E-cadherin receptors. The Langerhans cells play a significant role in immune reactions of the skin, and the antigen-presenting capabilities of these cells diminish upon ultraviolet exposure.

Merkel Cells

Merkel cells are also found in the basal layer with large numbers in touch-sensitive sites such as the fingertips and lips. They are closely associated with cutaneous nerves and seem to be involved in light touch sensation. Merkel cells are joined to keratinocytes by desmosomal junctions.

1.1.2 Dermis: The Inner Layer of the Skin

The thick layer of connective tissue underneath the epidermis is the dermis. It has four major cell types: fibroblasts, monocytes, macrophages, and dermal dendrocytes [7]. Apart from these cells, the dermis is mostly composed of collagen, elastin, and fibrillin that provide the skin its pliability, elasticity, and tensile strength. The dermis is tightly connected to the epidermis through a basement membrane and interacts with the epidermis in maintaining the properties of both skin layers.

The dermis is composed of two layers: the stratum papillare and stratum reticulare. The stratum papillare connects with the epidermis and projects its fingerlike papillae into the epidermis, giving the dermal–epidermal junction an undulating appearance. The stratum reticulare is a thick layer composed of dense connective tissue with a high concentration of collagenous, elastic, and reticular fibers.

The dermis maintains the flexibility and structure of the skin and assists the epidermis in protecting the body from mechanical injury and in thermoregulation. It also contains nerve endings (mechanoreceptors) that provide sensation to touch and heat, as well as sweat glands, sebaceous glands, apocrine glands, hair follicles, and lymphatic and blood vessels.

1.1.4 Hypodermis

The hypodermis is the deepest layer of the skin beneath the dermal layer. Adipocytes are the primary cells in the hypodermis, but fibroblast and macrophages are also present. The boundary between the deep reticular dermis and the hypodermis is an abrupt transition from a predominantly fibrous dermal connective tissue to an adipose tissue-rich subcutaneous region. This layer is mainly responsible for lipid storage and serves as an energy supply reserve, cushions and protects the skin, and allows for its mobility over underlying structures.

1.2 Appendages Present in the Skin

Apart from various layers, skin also contains several appendages, including sweat glands (eccrine and apocrine glands), sebaceous glands, nails, hair, and hair follicles. Here, we discuss in brief the structure and function of hair and hair follicles, as the role of miRNAs in hair follicle biology has been extensively elucidated by studies investigating the role of miRNAs in skin.

1.2.1 Hair and Hair Follicles

Hairs are filamentous structures that grow from a follicle in the dermal layer of the skin. Hair plays a role in the regulation of body temperature and in protection, insulation, specialized sensation, and coloration of the skin. The hair can be divided into two parts: the hair shaft and the hair follicle. The hair shaft (or the hair proper) grows within an epithelial channel formed by follicular keratinocytes, called the hair follicle, which is embedded in the dermis.

1.2.2 Structure of the Hair Follicle

The hair follicle has two main parts:

1. The permanent hair follicle: This is the upper two-thirds of the hair follicle, which remains at the end of the catagen phase of the hair cycle
2. The transient lower third of the follicle: This part of the follicle is destroyed upon cessation of hair growth upon entry of the follicle into catagen (see below)

The follicle wall itself has two layers: the outer root sheath that extends the entire length of the follicle and the inner root sheath that envelops the hair shaft to the point of its emergence on the skin surface. The outer root sheath of a hair follicle is continuous with the epidermis. The sebaceous glands, which produce and secrete hair-lubricating natural oils (sebum), are inserted into the root sheath near the base of the epidermis. The bulge, which is a thickened portion of the hair follicle containing follicular stem cells, is located at an oblique angle to the epidermis such that the hair forms to lie at an angle to the surface. These stem cells, which are capable of both initiating hair follicle growth during anagen and regrowing damaged epidermis, give rise to hair progenitor cells that proliferate and differentiate into different layers of the hair follicle and also form the hair shaft [8]. The bulb is the lowermost part of the hair, is shaped like an inverted cup, and consists of an aggregate of epithelial cells. The bulb, as well as the outer root sheath, is surrounded by a basement membrane upon which the basal cells rest. The bulb also partially surrounds the cluster of specialized fibroblasts of the dermal papilla. The dermal papilla and the dermal sheath are derived from progenitor cells of the dermis. At the end of the hair growth cycle, the lower follicle and bulb regress during catagen, but the cells of the dermal papilla do not disappear, suggesting that they may have a primary role in hair formation through an inductive process.

1.2.3 The Hair Growth Cycle

Unlike other parts of the skin, the hair follicle undergoes repeated cycles of growth and regression. The three phases of the hair cycle are:

1. Anagen—This is the growth phase, which can vary from 2 to 6 years, and determines the length of the hair. At any time point, approximately 85 % of all hairs are in anagen (the growth phase).
2. Catagen—This is a brief period (1–2 weeks) of a transitional phase during which the bulb and the transient part of the hair follicle involutes, and the dermal papilla breaks away to remain below in the dermis.
3. Telogen—The resting phase follows catagen. During this phase, which normally lasts about 1–3 months, hair is not produced. Thus, the hair does not elongate but does remain attached to the follicle, while the dermal papilla remains in a resting phase in the dermis. Approximately 10–15 % of all hairs are in this phase at any one time.

The hair follicle reenters the anagen phase at the end of telogen. At this time, the dermal papilla and the base of the follicle join together once more and begin the formation of a new hair as the growth cycle starts all over again.

There are various types of tumors originating from the hair follicles, including tricholemmona, trichomatricoma, hamartoma, trichoepitheliomas, trichoblastoma, etc. A variety of morphological evidence has suggested that many basal cell carcinomas originate directly from the follicular outer root sheath and, possibly, directly from the bulge [9, 10].

2 Types of Skin Cancer

Skin cancers are mainly divided into melanoma and nonmelanoma skin cancer. Nonmelanoma skin cancer arise from the predominant cells of the epidermis, the keratinocytes, whereas melanomas arise from melanocytes that provide pigment to skin and hair. Nonmelanoma skin cancer is the most common cancer in the world, with approximately 3.5 million new diagnoses each year in the USA. Indeed, skin cancers account for nearly half of all cancers in the USA (American Cancer Society website at http://www.cancer.org/cancer/cancercauses/sunanduvexposure/skin-cancer-facts). Of concern, the incidence of both melanoma and nonmelanoma skin cancers has been rising every year (http://www.skincancer.org/skin-cancer-information/skin-cancer-fact).

2.1 *Melanomas*

Melanomas are less common compared to other types of skin cancer but they are more dangerous, particularly if not detected early, and are responsible for the majority of deaths from skin cancer. Melanomas represent about 2% of all skin cancers (about 76,250 new cases were expected to be diagnosed in 2012) and contribute to more than 80% of the deaths from skin cancer (based on statistics reported by the American Cancer Society). The main risk factor for developing melanoma is sun exposure. Although old age is also a risk factor for melanoma, it occurs in young people as well, unlike most other common skin cancers.

2.2 *Nonmelanoma Skin Cancers*

2.2.1 Basal Cell Carcinoma

These cancers share features with the cells (keratinocytes) in the basal layer of the epidermis. About eight out of ten skin cancers are basal cell carcinomas (see American Cancer Society's web page). Cumulative sun exposure is the most common risk factor, and basal cell carcinoma usually develops in areas that are repeatedly exposed to sun and in older individuals. This carcinoma is rarely fatal, but if left untreated, it can grow into nearby areas and invade the bone or other tissues beneath the skin. Treatment generally involves surgical resection of the lesion, but removal often results in significant morbidity due to the location and growth into underlying tissues.

2.2.2 Squamous Cell Carcinoma

Squamous cell carcinomas are the most common skin cancer after basal cell carcinoma. As with basal cell carcinoma, this carcinoma is also commonly found on areas exposed to the sun. Squamous cell carcinomas tend to be more aggressive than basal cell cancers and carry a significant risk of metastasis, often spreading to lymph nodes and/or distant parts of the body [11].

2.2.3 Less Common Types of Skin Cancer

Other nonmelanoma skin cancers together account for less than 1% of these cancers (http://www.cancer.org/Cancer/SkinCancer-BasalandSquamousCell/DetailedGuide/skin-cancer-basal-and-squamous-cell-what-is-basal-and-squamous-cell) and include the following:

a. Merkel cell carcinoma (develops from Merkel cells)
b. Kaposi sarcoma (develops from lymphatic endothelial or vascular endothelial cells in the dermis)
c. Cutaneous (skin) lymphoma (develops from skin lymphocytes)
d. Skin adnexal tumors (develop from hair follicles or sweat glands)
e. Various types of sarcomas (develop from connective tissue cells)

3 Cancer as a Disease of Multiple Cellular Processes

Although cancer is typically thought of as a disease of hyperproliferation, multiple cell processes can be dysregulated and contribute to tumorigenesis. In addition to excessive proliferation, aberrant or absent apoptosis, abnormal morphogenesis, aberrant angiogenesis, absent senescence, abnormal migration and/or metastatic capacity, aberrant or absent differentiation, and genetic instability can all contribute to the development of cancer (reviewed in [12]). Furthermore, tumors can form when transformed cells escape immune surveillance. Accumulating evidence in the skin points to a role of miRNAs in regulating most of these processes in this organ, as described below. Thus, it seems likely that defects in miRNAs can and do contribute to the various types of skin cancer. Reports in which miRNA expression patterns are examined in normal skin and skin cancers are beginning to enter the literature and indicate that this, indeed, is the case.

4 MicroRNAs in the Skin

A large number of microRNAs have been identified in various tissues and organs and are described using a numbering system. Of the characterized miRNAs (or miRs), many are shown to be expressed in the skin or in cells derived from the skin. These miRNAs have been profiled in multiple skin cell types and tissues, including melanoma cells, normal human skin, psoriatic lesions, and artificial human epidermis, as well as mouse skin. Melanoma in particular has been rather intensively studied, and data indicate a likely role of miRNAs in various aspects of melanoma (see below). In addition, transgenic mouse models lacking Dicer and Drosha, the two RNases required for the production of miRNAs, specifically in the epidermis, either in utero or as adult animals, have been generated and characterized. While this approach does not allow the identification of the miRNA(s) involved in a particular cell response or phenotype observed, it is ideal for demonstrating the potential role of miRNAs in general in the epidermis. As described below, these studies have demonstrated a key involvement of miRNAs in multiple processes in the skin. Other studies in which the miRNA expression pattern of nonmelanoma skin cancers is profiled and the levels of the identified miRNAs are manipulated in skin cells have also provided evidence of a role of miRNAs in the development and progression of nonmelanoma skin cancers.

4.1 Role of MicroRNAs in Melanoma

There are numerous data in the literature to support the role of miRNAs in the development of melanoma as well as the regulation of its progression and metastasis, and, indeed, a melanoma miRNA "signature" has been proposed. MiRNAs implicated in melanoma development, progression, and/or metastasis include those that are upregulated, such as miR-146a, -155 [13], -214 [14], -221, -222 [15], and others, as well as some that are downregulated, such as miR-200c [13], let-7a, and let-7b [15]. Segura et al. [16] also identified a set of six miRNAs (miR-150, -342-3p, -455–3p, -145, -155, and -497) that predicted post-recurrence survival in melanoma with an approximate estimated accuracy of 80%. Several excellent reviews have been written concerning the role of miRNAs in melanoma (e.g., [2, 14, 15, 17–19]), and we will instead focus on the role of miRNAs in epidermal keratinocytes, skin structure, and the nonmelanoma skin cancers.

4.2 Role of MicroRNAs in Skin Structure and Function

Several years ago, Fuchs and colleagues and the Millar laboratory both demonstrated the importance of Dicer, the enzyme that processes pre-miRNAs to mature miRNAs, to epidermal structure and function. In 2006, Yi et al. [20] generated an epidermal-specific deletion of the Dicer gene using a keratin 14 promoter-driven Cre recombinase to delete Dicer early in epidermal development (in skin epithelial progenitor cells) and observed a profound phenotype. The knockout mice exhibited abnormal hair follicle development, resulting in evagination rather than invagination of the hair follicles and the formation of cyst-like structures that disturbed the architecture of the epidermis. The distortion from the malformed hair follicles also appeared to result in barrier disruption, and the pups wound up losing weight and dying approximately 4 days after birth from apparent dehydration [20]. These authors observed no obvious differences in the interfollicular epidermal structure otherwise, suggesting that the primary defect was in the regulation of hair follicle growth.

At about the same time, a similar epidermal-specific Dicer knockout mouse model (using a keratin 14 promoter-driven Cre recombinase as well) was generated by Sarah Millar's group [21]. These authors also found disordered hair follicle morphogenesis and poor survival of the animals, with stunting of the pups' growth and alopecia (lack of hair growth). However, in these studies the authors noted an effect on the interfollicular epidermis, which showed some expansion in the knockout mice [21]. Millar and colleagues also created an inducible epidermal-specific knockout mouse model to determine the effect of deletion of Dicer to decrease miR levels in adult animals [22]. In this mouse model, they found that deletion of Dicer in the anagen phase of the hair cycle resulted in hair shaft defects and hair loss that was progressive (in an anterior to posterior direction) and permanent. Dicer was not required for the maintenance of hair follicles in telogen, but the loss of Dicer during anagen resulted in a failure of catagen and subsequent follicular degradation. In addition, the interfollicular epidermis showed delayed thickening and hyperproliferation [22].

The Fuchs laboratory also examined the effect of epidermal-specific ablation of the *DGCR8* gene, the product of which is a key component of the microprocessor complex required for the ultimate generation of mature miRNAs [23]. Although Dicer is known as the RNase III enzyme that processes pre-miRNA into mature miRNAs in the cytoplasm, and its knockout therefore should reduce miRNA levels globally, Dicer has also been reported to play a role in heterochromatin silencing and the generation of small RNAs other than miRNAs. In contrast, *DGCR8* recognizes the pri-miRNA hairpin loop and is thought to be specific for miRNA production. A mouse model in which *DGCR8* is ablated specifically in the epidermis (using a keratin 14 promoter-driven Cre recombinase) exhibits a phenotype similar to that observed in the Dicer knockout mice, suggesting that the effects of ablation of both genes is the result of inhibited miRNA production rather than to some miRNA-independent function of Dicer [23].

Similarly, Millar and colleagues have generated an epidermal-specific Drosha knockout mouse in adult animals [22]. Drosha, like Dicer, is an RNase III enzyme that is necessary for the generation of miRNAs, converting pri-miRNAs to shorter pre-miRNAs. Also similarly to Dicer, the ablation of the Drosha gene results in a phenotype in the knockout mice characterized by a failure of hair follicles to enter catagen and eventual progressive and permanent hair loss. In fact, the adult Drosha knockout was indistinguishable from the adult Dicer knockout [22], suggesting that the phenotype observed for both models is the result of inhibition of miRNA generation rather than any other function of the enzymes.

Although these studies demonstrate a critical role of miRNAs in regulating skin structure and function, they do not identify particular miRNAs that may be of importance in these processes. Researchers have approached the question of the identity of miRNAs regulating a given cell response largely by performing miRNA microarray analyses on epidermal cells undergoing alterations such as differentiation or senescence or on normal versus tumor tissue or stem versus non-stem cells. Once such a miRNA is identified, its involvement in a particular cell response can be determined by overexpression of anti-miRs, called antagomirs, or miR mimetics and monitoring the effects of these manipulations on the response. Verification of a role of a specific miR on a cell response then raises the question as to mechanism of the miR's effect, i.e., the targets downregulated by the miR. Identification of the targets of the miR usually relies at least partially on the use of miR prediction algorithms for matching the miR seed sequence (the nucleotide sequence that binds to a complementary sequence in the 3′-UTR of the target gene) to potential regulated genes. Validation of these potential targets requires that overexpression of miR mimetics should decrease and of miR antagomirs should increase the mRNA and/or protein expression of the suspected target gene. An additional approach to confirm that the gene is directly modulated by the miR involves creating a reporter construct in which the 3′-UTR of the gene of interest is fused to the gene for luciferase. If the gene is a direct target, transfection of this reporter construct together with the miR mimetic should result in decreased luciferase activity. Using these multiple approaches, it is possible to demonstrate that a specific miRNA mediates a certain cell response, at least in part, by modulating the levels of a particular target gene product.

Recent studies using these methods have begun to demonstrate roles for certain miRs in the skin, as well as possible mechanisms by which these miRs may affect skin structure and function (i.e., the proteins targeted). One such miR is miR-34, in particular miR-34a and c, which appear to suppress p63-mediated cell cycle progression in epidermal keratinocytes. p63 is a transcription factor of the p53 family, which is expressed in the stratum basale and plays a role in maintaining the proliferative and regenerative ability of the epidermis (reviewed in [24, 25]). Antonini et al. [26] demonstrated that p63 mediates cell cycle progression at least in part by repressing miR-34a and c. Knockdown of p63 results in upregulation of miR-34 and decreased levels of proliferative markers such as cyclin D1. Indeed, miR-34 appears to be a direct transcriptional target of p63. Preventing the increase in miR-34 induced by p63 knockdown using antagomirs also inhibited the reduction in proliferation markers observed in response to p63 loss [26].

In addition to regulating miRNA expression, p63 can also be regulated miRNA expression. Thus, Rivetti di Val Cervo et al. [27] found that human keratinocyte cell senescence was associated with the upregulation of several miRNAs, including miR-138, -181a, -181b, and -130b. Indeed, overexpression of miR-138 or miR-181a or b, but not miR-130b, induced cell senescence, in part by decreasing sirtuin-1, a histone deacetylase associated with longevity [27]. On the other hand, miR-130b reduced the levels of ΔNp63, and, conversely, ΔNp63 inhibited the expression of miR-130b, as well as miR-138, -181a, and -181b, by binding to p63 response elements in close proximity to their genomic loci. Interestingly, these authors also found that miR-138 mRNA levels were significantly elevated, and miR-130b, -181a, and -181b tended to be increased, in skin biopsies obtained from aged (older than 60) versus young individuals (less than 10 years of age) [27]. Thus, there seems to be a feedback loop established between p63 and particular miRNAs, which allows p63 to regulate keratinocyte regenerative capacity and senescence through modulating miRNA expression, and this loop may impact the skin-aging process.

MiR-203 is another miRNA that apparently regulates the stem cell character of epidermal progenitor cells, as shown by Fuchs and colleagues [28]. These authors generated a transgenic mouse model in which miR-203 was overexpressed in the epidermis under the control of the keratin 14 promoter. The miR-203 transgenic mice exhibited a thinner epidermis as a result of decreased proliferation compared to wild-type animals. In addition, p63 levels were reduced and there seemed to be basal cell depletion. Conversely, antagomirs to miR-203 expressed in the epidermis increased epidermal proliferation in vivo [28] and partially blocked the inhibition of proliferation [28] or stimulation of differentiation [29] induced by an elevated calcium level in vitro. One mechanism by which miR-203 may exert this capacity to downregulate the proliferative capacity of epidermal cells is through an ability to decrease p63 levels directly [28], as shown by Lena et al. [30]. These authors demonstrated similar findings as those of Yi et al. [28], in that miR-203 expression was regulated by differentiation. Importantly, these authors showed that miR-203 directly regulated p63 expression and appeared to act as a switch between proliferative basal progenitors and terminally differentiating suprabasal cells. Indeed, miR-203 and p63 mRNA levels were inversely correlated during differentiation of murine embryonic stem cells into epidermal progenitor (keratin 14-positive) cells [30]. However, these authors observed that miR-203 alone was not sufficient to induce keratinocyte differentiation.

These results in the mouse were confirmed in human keratinocytes in vitro or freshly isolated from patient skin [31]. Thus, miR-203 was found to be upregulated upon induction of human keratinocyte differentiation in vitro by exposure to an elevated medium calcium level for a shorter (3 days) or longer (7 days) time period. Likewise, keratinocytes were freshly isolated from human skin and separated by their speed of adherence to collagen into rapidly adherent (stem cells), slowly adherent (transient amplifying), and non-adherent (suprabasal differentiated) cells. In the differentiated cells, miR-203 was observed to be more highly expressed than in either the stem or transit amplifying cells, and in epidermis in situ, miR-203 (as well as miR-23b) was expressed almost exclusively in the suprabasal layers [31]. Thus, these results are consistent with a role for miR-203 in triggering or mediating differentiation. Also in this report, a total of eight miRs (miR-203, -95, -210, -224, -26a, -200a, -27b, and -328) were upregulated under all differentiation conditions and one (miR-376a) was downregulated under these conditions. The proposed mechanism by which the identified miRNAs might function to regulate differentiation involved the endothelin A receptor, eyes absent homolog 4 (EYA4), and ganglioside-induced differentiation-associated protein-1 (GDAP1), since several of the upregulated miRs suppressed the 3′-UTR-induced luciferase reporter activity of these genes [31].

In regard to miR-203, Sonkoly et al. [29] found that the differentiation-dependent upregulation of this miRNA can be mediated by the activation of protein kinase C. In addition to an ability of elevated calcium levels to raise miR-203 levels, these authors showed that treatment of human keratinocytes with 1,25-dihydroxyvitamin D_3 or 12-O-tetradecanoylphorbol 13-acetate (TPA), both of which are known inducers of differentiation (reviewed in [32]), resulted in significant increases in miR-203 expression. Similarly, cell confluence, which also promotes differentiation [33], enhanced miR-203 levels, whereas exposure to keratinocyte growth factor (KGF) or epidermal growth factor, which promotes proliferation and/or inhibits differentiation (reviewed in [34–36]), decreased miR-203 expression. Finally, the effects of TPA on miR-203 levels could be blocked by inhibitors of protein kinase C [29].

Another miRNA found to play a role in regulating keratinocyte differentiation is miR-24. This miRNA is upregulated in vitro upon calcium-induced differentiation, and in epidermis, in situ miR-24 is expressed in the suprabasal layers [37]. Overexpression of miR-24 in keratinocytes in vitro was sufficient to promote keratinocyte differentiation and inhibit proliferation. Furthermore, expression of miR-24 under control of the keratin 5 promoter in transgenic mice in vivo resulted in accelerated keratinocyte differentiation (i.e., elevated levels of the spinous layer marker keratin 10 in basal keratinocytes), decreased proliferation, and a thinner epidermis [37]. On the other hand, injection of an anti-miR-24 antagomir into newborn mice increased keratinocyte proliferation. Targets of miR-24 were identified and many of them appeared to affect the actin cytoskeleton. Three in particular, p21-activated kinase 4 (PAK4), tyrosine kinase substrate 5 (TKS5), and Rho GTPase-activating protein 19 (ArhGAP19), were shown to be downregulated by miR-24, to affect actin cytoskeleton rearrangement, and to promote keratinocyte differentiation. Small interfering RNA-mediated knockdown of PAK4, TKS5, or ArhGAP19 reproduced the effects of miR-24 overexpression, whereas expression of miR-24-resistant PAK4, TKS5, or ArhGAP19 blocked the effects of miR-24 overexpression [37]. MiR-24 also targets c-Myc and E2F2, an action that appears to mediate the consequences of miR-24 expression on keratinocyte proliferation but not on differentiation or actin dynamics [37]. Thus, miR-24 seems to mediate keratinocyte differentiation through its effects on the actin cytoskeleton.

Other targets of miRNA regulation in keratinocytes include specialized adenine and thymine-rich binding protein 1 (SATB1), a protein that helps to integrate a higher-order chromatin structure and remodeling with gene transcription, and cyclin-dependent kinase-6 (CDK6), which phosphorylates and regulates the retinoblastoma (Rb) protein, thereby controlling cell cycle progression (reviewed in [38]). The level of these two proteins is reduced by miR-191, the expression of which is upregulated in senescent keratinocytes [39]. Indeed, overexpression of miR-191 itself inhibits keratinocyte proliferation and induces keratinocyte senescence, and likewise, silencing of SATB1 or CDK6 with siRNA reproduces the effect of miR-191 overexpression to inhibit proliferation and promote senescence [39]. Thus, in summary, miR-191 can regulate both proliferation and senescence in keratinocytes by modulating SATB1 and CDK6 levels.

4.3 Role of MicroRNAs in Epidermal Tumorigenesis and Nonmelanoma Skin Cancer

Another miRNA of importance in skin is miR-21, which has been reported to play a role in keratinocyte proliferation and migration in response to bone morphogenetic protein-4 (BMP4), as well as in epidermal tumorigenesis. Thus, Ahmed et al. [40] showed that miR-21 expression is inhibited by BMP4 concomitant with the ability of this growth factor to increase the expression of phosphatase and tensin homolog (PTEN, an inhibitor of the Akt pathway), programmed cell death 4 (PDCD4, an inhibitor of the activation of c-Jun N-terminal kinase or JNK), tissue inhibitor of metalloproteinases 3 (TIMP3) and tropomyosin 1 (TPM1), and to inhibit proliferation and migration. Importantly, a miR-21 mimic prevented the BMP4-induced increase in the expression of the above four genes as well as the inhibition of keratinocyte proliferation. In addition, the miR-21 mimic also significantly enhanced cell migration and blocked the capacity of BMP4 to delay scratch wound healing, indicating that miR-21 mediates BMP4's effects on proliferation and migration in keratinocytes.

Consistent with these in vitro results, Ma et al. [41] have demonstrated that ablation of the *miR-21* gene in a mouse model resulted in reduced proliferation and increased apoptosis in the epidermis. The miR-21 knockout mice also exhibited reduced susceptibility to a two-stage epidermal tumorigenesis protocol. In this procedure, mouse skin is treated with carcinogenic 7,12-dimethylbenz(a)anthracene (DMBA) to initiate (i.e., induce DNA mutations in) epidermal keratinocytes, followed by repeated application of a tumor promoter, such as the phorbol ester TPA, to elicit tumorigenesis. In this model, the application of DMBA alone results in the formation of few or no tumors, whereas repeated treatment with phorbol ester alone induces the formation of papillomas that rapidly regress upon termination of TPA exposure. Using this protocol, Ma et al. [41] observed fewer tumors per mouse and delayed tumor formation in the miR-21 knockout relative to wild-type mice. At the molecular level, these knockout keratinocytes also showed increased levels of PTEN, PDCD4, and Sprouty-1 and -2 (inhibitors of Raf-1 upstream of activation of extracellular signal-regulated kinase-1/2 or ERK-1/2), as well as decreased ERK-1/2, Akt, and JNK phosphorylation/activation [41]. It should be noted that BMP4, which as described above decreases miR-21 expression [40], also inhibits tumor formation in the DMBA/TPA two-stage carcinogenesis model [40], providing an additional link between miR-21 and epidermal tumorigenesis.

Increased miR-21 expression has also been found to be associated with squamous cell carcinoma in mice and humans [42–44]. One likely additional target of miR-21 is grainyhead-like 3 (GRHL3), a transcription factor and tumor suppressor. Darido et al. [45] have shown that epidermal-specific GRHL3 knockout mice exhibit thickened, hyperproliferative epidermis, as well as accelerated tumor formation and increased tumor and carcinoma numbers than the wild-type controls. Interestingly, these GRHL3 knockout mice even demonstrate tumor formation with TPA application alone. These mice show increased Akt activation and reduced PTEN levels, and heterozygous loss of PTEN accelerated carcinoma formation further in the GRHL3 knockout, PTEN heterozygous mice. A microarray analysis of miRNAs in two human squamous cell carcinoma cell tumors versus the adjacent normal epidermis showed a profound upregulation of miR-21 expression, which was confirmed by real-time reverse transcription polymerase chain reaction (RT-PCR) [45]. Overexpression of miR-21 decreased GRHL3 expression and levels, with an accompanying decrease in PTEN expression and levels. In contrast, an antagomir to miR-21 increased GRHL3 mRNA and protein expression as well as PTEN levels and reduced proliferation in a squamous cell carcinoma cell line [45].

Additional studies from the Andersen laboratory suggest that miR-21 is normally directly transcriptionally repressed by GRHL3 [46]. However, using miR-21 knockdown in normal keratinocytes, these authors found no effect of miR-21 knockdown on proliferation or differentiation in normal cells. On the other hand, manipulation of miR-21 levels altered the levels of MutS protein homolog 2 (MSH-2), a DNA mismatch repair enzyme, with overexpression decreasing and knockdown increasing MSH-2 protein expression. Importantly, this effect was more pronounced in transformed or tumor keratinocytes compared to normal cells. GRHL3 expression and levels were found to be decreased in human squamous cell carcinoma (10 samples in Darido et al. [46] and 60 samples in Bhandari et al. [46]), as well as squamous cell carcinomas generated in the DMBA/TPA protocol in mouse skin [46].

Other miRNAs found to be altered in human squamous cell carcinoma include miR-203 and miR-184, which are downregulated and upregulated, respectively [42]. On the other hand, miR-205 expression was unchanged in squamous cell carcinoma compared to normal skin [42]. MiR-205 is of interest because of the report from Lavker and colleagues [47] that this miRNA can promote keratinocyte migration by downregulating the lipid phosphatase SHIP2, which restricts motility, increasing cell adhesion to collagen and regulating the actin cytoskeleton. Another microarray study (verified by quantitative RT-PCR) revealed significantly increased mRNA levels of miR-135b, miR-424, and miR-766 and decreased miR-30a and a*, miR-378, miR-145, miR-140-3p, and miR-26a in cutaneous squamous cell carcinoma. In yet another report [48], miR-361-5p was demonstrated to be decreased in human squamous cell carcinoma versus normal epidermis. This decrease in miR-361-5p was inversely correlated with the levels of vascular endothelial cell growth factor (VEGF), and indeed, the VEGF downstream 3′-UTR possesses a miRNA recognition element that could be repressed by miR-361-5p [48]. This effect on VEGF, a growth factor known to promote angiogenesis, could potentially represent one mechanism by which downregulated miR-361-5p could underlie the development of squamous cell carcinoma.

Cutaneous squamous cell carcinoma and squamous cell carcinoma cell lines also exhibited downregulated miR-124 and -214 expression. These miRNAs normally decrease the levels of ERK-1 and ERK-2, which mediate abnormal cell proliferation in the squamous cell carcinoma cell lines. Re-expression of miR-124 and -214 in these cell lines inhibited proliferation by normalizing ERK-1/2 levels. In addition, miR-125b may play a role in the generation of squamous cell carcinoma. The expression of this miR, in addition to 53 others, was found to be decreased in squamous cell carcinoma compared to healthy skin by microarray analysis. (Upregulated miRs identified in this study were miR-31, miR-135b, miR-21, and miR-223.) Overexpression of miR-125b repressed the proliferation, colony formation, migration, and invasive capacity of squamous carcinoma cells. The effect on invasion is presumably related to the ability of miR-125b to downregulate matrix metalloproteinase-13 (MMP13) [43], which as an MMP can promote invasiveness by degrading the extracellular matrix that would otherwise limit cell migration. Indeed, knockdown of MMP13 reproduced the phenotype caused by overexpression of miR-125b [43], suggesting the importance of both the miRNA and MMP13 in squamous cell carcinoma.

Using a microarray analysis (with a subset of miRNAs verified by quantitative RT-PCR), another study examined the expression of miRNAs in basal cell carcinoma compared with non-lesional skin [49]. Eighteen miRNAs were identified with significant upregulation, as well as ten miRNAs that showed a significant decrease. A computer analysis predicted that these miRNAs should target many tumor-promoting pathways, such as ERK-1/2 and the Hedgehog signaling pathway, suggesting that miRNAs may be involved in the development of basal cell carcinoma in addition to squamous cell carcinoma.

Another microarray study analyzed the miRNA expression profile in two subtypes of basal cell carcinoma, nodular and more aggressive infiltrative basal cell carcinoma, to determine whether there might be miRNA differences that potentially distinguish these two subtypes (since no such mRNA differences have been found). Heffelfinger et al. [50] identified 20 miRNAs by sequencing, that showed differential expression in nodular versus infiltrative basal cell carcinoma. Six of these were assayed by quantitative RT-PCR and analyzed for a second ("replication") set of nodular and infiltrative basal cell carcinoma. Of these, five were validated by quantitative RT-PCR but only three of the five were significant among all tumors analyzed. One, in particular miR-183, was found to distinguish between the two tumor subtypes with no overlap. The prediction of potential miR-183 targets indicated a likely role of this miRNA in regulating cell motility and invasiveness [50], as has been seen in other cancer cell types and consistent with its differential expression in nodular (higher) versus infiltrative (lower) basal cell carcinoma.

Thus, a variety of studies have indicated a role of miRNAs in the development of nonmelanoma skin cancer, and there appear to be a number of different miRNAs that may be involved. This idea is further supported by the finding that the miRNA-processing enzymes Dicer and Drosha are altered in skin cancer [51]. Thus, Drosha levels are significantly increased in basal and squamous cell carcinoma compared to normal skin from the same individuals, whereas Dicer expression is decreased in basal cell carcinoma and increased in squamous cell carcinoma [51]. This dysregulation of Dicer and Drosha expression could contribute to the changes in miRNA levels that have been documented to occur in nonmelanoma skin cancer.

In summary, then, accumulating data point to an important role of miRNAs in various processes in the skin. Returning to the hallmarks of cancer cells discussed earlier, in epidermal and/or follicular keratinocytes, miRNAs have been found to regulate the processes of, or proteins involved in: proliferation [26, 28, 30, 43, 52–54], apoptosis [21], morphogenesis [20–23, 55], senescence [27, 39], migration (cell motility) and/or invasion [37, 43, 47, 52], differentiation [28, 29, 56], angiogenesis (e.g., VEGF [48]), and genome stability (e.g., MSH2 [46]), as summarized in Table 13.2. Furthermore, studies in other systems indicate that miRNAs are also involved in regulating the immune system (reviewed in [3, 57]). Thus, miRNAs can regulate all of the cellular processes known to be dysregulated in cancer. It is then not surprising that accumulating data are indicating a role for these small noncoding RNAs in tumorigenesis in the skin, and it seems likely that additional studies will provide further evidence of their importance in skin cancer development, progression, and metastasis.

Table 13.2 MicroRNAs of importance in the skin

MicroRNA	Keratinocyte response regulated	miR Target(s)	Reference(s)
miR-34a,c	Proliferation and cell cycle progression	p63	[26]
miR-138,-181a,b	Senescence	Sirtuin-1	[27]
miR-130b	Proliferation	ΔNp63	[27]
miR-17,-20b,-106a,-30a, -143,-455-3p	Differentiation	Rb p21 MAPK 1, 8, and 9	[56]
miR-203	Proliferation	p63	[30, 28]
miR-203,-95,-210,-224,-26a, -200a,-27b,-328	Differentiation	EYA4 Endothelin A receptor GDAP1	[29, 31]
miR-24	Differentiation, proliferation, migration (actin dynamics)	PAK4 TKS5 ArhGAP19 c-myc E2F2	[37]
miR-191	Senescence, proliferation	SATB1 CDK6	[30]
miR-21	Migration, proliferation, apoptosis, genomic stability	PTEN PDCD4 TIMP3 TPM1 GRHL3 MSH-2 SPRY1 and 2	[40, 41, 45, 46]
miR-205	Migration	SHIP2	[47]
miR-361-5p	Angiogenesis	VEGF	[48]
miR-124,-214	Proliferation	ERK-1/2	[54]
miR-125b	Proliferation, migration	MMP13	[53]

References

1. Sand M, Sand D, Altmeyer P, Bechara FG. MicroRNA in non-melanoma skin cancer. Cancer Biomark. 2012;11:253–7
2. Schneider MR. MicroRNAs as novel players in skin development, homeostasis and disease. Br J Dermatol. 2012;166:22–8
3. Sonkoly E, Stahle M, Pivarcsi A. MicroRNAs: novel regulators in skin inflammation. Clin Exp Dermatol. 2008;33:312–5
4. Lavker RM, Sun TT. Epidermal stem cells. J Invest Dermatol. 1983;81:121–7
5. Tsuruta D, Hopkinson SB, Jones JC. Hemidesmosome protein dynamics in live epithelial cells. Cell Motil Cytoskeleton. 2003;54:122–34
6. Bergstresser PR, Taylor JR. Epidermal 'turnover time'—a new examination. Br J Dermatol. 1977;96:503–9
7. Sueki H, Whitaker D, Buchsbaum M, Murphy GF. Novel interactions between dermal dendrocytes and mast cells in human skin. Implications for hemostasis and matrix repair. Lab Invest. 1993;69:160–72
8. Cotsarelis G, Sun TT, Lavker RM. Label-retaining cells reside in the bulge area of pilosebaceous unit: implications for follicular stem cells, hair cycle, and skin carcinogenesis. Cell. 1990;61:1329–37

9. Donovan J. Review of the hair follicle origin hypothesis for basal cell carcinoma. Dermatol Surg. 2009;35:1311–23
10. Schmidt KT, Ma A, Goldberg R, Medenica M. Multiple adnexal tumors and a parotid basal cell adenoma. J Am Acad Dermatol. 1991;25:960–4
11. Nuno-Gonzalez A, Vicente-Martin FJ, Pinedo-Moraleda F, Lopez-Estebaranz JL. High-risk cutaneous squamous cell carcinoma. Actas Dermosifiliogr. 2012;103:567–78
12. Kopnin BP. Targets of oncogenes and tumor suppressors: key for understanding basic mechanisms of carcinogenesis. Biochemistry (Mosc). 2000;65:2–27
13. Philippidou D, Schmitt M, Moser D, Margue C, Nazarov PV, Muller A, Vallar L, Nashan D, Behrmann I, Kreis S. Signatures of microRNAs and selected microRNA target genes in human melanoma. Cancer Res. 2010;70:4163–73
14. Bar-Eli M. Searching for the 'melano-miRs': miR-214 drives melanoma metastasis. EMBO J. 2011;30:1880–1
15. Glud M, Gniadecki R. MicroRNAs in the pathogenesis of malignant melanoma. J Eur Acad Dermatol Venereol. 2013;27:142–50
16. Segura MF, Belitskaya-Levy I, Rose AE, Zakrzewski J, Gaziel A, Hanniford D, Darvishian F, Berman RS, Shapiro RL, Pavlick AC, Osman I, Hernando E. Melanoma MicroRNA signature predicts post-recurrence survival. Clin Cancer Res. 2010;16:1577–86
17. Bell RE, Levy C. The three M's: melanoma, microphthalmia-associated transcription factor and microRNA. Pigment Cell Melanoma Res. 2011;24:1088–106
18. Segura MF, Greenwald HS, Hanniford D, Osman I, Hernando E. MicroRNA and cutaneous melanoma: from discovery to prognosis and therapy. Carcinogenesis. 2012;33:1823–32
19. Sand M, Gambichler T, Sand D, Skrygan M, Altmeyer P, Bechara FG MicroRNAs and the skin: tiny players in the body's largest organ. J Dermatol Sci. 2009;53:169–75
20. Yi R, O'Carroll D, Pasolli HA, Zhang Z, Dietrich FS, Tarakhovsky A, Fuchs E. Morphogenesis in skin is governed by discrete sets of differentially expressed microRNAs. Nat Genet. 2006;38:356–362
21. Andl T, Murchison EP, Liu F, Zhang Y, Yunta-Gonzalez M, Tobias JW, Andl CD, Seykora JT, Hannon GJ, Millar SE. The miRNA-processing enzyme dicer is essential for the morphogenesis and maintenance of hair follicles. Curr Biol. 2006;16:1041–9
22. Teta M, Choi YS, Okegbe T, Wong G, Tam OH, Chong MM, Seykora JT, Nagy A, Littman DR, Andl T, Millar SE. Inducible deletion of epidermal Dicer and Drosha reveals multiple functions for miRNAs in postnatal skin. Development. 2012;139:1405–16
23. Yi R, Pasolli HA, Landthaler M, Hafner M, Ojo T, Sheridan R, Sander C, O'Carroll D, Stoffel M, Tuschl T, Fuchs E. DGCR8-dependent microRNA biogenesis is essential for skin development. Proc Natl Acad Sci U S A. 2009;106:498–502
24. King KE, Weinberg WC. p63: defining roles in morphogenesis, homeostasis, and neoplasia of the epidermis. Mol Carcinog. 2007;46:716–24
25. Dai X, Segre JA. Transcriptional control of epidermal specification and differentiation. Curr Opin Genet Dev. 2004;14:485–91
26. Antonini D, Russo MT, De Rosa L, Gorrese M, Del Vecchio L, Missero C. Transcriptional repression of miR-34 family contributes to p63-mediated cell cycle progression in epidermal cells. J Invest Dermatol. 2010;130:1249–57
27. Rivetti di Val Cervo P, Lena AM, Nicoloso M, Rossi S, Mancini M, Zhou H, Saintigny G, Dellambra E, Odorisio T, Mahe C, Calin GA, Candi E, Melino G. p63-microRNA feedback in keratinocyte senescence. Proc Natl Acad Sci U S A. 2012;109:1133–8
28. Yi R, Poy MN, Stoffel M, Fuchs E. A skin microRNA promotes differentiation by repressing 'stemness'. Nature. 2008;452:225–229
29. Sonkoly E, Wei T, Pavez Lorie E, Suzuki H, Kato M, Torma H, Stahle M, Pivarcsi A. Protein kinase C-dependent upregulation of miR-203 induces the differentiation of human keratinocytes. J Invest Dermatol. 2010;130:124–34
30. Lena AM, Shalom-Feuerstein R, Rivetti di Val Cervo P, Aberdam D, Knight RA, Melino G, Candi E. miR-203 represses 'stemness' by repressing DeltaNp63. Cell Death Differ. 2008;15:1187–95

31. Hildebrand J, Rutze M, Walz N, Gallinat S, Wenck H, Deppert W, Grundhoff A, Knott A. A comprehensive analysis of microRNA expression during human keratinocyte differentiation in vitro and in vivo. J Invest Dermatol. 2011;131:20–9
32. Bollinger Bollag W, Bollag RJ. 1,25-Dihydroxyvitamin D(3), phospholipase D and protein kinase C in keratinocyte differentiation. Mol Cell Endocrinol. 2001;177:173–82
33. Poumay Y, Pittelkow MR. Cell density and culture factors regulate keratinocyte commitment to differentiation and expression of suprabasal K1/K10 keratins. J Invest Dermatol. 1995;104:271–6
34. Werner S. Keratinocyte growth factor: a unique player in epithelial repair processes. Cytokine Growth Factor Rev. 1998;9:153–65
35. Hashimoto K. Regulation of keratinocyte function by growth factors. J Dermatol Sci. 2000;24 Suppl 1:46–50
36. Pastore S, Mascia F, Mariani V, Girolomoni G. The epidermal growth factor receptor system in skin repair and inflammation. J Invest Dermatol. 2008;128:1365–74
37. Amelio I, Lena AM, Viticchie G, Shalom-Feuerstein R, Terrinoni A, Dinsdale D, Russo G, Fortunato C, Bonanno E, Spagnoli LG, Aberdam D, Knight RA, Candi E, Melino G. miR-24 triggers epidermal differentiation by controlling actin adhesion and cell migration. J Cell Biol. 2012;199:347–63
38. Coqueret O. Linking cyclins to transcriptional control. Gene. 2002;299:35–55
39. Lena AM, Mancini M, Rivetti di Val Cervo P, Saintigny G, Mahe C, Melino G, Candi E. MicroRNA-191 triggers keratinocytes senescence by SATB1 and CDK6 downregulation. Biochem Biophys Res Commun. 2012;423:509–514
40. Ahmed MI, Mardaryev AN, Lewis CJ, Sharov AA, Botchkareva NV. MicroRNA-21 is an important downstream component of BMP signalling in epidermal keratinocytes. J Cell Sci. 2011;124:3399–404
41. Ma X, Kumar M, Choudhury SN, Becker Buscaglia LE, Barker JR, Kanakamedala K, Liu MF, Li Y. Loss of the miR-21 allele elevates the expression of its target genes and reduces tumorigenesis. Proc Natl Acad Sci U S A. 2011;108:10144–9
42. Dziunycz P, Iotzova-Weiss G, Eloranta JJ, Lauchli S, Hafner J, French LE, Hofbauer GF Squamous cell carcinoma of the skin shows a distinct microRNA profile modulated by UV radiation. J Invest Dermatol. 2010;130:2686–9
43. Xu N, Zhang L, Meisgen F, Harada M, Heilborn J, Homey B, Grander D, Stahle M, Sonkoly E, Pivarcsi A. MicroRNA-125b down-regulates matrix metallopeptidase 13 and inhibits cutaneous squamous cell carcinoma cell proliferation, migration, and invasion. J Biol Chem. 2012;287:29899–908
44. Sand M, Skrygan M, Georgas D, Sand D, Hahn SA, Gambichler T, Altmeyer P, Bechara FG. Microarray analysis of microRNA expression in cutaneous squamous cell carcinoma. J Dermatol Sci. 2012;68:119–26
45. Darido C, Georgy SR, Wilanowski T, Dworkin S, Auden A, Zhao Q, Rank G, Srivastava S, Finlay MJ, Papenfuss AT, Pandolfi PP, Pearson RB, Jane SM. Targeting of the tumor suppressor GRHL3 by a miR-21-dependent proto-oncogenic network results in PTEN loss and tumorigenesis. Cancer Cell. 2011;20:635–48
46. Bhandari A, Gordon W, Dizon D, Hopkin AS, Gordon E, Yu Z, Andersen B. The Grainyhead transcription factor Grhl3/Get1 suppresses miR-21 expression and tumorigenesis in skin: modulation of the miR-21 target MSH2 by RNA-binding protein DND1. Oncogene. 2012;32(12):1497–507
47. Yu J, Peng H, Ruan Q, Fatima A, Getsios S, Lavker RM. MicroRNA-205 promotes keratinocyte migration via the lipid phosphatase SHIP2. FASEB J. 2010;24:3950–9
48. Kanitz A, Imig J, Dziunycz PJ, Primorac A, Galgano A, Hofbauer GF, Gerber AP, Detmar M. The Expression Levels of MicroRNA-361–5p and Its Target VEGFA Are Inversely Correlated in Human Cutaneous Squamous Cell Carcinoma. PLoS One. 2012;7:e49568
49. Sand M, Skrygan M, Sand D, Georgas D, Hahn SA, Gambichler T, Altmeyer P, Bechara FG. Expression of microRNAs in basal cell carcinoma. Br J Dermatol. 2012;167:847–55

50. Heffelfinger C, Ouyang Z, Engberg A, Leffell DJ, Hanlon AM, Gordon PB, Zheng W, Zhao H, Snyder MP, Bale AE. Correlation of global microRNA expression with basal cell carcinoma subtype. G3 (Bethesda). 2012;2:279–86
51. Sand M, Gambichler T, Skrygan M, Sand D, Scola N, Altmeyer P, Bechara FG. Expression levels of the microRNA processing enzymes Drosha and dicer in epithelial skin cancer. Cancer Invest. 2010;28:649–53
52. Viticchie G, Lena AM, Latina A, Formosa A, Gregersen LH, Lund AH, Bernardini S, Mauriello A, Miano R, Spagnoli LG, Knight RA, Candi E, Melino G. MiR-203 controls proliferation, migration and invasive potential of prostate cancer cell lines. Cell Cycle. 2011;10:1121–31
53. Xu N, Brodin P, Wei T, Meisgen F, Eidsmo L, Nagy N, Kemeny L, Stahle M, Sonkoly E, Pivarcsi A. MiR-125b, a microRNA downregulated in psoriasis, modulates keratinocyte proliferation by targeting FGFR2. J Invest Dermatol. 2011;131:1521–9
54. Yamane K, Jinnin M, Etoh T, Kobayashi Y, Shimozono N, Fukushima S, Masuguchi S, Maruo K, Inoue Y, Ishihara T, Aoi J, Oike Y, Ihn H. Down-regulation of miR-124/-214 in cutaneous squamous cell carcinoma mediates abnormal cell proliferation via the induction of ERK. J Mol Med (Berl). 2013;91:69–81
55. Bostjancic E, Glavac D. Importance of microRNAs in skin morphogenesis and diseases. Acta Dermatovenerol Alp Panonica Adriat. 2008;17:95–102
56. Wu N, Sulpice E, Obeid P, Benzina S, Kermarrec F, Combe S, Gidrol X. The miR-17 family links p63 protein to MAPK signaling to promote the onset of human keratinocyte differentiation. PLoS One. 2012;7:e45761
57. Contreras J, Rao DS. MicroRNAs in inflammation and immune responses. Leukemia. 2012;26:404–13

Chapter 14
MicroRNAs in the Development and Progression of Prostate Cancer

Nagalakshmi Nadiminty, Ramakumar Tummala and Allen C. Gao

Abstract Prostate cancer is one of the cancers with high incidence and mortality rates among US men. Even though prostate-specific antigen screening and other diagnostic approaches have improved patient prognosis, an urgent need for molecularly-based biomarkers of progression persists. MicroRNA (miRNA) dysregulation can have profound consequences as the loss of tumor-suppressive miRNAs enhances the expression of target oncogenes, while the upregulation of oncogenic miRNAs represses the target tumor suppressor genes. The realization of the importance of miRNAs in biological processes has led to a quest to understand the molecular mechanisms regulating miRNAs using a variety of model systems and to entertain the possibility of using miRNA antagonists or mimics for anticancer therapy. The promise of miRNAs as diagnostic and prognostic markers will also need to be realized by using validated datasets and standardized methodologies which will give us a way to compare and verify expression profiles. Here, we discuss the past and current studies which have led us to this point as miRNA-based therapeutics make their way into clinical trials.

Keywords MicroRNAs · Prostate cancer · CRPC · Progression · Androgen receptor · Biomarkers · Circulating miRNA · Clinical trials · p53 · Epigenetics

1 Introduction

Prostate cancer (CaP) ranks first in incidence and second in cancer-related mortality rates among males in the US [1]. There were 241,740 cases and 28,170 deaths in 2012, which makes it the second deadliest after lung cancer. Androgen deprivation and radiation constitute the first-line therapy against androgen-dependent as well as castration-resistant prostate cancer (CRPC). Even though most patients who fail androgen deprivation therapy (ADT) go on to develop CRPC, the majority of patients

A. C. Gao (✉) · N. Nadiminty · R. Tummala
Department of Urology and UC Davis Comprehensive Cancer Center,
University of California Davis Medical Center,
4645 2nd Ave, Research III, Suite 1300, Sacramento, CA 95817, USA
e-mail: acgao@ucdavis.edu

N. Nadiminty.
e-mail: nnadiminty@ucdavis.edu

S. R. Singh, P. Rameshwar (eds.), *MicroRNA in Development and in the Progression of Cancer*, DOI 10.1007/978-1-4899-8065-6_14,

continue to live with indolent CaP. There is an urgent need to predict which patients may fail ADT and which may never progress to CRPC. Existing diagnostic methods and prognostic indicators have improved risk stratification, so that patients who are at higher risk of progression are treated more aggressively, but the problems of overtreatment and subsequent side effects associated with treatment continue to be significant stumbling blocks. By itself, the Gleason score remains the best predictor of recurrence, progression, and death [2]. Additional prognostic modalities such as the Cancer of the Prostate Risk Assessment (CAPRA) score have improved outcomes, but molecularly-based biomarkers have been in short supply. One of the areas in which molecular biomarkers can best help is the characterization of biopsy specimens into indolent and aggressive cancers, and this is where small noncoding RNAs such as microRNAs (miRNAs) enter the picture.

miRNAs are small non-protein-coding RNA molecules that regulate gene expression by binding to the 5′- or 3′-untranslated regions (UTRs) of target transcripts, inhibiting their translation or inducing degradation. miRNA genes are frequently located in fragile sites, minimal regions of loss of heterozygosity, and minimal regions of amplification [3]. Oncogenic miRNAs are associated with regions of amplification, while tumor suppressor miRNAs are associated with frequently deleted chromosomal regions. Single nucleotide polymorphisms in miRNA genes can also be predictive of cancer recurrence and survival [4]. miRNA genes can be mono- or polycistronic, and can be intergenic, intronic, or exonic [5]. When located in exons of protein-coding genes, miRNAs frequently share the promoter of the host gene and are cotranscribed with it. The genes encoding miRNAs are transcribed in much the same way as protein-coding genes are, and produce a primary transcript of several hundred nucleotides in length. After multiple stages of processing in both the cytoplasm and nucleus, the mature 21–23-nucleotide (nt) miRNA is generated. Mature miRNAs are incorporated into RNA-induced silencing complexes (RISCs), and bind with imperfect complementarity to specific seed sequences in their target genes. Recent estimates predict that approximately 30 % of all genes are regulated by miRNAs where each miRNA can target multiple genes, and each messenger RNA (mRNA) can be targeted by multiple miRNAs, generating a complex network of gene regulation which can have a profound impact on cellular programs. These layers of complexity are required since miRNAs can be double-edged swords due to the immense variety of possible downstream effects. miRNAs are well on the way to becoming the central focus of many fields of research in oncology, as evidenced by the exponential growth in number of publications since their discovery a decade ago.

2 miRNAs Deregulated in CaP

A comprehensive literature search using the keywords "miRNAs and prostate cancer" reveals >350 studies which focused on examining the role of miRNAs in CaP. A large number of miRNAs have already been explored in CaP with regard to their biological function, expression profile in cell lines and clinical samples, prognostic/

diagnostic ability, and population dynamics. Results from most of these studies to date have been summarized in Table 14.1. It is interesting to note that most of the miRNAs studied so far are down-modulated in CaP, denoting their role as primarily tumor suppressors, whereas the number of oncogenic miRNAs, commonly referred to as oncomiRs, is only a handful. This makes the list of miRNAs that may potentially be used as markers of progression and recurrence short, since upregulated genes are considered to be more reliable biomarkers compared to downregulated genes. Several miRNAs target the androgen receptor (AR), which is an important survival factor for prostatic tissues, both in benign stages and in neoplastic stages. The list of miRNAs that target the AR is similarly short, which is conducive to developing strategies to modulate AR expression and functions.

3 Importance of the miRNA–AR Axis in CaP

Androgen signaling and AR expression levels are critical in the carcinogenesis and survival of CaP, even in hormone-refractory stages. Activation of target genes by AR is required for the normal functioning of the prostate gland as well as for the progression of CaP. The overexpression and aberrant ligand-independent activation of AR have been implicated in the development of CRPC. AR signaling aberrations and miRNAs appear to be closely linked to the progression of CaP, either by regulation of AR signaling by miRNAs or by androgen-dependent regulation of miRNA expression. The disruption of this AR–miRNA axis may contribute to the development of CaP. AR-targeting miRNAs presumably maintain expression of AR at optimal levels. The loss of AR-targeting miRNAs can potentially lead to elevated levels of AR expression and contribute to the development of CRPC. In addition, shortening of AR 3′ UTR resulting from alternative splicing or alternative polyadenylation may lead to the loss of miRNA binding sites, which would potentially disrupt miRNA-dependent repression of AR leading to AR overexpression [6].

A systematic analysis performed to identify potential AR-targeting miRNAs revealed that miRs-135b, -185, -297, -299-3p, -34a, -34c, -371-3p, -421, -449a, -449b, -634, -654-5p, and -9 were found to directly target a longer 3′UTR than previously used by most target prediction algorithms [7]. Of these, miRs-34a and -34c are of interest due to their close association with the tumor suppressor p53. It has been postulated that miRs of the miR-34 family are responsible for most of the functions of p53 [8], even though miR-34-deficient mouse models have failed to corroborate these results [9]. Studies from our laboratory have demonstrated that miRNAs which target the AR indirectly may also play an important role in CaP progression. We showed that miR-let-7c is underexpressed in CRPC and indirectly regulates the expression of AR via modulation of c-Myc, one of the transcription factors binding to AR promoter [10, 11]. In a study which examined the regulation of miRNAs by androgen, 17 miRNAs were > 1.5-fold upregulated or downregulated upon dihydrotestosterone (DHT) treatment in CaP cell lines, and 42 after castration in AR-positive xenografts. Only four miRNAs (miRs-10a, -141, -150*, and -1225-5p) were found to be regulated by androgen in both

Table 14.1 MiRNAs differentially expressed in prostate cancer

miRNA	Expression	Target	Altered function	References
let-7 family	Downregulated	Ras, c-Myc, Cyclin D1, Cdc25A, EZH2, PBX3, Lin28	Induces apoptosis, inhibits proliferation, regulates AR signaling	[10, 11, 39, 56]
miR-1	Downregulated	SLUG, PNP, FN1, LASP1, PTMA, BRCA1, CHK1, MCM7	Inhibits invasion, proliferation, EMT, tumorigenesis, prognostic marker	[57–59]
miR-7	Downregulated	ERBB2	Inhibits cell proliferation, tumor progression	[60]
miR-9	Downregulated	AR	Inhibits androgen-induced proliferation	[7]
miR-10a	Upregulated	HOXA1	Affects gene expression, cell differentiation	[61]
miR-15a	Downregulated	FGF2, FGFR1, CCND1, WNT3A	Reduces tumor-supportive ability of stromal cells	[26, 62]
miR-16	Downregulated	FGF2, FGFR1, CDK1, CDK2, CCND1, WNT3A	Reduces tumor-supportive ability of stromal cells; inhibits proliferation, inhibits growth of metastases	[26, 62, 63]
miR-17*	Downregulated	Mitochondrial antioxidant enzymes	Suppresses tumorigenesis	[64]
miR-17-3p	Downregulated	Vimentin	Reduces tumor growth; putative tumor suppressor	[65]
miR-20	Upregulated	VEGFA, CDKA1A, NCOA3, HIF1A, CAV1	High in CaP; biomarker for CaP	[66]
miR-20a	Upregulated	CX43	Promotes proliferation, tumor growth; high in CaP; increased in high Gleason grade	[67, 68]
miR-21	Upregulated	RECK, MARCKS, PDCD4, PTEN, TPM1, SPRY2, TIMP3	Oncogenic; increases invasiveness; resistance to apoptosis	[17, 18, 34]
miR-22	Downregulated	PTEN	Induces apoptosis and inhibits metastasis	[13, 69, 70]
miR-23a/b	Downregulated	Rac1, PRDXIII	Suppresses metastasis and response to hypoxia	[71, 72]
miR-24	Upregulated	FAF1	Inhibits apoptosis	[16]
miR-25	Upregulated	PTEN	Induces cell proliferation	[16, 73]
miR-26a	Up/downregulated	PLAG1, EZH2	Influences apoptosis, proliferation, invasion	[13, 73, 74]
miR-27a	Upregulated	Prohibitin	AR-regulated; induces AR target genes; induces cell growth	[75]
miR-27b	Downregulated	Rac1, CYP1B1, NOTCH1	Inhibits proliferation, regulates hormone metabolism	[13, 71, 74]
miR-29a/b	Downregulated	VEGFA, hnRNP-K, DKK1, sFRP2	Reduces cell proliferation; regulates cell differentiation and immune response	13, 76]

Table 14.1 (continued)

miRNA	Expression	Target	Altered function	References
miR-30b/c	Up/downregulated	BCL-9, MTA1, Snail1, GalNAc	Biomarker, influences metastasis	[13, 77]
miR-30d	Downregulated	GRP78	Induces apoptosis	[78]
miR-31	Downregulated	BCL-2L2, E2F6	Promotes apoptosis, inhibits proliferation	[79, 80]
miR-32	Upregulated	BTG2, Bim, C9orf5	AR-regulated; reduces apoptosis; short progression-free survival	[16, 81]
miR-34a	Downregulated	c-Myc, BCL-2, SIRT-1, E2F3, MET, CDK4–6, DLL1, CD44	Suppresses malignancy; inhibits proliferation and survival and metastasis	[16, 82–86]
mir-34b	Upregulated	CDK6, CREB, c-Myc, MET	Affects cell cycle and proliferation	[16, 83]
miR-34c	Downregulated	CDK6, MET, c-Myc, E2F3	Inhibits proliferation	[16, 83]
miR-92	Downregulated	Bim	Induces apoptosis	[13, 73, 87]
miR-96	Upregulated	FOXO1, hZIPs	Affects apoptosis	[49, 50]
miR-100	Up/downregulated	PSA, SMARCA5, SMARCD1, Ras, c-Myc	Independent predictor of biochemical recurrence; decreases proliferation	[88, 89]
miR-101	Downregulated	COX-2, EZH2	Genomic loss during progression; Suppresses growth, invasion	[20, 90]
miR-106a	Upregulated	RB1	Inhibits apoptosis	[73]
miR-106b	Upregulated	PTEN	Proto-oncogenic; cooperates with host gene MCM7 in transformation	[69]
miR-107	Downregulated	Granulin	Inhibits proliferation	[73, 91]
miR-125a	Up/downregulated	ERBB2, ERBB3	Cell proliferation, apoptosis	[13, 16, 92]
miR-125b	Upregulated	BAK1, p53, PUMA	AR-regulated; inhibits apoptosis, promotes castration resistance	[14, 15, 48]
miR-126	Downregulated	CRK, Spred1, PIK3R2/p85-beta	Inhibits proliferation, invasion, progression	[74, 93]
miR-126*	Downregulated	Prostein	Inhibits metastasis	[93, 94]
miR-128a	Downregulated	GOLM1, PHB, TROVE2, TMSB10	Inhibits invasion, progression	[16, 95]
miR-130a	Downregulated	CCNB1, ROCK1, GTF2H1, STX6	Impairs tumor growth; induces cell cycle arrest	[96]
miR-132	Downregulated	TALIN-2, HBEGF	Induces cell death	[97]
miR-133a	Downregulated	EGFR, PNP1	Inhibits proliferation, invasion	[59, 98]
miR-133b	Downregulated	FAIM	Inhibits proliferation, metabolic activity	[99]
miR-135b	Downregulated	AR	Inhibits androgen-induced proliferation	[7]

Table 14.1 (continued)

miRNA	Expression	Target	Altered function	References
miR-141	Upregulated	Clock, SHP	Activates AR and metastases	[13, 100–102]
miR-143	Downregulated	KRAS, ERK5, MYO6, KLK4, KLK10	Inhibits progression, proliferation, migration, invasion, enhances docetaxel sensitivity	[36, 103–106]
miR-145	Downregulated	SWAP70, FSCN1, BNIP3, TNFSF10, MYO6, Myc	Inhibits proliferation, invasion, tumorigenesis, migration, induces apoptosis	[31, 35, 36, 103, 107–109]
miR-146a	Downregulated	EGFR, MMP-2, ROCK1, CXCR4	Inhibits invasion, proliferation, metastasis	[110, 111]
miR-148a	Up/downregulated	CAND1, MSK1	Regulated by androgen, inhibits growth, migration, invasion, increases paclitaxel sensitivity	[112, 113]
miR-153	Upregulated	PTEN	Promotes proliferation	[114]
miR-181a	Downregulated	GRP78	Induces apoptosis	[78]
miR-181a-1	Upregulated	RB1, RBAK	Induces tumor progression	[16, 74]
miR-182	Upregulated	hZIP1	Regulates zinc homoeostasis, inhibits apoptosis	[16, 49, 50]
miR-182-5p	Upregulated	Unknown	High in high-grade CaP	[115]
miR-183	Upregulated	hZIP1	Regulates zinc homoeostasis	[50]
miR-185	Downregulated	AR	Inhibits androgen-induced proliferation	[7]
miR-193b	Downregulated	Unknown	Reduces growth, putative tumor suppressor	[116]
miR-194	Upregulated	DNMT3a, MeCP2	Induces genomic instability	[117]
miR-195	Up/downregulated	CDK4, GLUT3, WEE1, CDK6, Bcl-2	Cell cycle, proliferation, apoptosis	[13, 16, 74]
miR-200a/b	Up/downregulated	SLUG, PDGF-D, NOTCH1, Lin28B	Regulates EMT, cell growth	[37, 38, 58, 118]
miR-200c	Upregulated	SEC23A, JAGGED1	Induces cell growth and metastasis, inhibits apoptosis	[119, 120]
miR-203	Downregulated	CKAP2, LASP1, BIRC5, WASF1, ASAP1, RUNX2, survivin	Inhibits proliferation, migration, invasion, EMT	[96, 121, 122]
miR-204	Upregulated	PDEF	Increases cell growth	[123]
miR-205	Downregulated	VEGFA, HRAS, KLK2, NCOR2, E2F6, PKCepsilon	Promotes apoptosis, MET, inhibits proliferation	[76, 79, 96, 124–126]
miR-210	Upregulated	EFNA3, MNT, HOXA1, APC, ELK3	Induces hypoxia, proliferation and migration	[13, 127]
miR-214	Upregulated	EZH2, N-Ras, PTEN	Induces proliferation and cell cycle	[73]
miR-218	Upregulated	RAS, c-Myc, SMARCA5	Induces cell proliferation	[128]

Table 14.1 (continued)

miRNA	Expression	Target	Altered function	References
miR-221	Up/downregulated	ARHI, c-Kit, p27kip1	Oncogenic	[129–133]
miR-222	Up/downregulated	ARHI, c-Kit, p27kip1	Oncogenic	[80, 129–133]
miR-223	Downregulated	NFI-A	Induces differentiation	[73]
miR-224	Up/downregulated	KLK1, API-5	Influences progression-free survival	[70, 74, 134]
miR-296	Downregulated	HMGA1	Inhibits growth and invasion	[135]
miR-297	Downregulated	AR	Inhibits androgen-induced proliferation	[7]
miR-299-3p	Downregulated	AR	Inhibits androgen-induced proliferation	[7]
miR-301a	Downregulated	FOXF2, BBC3, PTEN, COL2A1	Inhibits proliferation	[132]
miR-320	Downregulated	β-catenin, ETS2	Inhibits progression	[25, 136, 137]
miR-330	Downregulated	E2F1	Inhibits growth, induces apoptosis	[138]
miR-331-3p	Downregulated	DOHH, ERBB2, KLK4, KLK10, EGFR, HER2	Inhibits AR and Akt signaling, inhibits proliferation	[21, 60, 106, 139]
miR-345	Upregulated	BAG3	Induces invasion, metastasis	[13]
miR-370	Upregulated	FOXO1	Induces proliferation	[140]
miR-371-3p	Downregulated	AR	Inhibits androgen-induced proliferation	[7]
miR-373	Downregulated	CD44	Suppress CD44 translation, induce invasion	[141]
miR-375	Upregulated	SEC23A	Stimulates proliferation; serum levels predict high risk for progression	[77, 100, 119]
miR-421	Downregulated	AR	Inhibits androgen-induced proliferation	[7]
miR-449a/b	Downregulated	AR, HDAC-1, Cyclin D1	Induces senescence and growth arrest	[7, 142, 143]
miR-488*	Downregulated	AR	Inhibits proliferation, induces apoptosis	[144]
miR-521	Upregulated	CSA	Influences DNA repair	[22]
miR-616	Upregulated	TFPI-2	Induces castration resistance	[145]
miR-634	Downregulated	AR	Inhibits androgen-induced proliferation	[7]
miR-642-5p	Downregulated	DOHH	Regulates eIF5A activity, inhibits cell proliferation	[139]
miR-654-5p	Downregulated	AR	Inhibits androgen-induced proliferation	[7]
miR-708	Downregulated	CD44, Akt2	Decreases tumorigenicity, regression of established tumors	[146]

cell lines and xenografts. Of these, miR-141 was found to be expressed more in CaP and CRPC compared to benign prostate hyperplasia [12]. A profiling study analyzed cell lines, xenograft models, and patient samples to establish correlations between AR expression and miRNA signature [13]. AR signaling results in upregulation of miR-125b, which acts as an oncogene in CaP [14]. miR-125b is overexpressed in AR-positive cell lines compared to AR-negative ones and is overexpressed in the majority of CaP patient samples compared to benign prostate tissue [15]. miR-338 has also been shown to be induced by androgen activation in LNCaP cells [16]. AR has been found to induce the oncomiR miR-21 in LNCaP and LAPC-4 cell lines. Elevation of miR-21 was further demonstrated to promote tumor growth and castration resistance in a LNCaP mouse xenograft model [17, 18]. To complicate the role of AR-induced miRNA in CaP, other studies find that AR has anticancer interactions with miRNA. AR is essential to p53-induced apoptosis, which is mediated by miR-34a/c [19], and also induces expression of the antiproliferative miR-101 [20]. Finally, AR signaling was also shown to be indirectly regulated by miR-331-3p in LNCaP cells, suggesting yet another role for miRNA in the AR signaling pathway [21]. Since the only function of miRs is to bind to 3′ UTR and inhibit translation of target genes, androgens may operate via induction of miRs to inhibit repressors of AR function. In concordance, knockdown of DICER in LNCaP cells and in tissues in mice induced the expression of co-repressors, NCoR and SMRT. These studies demonstrate a feedback loop between miRs, co-repressors, and AR and the imperative role of miRs in AR function.

4 miRNAs in Radiation Response

Ionizing radiation activates a multitude of survival and death signaling pathways. Previous studies demonstrate that radiation induces changes in a large number of genes, which are involved in DNA repair/synthesis, stress response, and cell cycle control. However, the role of miRNAs and how they integrate into the radiation signaling pathways is largely unknown. About 60 % of cancer patients receive radiation treatment, and while it is very effective, some patients may benefit from identification of novel radiosensitizers. It is expected that radiosensitizers should reduce the dose or frequency of radiation treatment and improve disease outcome in patients. In one study, global miRNA profiling was performed to determine important miRNAs in radiation stress response. The study found that of the 330 miRNAs analyzed, 10 miRNAs were significantly downregulated while 5 were significantly upregulated following irradiation in LNCaP and C4-2 CaP cells. They also found that miR-521 played a major role in modulating the radiation sensitivity of CaP cells [22]. Another study performed miRNA array and found that of 132 cancerous miRNAs examined, 10 miRNAs were significantly upregulated by irradiation in LNCaP cells. They also showed that miR-106b induced radioresistance in LNCaP cells by inhibiting the radiation-induced increase in p21 [23]. The identities of the miRNAs found to be modulated by irradiation from both studies are completely different, despite the use of identical experimental systems. One of the possible reasons may be that rapidly occurring changes in miRNA expression profiles may not be detected in samples collected after longer periods of time after irradiation. One of these studies used samples collected 4 h after irradiation, while another collected samples 24 h after irradiation. These results exemplify the care to be taken when radiation response in miRNA expression profiles is examined and the need for standardization of duration and amount of radiation applied. Another study compared single-dose radiation to fractionated radiation and found that fractionated radiation alters more miRNAs than a single dose. Some miRNAs were altered to a similar extent in both p53-positive and p53-null cells, indicating that p53 may not be the sole determinant of radiation response in CaP cells [24].

5 miRNAs in Other Signaling Pathways

miRNAs targeting other important cellular signaling pathways such as Wnt, NF-κB, p53, and cytokine signaling have also been described in CaP. MiR-320 downregulates Wnt signaling by directly targeting β-catenin and suppresses stem cell-like characteristics of CaP cells, possibly reorienting them towards a more differentiated phenotype [25]. The miR-15a-16-1 cluster targets Wnt3a and CCND1 in addition to other oncogenic targets and inhibits proliferation, survival, and invasion of CaP cells [26], indicating that coordinated inhibition of multiple signaling pathways may elicit stronger responses in terms of CaP cell survival. Very few miRNAs targeting either the classical or the alternative NF-κB pathways have been studied in CaP. Recent studies have identified miR-181b as being overexpressed in prostate carcinomas and that it targets CYLD, a known tumor suppressor and inhibitor of classical NF-κB signaling. miR-181b is part of a feedback loop involving STAT3, CYLD, IL-6, and NF-κB and participates in an epigenetic circuit to promote cell transformation [27]. Similarly, miR-21, which is induced by STAT3 and targets PTEN, is involved in induction of NF-κB activation by a positive feedback loop [27]. In addition, miR-21 has been universally reported to be overexpressed in human carcinomas including carcinoma of the prostate [18, 28]. A few recent reports have contradicted the oncogenic role of miR-21 in CaP [29], which goes to show that miRNAs have cell type- and tissue-specific roles in human cancers, and great care should be exercised in extending findings from one human cancer to another.

Several miRNAs have been shown to be downstream of the tumor suppressor p53, which include miR-34 family, miR-192/215, and miR-145—known transcriptional targets of p53. Most of these are downregulated in CaP, which demonstrates their importance in the tumor suppressor network spearheaded by p53. The miR-34 family of miRNAs and miR-145 are also regulated by methylation of promoters [30, 31]. In addition, p53 has been shown to be regulated by a few oncogenic miRNAs, primarily by miR-125b in CaP, which binds to the 3′UTR of p53 and regulates its expression [32]. miR-125b is an androgen-induced miRNA [15], implying that activation of AR signaling represses p53 and its related tumor-suppressive functions.

Cytokine signaling is also modulated by miRNAs in CaP. For example, IL-6 is regulated by the let-7 family, which in turn is regulated by a feedback loop involving Lin28, STAT3, and NF-κB [33]. IL-6 signaling results in activation of STAT3, which in turn induces the oncomiR, miR-21 [27], which had been shown to be overexpressed in CaP and to mediate proliferation, invasion, and metastasis [17, 18, 34]. The exact mechanisms by which miR-21 operates in CaP are still debated [29] and need to be examined using in vitro and in vivo models. miR-145, which is underexpressed in CaP due to methylation and p53 gene mutation [31], induces expression of pro-apoptotic genes such as TNFSF10 and IL-24 [35]. Epithelial-to-mesenchymal transition (EMT) is defined as the process by which epithelial cells acquire mesenchymal characteristics and transition to an elongated fibroblastic phenotype, thus acquiring attributes such as increased cell motility and invasion. Several miRNAs such as miR-200/200b, miR-143, miR-145, and the let-7 family have been shown to control EMT in CaP [36–39]. These findings collectively indicate that miRNAs occupy a critical niche in cell survival processes and fine-tune the survival and death signals via cross talk with the tumor microenvironment.

6 Epigenetic Regulation of CaP-Related miRNAs

Deregulated expression of miRNAs in cancer cells can result due to genomic abnormalities such as chromosomal rearrangements, genomic amplifications, and deletions of miRNA genes, and also due to altered transcriptional and posttranscriptional control of miRNA expression. In addition, epigenetic changes, such as methylation of CpG islands in the promoter regions of miRNA genes, can alter miRNA expression in cancer cells. An extensive analysis of miRNA genes shows that ~50% are associated with CpG islands suggesting their possible regulation by the DNA methylation machinery. In CaP, miRNA deregulation affects epigenetic reprogramming, blockade of apoptosis, promotion of cell cycle, migration, and invasion and is an alternative mechanism to sustain castration-resistant growth. Although several miRNAs have been reported to be regulated epigenetically in CaP, only a few have been experimentally proven to contribute to the disease. Of the miRNAs that are involved in the epigenetic process in CaP, three distinct types can be described: (1) miRNAs that regulate genes of the epigenetic machinery such as miR-101 and miR-449a, which regulate EZH2 and HDAC1, respectively; (2) miRNAs that are epigenetically regulated, such as miR-1 (MCM7, BRCA1), miR-200c/141 cluster (ZEB1/ZEB2), miR-132 (TALIN-2), miR-205 (ZEB1/ZEB2), miR-126 (DNMT1), miR-193b, miR-196b (target(s) unknown), miR-145 (TNFSF10), miR-34 family (SIRT1/CD44), and miR-21 (multiple targets); (3) miRNAs that are involved in the epigenetic regulation of AR, such as miR-34 family, miR-141, miR-494, and miR-29a/b/c (for a complete review see [40]). In addition, AR-regulated miRNAs such as miR-125b and miR-21 have also been shown to be epigenetically regulated [15, 18]. Thus, epigenetic regulation of miRNAs adds an extra degree of complexity to the picture which needs to be further elucidated to fully realize the potential of manipulating the epigenetic machinery for therapeutic purposes.

7 Circulating miRNAs in CaP

Recent evidence suggests that miRNA profiles from tissue sources as well as circulating body fluids may be good tools for prognostic and diagnostic purposes. miRNA profiles not only can distinguish between tumors of different developmental origin but also possess other prerequisites to be considered useful noninvasive biomarkers. First, they are exceptionally stable in a wide variety of clinical samples such as formalin-fixed paraffin-embedded tissues, blood, serum, and urine [41]. Second, they can be quantitatively measured reliably in small amounts of samples by real-time quantitative reverse transcription polymerase chain reaction (RT-PCR) and are amenable to high-throughput strategies [42]. Third, they are resistant to endogenous ribonuclease activity as well as variations in temperature and pH [43]. In addition, they are highly conserved among different species making the use of animals for preclinical studies feasible. A circulating tumor biomarker should also be able to detect a tumor before it can easily be detected by other means, and this is one area in which application of miRNAs has not been explored. The first studies that demonstrated that circulating cell-free miRNA profiles in body fluids are altered in response to different malignancies [44, 45] brought the exciting and limitless possibilities of circulating miRNAs to the fore. Since then, several preclinical studies have analyzed the sensitivity and specificity of cell-free miRNAs as biomarkers. Some of these studies are summarized in Table 14.2.

The mechanisms involved in the release of miRNAs into circulation have been under debate. The theory that miRNAs are passively released into extracellular spaces is being increasingly challenged by recent evidence that shows that miRNAs are released within microvesicles or endosomes and sometimes as Ago2-coupled complexes. These miRNAs may constitute a distinct miRNA profile of that particular tumor type and assist in prognosis. But confounding factors such as contamination of circulating miRNAs by cellular miRNAs and by erythrocyte miRNAs released due to hemolysis still exist. Similarly, even among the limited number of studies which analyzed prognostic indicators of miRNA profiles for CaP, the inconsistencies in sample selection, sample collection, methods of extraction of miRNAs, experimental platforms used, and ignorance of cellular origin make it difficult to effectively compare the results and draw conclusions about the efficacy of a particular miRNA or a panel of miRNAs in risk stratification or prognosis. Large-scale clinical studies with rigorous controls and an internationally established code for sample selection and collection are needed before the promise of miRNAs as circulating biomarkers can be realized in the clinic.

Table 14.2 Circulating miRNAs in prostate cancer

Body fluid	Sample size	Methodology	Findings	References
Plasma	25 patients (metastatic CaP), 25 healthy controls	qRT-PCR (6 miRNAs)	miR-141 levels differentiate CaP patients from healthy controls	[45]
	21 patients (metastatic CaP)	qRT-PCR (miR-141)	miR-141 levels correlated with PSA and with progression	[147]
	51 patients (18 localized, 8 local advanced, 25 metastatic), 20 healthy controls	qRT-PCR (miR-21, miR-221, and miR-141)	miR-21 and miR-221 levels higher in CaP compared to healthy controls; miR-21, miR-221, and miR-141 higher in metastatic vs. localized disease	[148]
	82 patients of stage 2–3; with risk stratification	qRT-PCR (miR-20a, miR-21, miR-145, and miR-221)	miR-20a levels higher in stage 3 compared to lower stages; miR-20a and miR-21 levels higher in high-risk patients; all four could distinguish high risk from low risk	[149]
Serum	6 patients (stages 2–4 CaP), 8 healthy controls	Custom microarray (547 miRNAs)	15 miRNAs elevated in CaP patients	[150]
	56 patients (20 localized CaP, 20 androgen-dependent CaP, 10 CRPC), 6 BPH controls	qRT-PCR (miR-21)	miR-21 levels higher in CRPC compared to BPH; associated with docetaxel resistance in CRPC	[151]
	29 patients (9 low risk, 11 intermediate risk, 9 high risk), 9 healthy controls	Multiplex qRT-PCR (677 miRNAs)	10 miRNAs differentially expressed in CaP; 7 miRNAs correlated with risk groups	[152]
	7 high-grade and 14 low-grade patients (profiling); 116 patients of various grades (validation)	qRT-PCR (667 miRNAs)	miR-141, miR-200b and miR-375 elevated in high-grade patients and correlated with clinicopathological parameters	[101]
	45 patients (37 localized, 8 metastatic), 18 BPH controls, 20 healthy controls	qRT-PCR (5 miRNAs)	miR-26a, miR-195, and let-7i levels higher in CaP compared to BPH	[153]
	14 TRAMP mice and 14 healthy controls (profiling); 25 metastatic CaP and 25 healthy controls (validation)	Affymetrix microarray, qRT-PCR (609 murine miRNAs, 10 human miRNAs)	miR-141, miR-298, miR-346, and miR-375 levels higher in CaP	[154]

Table 14.2 (continued)

Body fluid	Sample size	Methodology	Findings	References
	28 low-risk, 30 high-risk localized CaP and 26 metastatic CRPC	TaqMan miRNA microarray, qRT-PCR	miR-375, miR-387*, and miR-141 higher in CRPC compared to low-risk localized CaP; miR-409-3p lower in CRPC	[155]
Plasma and serum	78 patients (various grades, 15 with diagnosed metastases) and 28 healthy controls for profiling; 119 patients (47 recurrent after radical prostatectomy and 72 nonrecurrent)	qRT-PCR (742 miRNAs)	12 miRNAs altered in CaP compared to healthy controls; 16 miRNAs altered in metastatic vs. localized	[156]

8 miRNAs as Biomarkers, Prognostic Markers, and/or Therapeutic Targets

Numerous studies have described the potential for a particular miRNA or a panel of miRNAs to be used as biomarkers or prognostic markers in CaP. In addition, some studies have attempted to antagonize the functions of miRNAs with a view to using them as therapeutic targets. miR-21, which is one of the few oncomiRs to be described in CaP, is an example of an miRNA that may serve as a biomarker as well as a therapeutic target. miR-21 targets several mRNAs, such as MARCKS [34], RECK [17], and PDCD-4 [46] and is postulated as an independent predictor of biochemical recurrence and as a potential therapeutic target [47]. Similarly, miR-125b, which targets p53 and other molecules in the p53 pathway, such as BAK1 [15, 48], may serve as a biomarker of castration resistance, tumor stage, and perineural invasion as well as a therapeutic target. Another miRNA that is overexpressed in CaP, miR-96 (FOXO1, hZIPs), may be a prognostic marker of biochemical progression and tumor recurrence [49, 50]. Other miRNAs that have been shown to be downregulated in CaP and that are implicated in prognosis are miR-331-3p, miR-146, miR-1, miR-143, miR-145, miR-34 family, miR-200 family, let-7 family, miR-1, etc. This preclinical evidence needs to be corroborated by clinical studies, and although it is patently obvious that miRNAs could help classify CaP progression and recurrence, their potential is still far away from clinical application.

Different approaches are being developed to achieve gain or loss of miRNA functions. Restoring the functions of tumor suppressor miRNAs which have been repressed can be achieved by adeno-associated viruses, lentiviruses, cationic liposomes, or polymer-based nanoparticle formulations [51]. On the other hand, antagonizing functions of oncomiRs can be achieved by introduction of antagomiRs, oligonucleotides which inhibit target pairing competitively [52], or by miRNA "sponges," which have been designed to carry multiple binding sites for several endogenous miRNAs [53]. Some small molecules such as azobenzene, which blocks miR-21 function, [54] are also being explored as potential inhibitors of miRNA function.

9 miRNAs in Clinical Trials

miRNA therapeutics that are in preclinical development include miR-208/499 in chronic heart failure, miR-195 in post-myocardial infarction remodeling, and let-7 for non-small cell lung cancer. Some of these therapeutics may reach clinical trial stages in the not-so-distant future. There are currently a number of companies which have miRNA therapeutics programs with the most successful being miravirsen (SPC3649), an miR-122 inhibitor (Santaris Pharma), which is in phase II studies for treatment of hepatitis C. Miravirsen was the first miRNA-targeted drug to enter clinical trials. Recently, MGN-4893 (Miragen Therapeutics) that targets miR-451 was given orphan drug status by the US Food and drug Administration (FDA) to treat polycythemia vera, a myeloproliferative disorder characterized by an overabundance of blood cells and platelets in the body.

Although there are currently no clinical trials that use miRNAs as a treatment option for CaP, recent successes by Mirna Therapeutics and researchers at MD Anderson Cancer Center in inhibiting CaP tumor growth, decreasing lung metastasis, and extending survival in mice using liposome-based systemic delivery of miR-34a to suppress the adhesion molecule CD44 are promising. This group hopes to advance miR-34a as a treatment option for CaP patients. Currently, there are a few observational clinical trials to study miRNAs in CaP (www.clinicaltrials.gov) as detailed in Table 14.3.

10 Challenges in Using miRNA-Based Therapeutics

miRNAs are naturally occurring molecules, and distinct advantages of using miRNAs as therapeutic agents over currently used conventional drugs are apparent. These include their broad specificity, which would be a disadvantage with other therapies but is a distinct advantage with miRNA-based therapeutics. Consequently, miRNA-based therapeutics can target multiple genes in one or multiple pathways concurrently. In addition, tumor-suppressive miRNAs can be used to cooperatively target one or multiple target genes. Another advantage of using miRNAs as drugs is their small size, which makes them less antigenic than protein- or oligosaccharide-based gene replacement strategies. But as with other kinds of therapeutic agents, there are several challenges associated with using miRNAs as therapeutics; a major hurdle is the mode of delivery. Even though viral-mediated delivery systems (adenoviral, lentiviral) have shown promise in preclinical studies, they are not likely to be extrapolated to human use. Other strategies such as liposome- or nanoparticle- mediated delivery or conjugation to cell-penetrating peptides may be plausible. Adjuvant carrier systems which can in-

Table 14.3 Current miRNA-related clinical trials in CaP

Trial title	Study type	Institution	Trial identifier
MicroRNA expression profiles in high-risk prostate cancer	Observational; to study whether miRNA profiles correlate with disease outcome	Würzburg University Hospital, Germany	NCT01220427
Molecular correlates of sensitivity and resistance to therapy in prostate cancer	Observational; to study differences in miRNA profiles in order to discover new biomarkers and drug targets	University of Washington	NCT01050504
Trial of vaccine therapy in curative resected prostate cancer patients using autologous dendritic cells loaded with mRNA from primary prostate cancer tissue, hTERT, and Survivin	Treatment; secondary objective	Rikshospitalet University Hospital, Norway	NCT01197625
Phase II randomized study of combined androgen deprivation comprising Bicalutamide and Goserelin or Leuprolide Acetate with versus without Cixutumumab in patients with newly diagnosed hormone-sensitive metastatic prostate cancer	Biomarker/laboratory analysis, treatment; secondary objective	Saint Anthony's Hospital at Saint Anthony's Health Center, Illinois	NCT01120236

crease the stability of miRNAs in the cellular microenvironment and enhance uptake by target tissues need to be invented. Another major hurdle is the concern that delivery of exogenous miRNAs or their mimics may overwhelm the cellular RISC system and interfere with the processing of endogenous miRNAs. Other concerns include off-target effects, toxicity, and possible liver damage [55]. Population-based variation in miRNA expression profiles is another major challenge. Dosage and combinations of miRNAs for each type of cancer need to be established taking into account the gender, race, and environmental conditions of each patient. These are difficult but not insurmountable obstacles, and with the current pace in discovery and application of miRNA-based therapeutics, their resolution would not be too far away in the future.

11 Conclusions and Perspectives

Research in the last decade, since the discovery of miRNAs, has suggested that an intimate relationship exists between CaP and miRNA profiles making these discoveries of strong prognostic and therapeutic importance. The field is clearly promising and exciting but further accurate dissection of the mechanistic aspects is absolutely necessary to determine the specific roles of individual miRNAs and collective impact of a particular miRNA profile signature on disease outcome and progression. Once this knowledge is obtained, it would become easier to develop therapeutic approaches to target a specific miRNA or a set of miRNAs to achieve a desired outcome. At the same time, even though several studies have demonstrated the utility of miRNA profiling in predicting clinical outcome, the findings need to be validated and consistency needs to be improved. In conclusion, miRNAs represent valuable prognostic and therapeutic tools which may prove to be essential weapons in the fight against CaP progression, and it is up to the research community to come up with innovative and reliable techniques to utilize them effectively.

References

1. Siegel R, Naishadham D, Jemal A. Cancer statistics, 2012. CA Cancer J Clin. 2012;62(1):10–29.
2. Lucas SM, Heath EI. Current challenges in development of differentially expressed and prognostic prostate cancer biomarkers. Prostate Cancer. 2012;2012(640968).
3. Calin GA, et al. Human microRNA genes are frequently located at fragile sites and genomic regions involved in cancers. Proc Natl Acad Sci U S A. 2004;101(9):2999–3004.
4. Lin J, et al. Genetic variations in microRNA-related genes are associated with survival and recurrence in patients with renal cell carcinoma. Carcinogenesis. 2010;31(10):1805–12.
5. Long Y-S, et al. Identification of the transcriptional promoters in the proximal regions of human microRNA genes. Mol Biol Rep. 2010;38(6):4153–7.
6. Sikand K, Barik S, Shukla GC. MicroRNAs and androgen receptor 3′ untranslated region: a missing link in castration-resistant prostate cancer? Mol Cell Pharmacol. 2011;3(3):107–13.
7. Östling P, et al. Systematic analysis of microRNAs targeting the androgen receptor in prostate cancer cells. Cancer Res. 2011;71(5):1956–67.
8. He L, et al. A microRNA component of the p53 tumour suppressor network. Nature. 2007;447(7148):1130–4.
9. Jain AK, Barton MC. Unmet expectations: miR-34 plays no role in p53-mediated tumor suppression in vivo. PLoS Genet. 2012;8(7):e1002859.
10. Nadiminty N, et al. MicroRNA let-7c is downregulated in prostate cancer and suppresses prostate cancer growth. PLoS One. 2012;7(3):e32832.
11. Nadiminty N, et al. MicroRNA let-7c suppresses androgen receptor expression and activity via regulation of myc expression in prostate cancer cells. J Biol Chem. 2012;287(2):1527–37.
12. Waltering KK, et al. Androgen regulation of micro-RNAs in prostate cancer. Prostate. 2011;71(6):604–14.
13. Porkka KP, et al. MicroRNA expression profiling in prostate cancer. Cancer Res. 2007;67(13):6130–5.
14. Shi X-B, et al. miR-125b promotes growth of prostate cancer xenograft tumor through targeting pro-apoptotic genes. Prostate. 2011;71(5):538–49.
15. Shi X-B, et al. An androgen-regulated miRNA suppresses Bak1 expression and induces androgen-independent growth of prostate cancer cells. Proc Natl Acad Sci U S A. 2007;104(50):19983–8.
16. Ambs S, et al. Genomic profiling of microRNA and messenger RNA reveals deregulated microRNA expression in prostate cancer. Cancer Res. 2008;68(15):6162–70.
17. Reis S, et al. miR-21 may acts as an oncomir by targeting RECK, a matrix metalloproteinase regulator, in prostate cancer. BMC Urol. 2012;12(1):14.
18. Ribas J, et al. miR-21: an androgen receptor regulated microRNA that promotes hormone-dependent and hormone-independent prostate cancer growth. Cancer Res. 2009;69(18):7165–9.

19. Rokhlin OW, et al. MicroRNA-34 mediates AR-dependent p53-induced apoptosis in prostate cancer. Cancer Biol Ther. 2008;7(8):1288–96.
20. Cao P, et al. MicroRNA-101 negatively regulates Ezh2 and its expression is modulated by androgen receptor and HIF-1alpha/HIF-1beta. Mol Cancer. 2010;9(1):108.
21. Epis MR, et al. miR-331-3p regulates ERBB-2 expression and androgen receptor signaling in prostate cancer. J Biol Chem. 2009;284(37):24696–704.
22. Josson S, et al. Radiation modulation of microRNA in prostate cancer cell lines. Prostate. 2008;68(15):1599–606.
23. Li B, et al. Down-regulation of microRNA 106b is involved in p21-mediated cell cycle arrest in response to radiation in prostate cancer cells. Prostate. 2011;71(6):567–74.
24. John-Aryankalayil M, et al. Fractionated radiation alters oncomir and tumor suppressor miRNAs in human prostate cancer cells. Radiat Res. 2012;178(3):105–17.
25. Hsieh IS, et al. MicroRNA-320 suppresses the stem cell-like characteristics of prostate cancer cells by downregulating the Wnt/beta-catenin signaling pathway. Carcinogenesis. 2013;34(3):530–8.
26. Bonci D, et al. The miR-15a-miR-16-1 cluster controls prostate cancer by targeting multiple oncogenic activities. Nat Med. 2008;14(11):1271–7.
27. Iliopoulos D, et al. STAT3 activation of miR-21 and miR-181b-1 via PTEN and CYLD are part of the epigenetic switch linking inflammation to cancer. Mol Cell. 2011;39(4):493–506.
28. Ma X, et al. MicroRNAs in NF-kB signaling. J Mol Cell Biol. 2011;3(3):159–66.
29. Folini M, et al. miR-21: an oncomir on strike in prostate cancer. Mol Cancer. 2010;9(1):12.
30. Lodygin D, et al. Inactivation of miR-34a by aberrant CpG methylation in multiple types of cancer. Cell Cycle. 2008;7(16):2591–600.
31. Suh SO, et al. MicroRNA-145 is regulated by DNA methylation and p53 gene mutation in prostate cancer. Carcinogenesis. 2011;32(5):772–8.
32. Le MTN, et al. Conserved regulation of p53 network dosage by microRNA–125b occurs through evolving miRNA–target gene pairs. PLoS Genet. 2011;7(9):e1002242.
33. Iliopoulos D, Hirsch HA, Struhl K. An epigenetic switch involving NF-κB, Lin28, Let-7 microRNA, and IL6 links inflammation to cell transformation. Cell. 2009;139(4):693–706.
34. Li T, et al. MicroRNA-21 directly targets MARCKS and promotes apoptosis resistance and invasion in prostate cancer cells. Biochem Biophys Res Commun. 2009;383(3):280–5.
35. Zaman MS, et al. The functional significance of microRNA-145 in prostate cancer. Br J Cancer. 2010;103(2):256–64.
36. Peng X, et al. Identification of miRs-143 and -145 that is associated with bone metastasis of prostate cancer and involved in the regulation of EMT. PLoS One. 2011;6(5):e20341.
37. Kong D, et al. Epithelial to mesenchymal transition is mechanistically linked with stem cell signatures in prostate cancer cells. PLoS One. 2010;5(8):e12445.
38. Kong D, et al. miR-200 regulates PDGF-D-mediated epithelial-mesenchymal transition, adhesion, and invasion of prostate cancer cells. Stem Cells. 2009;27(8):1712–21.
39. Kong D, et al. Loss of let-7 upregulates EZH2 in prostate cancer consistent with the acquisition of cancer stem cell signatures that are attenuated by BR-DIM. PLoS One. 2012;7(3):e33729.
40. Paone A, Galli R, Fabbri M. MicroRNAs as new characters in the plot between epigenetics and prostate cancer. Front Genet. 2011;2:62.
41. Xi Y, et al. Systematic analysis of microRNA expression of RNA extracted from fresh frozen and formalin-fixed paraffin-embedded samples. RNA. 2007;13(10):1668–74.
42. Tam W. The emergent role of microRNAs in molecular diagnostics of cancer. J Mol Diagn. 2008;10(5):411–4.
43. Kelly BD, et al. A review of expression profiling of circulating microRNAs in men with prostate cancer. BJU Int. 2012;111(1):17–21.
44. Lawrie CH, et al. Detection of elevated levels of tumour-associated microRNAs in serum of patients with diffuse large B-cell lymphoma. Br J Haematol. 2008;141(5):672–5.
45. Mitchell PS, et al. Circulating microRNAs as stable blood-based markers for cancer detection. Proc Natl Acad Sci U S A. 2008;105(30):10513–18.

46. Lu Z, et al. MicroRNA-21 promotes cell transformation by targeting the programmed cell death 4 gene. Oncogene. 2008;27(31):4373–9.
47. Li T, et al. miR-21 as an independent biochemical recurrence predictor and potential therapeutic target for prostate cancer. J Urol. 2012;187(4):1466–72.
48. Shi X-B, Tepper CG, deVere White RW. Cancerous miRNAs and their regulation. Cell Cycle. 2008;7(11):1529–38.
49. Schaefer A, et al. MicroRNAs and cancer: current state and future perspectives in urologic oncology. Urol Oncol: Semin Ori Invest. 2010;28(1):4–13.
50. Mihelich BL, et al. miR-183-96-182 cluster is overexpressed in prostate tissue and regulates zinc homeostasis in prostate cells. J Biol Chem. 2011;286(52):44503–11.
51. Bhardwaj A, Singh S, Singh AP. MicroRNA-based cancer therapeutics: big hope from small RNAs. Mol Cell Pharmacol. 2010;2(5):213–9.
52. Krutzfeldt J, et al. Silencing of microRNAs in vivo with 'antagomirs'. Nature. 2005;438(7068):685–9.
53. Ebert MS, Neilson JR, Sharp PA. MicroRNA sponges: competitive inhibitors of small RNAs in mammalian cells. Nat Meth. 2007;4(9):721–6.
54. Gumireddy K, et al. Small-molecule inhibitors of microRNA miR-21 function. Angew Chem Int Ed. 2008;47(39):7482–4.
55. Grimm D, et al. Fatality in mice due to oversaturation of cellular microRNA/short hairpin RNA pathways. Nature. 2006;441(7092):537–41.
56. Ramberg H, et al. Regulation of PBX3 expression by androgen and Let-7d in prostate cancer. Mol Cancer. 2011;10(1):50.
57. Hudson RS, et al. MicroRNA-1 is a candidate tumor suppressor and prognostic marker in human prostate cancer. Nucleic Acids Res. 2011;40(8):3689–703.
58. Liu YN, et al. MiR-1 and miR-200 inhibit EMT via slug-dependent and tumorigenesis via slug-independent mechanisms. Oncogene. 2013;32(3):296–306.
59. Kojima S, et al. Tumour suppressors miR-1 and miR-133a target the oncogenic function of purine nucleoside phosphorylase (PNP) in prostate cancer. Br J Cancer. 2012;106(2):405–13.
60. Wu W, et al. MicroRNA regulation of growth factor receptor signaling in human cancer cells, in microRNA and cancer. Humana Press; 2011. pp. 147–63.
61. Garzon R, et al. MicroRNA fingerprints during human megakaryocytopoiesis. Proc Natl Acad Sci U S A. 2006;103(13):5078–83.
62. Musumeci M, et al. Control of tumor and microenvironment cross-talk by miR-15a and miR-16 in prostate cancer. Oncogene. 2011;30(41):4231–42.
63. Takeshita F, et al. Systemic delivery of synthetic microRNA-16 inhibits the growth of metastatic prostate tumors via downregulation of multiple cell-cycle genes. Mol Ther. 2009;18(1):181–7.
64. Xu Y, et al. miR-17* suppresses tumorigenicity of prostate cancer by inhibiting mitochondrial antioxidant enzymes. PLoS One. 2010;5(12):e14356.
65. Zhang X, et al. MicroRNA-17-3p is a prostate tumor suppressor in vitro and in vivo, and is decreased in high grade prostate tumors analyzed by laser capture microdissection. Clin Exp Metastasis. 2009;26(8):965–79.
66. Liu DF, et al. MicroRNA expression profile analysis revelas diagnostic marker for human prostate cancer. Asian Pac J Cancer Prev. 2012;13(7):3313–7.
67. Li X, et al. Suppression of CX43 expression by miR-20a in the progression of human prostate cancer. Cancer Biol Ther. 2012;13(10):890–8.
68. Pesta M, et al. Importance of miR-20a expression in prostate cancer tissue. Anticancer Res. 2010;30(9):3579–83.
69. Poliseno L, et al. Identification of the miR-106b~25 microRNA cluster as a proto-oncogenic PTEN-targeting intron that cooperates with its host gene MCM7 in transformation. Sci Signal. 2010;3(117):ra29.
70. Wang L, et al. Gene networks and microRNAs implicated in aggressive prostate cancer. Cancer Res. 2009;69(24):9490–7.

71. Ishteiwy RA, et al. The microRNA-23b/-27b cluster suppresses the metastatic phenotype of castration-resistant prostate cancer cells. PLoS One. 2012;7(12):e52106.
72. He HC, et al. MicroRNA-23b downregulates peroxiredoxin III in human prostate cancer. FEBS Lett. 2012;586(16):2451–8.
73. Volinia S, et al. A microRNA expression signature of human solid tumors defines cancer gene targets. Proc Natl Acad Sci U S A. 2006;103(7):2257–61.
74. Prueitt RL, et al. Expression of microRNAs and protein-coding genes associated with perineural invasion in prostate cancer. Prostate. 2008;68(11):1152–64.
75. Fletcher CE, et al. Androgen-regulated processing of the oncomir MiR-27a, which targets Prohibitin in prostate cancer. Hum Mol Genet. 2012;21(14):3112–27.
76. Szczyrba J, et al. Identification of ZNF217, hnRNP-K, VEGF-A and IPO7 as targets for microRNAs that are downregulated in prostate carcinoma. Int J Cancer. 2013;132(4):775–84.
77. Chen Z-H, et al. A panel of five circulating microRNAs as potential biomarkers for prostate cancer. Prostate. 2012;72(13):1443–52.
78. Su SF, et al. miR-30d, miR-181a and miR-199a-5p cooperatively suppress the endoplasmic reticulum chaperone and signaling regulator GRP78 in cancer. Oncogene. 2012;32:4694–4701.
79. Bhatnagar N, et al. Downregulation of miR-205 and miR-31 confers resistance to chemotherapy-induced apoptosis in prostate cancer cells. Cell Death and Dis. 2010;1: e105.
80. Fuse M, et al. Tumor suppressive microRNAs (miR-222 and miR-31) regulate molecular pathways based on microRNA expression signature in prostate cancer. J Hum Genet. 2012;57(11):691–9.
81. Jalava SE, et al. Androgen regulated miR-32 targets BTG2 and is overexpressed in castration-resistant prostate cancer. Oncogene. 2012;31(41):4460–71.
82. Yamamura S, et al. MicroRNA-34a modulates c-Myc transcriptional complexes to suppress malignancy in human prostate cancer cells. PLoS One. 2012;7(1):e29722.
83. Hermeking H. The miR-34 family in cancer and apoptosis. Cell Death Differ. 2009;17(2):193–9.
84. Fujita Y, et al. Effects of miR-34a on cell growth and chemoresistance in prostate cancer PC3 cells. Biochem Biophys Res Commun. 2008;377(1):114–9.
85. Kojima K, et al. MiR-34a attenuates paclitaxel-resistance of hormone-refractory prostate cancer PC3 cells through direct and indirect mechanisms. Prostate. 2010;70(14):1501–12.
86. Liu C, et al. The microRNA miR-34a inhibits prostate cancer stem cells and metastasis by directly repressing CD44. Nat Med. 2011;17(2):211–5.
87. Tsuchida A, et al. miR-92 is a key oncogenic component of the miR-17-92 cluster in colon cancer. Cancer Sci. 2011;102(12):2264–71.
88. Leite KR, et al. MicroRNA-100 expression is independently related to biochemical recurrence of prostate cancer. J Urol. 2011;185(3):1118–22.
89. Sun D, et al. miR-99 family of microRNAs suppresses the expression of prostate-specific antigen and prostate cancer cell proliferation. Cancer Res. 2011;71(4):1313–24.
90. Hao Y, et al. Enforced expression of miR-101 inhibits prostate cancer cell growth by modulating the COX-2 pathway in vivo. Cancer Prev Res. 2011;4(7):1073–83.
91. Wang W-X, et al. Dysregulation of the Mitogen Granulin in human cancer through the miR-15/107 microRNA gene group. Cancer Res. 2010;70(22):9137–42.
92. Scott GK, et al. Coordinate suppression of ERBB2 and ERBB3 by enforced expression of micro-RNA miR-125a or miR-125b. J Biol Chem. 2007;282(2):1479–86.
93. Watahiki A, et al. MicroRNAs associated with metastatic prostate cancer. PLoS One. 2011;6(9):e24950.
94. Musiyenko A, Bitko V, Barik S. Ectopic expression of miR-126*, an intronic product of the vascular endothelial EGF-like 7 gene, regulates prostein translation and invasiveness of prostate cancer LNCaP cells. J Mol Med. 2008;86(3):313–22.
95. Khan AP, et al. Quantitative proteomic profiling of prostate cancer reveals a role for miR-128 in prostate cancer. Mol Cell Proteomics. 2010;9(2):298–312.
96. Boll K, et al. MiR-130a, miR-203 and miR-205 jointly repress key oncogenic pathways and are downregulated in prostate carcinoma. Oncogene. 2013;32(3):277–85.

97. Formosa A, et al. DNA methylation silences miR-132 in prostate cancer. Oncogene. 2013;32(1):127–34.
98. Tao J, et al. microRNA-133 inhibits cell proliferation, migartion and invasion in prostate cancer cells by targeting the epidermal growth factor receptor. Oncol Rep. 2012;27(6):1967–75.
99. Patron JP, et al. Mir-133b targets antiapoptotic genes and enhances death receptor-induced apoptosis. PLoS One. 2012;7(4):e35345.
100. Brase JC, et al. Circulating miRNAs are correlated with tumor progression in prostate cancer. Int J Cancer. 2011;128(3):608–16.
101. Brase J, et al. Serum microRNAs as non-invasive biomarkers for cancer. Mol Cancer. 2011;9(1):306.
102. Xiao J, et al. miR-141 modulates androgen receptor transcriptional activity in human prostate cancer cells through targeting the small heterodimer partner protein. Prostate. 2012;72(14):1514–22.
103. Huang S, et al. miR-143 and miR-145 inhibit stem cell characteristics of PC-3 prostate cancer cells. Oncol Rep. 2012;28(5):1831–7.
104. Clape C, et al. miR-143 interferes with ERK5 signaling and abrogates prostate cancer progression in mice. PLoS One. 2009;4(10):e7542.
105. Xu B, et al. miR-143 decreases prostate cancer cells proliferation and migration and enhances their sensitivity to docetaxel through suppression of KRAS. Mol Cell Biochem. 2011;350(1-2):207–13.
106. White NM, et al. The miRNA-kallikrein axis of interaction: a new dimension in the pathogenesis of prostate cancer. Biol Chem. 2012;393(5):379–89.
107. Chiyomaru T, et al. SWAP70, actin-binding protein, function as an oncogene targeting tumor-suppressive miR-145 in prostate cancer. Prostate. 2011;71(14):1559–67.
108. Fuse M, et al. Restoration of miR-145 expression suppresses cell proliferation, migration and invasion in prostate cancer by targeting FSCN1. Int J Oncol. 2011;38(4):1093–101.
109. Chen X, et al. MicroRNA145 targets BNIP3 and suppresses prostate cancer progression. Cancer Res. 2010;70(7):2728–38.
110. Lin S-L, et al. Loss of mir-146a function in hormone-refractory prostate cancer. RNA. 2008;14(3):417–24.
111. Xu B, et al. MiR-146a suppresses tumor growth and progression by targeting EGFR pathway and in a p-ERK-dependent manner in castration-resistant prostate cancer. Prostate. 2012;72(11):1171–8.
112. Murata T, et al. miR-148a is an androgen-responsive microRNA that promotes LNCaP prostate cell growth by repressing its target CAND1 expression. Prostate Cancer Prostatic Dis. 2010;13(4):356–61.
113. Fujita Y, et al. MiR-148a attenuates paclitaxel resistance of hormone-refractory, drug-resistant prostate cancer PC3 cells by regulating MSK1 expression. J Biol Chem. 2010;285(25):19076–84.
114. Wu Z, et al. Upregulation of miR-153 promotes cell proliferation via downregulation of the PTEN tumor suppressor gene in human prostate cancer. Prostate. 2013;73(6):596–604.
115. Tsuchiyama K, et al. Expression of MicroRNAs associated with Gleason grading system in prostate cancer: miR-182-5p is a useful marker for high grade prostate cancer. Prostate. 2013;73(8):827–34.
116. Rauhala HE, et al. miR-193b is an epigenetically regulated putative tumor suppressor in prostate cancer. Int J Cancer. 2010;127(6):1363–72.
117. Tong AW, et al. MicroRNA profile analysis of human prostate cancers. Cancer Gene Ther. 2008;16(3):206–16.
118. Barron N, et al. Biochemical relapse following radical prostatectomy and miR-200a levels in prostate cancer. Prostate. 2012;72(11):1193–9.
119. Szczyrba J, et al. Downregulation of Sec23A protein by miRNA-375 in prostate carcinoma. Mol Cancer Res. 2011;9(6):791–800.
120. Vallejo DM, Caparros E, Dominguez M. Targeting notch signalling by the conserved miR-8/200 microRNA family in development and cancer cells. EMBO J. 2011;30(4):756–69.

121. Viticchiè G, et al. MiR-203 controls proliferation, migration and invasive potential of prostate cancer cell lines. Cell Cycle. 2011;10(7):1121–31.
122. Saini S, et al. Regulatory role of mir-203 in prostate cancer progression and metastasis. Clin Cancer Res. 2011;17(16):5287–98.
123. Turner DP, et al. Mechanisms and functional consequences of PDEF protein expression loss during prostate cancer progression. Prostate. 2011;71(16):1723–35.
124. Tucci P, et al. Loss of p63 and its microRNA-205 target results in enhanced cell migration and metastasis in prostate cancer. Proc Natl Acad Sci U S A. 2012;109(38):15312–17.
125. Majid S, et al. MicroRNA-205-directed transcriptional activation of tumor suppressor genes in prostate cancer. Cancer. 2010;116(24):5637–49.
126. Gandellini P, et al. miR-205 Exerts Tumor-Suppressive Functions in Human Prostate through Down-regulation of Protein Kinase Cε. Cancer Res. 2009;69(6):2287–95.
127. Chan YC, et al. miR-210: the master hypoxamir. Microcirculation. 2012;19(3):215–23.
128. Leite KR, et al. Change in expression of miR-let7c, miR-100, and miR-218 from high grade localized prostate cancer to metastasis. Urol Oncol: Semin Ori Invest. 2011;29(3):265–9.
129. Sun T, et al. The altered expression of MiR-221/-222 and MiR-23b/-27b is associated with the development of human castration resistant prostate cancer. Prostate. 2012;72(10):1093–103.
130. Galardi S, et al. miR-221 and miR-222 expression affects the proliferation potential of human prostate carcinoma cell lines by targeting p27Kip1. J Biol Chem. 2007;282(32):23716–24.
131. Lin D, et al. The expression and clinical significance of GTP-binding RAS-like 3 (ARHI) and MicroRNA 221 and 222 in prostate cancer. J Int Med Res. 2011;39(5):1870–5.
132. Spahn M, et al. Expression of microRNA-221 is progressively reduced in aggressive prostate cancer and metastasis and predicts clinical recurrence. Int J Cancer. 2010;127(2):394–403.
133. Sun T, et al. The Role of microRNA-221 and microRNA-222 in androgen-independent prostate cancer cell lines. Cancer Res. 2009;69(8):3356–63.
134. Mavridis K, Stravodimos K, Scorilas A. Downregulation and prognostic performance of microRNA 224 expression in prostate cancer. Clin Chem. 2013;59(1):261–9.
135. Wei J-J, et al. Regulation of HMGA1 expression by microRNA-296 affects prostate cancer growth and invasion. Clin Cancer Res. 2011;17(6):1297–305.
136. Szczyrba J, et al. The microRNA profile of prostate carcinoma obtained by deep sequencing. Mol Cancer Res. 2010;8(4):529–38.
137. Bronisz A, et al. Reprogramming of the tumour microenvironment by stromal PTEN-regulated miR-320. Nat Cell Biol. 2011;14(2):159–67.
138. Lee KH, et al. MicroRNA-330 acts as tumor suppressor and induces apoptosis of prostate cancer cells through E2F1-mediated suppression of Akt phosphorylation. Oncogene. 2009;28(38):3360–70.
139. Epis MR, et al. Regulation of expression of deoxyhypusine hydroxylase (DOHH), the enzyme that catalyzes the activation of eIF5A, by miR-331-3p and miR-642-5p in prostate cancer cells. J Biol Chem. 2012;287(42):35251–9.
140. Wu Z, et al. Upregulation of microRNA-370 induces proliferation in human prostate cancer cells by downregulating the transcription factor FOXO1. PLoS One. 2012;7(9):e45825.
141. Yang K, Handorean AM, Iczkowski KA. MicroRNAs 373 and 520c are downregulated in prostate cancer, suppress CD44 translation and enhance invasion of prostate cancer cells in vitro. Int J Clin Exp Pathol. 2009;2(4):361–9.
142. Noonan EJ, et al. miR-449a causes Rb-dependent cell cycle arrest and senescence in prostate cancer cells. Oncotarget. 2010;1(5):349–58.
143. Noonan EJ, et al. miR-449a targets HDAC-1 and induces growth arrest in prostate cancer. Oncogene. 2009;28(14):1714–24.
144. Sikand K, et al. miR 488* inhibits androgen receptor expression in prostate carcinoma cells. Int J Cancer. 2011;129(4):810–9.
145. Ma S, et al. MicroRNA-616 induces androgen-independent growth of prostate cancer cells by suppressing expression of tissue factor pathway inhibitor TFPI-2. Cancer Res. 2011;71(2):583–92.

146. Saini S, et al. miRNA-708 control of CD44 + prostate cancer-initiating cells. Cancer Res. 2012;72(14):3618–30.
147. Gonzales JC, et al. Comparison of circulating microRNA 141 to circulating tumor cells, lactate dehydrogenase, and prostate-specific antigen for determining treatment response in patients with metastatic prostate cancer. Clin Genitourin Cancer. 2011;9(1):39–45.
148. Yaman Agaoglu F, et al. Investigation of miR-21, miR-141, and miR-221 in blood circulation of patients with prostate cancer. Tumor Biol. 2011;32(3):583–8.
149. Shen J, et al. Dysregulation of circulating microRNAs and prediction of aggressive prostate cancer. Prostate. 2012;72(13):1469–77.
150. Lodes MJ, et al. Detection of cancer with serum miRNAs on an oligonucleotide microarray. PLoS One. 2009;4(7):e6229.
151. Zhang H-L, et al. Serum miRNA-21: elevated levels in patients with metastatic hormone-refractory prostate cancer and potential predictive factor for the efficacy of docetaxel-based chemotherapy. Prostate. 2010;71(3):326–31.
152. Moltzahn F, et al. Microfluidic-based multiplex qRT-PCR identifies diagnostic and prognostic microRNA signatures in the sera of prostate cancer patients. Cancer Res. 2011;71(2):550–60.
153. Mahn R, et al. Circulating microRNAs (miRNA) in serum of patients with prostate cancer. Urology. 2011;77(5):1265.e9–1265.e16.
154. Selth LA, et al. Discovery of circulating microRNAs associated with human prostate cancer using a mouse model of disease. Int J Cancer. 2011;131(3):652–61.
155. Nguyen HC, et al. Expression differences of circulating microRNAs in metastatic castration resistant prostate cancer and low-risk, localized prostate cancer. Prostate. 2013;73(4):346–54.
156. Bryant RJ, et al. Changes in circulating microRNA levels associated with prostate cancer. Br J Cancer. 2012;106(4):768–74.

Chapter 15
MicroRNAs in Development and Progression of Ovarian Cancer

Monika Sangwan and Neetu Dahiya

Abstract Ovarian cancer is one of the most fatal gynecological malignancies responsible for the death of more than 14,000 patients in the USA annually. Even after advancements in treatment and diagnostic approaches, the 5-year survival of ovarian cancer patients is less than 30%. The development of early detection methods combined with personalized medicine is crucial for the effective treatment and improvement of overall survival of ovarian cancer patients. Altered expression of microRNAs (miRNAs) and their multifaceted biological activities in cancer make them important candidates for diagnosis, prognosis, and therapy of cancer. The presence of miRNA in body fluids such as blood and urine provides a unique opportunity for developing noninvasive methods for screening, evaluating drug response, and detecting recurrence of the disease. This chapter presents an overview of miRNA alterations involved in ovarian cancer development, progression, and drug resistance.

Keywords Ovarian cancer · miRNA · Development · Progression · Metastasis

1 Introduction

Ovarian cancer is the leading cause of death among gynecological malignancies. Three main factors, including diagnosis at later stage, development of drug resistance, and higher recurrence rate are responsible for the high mortality rate of this disease. A number of studies have been performed to identify novel biomarkers for early detection and therapeutic applications with little or no success. The discovery of a biomarker or set of biomarkers for epithelial ovarian cancer (EOC) has been further complicated due to the poor understanding of its origin. Earlier, EOC was supposed to originate from ovarian surface epithelium; however, recent studies have supported a nonovarian origin of ovarian cancer [25].

N. Dahiya (✉)
Department of Obstetrics and Gynecology, University of Virginia,
Charlottesville, VA 22908, USA
e-mail: ineetudahiya@yahoo.com

M. Sangwan
University of Toronto, Genetics and Development Division, 399 Bathurst st,
Toronto, Ontario M5T2S8, Canada

S. R. Singh, P. Rameshwar (eds.), *MicroRNA in Development and in the Progression of Cancer,* DOI 10.1007/978-1-4899-8065-6_15,

Despite improvements in better diagnostics and clinical methods, poor understanding of the mechanism of development and progression of ovarian cancer is a major hurdle in devising an effective strategy for the treatment of this cancer. The discovery of microRNAs (miRNAs) and their potential role in human diseases have opened a new possibility for understanding the underlying mechanism of the disease's development and possible therapeutic interventions for effective treatment. Several research groups, including ours, have examined altered expression of miRNAs in ovarian cancer and studied the role of miRNA targets in ovarian carcinogenesis [11, 16, 21, 32, 53]. This book chapter presents a summary of up-to-date knowledge on miRNAs in ovarian cancer development, progression, and drug resistance.

2 miRNA Expression in Ovarian Cancer

A number of investigations have been carried out to understand the expression pattern of miRNAs, their potential targets, and mechanisms of action in ovarian cancer pathogenesis. Dahiya et al. [11] identified the differential expression of 70 miRNAs in EOC compared to normal human ovarian surface epithelial cell line (HOSE B). Among 14 miRNAs commonly deregulated in ovarian cancer cell lines and tumor tissue samples, 3 miRNAs (miR-221, miR-146b, and miR-508) were upregulated and 11 miRNAs (let-7f, miR-106b, miR-134, miR-155, miR-21, miR-346, miR-422a, miR-424, miR-519a, miR-648, and miR-662) were downregulated. Transcriptional targets of selected miRNAs were determined after transfection of precursor miRNAs in ovarian cancer cell lines followed by microarray analysis. Comparison of altered messenger RNAs (mRNAs) with predicted targets for the selected miRNAs showed very little overlap, suggesting the importance of experimental approaches for identifying novel miRNA targets. Zhang et al. [53] applied multiple approaches to explore miRNA expression pattern in EOC. Thirty-five miRNAs were differentially expressed in cancer vs. immortalized normal ovarian surface epithelial cells, of which 31 miRNAs were downregulated. Comparison of late-stage vs. early-stage cancer identified 13 downregulated miRNAs, suggesting a tumor-suppressing role of these miRNAs in ovarian cancer progression. To understand the contribution of epigenetic modifications in miRNA deregulation in ovarian cancer, Zhang et al. [53] treated five EOC cell lines with DNA demethylating agents 5-aza-2′-deoxycytidine and histone deacetylase inhibitor. Sixteen miRNAs downregulated in EOC showed increased expression after treatment, suggesting an important role of epigenetic silencing in miRNA deregulation in EOC [53].

Iorio et al. [21] performed miRNA microarrays on 89 samples (15 normal ovarian samples, 69 ovarian malignant tumors, and 5 ovarian carcinoma cell lines), and identified miR-200a, miR-200b, miR-200c, and miR-141 as highly upregulated, and miR-199a, miR-140, miR-145, and miR-125b1 as highly downregulated in cancer compared to normal ovarian control samples. The deregulation of some of the miRNAs was associated with epigenetic modification, as treatment with demethylating agent 5-aza-2′-deoxycytidine induced expression of some of the miRNAs. Nam et al. [32] also found upregulation of miR-200a, miR-200b, miR-200c, and miR-141 and downregulation of miR-145, miR-199a, and miR-125b. In addition to the above studies, increased expression of miR-200 family members in ovarian cancers has been reported in several other research studies [45, 47].

Although miRNA microarray provides a great platform for identifying miRNAs involved in ovarian cancer development and progression, the technique is limited for known miRNAs only. In order to identify known and novel miRNAs which could be potential diagnostic or therapeutic markers, Wyman et al. [45] performed next-generation sequencing on small-RNA complementary DNA (cDNA) libraries prepared from RNA extracted from EOC specimens. The authors identified six novel (miR-2114, miR-449c, miR-2115, miR-2116, miR-2117, and miR-548q) and 39 already annotated miRNAs significantly altered in cancer vs. normal HOSE B cell lines. Some miRNAs were able to distinguish between different ovarian cancer subtypes such as miR-449a (serous specific), miR-499-5p/miR-375/miR-196a/miR-196b/miR-182 (endometroid specific), and miR-486-5p/miR-144/miR-30a/miR-199a-5p (clear cell specific).

Among the four main subtypes of ovarian cancer, i.e., serous, mucinous, endometroid, and clear cell carcinoma, although serous cancer is the most common cancer, clear cell carcinoma, which represents only ~6 % of ovarian cancer is of major concern due to worse prognosis than serous cancers. Nagaraja et al. [31] applied the Illumina next-generation sequencing technique to study miRNA expression in clear cell ovarian cancer cell lines and short-term cultures of HOSE B. A total of 54 differentially expressed miRNAs were identified in cancer vs. normal cell line: 21 miRNAs were upregulated, and 33 miRNAs were downregulated in cancer. miR-100 was most significantly downregulated in clear cell cancer cells, and forced expression of miR-100 in ovarian cancer cell line OVSAYO resulted in decreased expression of mTOR protein levels. Downregulation of mTOR protein also reduced the expression of downstream targets 4EBP1 and P70S6. In addition to finding miR-100-mediated regulation of mTOR and its downstream targets, the authors also identified a novel role of miR-100 in the sensitization of OVSAYO cells to the rapamycin analog RAD001. This study provides an alternative mechanism for overcoming chemoresistance of clear cell cancer cells to RAD001.

To further extend the studies beyond miRNA expression pattern and miRNA target determination, Kim et al. [22] performed miRNA microarray analysis on 103 fresh benign, borderline, and malignant ovarian tumor samples and correlated miRNA expression with clinicopathological features. Among several deregulated miRNAs in cancer, the expression of three miRNAs, namely miR-519a, miR-153, and miR-485-5p, was correlated with clinicopathological parameters. Reduced expression of miR-153 and miR-485-5p was associated with International Federation of Gynecology and Obstetrics (FIGO) grade 3. Reduced expression of miR-153 and increased expression of miR-519a were associated with advanced clinical stage. Higher expression of miR-519a was also associated with poor progression-free survival. This study provides a novel role of these three miRNAs as diagnostic and prognostic biomarkers for ovarian cancer.

Since ovarian cancer is mostly diagnosed at later stages of the disease, most of the ovarian cancer biomarker studies have been conducted on late-stage tumors. There is a dire need to understand molecular changes occurring at early stages during the development of ovarian cancer. Marchini et al. [30] did miRNA profiling of 144 FIGO stage I EOC samples isolated from patients treated with a platinum-based regimen. A set of 34 miRNAs associated with progression-free survival was able to divide patients into two clusters. Further analysis using quantitative polymerase chain reaction (qPCR) identified the upregulation of miR-214, miR-199a-3p, miR-199a-5p, miR-145, miR-200b, and miR-143 and the downregulation of miR-30a, miR-30a*, miR-30d, miR-200c, and miR-20a in tumor tissue samples from relapsers compared to nonrelapsers. Examination of miRNA expression pattern with patient survival identified an association of high levels of miR-199a-3p and miR-199a-5p with lower overall and progression-free survival.

3 miRNA Target Genes Often Deregulated in Ovarian Cancer

In addition to high-throughput miRNA profiling studies, several authors attempted to identify miRNAs regulating genes or pathways with known functions in ovarian cancer pathogenesis. Deregulation of kallikrein-related peptidases has been reported in several cancers, and the altered expression of six kallikrein members, namely KLK4, KLK5, KLK6, KLK7, KLK10, and KLK15, was found to be associated with poor prognosis of ovarian cancer [51]. To gain further insight into the regulation of KLK10, which has been found to be upregulated in ovarian cancer, White et al. [42] examined miRNA-mediated regulation of KLK10. Correlation of miRNA:mRNA expression in ovarian cancer patients identified several potential miRNAs targeting KLK10. Based on prediction analysis, three miRNAs (let-7f, miR-224, and miR-516a) targeting the 3′UTR of KLK10 were selected. Modulation of miRNA expression levels led to downregulation of KLK10 protein, validating miRNA-mediated regulation of KLK10 in ovarian cancers.

Another gene important in ovarian cancer is *Six1,* a homeobox protein frequently deregulated in cancer. To understand the regulation of *Six1* in ovarian cancers, Imam et al. [20] applied prediction algorithms to identify potential miRNAs regulating *Six1.* miR-185 was predicted by seven different algorithms (Target Scan, miRanda, mirTarget2, miTarget, PITA, RNA22, and RNA hybrid) and inversely correlated with *Six1* expression in multiple cancers. Further analysis indicated that miR-185 regulated *Six1* at the mRNA and protein levels. Downregulation of *Six1* by miR-185 resulted in a significant effect on colony formation in vitro and inhibited tumor growth in vivo.

Amyloid precursor proteins (APPs) are known for their role in Alzheimer's disease; however, recently APPs have been found to regulate cell adhesion, motility, and proliferation. Fan et al. [14] identified the role of APP in the regulation of

Table 15.1 List of miRNAs significantly altered in ovarian cancer and their validated targets

Name of miRNA	Validated target	Reference
miR-214	PTEN	[42]
miR-15a, miR-16	BMI1	[2]
miR-125a	ARID3B	[12]
miRNA-200	ZEB1, ZEB2	[1]
miR-187	DAB2	[6]
miR-20a	APP	[14]
miR-125	BCL3	[18]
miR-185	SIX1	[20]
miR-125b	BAK-1	[24]
let-7f, miR-224, miR-516a	KLK10	[42]
miR-128, miR-152	CSF1	[43]

proliferation and invasion in OVCAR3 cell line via miR-20a-mediated degradation by binding to 3′UTR. Overexpression of miR-20 increased long-term proliferation and invasion of ovarian cancer cells OVCAR3. Bhattacharya et al. [2] applied computational prediction methods along with literature survey to find miRNAs targeting BMI-1, which is upregulated in epithelial malignancies, including ovarian cancers. miR-15a and miR-16 were selected as putative miRNAs interacting with BMI-1. An inverse correlation was observed between BMI-1 and the expression of miR-15 and miR-16 in ovarian cancer samples. Ovarian cancer samples expressing a high level of miR-15a and miR-16 showed low levels of BMI-1, and low BMI-1-expressing samples showed high levels of miR-15a and miR-16. A list of miRNAs altered in ovarian cancer and their validated targets is given in Table 15.1.

4 Regulation of miRNA Expression in Ovarian Cancer

Although a number of studies have been conducted to determine differential expression of miRNA in ovarian cancer, the transcriptional or posttranscriptional regulation of miRNA expression in ovarian cancer is not very well characterized. Recently, p53 has been identified as a novel regulator of miRNAs, either regulating miRNA transcription or modulating miRNA functions. p53 is a very important molecule from the ovarian cancer standpoint. More than 90 % of ovarian cancers exhibit p53 mutations, and it plays a crucial role in ovarian cancer pathogenesis. Corney et al. [9] reported p53-mediated regulation of miR-34, which is downregulated in 100 % of EOCs with p53 mutation and in 93 % EOC expressing wild-type p53. Downregulation of miR-34b*/c was more pronounced in stage 4 tumors compared to stage 3 tumors. To find the functional significance of miR-34 in EOC, the authors modulated miRNA levels and found an inhibitory effect of miR-34 on proliferation, motility, and invasion, which were, in part, regulated via changes in mesenchymal epithelial transition factor (MET) levels [9]. In addition to p53 regulation, promoter methylation and copy number change were also associated with reduced expression of miR-34a in EOC samples. Another p53-regulated miRNA, miR-31, which is downregulated in EOC, was identified by Creighton et al. [10]. The authors identified a p53-dependent role of miR-31 in ovarian cancer proliferation and apoptosis regulation. Forced expression of miR-31 in cells expressing dysfunctional p53 inhibited proliferation and induced apoptosis, but cells expressing functional p53 were unaffected by miR-31 levels. The genomic deletion at 9p21 was also suggested as a factor responsible for reduced expression of miR-31 in ovarian cancer. Association of miR-31 expression with alterations in the p53 pathway and functions suggests potential applications of miR-31-based therapies in patients with a nonfunctional p53 pathway.

Transcriptional regulation of miR-200 family members was studied in detail by Knouf et al. [23]. The authors applied the chromatin immunoprecipitation approach to identify molecules associated with the promoter region of these miRNAs and found an important role of p73 and p63 in the transcriptional regulation of miR-200a, b, and miR-429.

Yin et al. [50] identified TWIST1-mediated regulation of miR-199a and miR-214 in CD44+ ovarian cancer stem cells. TWIST1 also regulates transition of TypeI/CD44+ cells to Type II/CD44− cells through the regulation of IKKβ/NF-κB and PTEN/AKT pathways. Further analysis identified TWIST-mediated regulation of IKKβ/NF-κB and PTEN/AKT pathways via regulation of miRNAs miR-199a and miR-214. Another regulator of NF-κB and AKT is the adenoviral type 5 E1A (E1A). E1A is an inhibitor of metastasis, and E1A-based therapies are under trial for cancer treatment; however, the mechanism of metastasis inhibition is not very clear. Su et al. [39] identified the role of E1A in the regulation of miR-520h, which ultimately inhibited expression of TWIST and downstream effectors NF-κB and AKT.

5 miRNA Mutations/Polymorphism in Ovarian Cancer

To understand the role of genetic variations in miRNA genes in familial ovarian cancer, Shen et al. [36] studied miRNAs targeting key ovarian cancer genes. Thirty miRNAs were selected based on the miRNA:mRNA interactions using computational algorithm. Among seven, novel genetic variations were identified for four miRNAs: one was found in miR-17 (C/T) and miR-188 (T/C), two in miR-29b-2 (C/T, A/T), and three in miR-191 (C/T, C/G, C/A). Further analysis of the RNA secondary structure for three variants of miR-191 genes showed a slight change in the secondary structure of the C/A variant with reduced stability of the A allele compared to the C allele. Transfection of precursor miRNA in SKOV3 cell lines showed about twofold higher expression of the C allele compared to the A allele.

Pastrello et al. [33] investigated miRNA expression in BRCA1/BRCA2-negative familial breast and ovarian cancer patients. Three miRNAs, namely miR-146a, miR-17, and miR-369, were selected based on their ability to target the *BRCA1/2* gene and genetic localization in chromosomal regions deleted in sporadic and familial breast and ovarian cancers. The genetic variations in these three miRNAs were not associated with predisposition to breast and ovarian cancer; however, a single nucleotide polymorphism (SNP) in miR-146a was found to be associated with early onset of the disease. In another study, a G-to-C polymorphism in the precursor miR-146a gene was found associated with age of diagnosis in breast cancer and ovarian cancer patients. The authors hypothesized that the presence of a variant allele predisposes patients to early onset of cancer due to high levels of mature miR-146a [33].

6 Role of miRNA in Ovarian Cancer Metastasis

Among a number of miRNAs involved in ovarian cancer pathogenesis, members of the miR-200 family play a critical role in ovarian cancer progression and metastasis. The miR-200 family members, which include miR-200a, miR-200b, miR-200c, miR-141, and miR-429, have been reported to regulate epithelial-to-mesenchymal transition (EMT), a crucial event for metastasis. The miRNA-200 family regulates the zinc finger transcriptional repressors ZEB1 and ZEB2, which are known to promote EMT via downregulation of E-cadherin and Lgl-2 [1].

The authors proposed a double-negative feedback loop between miR-200 and ZEB, which regulates mesothelial-to-epithelial transition during ovarian cancer progression. An inverse correlation was found between the expression of miR-200 and ZEB, showing higher expression of ZEBs (ZEB1 and ZEB2) and lower expression of miR-200 in normal HOSE cell line and reverse expression pattern in cancer cells. Another important regulator of EMT is the epidermal growth factor receptor (EGFR), which is known to mediate EMT via regulation of E-cadherin in ovarian cancer cells. Dahl et al. [12] identified negative regulation of EMT by miR-125. The role of miR-125a in the negative regulation of EMT is further validated by the forced expression of miR-125a, which induced mesenchymal-to-epithelial conversion of highly invasive ovarian cancer cells. The authors identified indirect regulation of miR-125 transcription by EGFR via regulation of transcription factor PEA3. To gain further insight into the mechanism of miR-125a-mediated MET, the authors applied miRNA target prediction approaches to predict targets of miR-125a and identified ARID3B as a potential target. The potential role of ARID3B in MET is further strengthened by its known role in mesenchymal development during embryogenesis. The inverse correlation in the expression pattern of miR-125a and ARID3B in ovarian cancer cells further validates the findings. Role of miR-125 in ovarian cancer development was also studied by Guan et al. [18]. Modulation of the miR-125b level was associated with cellular proliferation of ovarian cancer cells in vitro and tumor growth in vivo. miRNA target prediction using miRTarAS identified BCL3 as a potential target of miR-125, which was validated very well in ovarian cancer cells. Modulation of miR-125b levels in SKOV3 and ES2, two different ovarian cancer cell lines, was correlated with BCL3 expression at the protein level [18]. Chao et al. [6] found upregulation of miR-187 in ovarian cancer cell lines compared to normal ovarian surface epithelial cell line. Recently, miR-187 has been reported to regulate EMT via direct regulation of Dab2 and subsequent deregulation of E-cadherin and vimentin expression. Inverse correlation between miR-187 and Dab2 expression in ovarian cancer also supports regulation of Dab2 by miR-187. Downregulation of Dab2 has been reported in a variety of cancers, which suggests a tumor suppressor function; however, the Dab2-mediated induction of EMT suggests dual functions of Dab2 in tumor progression [6].

7 miRNA in Drug Resistance

Platinum-based regimens are the most effective chemotherapeutic drugs used for the treatment of ovarian cancer. The standard chemotherapeutic approach for ovarian cancer treatment involves a combination of a platinum compound, such as cisplatin or carboplatin, with a taxane, such as paclitaxel or docetaxel. Cisplatin is one of the best platinum adducts used for treating ovarian cancer. However, most of the patients treated with cisplatin and other platinum-based drugs develop drug resistance. For effective treatment, it is important to overcome the drug resistance developed in patients treated with cisplatin. Involvement of miRNAs in the development of drug resistance in ovarian cancer has been reported by several studies [8, 15, 24, 38, 46]. Increased expression of miR-125b was found to be associated with cisplatin resistance in ovarian cancer cell line C13, a cisplatin-resistant variant of cell line OV2008. Bcl2 antagonist killer 2 (Bak-1), a direct target of miR-125b, was downregulated in resistant cell lines, and forced downregulation of Bak-1 decreased cisplatin-induced apoptosis in OV2008 cell lines [24].

Most of the let-7 family members have been classified as tumor suppressors due to downregulation in cancers. It has been reported that let-7 expression improves response of cancer cells to platinum-based chemotherapy or radiation therapy [41]. Among 11 let-7 family members identified in humans, most of the let-7 members have been categorized as tumor suppressors except let-7a-3, which exhibits oncogenic functions [29]. Expression of let-7a was found to be associated with drug resistance in ovarian cancer patients. Patients with high expression of let-7a showed better survival compared to patients with low expression of let-7a when treated with cisplatin. However, upon administration of a combination treatment with cisplatin and paclitaxel, patients expressing high let-7a showed worst progression-free and overall survival. On the other hand, patients with low expression of let-7a showed better survival with combination therapy using cisplatin and paclitaxel compared to cisplatin or paclitaxel therapy [29].

To identify potential miRNAs in chemoresistance in EOC, Yang et al. [48] studied 69 late-stage EOC patient samples, which were either responsive or resistant to cisplatin therapy. A set of 34 miRNAs was identified in noncomplete response vs. complete response groups, of which 24 were overexpressed in the noncomplete response group and 10 were overexpressed in the complete response group. let-7i was the topmost downregulated miRNA in chemotherapy resistance group vs. chemotherapy responsive group [48].

To identify miRNAs involved in the development of drug resistance, Sorrentino et al. [38] performed miRNA expression profiling on A2780 wild-type ovarian cancer cell line and its cisplatin-resistant and paclitaxel-resistant counterparts. Six miRNAs, namely let-7e, miR-30c, miR-125b, miR-130a, and miR-335, were involved with chemoresistance, of which three miRNAs, namely miR-30c, miR-130a, and miR-335, were downregulated in all cell lines, whereas miR-125b was upregulated in all resistant cell lines except the paclitaxel-resistant cell line [38].

Another important modulator of drug resistance in ovarian cancer is miR-214, which has been found upregulated in late-stage and high-grade ovarian cancers. miR-214-induced cell survival and cisplatin resistance via downregulation of PTEN and activation of the AKT pathway. High expression of miR-214 in recurrent cancers compared to primary cancers treated with cisplatin suggested its direct role in the development of ovarian cancer drug resistance [47]. Low expression of miR-214 in primary cancer and inverse correlation with PTEN expression also suggest its potential application as a biomarker for cisplatin resistance in ovarian cancer patients.

The Nodal-Activin receptor-like kinase 7 (ALK7) pathway is also involved in chemosensitivity [45]. Ye et al. [48] observed an inverse correlation between miR-376 and ALK7 expression in serous ovarian cancer samples isolated from patients in the complete response group and incomplete response group. The incomplete response group exhibited higher expression of miR-376a than the complete response group. The direct role of miR-376 in drug response was confirmed by modulation of miRNA levels in cancer cells, which resulted in subsequent change in sensitivity to cisplatin [48].

Chen et al. [7] studied the expression pattern of IKKb and miR-199a in two types of EOCs categorized into type I and type II based on MyD88 expression. Type I cells express high IKKb and low miR-199a, which by TLR and TNF-alpha stimulation, activates NF-κB and makes cells more resistant to cytotoxic drugs. However, the similar mechanism was not observed in type II EOC cells, which express high miR-199a and low IKKb. Nam et al. [32] also found an association of low expression of miR-199a with the development of drug resistance in ovarian cancer.

Different patients respond differently to the same drug treatment. Based on the response, patients can be categorized into three different groups: (1) complete response, (2) partial response, (3) no response. Most of the patients who do not respond to the treatment exhibit inherent drug resistance. To understand the mechanism of inherent drug resistance, Boren et al. [3] performed miRNA profiling on 16 ovarian cancer cell lines and compared the miRNA expression pattern to drug sensitivity of these cell lines. Cells were treated with doxorubicin, gemcitabine, paclitaxel, docetaxel, topotecan, and cisplatin followed by cell proliferation assay. A total of 27 miRNAs were identified that were significantly associated with cells' response to chemotherapeutic treatments. In addition to some miRNAs associated with a specific drug, some miRNAs were associated with more than one drug, for example, miR-213, miR-181a, miR-181b (doxorubicin, gemcitabine), miR-99a and miR-514 (docetaxel, paclitaxel), miR-518-AS (docetaxel, topotecan), and miR-520f (docetaxel, cisplatin). To identify cisplatin-responsive genes, the authors did miRNA profiling in sensitive and resistant cell lines. A set of 81 miRNAs showing differential expression pattern in cisplatin-sensitive cell lines and its cisplatin-resistant counterparts were identified. Three miRNAs, namely miR-340, miR-381, and miR-520f, were also identified in the inherent resistance to platinum in 16 ovarian cancer cell lines [3].

Another important determinant for selecting a specific drug for treating patients is the length of disease-free survival time. Patients with long disease-free intervals between first-line platinum-based therapies can be retreated with platinum-based regimens as a second line of chemotherapy. However, patients with shorter disease-free intervals will most likely not respond well to the second-line platinum-based therapy, and therefore, are given an alternate chemotherapeutic regimen. In order to gain insight on the biomarkers that can be useful for predicting response to a second line of platinum-based chemotherapy, Eitan et al. [13] studied the miRNA expression pattern in stage I and stage III ovarian cancer patients treated with platinum-based chemotherapies. Nineteen miRNAs showed a differential expression pattern between stage I and stage II patient samples. Seven miRNAs were significantly differentially expressed in platinum-sensitive vs. platinum-resistant patient samples. Three miRNAs, namely miR-27a, miR-378, and miR-23a, also showed significant differences in sensitive and resistant samples from stage III patients. miR-449b was able to predict overall survival of patients where higher expression was associated with improved survival. On the other hand, higher expression of miR-21, miR-23a, miR-24-2*, and miR-27a were associated with poor prognosis.

One of the factors determining the response of patients to a particular drug depends on the expression of different proteins involved in an efflux of drugs from cancer cells. P-glycoprotein (P-gp), a member of adenosine triphosphate (ATP)-binding cassette transporters is involved in the efflux of drugs from cancer cells, has been found to be upregulated in ovarian cancer. Increased expression of P-gp was associated with the development of drug resistance in ovarian cancer [17]. To find the mechanism involved in P-gp-mediated drug resistance in ovarian cancer, Li et al. [27] examined the expression of miR-27a and P-gp in A2780 and A2780/Taxol cells and found increased expression of both miR-27 and P-gp in resistant A2780/Taxol cells compared to sensitive A2780 cell line. The direct role of miR-27 in drug response was further confirmed by modulation of miR-27a using an miRNA mimic which increased sensitivity of cells to Taxol drugs and also inhibited HIPK2; a Target scan predicted targets of miR-27a.

8 Functional Roles of miRNAs in Ovarian Cancer

The development and progression of cancer involve a number of cellular activities such as adhesion, apoptosis, invasion, motility, and survival. Recently, the potential of miRNAs in regulating these important cellular activities with respect to ovarian cancer pathogenesis has been reported by several authors [26, 43, 44, 49].

Overexpression of miR-200a, an important modulator of EMT, inhibited migration and invasion of cancer cells by direct regulation of ZEB2, a transcriptional repressor of E-cadherin [44]. Altered expression of miR-22, which regulates various cellular processes including cell growth, motility, apoptosis, and cell cycle have been implicated in different human cancers, including ovarian cancer [28, 34, 51] (Li et al. 2010). To understand the role of miR-22 in ovarian cancer metastasis, Li et al. [26] investigated miRNA expression in low-metastatic and high-metastatic cell lines derived from SKOV3. Among several differentially expressed miRNAs between low- and high-metastatic cell lines, miR-22 was selected for further analysis due to decrease in miR-22 expression in the late stage compared to the early stage of the cancer [26]. Introduction of miR-22 in SKOV3IP inhibited migration of cells, whereas treatment with an miR-22 inhibitor increased cancer cell migration suggesting an inverse correlation between miR-22 expression and cell migration in ovarian cancer cells.

ALK7 and its ligand Nodal are apoptotic inductors in ovarian cancer cells. Using computation prediction based on complementary sequences in 3′UTR of ALK7, miR-376 was identified as a potential miRNA targeting ALK7. A luciferase reporter assay also confirmed miR-376-mediated downregulation of ALK7. Modulation of miR-376 levels in ovarian cancer cells affected cellular proliferation, survival, and spheroid formation. Upregulation of miR-376 reduced the expression of ALK7 and subsequently increased the size of the spheroid, whereas inhibition using anti-miR-376 increased the expression of ALK7 and decreased the size of the spheroid. Inhibition of miR-376 also resulted in a low number of apoptotic cells, further supporting the role of ALK7 in regulation of apoptosis in ovarian cancer cells [48]. [43] reported the role of miR-128 and miR-152 in regulation of colony-stimulating factor 1 (CSF-1), which is involved in the progression of ovarian cancer via regulation of CSF-1R, a tyrosine kinase receptor [43]. In normal ovarian surface epithelium, there is very low expression of CSF-1; however, increased expression of CSF-1 and its receptor was associated with poor prognosis in ovarian cancer [5].

Shih et al [37] identified miRNA survival signature for advanced ovarian cancer. Twenty-nine miRNAs were found to be associated with overall survival. Negative correlation between miR-410 and miR-645 and overall survival was found in advanced stage ovarian cancer. The negative association was independent of tumor debulking and FIGO stage. Due to diagnosis at later stages of the disease, there are very few studies on early stages of ovarian cancer. Marchini et al. [30] analyzed 144 EOC stage I patients to identify miRNA survival signature. A set of 34 miRNAs was found to be associated with overall survival and 11 miRNAs (miR-214, miR-199a-3p, miR-199a-5p, miR-145, miR-200b, miR-30a, miR-30a*, miR-30d, miR-200c, miR-20a, and miR-143) showed differential expression between relapsers and nonrelapsers. Three miRNAs (miR-200c, miR-199a-3p, and miR-199a-5p) were associated with progression-free survival, overall survival, or both. However, downregulation of miR-200c was confirmed in relapsers compared to nonrelapsers, but upregulation of miR-199a-3 and miR-199-5p was not validated, suggesting the potential of miR-200c in predicting the survival of EOC stage I patients. Increased expression of miR-200a, b, c, miR-141, miR-18a, miR-93, and miR-429 and decreased expression of let-7b and miR-199a were correlated with decreased progression-free and overall survival of the patients [32]. A putative tumor suppressor region on chromosome 14, which harbors a cluster of miRNAs located in Dlk1–Gtl2 region domain, was identified in EOCs. Decreased expression of eight Dlk–Glt2 domain miRNAs is associated with high tumor proliferation and poor survival [53]. Decreased expression of let-7i was indicative of a shorter progression-free survival in comparison to patients with higher expression of let-7i, suggesting a strong role of let-7i as a therapeutic target or chemoresistant modulator in EOC treatment [48].

9 Circulating miRNAs in Ovarian Cancer

Detection of miRNAs in circulation has revolutionized their applications for developing noninvasive biomarkers for diagnosis, therapy, and clinical outcome of a treatment for human diseases. The technology is also promising for monitoring the recurrence of the disease. Taylor and Gercel-Taylor [40] identified increased expression of the same miRNAs that are overexpressed in cancers in exosomes isolated from patients' blood samples. Resnick et al. [35] applied the reverse transcriptase PCR (RTPCR) method to examine the potential of serum miRNAs as biomarkers for EOC. Twenty-one miRNAs were differentially expressed in serum isolated from cancer patients compared to serum from normal patients. Among eight miRNAs selected for further study, five miRNAs, namely miR-21, miR-92, miR-93, miR-126, and miR-29a, were significantly upregulated and three miRNAs, namely miR-155, miR-127, and miR-99b, were significantly downregulated in cancer patients' serum compared to normal controls.

To identify novel miRNA biomarkers for ovarian cancer screening and treatment, Hausler et al. [19] performed miRNA profiling on whole blood samples collected from ovarian cancer patients and healthy normal controls. Among 147 significantly altered miRNAs between cancer and normal, 4 miRNAs were identified after multiple comparisons by the Benjamini–Hochberg approach. Although the authors were able to identify an miRNA's signature in ovarian cancer patients, it was not sensitive and specific enough for the monitoring of disease progression. However, a combination of an miRNA signature with other markers could be used for improving existing screening methods.

10 Conclusion

In summary, we conclude that miRNAs are important regulators of ovarian cancer development and progression. Identification of miRNAs and their targets specific to different cancer subtypes, various stages, and grades will be crucial for understanding the disease pathogenesis. In addition, the involvement of miRNAs in the development of drug resistance, recurrence of the disease, and determination of survival makes them hold great potential for devising effective treatment strategies.

References

1. Bendoraite A, Knouf EC, Garg KS, Parkin RK, Kroh EM, O'Briant KC, Ventura AP, Godwin AK, Karlan BY, Drescher CW, Urban N, Knudsen BS, Tewari M. Regulation of miR-200 family microRNAs and ZEB transcription factors in ovarian cancer: evidence supporting a mesothelial-to-epithelial transition. Gynecol Oncol. 2010;116:117–25.
2. Bhattacharya R, Nicoloso M, Arvizo R, Wang E, Cortez A, Rossi S, Calin GA, Mukherjee P. MiR-15a and MiR-16 control Bmi-1 expression in ovarian cancer. Cancer Res. 2009;69:9090–5.
3. Boren T, Xiong Y, Hakam A, Wenham R, Apte S, Chan G, Kamath SG, Chen DT, Dressman H, Lancaster JM. MicroRNAs and their target messenger RNAs associated with ovarian cancer response to chemotherapy. Gynecol Oncol. 2009;113:249–55.
4. Chambers SK. Role of CSF-1 in progression of epithelial ovarian cancer. Future Oncol. 2009;5:1429–40.
5. Chambers SK, Kacinski BM, Ivins CM, Carcanqiu ML. Overexpression of epithelial macrophage colony-stimulating factor (CSF-1) and CSF-1 receptor: a poor prognostic factor in epithelial ovarian cancer, contrasted with a protective effect of stromal CSF-1. Clin Cancer Res. 1997;3:999–1007.
6. Chao A, Lin CY, Lee YS, Tsai CL, Wei PC, Hsueh S, Wu TI, Tsai CN, Wang CJ, Chao AS, Wang TH, Lai CH. Regulation of ovarian cancer progression by microRNA-187 through targeting Disabled homolog-2. Oncogene. 2010;31:764–75.
7. Chen R, Alvero AB, Silasi DA, Kelly MG, Fest S, Visintin I, Leiser A, Schwartz PE, Rutherford T, Mor G. Regulation of IKKbeta by miR-199a affects NF-kappaB activity in ovarian cancer cells. Oncogene. 2008;27:4712–23.
8. Cheng W, Liu T, Wan X, Gao Y, Wang H. MicroRNA-199a targets CD44 to suppress the tumorigenicity and multidrug resistance of ovarian cancer-initiating cells. FEBS J. 2012;279:2047–59.
9. Corney DC, Hwang CI, Matoso A, Vogt M, Flesken-Nikitin A, Godwin AK, Kamat AA, Sood AK, Ellenson LH, Hermeking H, Nikitin AY. Frequent downregulation of miR-34 family in human ovarian cancers. Clin Cancer Res. 2010;16:1119–28.
10. Creighton CJ, Fountain MD, Yu Z, Nagaraja AK, Zhu H, Khan M, Olokpa E, Zariff A, Gunaratne PH, Matzuk MM, Anderson ML. Molecular profiling uncovers a p53-associated role for microRNA-31 in inhibiting the proliferation of serous ovarian carcinomas and other cancers. Cancer Res. 2010;70:1906–15.
11. Dahiya N, Sherman-Baust CA, Wang TL, Davidson B, Shih IeM, Zhang Y, Wood W 3rd, Becker KG, Morin PJ. MicroRNA expression and identification of putative miRNA targets in ovarian cancer. PLoS One. 2008;3:e2436.
12. Dahl KD, Dahl R, Kruichak JN, Hudson LG. The epidermal growth factor receptor responsive miR-125a represses mesenchymal morphology in ovarian cancer cells. Neoplasia. 2009;11:1208–15.
13. Eitan R, Kushnir M, Lithwick-Yanai G, David MB, Hoshen M, Glezerman M, Hod M, Sabah G, Rosenwald S, Levavi H. Tumor microRNA expression patterns associated with resistance to platinum based chemotherapy and survival in ovarian cancer patients. Gynecol Oncol. 2009;114:253–9.
14. Fan X, Liu Y, Jiang J, Ma Z, Wu H, Liu T, Liu M, Li X, Tang H. miR-20a promotes proliferation and invasion by targeting APP in human ovarian cancer cells. Acta Biochim Biophys Sin (Shanghai). 2010;42:318–24.

15. Fu X, Tian J, Zhang L, Chen Y, Hao Q. Involvement of microRNA-93, a new regulator of PTEN/Akt signaling pathway, in regulation of chemotherapeutic drug cisplatin chemosensitivity in ovarian cancer cells. FEBS Lett. 2012;586:1279–86.
16. Giannakakis A, Sandaltzopoulos R, Greshock J, Liang S, Huang J, Hasegawa K, Li C, O'Brien-Jenkins A, Katsaros D, Weber BL, Simon C, Coukos G, Zhang L. miR-210 links hypoxia with cell cycle regulation and is deleted in human epithelial ovarian cancer. Cancer Biol Ther. 2008;7:255–64.
17. Gottesman MM, Ling V. The molecular basis of multidrug resistance in cancer: the early years of P-glycoprotein research. FEBS Lett. 2006;580:998–1009.
18. Guan Y, Yao H, Zheng Z, Qiu G, Sun K. MiR-125b targets BCL3 and suppresses ovarian cancer proliferation. Int J Cancer. 2011;128:2274–83.
19. Häusler SF, Keller A, Chandran PA, Ziegler K, Zipp K, Heuer S, Krockenberger M, Engel JB, Hönig A, Scheffler M, Dietl J, Wischhusen J. Whole blood-derived miRNA profiles as potential new tools for ovarian cancer screening. Br J Cancer. 2010;103:693–700.
20. Imam JS, Buddavarapu K, Lee-Chang JS, Ganapathy S, Camosy C, Chen Y, Rao MK. MicroRNA-185 suppresses tumor growth and progression by targeting the Six1 oncogene in human cancers. Oncogene. 2010;29:4971–9.
21. Iorio MV, Visone R, Di Leva G, Donati V, Petrocca F, Casalini P, Taccioli C, Volinia S, Liu CG, Alder H, Calin GA, Ménard S, Croce CM. MicroRNA signatures in human ovarian cancer. Cancer Res. 2007;67:8699–707.
22. Kim TH, Kim YK, Kwon Y, Heo JH, Kang H, Kim G, An HJ. Deregulation of miR-519a, 153, and 485–5p and its clinicopathological relevance in ovarian epithelial tumours. Histopathology. 2010;57:734–43.
23. Knouf EC, Garg K, Arroyo JD, Correa Y, Sarkar D, Parkin RK, Wurz K, O'Briant KC, Godwin AK, Urban ND, Ruzzo WL, Gentleman R, Drescher CW, Swisher EM, Tewari M. An integrative genomic approach identifies p73 and p63 as activators of miR-200 microRNA family transcription. Nucleic Acids Res. 2012;40:499–510.
24. Kong F, Sun C, Wang Z, Han L, Weng D, Lu Y, Chen G. miR-125b confers resistance of ovarian cancer cells to cisplatin by targeting pro-apoptotic Bcl-2 antagonist killer 1. J Huazhong Univ Sci Technol Med Sci. 2011;31:543–549.
25. Kurman RJ, Shih IeM. The origin and pathogenesis of epithelial ovarian cancer: a proposed unifying theory. Am J Surg Pathol. 2010;34:433–43.
26. Li J, Liang S, Yu H, Zhang J, Ma D, Lu X. An inhibitory effect of miR-22 on cell migration and invasion in ovarian cancer. Gynecol Oncol. 2010a;119:543–8.
27. Li Z, Hu S, Wang J, Cai J, Xiao L, Yu L, Wang Z. MiR-27a modulates MDR1/P-glycoprotein expression by targeting HIPK2 in human ovarian cancer cells. Gynecol Oncol. 2010b;119:125–30.
28. Ling B, Wang GX, Long G, Qiu JH, Hu ZL. Tumor suppressor miR-22 suppresses lung cancer cell progression through post-transcriptional regulation of ErbB3. J Cancer Res Clin Oncol. 2012;138:1355–61.
29. Lu L, Schwartz P, Scarampi L, Rutherford T, Canuto EM, Yu H, Katsaros D. MicroRNA let-7a: a potential marker for selection of paclitaxel in ovarian cancer management. Gynecol Oncol. 2011;122:366–71.
30. Marchini S, Cavalieri D, Fruscio R, Calura E, Garavaglia D, Nerini IF, Mangioni C, Cattoretti G, Clivio L, Beltrame L, Katsaros D, Scarampi L, Menato G, Perego P, Chiorino G, Buda A, Romualdi C, D'Incalci M. Association between miR-200c and the survival of patients with stage I epithelial ovarian cancer: a retrospective study of two independent tumour tissue collections. Lancet Oncol. 2011;12:273–85.
31. Nagaraja AK, Creighton CJ, Yu Z, Zhu H, Gunaratne PH, Reid JG, Olokpa E, Itamochi H, Ueno NT, Hawkins SM, Anderson ML, Matzuk MM. A link between mir-100 and FRAP1/mTOR in clear cell ovarian cancer. Mol Endocrinol 2010;24:447–63.
32. Nam EJ, Yoon H, Kim SW, Kim H, Kim YT, Kim JH, Kim JW, Kim S. MicroRNA expression profiles in serous ovarian carcinoma. Clin Cancer Res. 2008;14:2690–5.

33. Pastrello C, Polesel J, Della Puppa L, Viel A, Maestro R. Association between hsa-mir-146a genotype and tumor age-of-onset in BRCA1/BRCA2-negative familial breast and ovarian cancer patients. Carcinogenesis. 2010;31:2124–6.
34. Patel JB, Appaiah HN, Burnett RM, Bhat-Nakshatri P, Wang G, Mehta R, Badve S, Thomson MJ, Hammond S, Steeg P, Liu Y, Nakshatri H. Control of EVI-1 oncogene expression in metastatic breast cancer cells through microRNA miR-22. Oncogene. 2011;30:1290–301.
35. Resnick KE, Alder H, Hagan JP, Richardson DL, Croce CM, Cohn DE. The detection of differentially expressed microRNAs from the serum of ovarian cancer patients using a novel real-time PCR platform. Gynecol Oncol. 2009;112:55–9.
36. Shen J, DiCioccio R, Odunsi K, Lele SB, Zhao H. Novel genetic variants in miR-191 gene and familial ovarian cancer. BMC Cancer. 2010;10:47.
37. Shih KK, Qin LX, Tanner EJ, Zhou Q, Bisogna M, Dao F, Olvera N, Viale A, Barakat RR, Levine DA. A microRNA survival signature (MiSS) for advanced ovarian cancer. Gynecol Oncol. 2011;121:444–50.
38. Sorrentino A, Liu CG, Addario A, Peschle C, Scambia G, Ferlini C. Role of microRNAs in drug-resistant ovarian cancer cells. Gynecol Oncol. 2008;111:478–86.
39. Su JL, Chen PB, Chen YH, Chen SC, Chang YW, Jan YH, Cheng X, Hsiao M, Hung MC. Downregulation of microRNA miR-520h by E1A contributes to anticancer activity. Cancer Res. 2010;70:5096–108.
40. Taylor DD, Gercel-Taylor C. MicroRNA signatures of tumor-derived exosomes as diagnostic biomarkers of ovarian cancer. Gynecol Oncol. 2008;110:13–21.
41. Weidhaas JB, Babar I, Nallur SM, Trang P, Roush S, Boehm M, Gillespie E, Slack FJ. MicroRNAs as potential agents to alter resistance to cytotoxic anticancer therapy. Cancer Res. 2007;67:11111–6.
42. White NM, Chow TF, Mejia-Guerrero S, Diamandis M, Rofael Y, Faragalla H, Mankaruous M, Gabril M, Girgis A, Yousef GM. Three dysregulated miRNAs control kallikrein 10 expression and cell proliferation in ovarian cancer. Br J Cancer. 2010;102:1244–53.
43. Woo HH, László CF, Greco S, Chambers SK. Regulation of colony stimulating factor-1 expression and ovarian cancer cell behavior in vitro by miR-128 and miR-152. Mol Cancer. 2012;11:58–68.
44. Wu Q, Guo R, Lin M, Zhou B, Wang Y. MicroRNA-200a inhibits CD133/1 + ovarian cancer stem cells migration and invasion by targeting E-cadherin repressor ZEB2. Gynecol Oncol. 2011;122:149–54.
45. Wyman SK, Parkin RK, Mitchell PS, Fritz BR, O'Briant K, Godwin AK, Urban N, Drescher CW, Knudsen BS, Tewari M. Repertoire of microRNAs in epithelial ovarian cancer as determined by next generation sequencing of small RNA cDNA libraries. PLoS One 2009;4:e5311.
46. Xu G, Zhong Y, Munir S, Yang BB, Tsang BK, Peng C. Nodal induces apoptosis and inhibits proliferation in human epithelial ovarian cancer cells via activin receptor-like kinase 7. Clin Endocrinol Metab. 2004;89:5523–34.
47. Yang H, Kong W, He L, Zhao JJ, O'Donnell JD, Wang J, Wenham RM, Coppola D, Kruk PA, Nicosia SV, Cheng JQ. MicroRNA expression profiling in human ovarian cancer: miR-214 induces cell survival and cisplatin resistance by targeting PTEN. Cancer Res. 2008a;68:425–33.
48. Yang N, Kaur S, Volinia S, Greshock J, Lassus H, Hasegawa K, Liang S, Leminen A, Deng S, Smith L, Johnstone CN, Chen XM, Liu CG, Huang Q, Katsaros D, Calin GA, Weber BL, Bützow R, Croce CM, Coukos G, Zhang L. MicroRNA microarray identifies Let-7i as a novel biomarker and therapeutic target in human epithelial ovarian cancer. Cancer Res. 2008b;68:10307–14.
49. Ye G, Fu G, Cui S, Zhao S, Bernaudo S, Bai Y, Ding Y, Zhang Y, Yang BB, Peng C. MicroRNA 376c enhances ovarian cancer cell survival by targeting activin receptor-like kinase 7: implications for chemoresistance. J Cell Sci. 2011;124:359–68.

50. Yin G, Chen R, Alvero AB, Fu HH, Holmberg J, Glackin C, Rutherford T, Mor G. TWISTing stemness, inflammation and proliferation of epithelial ovarian cancer cells through MIR199A2/214. Oncogene. 2010;29:3545–53.
51. Yousef GM, Polymeris ME, Yacoub GM, Scorilas A, Soosaipillai A, Popalis C, Fracchioli S, Katsaros D, Diamandis EP. Parallel overexpression of seven kallikrein genes in ovarian cancer. Cancer Res. 2003;63:2223–7.
52. Zhang J, Yang Y, Yang T, Liu Y, Li A, Fu S, Wu M, Pan Z, Zhou W. MicroRNA-22, downregulated in hepatocellular carcinoma and correlated with prognosis, suppresses cell proliferation and tumourigenicity. Br J Cancer. 2010;103:1215–20.
53. Zhang L, Volinia S, Bonome T, Calin GA, Greshock J, Yang N, Liu CG, Giannakakis A, Alexiou P, Hasegawa K, Johnstone CN, Megraw MS, Adams S, Lassus H, Huang J, Kaur S, Liang S, Sethupathy P, Leminen A, Simossis VA, Sandaltzopoulos R, Naomoto Y, Katsaros D, Gimotty PA, DeMichele A, Huang Q, Bützow R, Rustgi AK, Weber BL, Birrer MJ, Hatzigeorgiou AG, Croce CM, Coukos G. Genomic and epigenetic alterations deregulate microRNA expression in human epithelial ovarian cancer. Proc Natl Acad Sci. 2008;105:7004–9.

Part III
Role of MicroRNA and Potential Cancer Therapeutics

Chapter 16
MicroRNA and Cancer Drug Resistance

Daohong Chen and Harikrishna Nakshatri

Abstract The last decade witnessed exponential growth in our understanding of the role of microRNAs (miRNAs) in cancer. In addition to a clearly defined role in cancer initiation and progression, miRNAs are now believed to determine sensitivity to therapy. The post-genomic era has seen rapid growth in our understanding of drug resistance mechanisms at the genomic level including genomic aberrations involving miRNA clusters and in the development of targeted therapies. Aberrant activity of a single miRNA can influence multiple signaling pathways associated with therapeutic response because it can target multiple proteins. Although it is difficult to pinpoint a single downstream effector of a miRNA, studies focusing on known oncogenes and tumor suppressors targeted by miRNA as well as advanced bioinformatics capabilities have enabled the discovery of integrated mRNA–miRNA–protein circuitry in cancer cells that govern multiple aspects of cancer including drug sensitivity. These studies have provided evidence for specific miRNAs targeting signaling molecules involved in drug transport, drug metabolism, synthesis of ligands for receptors, drug-induced DNA damage response and apoptotic pathways, and growth factor receptors/kinases/phosphatases that form the backbone of targeted therapies. Extensive knowledge of miRNA expression pattern and targets has allowed clinical translation of miRNAs as prognostic and predictive markers of therapies. miRNAs expressed in cancer cells govern signaling not only in cancer cells but also in neighboring cells and distant organs because they are incorporated into secretory microvesicles, remain stable in body fluids, and cross plasma membrane or cell–cell junctions. These properties of miRNAs have generated considerable interest in developing means to restore "normal miRNA patterns" in cancer through therapeutic approaches. Indeed, miRNA-based therapies are already in phase II clinical trial for hepatitis C infection, confirming feasibility of this approach. Therefore, miRNA or miRNA antagomir-based therapies are likely the next-revolutionary therapeutic approach to combat cancer.

H. Nakshatri (✉)
Biochemistry & Molecular Biology, Indiana University School of Medicine, C218E,
980 West Walnut St, Indianapolis, IN 46202, USA
e-mail: hnakshat@iupui.edu

D. Chen · H. Nakshatri
Departments of Surgery, Indiana University School of Medicine,
Indianapolis, IN 46202, USA

S. R. Singh, P. Rameshwar (eds.), *MicroRNA in Development and in the Progression of Cancer*, DOI 10.1007/978-1-4899-8065-6_16,

Keywords MicroRNA · Drug transporters · Drug resistance · Apoptosis · Signaling pathways · Biomarkers · Therapy

1 Introduction

MicroRNA (miRNA) was originally discovered as short noncoding RNAs consisting of 19–25 nucleotides in *Caenorhabditis elegans* two decades ago [81]. Since then, more than 1,000 miRNAs have been identified among various species including mammals, and cumulated studies have demonstrated that miRNA plays a crucial role in regulation of numerous biological processes including response to therapies [73, 75, 147]. Within the past two decades, the number of publications on miRNAs has crossed 20,000. There are more than 500 publications on miRNAs and drug resistance in PuBMed.

RNA polymerase II transcribes intergenic or intragenic miRNAs as long primary transcripts with a hairpin structure (pri-miRNAs) in the nucleus. Most often, intragenic miRNAs are transcribed as part of the transcript corresponding to parental mRNA-encoding gene. The nuclear RNAse III Dorsa subsequently processes pri-miRNA into 70–100-nucleotide-long precursor molecule known as pre-miRNA [82]. Alternatively, pre-miRNA can be generated independently of Dorsa through the splicing machinery (miRtron pathway) [120]. Through an exportin-5-mediated mechanism, the pre-miRNAs are exported to the cytoplasm and then processed into double-strand RNA of approximately 22 nucleotides in length by Dicer (a cytoplasmic RNAse III)-containing protein complex. One of the resulting double miRNA strands is selected as a guide strand according to thermodynamic features and the complementary strand is degraded to generate single-stranded mature miRNA [56, 67].

Although the specific biological roles have so far been elucidated for only a small fraction of identified miRNAs through genomic approaches, the fundamental mechanistic mode of miRNA functioning is well established. The mature miRNA is incorporated into a molecular group known as RNA-induced silencing complex (RISC), and then the miRNA-containing RISC binds to target mRNAs. Consequently, mRNA is degraded if there is complete complementarity or fails to get translated into protein if there is partial complementarity [56, 67, 73]. Emerging data suggest that miRNA targeting can also go beyond mRNA to ribonucleoprotein and even the promoter regions of mRNA-encoding genes in a RISC-independent manner [4, 82]. To date, miRNAs have been computationally predicted to be able to target >60 % of the protein-coding genes that are involved in development, metabolism, cell proliferation, differentiation, and apoptosis [28]. Since these biological processing pathways significantly contribute to cancer progression and therapeutic response, it is understandable that miRNAs represent an important player in the regulation of cancer drug resistance. However, it is often difficult to pinpoint proteins downstream of a miRNA that mediate response to treatment because of multiple targets of a single miRNA and mRNA-containing

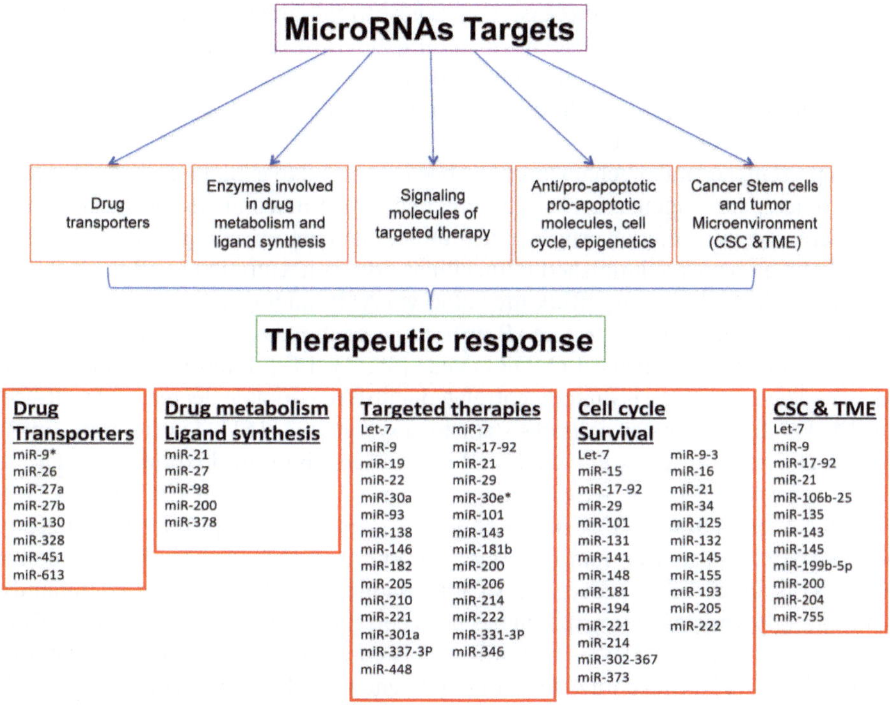

Fig. 16.1 Schematic view of miRNA targets in drug resistance mechanisms. miRNAs shown to participate in drug sensitivity are indicated

sequences that are complementary to multiple miRNAs. Moreover, there are significant data demonstrating miRNAs as a part of regulatory loop that maintains homeostasis. Therefore, these aspects of miRNA biology and functions have to be taken into consideration before labeling an miRNA as a oncomir or a tumor suppressor and hence targets for therapy. Figure 16.1 provides an overview of the topics that are discussed in this chapter.

2 MicroRNA Deregulation and Response to Therapy: Drug Transporters

Cancer therapy has witnessed a myriad of changes over the years including in traditional cytotoxic therapies and hormonal therapy. Furthermore, targeted therapies involving different combinations and sequence of administration of drugs based on genomic makeup of tumor on a case-by-case basis will likely become a common clinical practice. Effective cancer therapy has so far been able to eliminate macroscopic tumor either at primary site or at common distal sites. However, even if showing an impressive response to treatment initially, the majority of cancer patients will relapse since small cohorts of tumor cells can survive in cryptic anatomic loci and exhibit up to 90 % resistance to one or more therapeutic compounds [34, 73]. As such, drug resistance was and still is a major challenge in front of cancer treatment.

At present, several mechanisms have been proposed to explain drug resistance to chemotherapeutic agents, including deregulated pharmacokinetic processing such as decreased intracellular drug concentrations mediated by drug transporters and metabolic enzymes. Diminished intracellular drug concentration is often attributed to energy-dependent drug efflux mediated via ATP-binding cassette (ABC) transporters, in which P-glycoprotein (P-gp) is one of the classical members encoded by the multidrug-resistance 1 (MDR1) gene [38, 73, 141]. It has been reported that elevated levels of P-gp is directly correlated with a remarkable decrease of miR-451 in MCF-7 breast cancer cells resistant to doxorubicin [161]. Furthermore, miR-451-transfected MCF-7 cells exhibited higher sensitivity to doxorubicin than the vector control [72]. Likewise, breast cancer resistance protein (BCRP/ABCG2), which is a target of miR-328, shares drug substrates with P-gp [109]. Another group of the ABC transporter family termed multidrug resistance-associated proteins (MRPs) is involved in processing of glutathione (GSH)-conjugated compounds. Interestingly, resistance to cisplatin has been correlated with MRP2-mediated efflux of cisplatin–GSH complexes, and miR-489 is currently proposed as a regulator of MRP2 [113, 132]. Since miR-451, miR-328, and miR-489 downregulate the levels of proteins that pump chemotherapeutics out of cancer cells, one can anticipate levels of these miRNAs to be lower in cancer cells compared to normal cells. Indeed, downregulation of miR-451 in breast cancer is associated with endocrine resistance [5]. Similarly, miR-489 levels are lower in breast cancer cells that are resistant to the antiestrogen tamoxifen [100].

miRNAs can also regulate drug transporters indirectly by targeting transcription regulators of drug transporters. For example, peroxisome proliferator-activated receptor gamma (PPARγ) regulates the expression of ABCG1 [136]. Similarly, liver X receptor (LXR) controls the expression of ABCB1 and ABCC1. miR-27a, miR-27b, and miR-130 target PPARγ [68, 83, 84], whereas miR-613 targets LXRα [106]. LXR itself is involved in the expression of ABCA1 through miRNA as it suppresses the expression of miR-26, which targets ABCA1. Therefore, a complex interplay between miRNAs, drug transporters, and transcription regulators of drug transporters will ultimately determine drug availability in a tumor cell. Consequently, it is a combination of miRNAs but not a single miRNA's influence therapeutic response. To aid in the process of determining significance of overexpression or underexpression of multiple miRNAs in cancer, we have recently developed an online tool to evaluate prognostic relevance of miRNAs in multiple cancers [37]. Such tools may help in treatment decisions in future, and drugs that are subject to drug efflux pumps-mediated exclusion can be avoided based on miRNA expression pattern in tumors.

3 MicroRNAs and Enzymes Involved in Metabolizing Drugs and Synthesis of Ligands

Besides drug transporter molecules, enzymes involved in metabolic clearance of drugs can influence sensitivity to therapeutic molecules. Rapid drug metabolism resulting from miRNA-driven enzyme induction is correlated with decreased response to cancer treatment. For example, downregulation of miR-27 is linked to higher cytochrome P450 enzymes (CYP1B1/CYP3A4) and consequently enhanced doxorubicin metabolism [54, 72, 73]. Similarly, PPARα is an important transcription factor that regulates the expression of genes encoding xenobiotic and lipid-metabolizing enzymes, androstane receptor, CYP3A, and CYP2C [123]. miR-21 and miR-27b negatively regulate PPARα expression [66]. Therefore, miR-21 and miR-27b levels in tumors can influence the rate at which cancer cells clear drugs.

miRNAs can also play a role in controlling the expression levels of enzymes that provide active ligands to signaling pathway molecules. For example, aromatase (CYP19A1) converts testosterone to estrogen in postmenopausal women and drives the growth of estrogen receptor (ER)-positive breast cancers. Therefore, aromatase inhibitors are the standard breast cancer therapy. Elevated aromatase expression due to miRNA deregulation can potentially contribute to resistance to aromatase inhibitors. In this respect, miR-378 and miR-98 have been shown to target CYP19A1 [153]. Similarly, miR-200a controls progesterone receptor activity by regulating the expression of progesterone-metabolizing enzyme 20α-hydroxysteroid dehydrogenase (20α-HD) [148]. miR-200a directly represses STAT5a, a transcriptional repressor of 20α-HD, which leads to elevated 20α-HD levels, diminished progesterone levels, and reduced progesterone receptor activity.

4 miRNAs and Signaling Molecules of Targeted Therapy

In contemporary therapeutics of oncology, targeted treatment has been hailed as a major breakthrough and has been heralded as one of the successful outcomes of human genome project. Genomic imprints of tumors obtained through next-generation sequencing have allowed identification of driver mutations in cancers and development of molecules that selectively target genomic aberrations unique to the tumor, thereby minimizing side effects to normal tissues. Imatinib (Gleevec) is the first successful targeted therapy as it inhibits kinase activity of break point cluster–Abelson (BCR-ABL) fusion kinase that is expressed in chronic myeloid leukemia. Resistance to imatinib is associated with deregulated miRNAs. For example, 19-miRNA expression profile has been shown to predict response to imatinib [124]. Similarly, miR-138 is described as a predictor of response to imatinib as it represses BCR-ABL expression. Autophagy is a mechanism of imatinib resistance and miR-30a has been shown to enhance sensitivity to imatinib by blocking autophagy machinery [155].

ER is one of the oldest molecules of targeted therapy; Cole et al. first published in 1971 the use of tamoxifen as an antiestrogen to block the activity of ER in breast cancer [19]. More than 70 % of breast cancers express ER and tamoxifen has been effective in reducing recurrence of these tumors. Response of tumors to tamoxifen is dependent on the expression and activity of ER. Only those tumors that are dependent on estrogen working through ERα respond to tamoxifen. miR-101, miR-206, and miR-221/222 repress ER expression, and overexpression of these miRNAs is associated with resistance to antiestrogens [116]. This miRNA cluster is upregulated in ERα-negative breast cancer [71]. We and others have shown that estrogen-ERα-miRNAs form a regulatory loop where E2:ERα induces the expression of several miRNAs, which in turn regulate ERα levels or the levels of downstream effectors of ERα [6, 11].

The transcription coregulator AIB1 (also called SRC-3 or NCOA3), which is essential for transactivation by multiple transcription factors including ERα, is amplified or overexpressed in several cancers [92]. Co-overexpression/amplification of AIB1 and ERBB2 oncogene in ERα-positive breast cancer is associated with poor response to antiestrogens [105]. miR-206 targets not only ERα but also AIB1 [1]. Similarly, miR-17-5p regulates AIB1 levels [46]. miR-346 targets the receptor-interacting protein 140 (RIP140), a transcriptional corepressor downregulated in endocrine-resistant cell lines, and is required for E2-mediated repression of genes [139]. Collectively, miRNAs that regulate the expression of ERα and its co-regulatory molecules as well as miRNAs induced/repressed by E2 can influence response to antiestrogens. Unfortunately, these miRNAs cannot be exploited as therapeutic targets as a recent large-scale miRNA profiling of ERα-positive breast tumors failed to identify a specific set of miRNAs that can predict response to tamoxifen treatment [94].

The pathways of epidermal growth factor receptor (EGFR) or/and human epidermal growth factor 2 (ERBB2, also called HER2) are major therapeutic targets in modern oncology. Overexpression, mutational activation, and/or amplification of genes in EGFR/HER2 pathways are frequently observed in multiple cancers [52]. Several antibodies and small molecule inhibitors have been developed to inhibit EGFR/HER2 signaling cascade and have already delivered clinical benefits to cancer patients. Nevertheless, more than 50 % of cancer patients who receive EGFR/HER2-targeted therapies eventually fail to respond due to comprehensive drug resistance mechanisms including continued overexpression of the drug target and activation/mutation of downstream kinases/phosphatases [73, 115]. For example, PI3K/AKT/PTEN (phosphatidylinositide 3-kinase/protein kinase B/phosphatase and tensin homolog) pathway is one of the major effectors of EGFR/HER2 [52]. This pathway is frequently mutated in cancer and multiple miRNAs target this pathway. HER3, a dimerizing partner of HER2, is often upregulated in cells resistant to EGFR- and HER2-targeted therapies [126]. In general, resistance mechanisms to anticancer kinase inhibitors involve growth factor-driven activation of redundant growth factor receptors [149]. Therefore, resistance to EGFR/HER2-targeted therapies may involve multiple miRNAs that target EGFR/HER2, their downstream effectors, and other growth factor receptors that can activate downstream effectors of

EGFR/HER2. For example, miR-7 and miR-331-3p directly regulate the expression of EGFR and HER2, respectively [35]. miR-205 regulates HER3 expression [57]. miR-21 upregulation is associated with resistance to trastuzumab, a HER2-targeted therapy [36].

Multiple miRNAs have been shown to target PTEN, a tumor suppressor that negatively regulates PI3K/AKT pathway. miR-21, miR-221, miR-222, miR-301, miR-93, miR-214, miR-19, and miR-17-92 cluster target PTEN either directly or indirectly [32, 117, 130]. Elevated expression of any of these miRNAs in cancer can potentially dampen the effects of EGFR/HER2-targeted therapies by elevating basal PI3K/AKT activity levels. Downregulation of specific miRNAs can also contribute to drug resistance. For example, reduced expression of miR-22 leads to resistance to paclitaxel in cancer cells with p53 mutation, which correlates with lower PTEN levels [87]. Interestingly, we reported specific downregulation of miR-22 in metastatic cancer cells compared to parental tumor cells, which provides an explanation for resistance of metastatic tumors to therapies [110]. Apart from PI3K/AKT/PTEN, miRNAs targeting Ras pathway can influence response to EGFR-based therapies. For example, KRAS-interacting miR-143, which is frequently downregulated in colorectal cancer, may influence sensitivity to 5-fluorouracil (5-FU) and outcomes of EGFR-targeted agents. A significant lower progression-free rate upon anti-EGFR treatment was observed in the cancer patients with miR-143 downregulation [112].

The signal-induced latent transcription factors such as nuclear factor kappa B (NF-κB) and signal transducer and activator of transcription 3 (STAT3) are considered to play a major role in therapeutic resistance. Inhibitor-of-kappa B (IκB) sequesters NF-κB in the cytoplasm of a normal cell and growth factor- or cytokine-induced phosphorylation and degradation of IκB lead to nuclear translocation of NF-κB. A number of groups including our group have shown constitutive activation of NF-κB in cancer and activated NF-κB contributing to chemoresistance [62, 103]. More than 300 publications report an association between NF-κB and miRNA signaling. miRNAs acting either upstream or downstream of NF-κB have been shown modulate response to therapy. For example, miR-30e* suppresses the expression of IκBα to increase NF-κB activity [59]. miR-182 increases NF-κB activity by targeting cylindromatosis (CYLD), an NF-κB-negative regulator [133]. NF-κB/miR-448 autoregulatory loop has been shown to be involved in chemotherapy-induced epithelial-to-mesenchymal transition (EMT) and tumor progression. We had demonstrated NF-κB inducing the expression of zinc finger E-box binding homeobox 1 (ZEB1) and ZEB2, which are the major negative regulators of miR-200 family miRNAs with antitumor and chemosensitizing activities [15, 43, 86]. NF-κB is involved in induction of multiple miRNAs including miR-146, miR-143, miR-21, and miR-301a and repression of miR-29 and Let-7 [95]. Although NF-κB-mediated resistance to chemotherapy was initially thought to be due to NF-κB-dependent expression of anti-apoptotic proteins such as cIAP-1 and cIAP-2 [42, 74], miRNAs playing a role in NF-κB-mediated chemoresistance is another possibility.

STAT3 is another signal-activated transcription factor involved in chemoresistance and is a major transcription factor in the miRNA circuitry. miR-337-3p increases taxane sensitivity by targeting STAT3 [26]. Similarly, miR-9 reduces STAT3 levels and prevents mesenchymal differentiation of glioblastoma [69]. STAT3 increases the expression of miRNAs such as miR-21, miR-17, miR-181b, and miR-19a [156]. STAT3 also suppresses the expression of miR-200c, which plays a role in chemosensitivity [119]. STAT3-induced miR-17 plays a role in conferring resistance to MEK inhibitors by blocking the expression of pro-apoptotic BIM [21]. Thus, miRNAs that regulate and are regulated by STAT3 can potentially influence sensitivity to chemotherapy.

Hypoxia plays a significant role in determining sensitivity to chemotherapy. Apart from limited access to drugs due to limited blood supply, hypoxia-induced signals directly confer resistance to chemotherapy. The hypoxia-inducible transcription factor 1 (HIF1) transcriptionally upregulates oncogenic miR-210, which has multiple roles in cancer including chemoresistance [12]. For example, elevated circulating miR-210 levels correlate with resistance to trastuzumab therapy in breast cancer [60].

5 miRNA in Pro-Survival Response

Tumor growth is characterized by increased cell proliferation or/and diminished apoptosis. Many cytotoxic drugs such as doxorubicin and cisplatin exert their antitumor efficacy by damaging the genetic material of cells and then suppressing the capability of cancer cells to divide. Upon exposure to chemotherapeutic drugs, several cellular/molecular responses are activated including those involved in cell cycle, apoptosis, DNA repair, and epigenetic regulation. Cell death occurs when damage exceeds the cellular repair capabilities. Although the pro- and antigrowth programs are finely tuned and delicately balanced by miRNAs in normal cells, genomic abnormalities in cancer cells may shift the balance toward cell survival.

Cyclin-dependent kinases (CDKs) control cell cycle progression and are subjected to modulation by the tumor suppressors such as retinoblastoma protein and p53. Normally, p53 is able to detect DNA damage and then suppresses CDKs to halt cell cycle progression through activation of CDK inhibitors such as p21. Loss or mutation in p53 has been linked to diminished sensitivity to DNA damage and enhanced drug resistance [85]. One underlying mechanism of lower p53 in cancer cell is the upregulation of miR-125, which binds to the 3′-untranslated region (UTR) of p53 gene and blocks translation in cells that are resistant to doxorubicin [72, 79]. miR-17-92 cluster, which includes miR-17/20a, can also reduce overall p53 activity by targeting p21 [45]. As p21 normally interacts with CDK2 to prevent cells from entering the S-phase, loss of p21 enables cancer cells to progress through the cell cycle with DNA damage. In addition, the G1/S checkpoint is affected by several other miRNAs dysregulated in drug-resistant cancers, which include miRNA-221/222/214 targeting PTEN and miRNA-16/29/34 targeting CDK6 [72, 88, 91, 113].

Selective CDK4/6 inhibitor PD0332991 has shown clinical promise and is currently under phase I/II studies in multiple cancers including breast cancer [31]. Since these studies are in an early phase, it is not known whether resistance to this drug will eventually be observed. Nonetheless, miRNAs that target CDK4/6, p21, p27, and downstream effectors such as retinoblastoma protein and E2F family transcription factors will likely determine response to this targeted therapy. For example, miR-302 inhibits tumorigenicity by targeting CDK4/6 [89]. miR-302–367 cluster has been shown to have a tumor suppressor activity in glioma, cervical cancer, and sensitize breast cancer cells to radiation therapy [30]. Therefore, tumors with lower levels of miR-302 may show resistance to PD0332991.

Programmed cell death plays an important role in maintaining tissue homeostasis and diminished apoptosis is one of the hallmarks of cancer cells. Anticancer cytotoxic drugs are designed to reactivate apoptotic pathways by suppressing pro-survival signals such as MAPK/AKT. Several miRNAs, including miRNA-205/214/221/222, target PTEN which serves as a negative regulator of AKT, and their upregulation has been linked to DNA-damaging drug resistance [32, 40, 73]. Furthermore, as key downstream processors of AKT pathway, BCL2 protein family members are associated with miRNA-mediated drug resistance. Downregulation of miRNA-15/16/21/34 leads to elevated BCL2 levels and resistance to apoptosis-inducing compounds [16]. BCL2 family member MCL-1 has gained considerable attention lately for its role in drug resistance including resistance to BCL2 inhibitor ABT-263 [76]. miR-193b, miR-181, miR-125b, miR-101, and miR-29 target MCL-1 [13, 134, 145]. Bcl-xL, another BCL2 family member, also plays a significant role in chemoresistance. Let-7 family of miRNAs regulates Bcl-xL expression [131]. In this context, let-7 family members are frequently expressed at a lower level in tumors that are resistant to chemotherapy [135]. Furthermore, high let-7a level is associated with better response to EGFR-targeted therapies in colon cancer patients [121]. In addition, let-7a levels can be used as a marker for selecting ovarian cancer patients for paclitaxel- or cisplatin-based therapy [121]. However, a miRNA may not always function as an anti-apoptotic or pro-apoptotic miRNA in all cancers because the same miRNA can target both pro- and anti-apoptotic proteins. For example, although let-7 reduces anti-apoptotic Bcl-xL, it also targets caspase-3, the common effector of apoptotic signaling cascade. In this context, let-7 is upregulated in cyclophosphamide-resistant breast cancer [73, 140]. Similarly, miR-205 targets both oncogenic HER3 and the tumor suppressor PTEN [40, 57]. Furthermore, miRNA-145/155 has been shown to inhibit caspase-3 activity and prevent apoptosis in certain cancers [107]. However, miR-145 also has a pro-apoptotic role as a part of MDM2-p53 feedback loop [157].

Basal and drug-induced DNA repair activity in cancer cells plays a substantial role in resistance to genotoxic drugs such as alkylating agent, platinum compounds, and topoisomerase inhibitors. In this regard, enhanced nucleotide excision repair (NER) function protects cancer cells from drug-induced DNA damage. Furthermore, decrease in miR-373, a negative regulator of NER repair protein RAD23B, can augment DNA repair and in turn survival [118, 151]. Although efficient DNA repair activity mostly favors drug resistance, too little DNA repair activity is not always associated with favorable response to therapy. Single-base mispairs or DNA loop mismatch triggered by the drug treatment are normally removed through mismatch repair (MMR) mechanism, and defects in MMR facilitate drug resistance since mutations contribute to selecting more progressive malignant cells. In agreement with the note, abnormal expression of miR-21 and miR-141, which target two key components in MMR pathway, has been reported in drug-resistant cancer cell lines [143]. Additionally, variants of pri-miR-17 have been linked to reduced apoptotic response to therapeutic compounds in tumors with defects in breast cancer 1 (BRCA1), a major player in the repair of double-strand breaks and interstrand cross-links [128, 151].

In addition to various genetic alterations such as DNA mutations and copy number variations, epigenetic changes have emerged as significant contributors to cancer progression and drug resistance. Aberrant DNA methylation represents an epigenetic hallmark in tumor pathogenesis, and DNA methyltransferases (DNMTs) have been linked to cancer-specific DNA methylation. A number of DNMT inhibitors are in clinical use, particularly for hematologic malignancies and ovarian cancer. For example, decitabine has shown clinical activity in ovarian cancer and acute myeloid leukemia (AML) [61, 97]. miRNAs can potentially play a role in determining sensitivity to DNMT inhibitors. Recently, miR-29 and miR-148 were reported to play a role in upregulation of DNMTs in drug-resistant cancer cells [27, 29]. In addition, miR-132 and miR-194, which target methyl-binding protein MeCP2, were also elevated in cancer cells resistant to cisplatin and doxorubicin [72, 113]. In some instances, enhanced DNMT activity itself is responsible for downregulation of miRNAs in cancer, which can be reversed by DNMT inhibitors. For example, miR-9-3 and miR-193a are epigenetically silenced in small cell lung cancer, which can be reversed by DNMT inhibitors [44].

Cancer-associated epigenetic state can result from alterations in chromatin structure due to histone modification, which is mediated by chromatin-modifying enzymes such as histone acetyltransferases, histone deacetylases (HDAC), histone methyltransferases, and histone demethylases [24]. A wide array of miRNAs regulates the expression of these enzymes and differential expression of miRNAs targeting these enzymes in cancer cells has been observed [72, 113, 122]. The most prominent among them is miR-101, which targets histone methyltransferase EZH2 [144]. Genomic loss of miR-101 leading to upregulation of EZH2 is observed in prostate cancer. As inhibitors of EZH2 are just about to enter clinic, emergence of resistance to such drugs through miRNAs is a possibility [98]. miRNAs may play a similar role in determining sensitivity to clinically used HDAC inhibitors as these inhibitors significantly alter the expression of anti-apoptotic and pro-apoptotic molecules such as MCL-1, BIM, BMF, and NOXA [150].

6 miRNAs in Cancer Stem Cells and Tumor Microenvironment

A tumor in an organ represents a health disorder with heterogeneous morphology, molecular profiles, and therapeutic sensitivity [3, 111]. The current evidence indicates that within malignant tissues' heterogeneous population of cancer cells, a minor portion of cells embody all the properties of a tumor including resistance to chemotherapy and these cells are termed as cancer stem cells (CSCs) or tumor-initiating cells. CSCs are thought to be the core subset being capable of self-renewal, undergo differentiation, and generate heterogeneous tumor cells [18, 104]. Interestingly, miRNAs have been shown to regulate the major functional mediators of CSCs.

Since CSCs demonstrate reactivation of signaling pathways that control the embryonic and somatic stem cells, the studies on CSCs have been greatly inspired by the knowledge of normal stem cells. It was observed that mice deficient in Dicer-1, a cytoplasmic mediator in miRNA-generating machinery, died early in development due to diminished miRNAs and stem cells [25]. Conversely, miR-145 targets the pluripotency factors including OCT4, sex determining region Y-box 2 (SOX2), and KLF4. miR-145 inhibits the expression of pluripotency genes, and thereby the stem cell's self-renewal, and promotes lineage-restricted differentiation. Deficiency in miR-145 upregulates the expression of these pluripotency factors and suppresses differentiation [152]. Furthermore, miRNAs are involved in tissue specificity commitment. Co-transcription of miR-145 and miR-143 was observed in multipotent cardiac progenitors before being localized to smooth muscle cells. In addition, miR-106b-25 cluster and miR-204 regulate neuronal stem cell and mesenchymal progenitor, respectively [22], and overexpression of miR-106b-25 cluster in prostate cancer is associated with poor outcome [48]. Interestingly, the expression of miR-143 and miR-145 is downregulated in several cancers and this downregulation may potentially lead to increased CSCs and resistance to therapy [47]. Similar to miR-143 and miR-145, Let 7 and miR-200 family members have been shown to exhibit differentiation-promoting and anti-CSC properties in a variety of cancers [53, 114].

Several fundamental pathways associated with self-renewal and asymmetric division of embryonic stem cells are reactivated in CSCs, including Notch, Hedgehog, and Wnt/β-catenin signaling cascades [25]. Differential miRNA profiles between normal stem cells and CSCs have been reported and few of the differentially expressed miRNAs target the above three pathways [90]. miR-199b-5p suppresses the expression of HES1, a transcriptional factor downstream of Notch pathway, and its expression is reduced in metastatic cancers [33]. In contrast, miR-17-92 cluster is strikingly upregulated in lung cancer and functions as an oncogene via activating sonic hedgehog signaling [142]. Additionally, miR-135, which is highly expressed in colorectal tumors, functions as an oncogene by targeting the adenomatous polyposis coli (APC). Loss of APC leads to elevated β-catenin accumulation and enhanced self-renewal of CSCs [102]. In this regard, several miRNAs modulate and are modulated by Wnt signaling network in stem cells including CSCs [55, 159].

During tumor progression, cancer cells shift their epithelial phenotype to a more aggressive mesenchymal phenotype, which is termed EMT. EMT has been linked to cancer metastasis, drug resistance, and poor clinical outcomes due to shift in metabolic and signaling pathways [3, 138]. Cells that have undergone EMT share some of the properties of CSCs, as measured by tumor-generating capacity, sphere formation, and cell surface marker profile [41, 129, 137]. For example, in breast cancer, induction of EMT simultaneously converts the tumor cells into the CD44+/

CD24− CSC population [96]. Transforming growth factor beta (TGFβ), RAS, and Wnt signaling activities as well as various transcription factors in the EMT pathway can induce CSC [7, 96]. TGFβ activates downstream mediators of EMT through upregulation of miR-21 and miR-155 [8]. TGFβ also reduces E-cadherin levels, a hallmark of EMT, by inducing ZEB1 and ZEB2. ZEB1 and ZEB2 contribute to EMT by downregulating miR-200 family miRNAs [10]. In summation, miRNA–EMT nexus contributes to therapeutic resistance by enhancing CSC phenotype of cancer cells.

miRNAs not only function in cells that express them but also can influence gene expression in heterotypic cells as microvesicles/exosomes secreted by cells contain miRNAs, which are delivered to heterotypic cells. Consequently, miRNA expression pattern in tumor cells can influence tumor microenvironment and response to therapy. For example, tumor-secreted miR-9 activates JAK/STAT pathway in endothelial cells and promotes angiogenesis [160]. miRNAs secreted by macrophages enter cancer cells and change invasive behavior [154]. miRNAs released by cancer cells create metastatic niche in distant organs [39]. Overall, miRNAs in the cancer-derived microvesicles play vital roles in tumor invasion, metastasis, inflammation, self-renewal, and drug sensitivity.

7 miRNAs as Biomarkers of Therapeutic Response

Considering a wide range of effects attributed to miRNAs, it is not surprising that significant efforts have been placed to develop miRNAs as biomarkers. Tumor biomarkers are used to predict cancer status and therapeutic efficacy [49, 65]. Gene expression profiling analyses are increasingly being translated into prognostic/predictive tools for various tumor types in clinical settings and such analyses have already delivered comprehensive information on effective chemotherapy regimes [10, 108]. However, gene expression profiling studies are usually done with tumor tissues rather than body fluids and mRNA is significantly degraded in formalin-fixed paraffin-embedded (FFPE) tissues. In this regard, emerging miRNA assays open a new gateway for the biomarker world and hold the strength to solve challenging issues in the evaluation of cancer therapeutic response. It has been demonstrated that miRNAs are stable in FFPE tissues and are detectable in body fluids such as blood [14, 49]. Therefore, miRNA assays are likely superior to gene expression arrays in clinical setting. Indeed, miRNA profiling has been reported to generate more accurate information about tumor classification than cDNA arrays [50]. Interestingly, certain miRNAs have been proposed to predict the sensitivity of tumor cells to anticancer compounds, and thereby clinical outcomes. In particular, upregulation of miRNA-21 was found to play a role in mediating resistance to anti-HER2 therapy in breast cancer [36]. Likewise, let-7g and miR-181b miRNAs are significantly linked to 5-FU response in colorectal cancer cells [50]. In addition, let-7 levels predict response to cisplatin therapy in esophageal squamous cell carcinoma [135]. In contrast, downregulation of miR-21 and miR-200b enhanced sensitivity to gemcitabine in pancreatic cancer cells [146].

Since the first detection in the serum of the patients with B-cell lymphoma and other cancers in 2008 [78, 101], extracellular miRNAs have emerged as minimally invasive biomarkers of cancer progression and therapeutic response [125]. Impressively, miRNAs in body fluids are stable under tough conditions such as boiling, low/high pH, and survive well through multiple freeze–thaw cycles [9, 20]. There is convincing evidence for differential levels of specific miRNAs in the serum/plasma of cancer patients and normal controls [14, 20]. Serum miRNA profiling was shown to hold much higher sensitivity than traditional CA125 screening in early detection of ovarian cancer [9]. Persistent elevation of small RNA U6 was observed in the serum of breast cancer patients with or without active disease [2]. Furthermore, body fluid miRNAs can be utilized to predict/monitor therapeutic response of cancer. For example, increased serum miR-21 levels have been linked to docetaxel resistance in patients with castration-resistant prostate cancer [20]. This field is still in infancy and much more progress is expected in the coming years.

8 miRNA Therapy

miRNA as a therapy is emerging as one of the key branches of gene therapy, which hits disease targets through manipulating genetic materials (DNA or RNA). The concept of gene therapy was conceived four decades ago and the first human trial was practiced on a 4-year-old child with a genetic immune disorder termed severe combined immunodeficiency (SCID) in1990. Although being challenged by host immune response and molecular delivering issues, gene therapy is achieving some clinical success [64, 80]. The US Food and Drug Administration has already approved two RNA-based therapies: an antisense drug fomivirsen for cytomegalovirus-induced retinitis and an RNA aptamer pegaptanib for age-related macular degeneration [63, 70]. Being much smaller in molecular size than protein-coding genes and more effective in hitting the target than antisense approach, miRNAs are less likely to induce harmful immune response and have higher pharmacological potency as well as therapeutic efficacy [70]. Several RNA-based therapies are currently being tested in cancer patients [23]. Recent exciting progresses in selective drug delivery technologies such as miRNAs conjugated with a single-chain fragmented antibody targeting tumor cells and a positively charged peptide binding to RNA will likely accelerate clinical development of miRNA-based therapies [127].

Although miRNAs as a single agent may not prove to be effective in cancer therapy, it is conceivable that therapeutic miRNAs sensitize tumor cells to chemotherapy. Multidrug resistance is due to overexpression of ABC family transporters. Stem cell-enriched transcription factor SOX2 regulates the expression of these transporters [58]. SOX2 is a direct target of miR-9*. Overexpression of miR-9* in a chemotherapy-resistant glioma stem cell line inhibited SOX2 expression and in turn restored drug sensitivity [58]. Tamoxifen resistance is linked to enhanced expression of anti-apoptotic BCL2 protein resulting from decreased miR-15/16, and, accordingly, restoration of these miRNAs resensitized breast cancer cell to tamoxifen

[17]. miR-126 targets RAS and PI3K pathways and thereby sensitizes resistant lung cancer cells to Gefitinib (EGFR inhibitor) [158]. miR-31, which regulates 13 DNA repair-related genes and is significantly diminished in radio-resistant tumor cells, has been proposed as both a functional biomarker to predict response and a potential agent to improve the therapeutic efficacy of radiation [93]. In some instances, antagomirs can be used to sensitize cancer cells to chemotherapy. For example, resistance to 5-FU involves miR-21-dependent downregulation of core mismatch repair mutator genes, which could be reversed by miR-21 antagomirs [51, 99]. Antagonists of miR-122 are in phase II study for hepatitis C infection after successful preclinical studies in primates [77]. miRNA therapeutics has miR-34 and let-7 miRNAs in the preclinical stage for cancer. With improved nucleic acid chemistry, encapsulation, and targeting strategies, it will not be a surprise if miRNAs or miRNA antagomirs will become major weapons to contain or eliminate cancer within a decade.

References

1. Adams BD, Cowee DM, White BA. The role of miR-206 in the epidermal growth factor (EGF) induced repression of estrogen receptor-alpha (ERalpha) signaling and a luminal phenotype in MCF-7 breast cancer cells. Mol Endocrinol. 2009;23:1215–30.
2. Appaiah HN, Goswami CP, Mina LA, Badve S, Sledge GW Jr, Liu Y, Nakshatri H. Persistent upregulation of U6:SNORD44 small RNA ratio in the serum of breast cancer patients. Breast Cancer Res. 2011;13:R86. (BCR)
3. Badve S, Nakshatri H. Breast-cancer stem cells-beyond semantics. Lancet Oncol. 2012;13:e43–48.
4. Beitzinger M, Meister G. MicroRNAs: from decay to decoy. Cell. 2010;140:612–4.
5. Bergamaschi A, Katzenellenbogen BS. Tamoxifen downregulation of miR-451 increases 14-3-3zeta and promotes breast cancer cell survival and endocrine resistance. Oncogene. 2012;31:39–47.
6. Bhat-Nakshatri P, Wang GH, Collins NR, Thomson MJ, Geistlinger TR, Carroll JS, Brown M, Hammond S, Srour EF, Liu YL, et al. Estradiol-regulated microRNAs control estradiol response in breast cancer cells. Nucleic Acids Res. 2009;37:4850–61.
7. Bhat-Nakshatri P, Appaiah H, Ballas C, Pick-Franke P, Goulet R Jr, Badve S, Srour EF, Nakshatri H. SLUG/SNAI2 and tumor necrosis factor generate breast cells with CD44+/CD24- phenotype. BMC Cancer. 2010;10:411.
8. Blower PE, Verducci JS, Lin S, Zhou J, Chung JH, Dai Z, Liu CG, Reinhold W, Lorenzi PL, Kaldjian EP, et al. MicroRNA expression profiles for the NCI-60 cancer cell panel. Mol Cancer Ther. 2007;6:1483–91.
9. Brase JC, Wuttig D, Kuner R, Sultmann H. Serum microRNAs as non-invasive biomarkers for cancer. Mol Cancer. 2010;9:306.
10. Burk U, Schubert J, Wellner U, Schmalhofer O, Vincan E, Spaderna S, Brabletz T. A reciprocal repression between ZEB1 and members of the miR-200 family promotes EMT and invasion in cancer cells. EMBO Rep. 2008;9:582–9.
11. Castellano L, Giamas G, Jacob J, Coombes RC, Lucchesi W, Thiruchelvam P, Barton G, Jiao LR, Wait R, Waxman J, et al. The estrogen receptor-alpha-induced microRNA signature regulates itself and its transcriptional response. Proc Natl Acad Sci USA. 2009;106:15732–7.
12. Chan SY, Loscalzo J. MicroRNA-210: a unique and pleiotropic hypoxamir. Cell Cycle. 2010;9:1072–83.

13. Chen J, Zhang X, Lentz C, Abi-Daoud M, Pare GC, Yang X, Feilotter HE, Tron VA. miR-193b Regulates Mcl-1 in Melanoma. Am J Pathol. 2011;179:2162–8.
14. Cho WC. MicroRNAs: potential biomarkers for cancer diagnosis, prognosis and targets for therapy. Int J Biochem Cell Biol. 2010;42:1273–81.
15. Chua HL, Bhat-Nakshatri P, Clare SE, Morimiya A, Badve S, Nakshatri H. NF-kappaB represses E-cadherin expression and enhances epithelial to mesenchymal transition of mammary epithelial cells: potential involvement of ZEB-1 and ZEB-2. Oncogene. 2007;26:711–24.
16. Cimmino A, Calin GA, Fabbri M, Iorio MV, Ferracin M, Shimizu M, Wojcik SE, Aqeilan RI, Zupo S, Dono M, et al. miR-15 and miR-16 induce apoptosis by targeting BCL2. Proc Natl Acad Sci USA. 2005;102:13944–9.
17. Cittelly DM, Das PM, Salvo VA, Fonseca JP, Burow ME, Jones FE. Oncogenic HER2{Delta}16 suppresses miR-15a/16 and deregulates BCL-2 to promote endocrine resistance of breast tumors. Carcinogenesis. 2010;31:2049–57.
18. Clevers H. The cancer stem cell: premises, promises and challenges. Nat Med. 2011;17:313–9.
19. Cole MP, Jones CT, Todd ID. A new anti-oestrogenic agent in late breast cancer. An early clinical appraisal of ICI46474. Br J Cancer. 1971;25:270–5.
20. Cortez MA, Bueso-Ramos C, Ferdin J, Lopez-Berestein G, Sood AK, Calin GA. MicroRNAs in body fluids-the mix of hormones and biomarkers. Nat Rev Clin Oncol. 2011;8:467–77.
21. Dai B, Meng J, Peyton M, Girard L, Bornmann WG, Ji L, Minna JD, Fang B, Roth JA. STAT3 mediates resistance to MEK inhibitor through microRNA miR-17. Cancer Res. 2011;71:3658–68.
22. Das DK, Mukhopadhyay P. MicroRNA signatures of stem cells. Exp Clin Cardiol. 2011;16:e13–16.
23. Davidson BL, McCray PB Jr. Current prospects for RNA interference-based therapies. Nat Rev Genet. 2011;12:329–40.
24. Dawson MA, Kouzarides T. Cancer epigenetics: from mechanism to therapy. Cell. 2012;150:12–27.
25. DeSano JT, Xu L. MicroRNA regulation of cancer stem cells and therapeutic implications. AAPS J. 2009;11:682–92.
26. Du L, Subauste MC, DeSevo C, Zhao Z, Baker M, Borkowski R, Schageman JJ, Greer R, Yang CR, Suraokar M, et al. miR-337-3p and its targets STAT3 and RAP1A modulate taxane sensitivity in non-small cell lung cancers. PloS One. 2012;7:e39167.
27. Duursma AM, Kedde M, Schrier M, le Sage C, Agami R. miR-148 targets human DNMT3b protein coding region. RNA. 2008;14:872–7.
28. Ebert MS, Sharp PA. Roles for microRNAs in conferring robustness to biological processes. Cell. 2012;149:515–24.
29. Fabbri M, Garzon R, Cimmino A, Liu Z, Zanesi N, Callegari E, Liu S, Alder H, Costinean S, Fernandez-Cymering C, et al. MicroRNA-29 family reverts aberrant methylation in lung cancer by targeting DNA methyltransferases 3A and 3B. Proc Natl Acad Sci USA. 2007;104:15805–10.
30. Fareh M, Turchi L, Virolle V, Debruyne D, Almairac F, de-la-Forest Divonne S, Paquis P, Preynat-Seauve O, Krause KH, Chneiweiss H, et al. The miR 302-367 cluster drastically affects self-renewal and infiltration properties of glioma-initiating cells through CXCR4 repression and consequent disruption of the SHH-GLI-NANOG network. Cell Death Differ. 2012;19:232–44.
31. Flaherty KT, Lorusso PM, Demichele A, Abramson VG, Courtney R, Randolph SS, Shaik MN, Wilner KD, O'Dwyer PJ, Schwartz GK. Phase I, dose-escalation trial of the oral cyclin-dependent kinase 4/6 inhibitor PD 0332991, administered using a 21-day schedule in patients with advanced cancer. Clin Cancer Res. 2012;18:568–76.
32. Garofalo M, Di Leva G, Romano G, Nuovo G, Suh SS, Ngankeu A, Taccioli C, Pichiorri F, Alder H, Secchiero P, et al. miR-221 & 222 regulate TRAIL resistance and enhance tumorigenicity through PTEN and TIMP3 downregulation. Cancer Cell. 2009;16:498–509.

33. Garzia L, Andolfo I, Cusanelli E, Marino N, Petrosino G, De Martino D, Esposito V, Galeone A, Navas L, Esposito S, et al. MicroRNA-199b-5p impairs cancer stem cells through negative regulation of HES1 in medulloblastoma. PloS One. 2009;4:e4998.
34. Gilbert LA, Hemann MT. Chemotherapeutic resistance: surviving stressful situations. Cancer Res. 2011;71:5062–6.
35. Giles KM, Barker A, Zhang PM, Epis MR, Leedman PJ. MicroRNA regulation of growth factor receptor signaling in human cancer cells. Methods Mol Biol. 2011;676:147–63.
36. Gong C, Yao Y, Wang Y, Liu B, Wu W, Chen J, Su F, Yao H, Song E. Up-regulation of miR-21 mediates resistance to trastuzumab therapy for breast cancer. J Biol Chem. 2011;286:19127–37.
37. Goswami CP, Nakshatri H. PROGmiR: a tool for identifying prognostic miRNA biomarkers in multiple cancers using publicly available data. J Clin Bioinforma. 2012;2:23.
38. Gottesman MM, Ling V. The molecular basis of multidrug resistance in cancer: the early years of P-glycoprotein research. Febs Letters. 2006;580:998–1009.
39. Grange C, Tapparo M, Collino F, Vitillo L, Damasco C, Deregibus MC, Tetta C, Bussolati B, Camussi G. Microvesicles released from human renal cancer stem cells stimulate angiogenesis and formation of lung premetastatic niche. Cancer Res. 2011;71:5346–56.
40. Greene SB, Gunaratne PH, Hammond SM, Rosen JM. A putative role for microRNA-205 in mammary epithelial cell progenitors. J Cell Sci. 2010;123:606–18.
41. Gupta PB, Chaffer CL, Weinberg RA. Cancer stem cells: mirage or reality? Nat Med. 2009;15:1010–2.
42. Gyrd-Hansen M, Meier P. IAPs: from caspase inhibitors to modulators of NF-kappaB, inflammation and cancer. Nat Rev Cancer. 2010;10:561–74.
43. Heerdt BG, Houston MA, Mariadason JM, Augenlicht LH. Dissociation of staurosporine-induced apoptosis from G2-M arrest in SW620 human colonic carcinoma cells: initiation of the apoptotic cascade is associated with elevation of the mitochondrial membrane potential (deltapsim). Cancer Res. 2000;60:6704–13.
44. Heller G, Weinzierl M, Noll C, Babinsky V, Ziegler B, Altenberger C, Minichsdorfer C, Lang G, Dome B, End-Pfutzenreuter A, et al. Genome-wide miRNA expression profiling identifies miR-9-3 and miR-193a as targets for DNA methylation in non-small cell lung cancers. Clin Cancer Res. 2012;18:1619–29.
45. Hong L, Lai M, Chen M, Xie C, Liao R, Kang YJ, Xiao C, Hu WY, Han J, Sun P. The miR-17-92 cluster of microRNAs confers tumorigenicity by inhibiting oncogene-induced senescence. Cancer Res. 2010;70:8547–57.
46. Hossain A, Kuo MT, Saunders GF. Mir-17-5p regulates breast cancer cell proliferation by inhibiting translation of AIB1 mRNA. Mol Cell Biol. 2006;26:8191–201.
47. Huang S, Guo W, Tang Y, Ren D, Zou X, Peng X. miR-143 and miR-145 inhibit stem cell characteristics of PC-3 prostate cancer cells. Oncol Rep. 2012;28:1831–7.
48. Hudson RS, Yi M, Esposito D, Glynn SA, Starks AM, Yang Y, Schetter AJ, Watkins SK, Hurwitz AA, Dorsey TH, et al. MicroRNA-106b-25 cluster expression is associated with early disease recurrence and targets caspase-7 and focal adhesion in human prostate cancer. Oncogene. 2013;32:4139–47.
49. Hui A, How C, Ito E, Liu FF. Micro-RNAs as diagnostic or prognostic markers in human epithelial malignancies. BMC Cancer. 2011;11:500.
50. Hummel R, Hussey DJ, Haier J. MicroRNAs: predictors and modifiers of chemo- and radiotherapy in different tumour types. Eur J Cancer. 2010;46:298–311.
51. Hwang JH, Voortman J, Giovannetti E, Steinberg SM, Leon LG, Kim YT, Funel N, Park JK, Kim MA, Kang GH, et al. Identification of microRNA-21 as a biomarker for chemoresistance and clinical outcome following adjuvant therapy in resectable pancreatic cancer. PloS One. 2010;5:e10630.
52. Hynes NE, MacDonald G. ErbB receptors and signaling pathways in cancer. Curr Opin Cell Biol. 2009;21:177–84.
53. Iliopoulos D, Hirsch HA, Struhl K. An epigenetic switch involving NF-kappaB, Lin28, Let-7 microRNA, and IL6 links inflammation to cell transformation. Cell. 2009;139:693–706.

54. Ingelman-Sundberg M, Rodriguez-Antona C. Pharmacogenetics of drug-metabolizing enzymes: implications for a safer and more effective drug therapy. Philos Trans R Soc B-Biol Sci. 2005;360:1563–70.
55. Inui M, Martello G, Piccolo S. MicroRNA control of signal transduction. Nat Rev Mol Cell Biol. 2010;11:252–63.
56. Iorio MV, Croce CM. MicroRNA dysregulation in cancer: diagnostics, monitoring and therapeutics. A comprehensive review. EMBO Mol Med. 2012;4:143–59.
57. Iorio MV, Casalini P, Piovan C, Di Leva G, Merlo A, Triulzi T, Menard S, Croce CM, Tagliabue E. microRNA-205 regulates HER3 in human breast cancer. Cancer Res. 2009;69:2195–200.
58. Jeon HM, Sohn YW, Oh SY, Kim SH, Beck S, Kim S, Kim H. ID4 imparts chemoresistance and cancer stemness to glioma cells by derepressing miR-9*-mediated suppression of SOX2. Cancer Res. 2011;71:3410–21.
59. Jiang L, Lin C, Song L, Wu J, Chen B, Ying Z, Fang L, Yan X, He M, Li J, et al. MicroRNA-30e* promotes human glioma cell invasiveness in an orthotopic xenotransplantation model by disrupting the NF-kappaB/IkappaBalpha negative feedback loop. J Clin Invest. 2012;122:33–47.
60. Jung EJ, Santarpia L, Kim J, Esteva FJ, Moretti E, Buzdar AU, Di Leo A, Le XF, Bast RC Jr, Park ST, et al. Plasma microRNA 210 levels correlate with sensitivity to trastuzumab and tumor presence in breast cancer patients. Cancer. 2012;118:2603–14.
61. Kantarjian HM, Thomas XG, Dmoszynska A, Wierzbowska A, Mazur G, Mayer J, Gau JP, Chou WC, Buckstein R, Cermak J, et al. Multicenter, randomized, open-label, phase III trial of decitabine versus patient choice, with physician advice, of either supportive care or low-dose cytarabine for the treatment of older patients with newly diagnosed acute myeloid leukemia. J Clin Oncol. 2012;30:2670–7.
62. Karin M, Cao Y, Greten FR, Li ZW. NF-kappaB in cancer: from innocent bystander to major culprit. Nat Rev Cancer. 2002;2:301–10.
63. Kasinski AL, Slack FJ. Epigenetics and genetics. MicroRNAs en route to the clinic: progress in validating and targeting microRNAs for cancer therapy. Nat Rev Cancer. 2011;11:849–64.
64. Kay MA. State-of-the-art gene-based therapies: the road ahead. Nat Rev Genet. 2011;12:316–28.
65. Kelloff GJ, Sigman CC. Cancer biomarkers: selecting the right drug for the right patient. Nat Rev Drug Discov. 2012;11:201–14.
66. Kida K, Nakajima M, Mohri T, Oda Y, Takagi S, Fukami T, Yokoi T. PPARalpha is regulated by miR-21 and miR-27b in human liver. Pharm Res. 2011;28:2467–76.
67. Kim VN, Han J, Siomi MC. Biogenesis of small RNAs in animals. Nat Rev Mol Cell Biol. 2009;10:126–39.
68. Kim SY, Kim AY, Lee HW, Son YH, Lee GY, Lee JW, Lee YS, Kim JB. miR-27a is a negative regulator of adipocyte differentiation via suppressing PPARgamma expression. Biochem Biophys Res Commun. 2010;392:323–8.
69. Kim TM, Huang W, Park R, Park PJ, Johnson MD. A developmental taxonomy of glioblastoma defined and maintained by MicroRNAs. Cancer Res. 2011;71:3387–99.
70. Kole R, Krainer AR, Altman S. RNA therapeutics: beyond RNA interference and antisense oligonucleotides. Nat Rev Drug Discov. 2012;11:125–40.
71. Kondo N, Toyama T, Sugiura H, Fujii Y, Yamashita H. miR-206 expression is down-regulated in estrogen receptor alpha-positive human breast cancer. Cancer Res. 2008;68:5004–8.
72. Kovalchuk O, Filkowski J, Meservy J, Ilnytskyy Y, Tryndyak VP, Chekhun VF, Pogribny IP. Involvement of microRNA-451 in resistance of the MCF-7 breast cancer cells to chemotherapeutic drug doxorubicin. Mol Cancer Ther. 2008;7:2152–9.
73. Kutanzi KR, Yurchenko OV, Beland FA, Checkhun VF, Pogribny IP. MicroRNA-mediated drug resistance in breast cancer. Clin Epigenetics. 2011;2:171–85.
74. LaCasse EC, Baird S, Korneluk RG, MacKenzie AE. The inhibitors of apoptosis (IAPs) and their emerging role in cancer. Oncogene. 1998;17:3247–59.
75. Lagos-Quintana M, Rauhut R, Lendeckel W, Tuschl T. Identification of novel genes coding for small expressed RNAs. Science. 2001;294:853–8.

76. Lam LT, Lu X, Zhang H, Lesniewski R, Rosenberg S, Semizarov D. A microRNA screen to identify modulators of sensitivity to BCL2 inhibitor ABT-263 (navitoclax). Mol Cancer Ther. 2010;9:2943–50.
77. Lanford RE, Hildebrandt-Eriksen ES, Petri A, Persson R, Lindow M, Munk ME, Kauppinen S, Orum H. Therapeutic silencing of microRNA-122 in primates with chronic hepatitis C virus infection. Science. 2010;327:198–201.
78. Lawrie CH, Gal S, Dunlop HM, Pushkaran B, Liggins AP, Pulford K, Banham AH, Pezzella F, Boultwood J, Wainscoat JS, et al. Detection of elevated levels of tumour-associated microRNAs in serum of patients with diffuse large B-cell lymphoma. Br J Haematol. 2008;141:672–5.
79. Le MT, Teh C, Shyh-Chang N, Xie H, Zhou B, Korzh V, Lodish HF, Lim B. MicroRNA-125b is a novel negative regulator of p53. Genes Dev. 2009;23:862–76.
80. Lee B, Davidson BL. Gene therapy grows into young adulthood: special review issue. Hum Mol Genet. 2011;20:R1.
81. Lee RC, Feinbaum RL, Ambros V. The C-elegans heterochronic gene Lin-4 encodes small rnas with antisense complementarity to Lin-14. Cell. 1993;75:843–54.
82. Lee Y, Kim M, Han JJ, Yeom KH, Lee S, Baek SH, Kim VN. MicroRNA genes are transcribed by RNA polymerase II. Embo J. 2004;23:4051–60.
83. Lee EK, Lee MJ, Abdelmohsen K, Kim W, Kim MM, Srikantan S, Martindale JL, Hutchison ER, Kim HH, Marasa BS, et al. miR-130 suppresses adipogenesis by inhibiting peroxisome proliferator-activated receptor gamma expression. Mol Cell Biol. 2011;31:626–38.
84. Lee JJ, Drakaki A, Iliopoulos D, Struhl K. MiR-27b targets PPARgamma to inhibit growth, tumor progression and the inflammatory response in neuroblastoma cells. Oncogene. 2012;31:3818–25.
85. Lehmann BD, Pietenpol JA. Targeting mutant p53 in human tumors. J Clin Oncol. 2012;30:3648–50.
86. Leskela S, Leandro-Garcia LJ, Mendiola M, Barriuso J, Inglada-Perez L, Munoz I, Martinez-Delgado B, Redondo A, de Santiago J, Robledo M, et al. The miR-200 family controls beta-tubulin III expression and is associated with paclitaxel-based treatment response and progression-free survival in ovarian cancer patients. Endocr Relat Cancer. 2011;18:85–95.
87. Li J, Zhang Y, Zhao J, Kong F, Chen Y. Overexpression of miR-22 reverses paclitaxel-induced chemoresistance through activation of PTEN signaling in p53-mutated colon cancer cells. Mol Cell Biochem. 2011a;357:31–8.
88. Li LM, Hou DX, Guo YL, Yang JW, Liu Y, Zhang CY, Zen K. Role of microRNA-214-targeting phosphatase and tensin homolog in advanced glycation end product-induced apoptosis delay in monocytes. J Immunol. 2011b;186:2552–60.
89. Lin SL, Chang DC, Ying SY, Leu D, Wu DT. MicroRNA miR-302 inhibits the tumorigenecity of human pluripotent stem cells by coordinate suppression of the CDK2 and CDK4/6 cell cycle pathways. Cancer Res. 2010;70:9473–82.
90. Liu C, Tang DG. MicroRNA regulation of cancer stem cells. Cancer Res. 2011;71:5950–4.
91. Liu Q, Fu H, Sun F, Zhang H, Tie Y, Zhu J, Xing R, Sun Z, Zheng X. miR-16 family induces cell cycle arrest by regulating multiple cell cycle genes. Nucleic Acids Res. 2008;36:5391–404.
92. Lydon JP, O'Malley BW. Minireview: steroid receptor coactivator-3: a multifarious coregulator in mammary gland metastasis. Endocrinology. 2011;152:19–25.
93. Lynam-Lennon N, Reynolds JV, Marignol L, Sheils OM, Pidgeon GP, Maher SG. MicroRNA-31 modulates tumour sensitivity to radiation in oesophageal adenocarcinoma. J Mol Med (Berl). 2012;90:1449–58.
94. Lyng MB, Laenkholm AV, Sokilde R, Gravgaard KH, Litman T, Ditzel HJ. Global microRNA expression profiling of high-risk ER + breast cancers from patients receiving adjuvant tamoxifen mono-therapy: a DBCG study. PloS One. 2012;7:e36170.
95. Ma X, Becker Buscaglia LE, Barker JR, Li Y. MicroRNAs in NF-kappaB signaling. J Mol Cell Biol. 2011;3:159–66.

96. Mani SA, Guo W, Liao MJ, Eaton EN, Ayyanan A, Zhou AY, Brooks M, Reinhard F, Zhang CC, Shipitsin M, et al. The epithelial-mesenchymal transition generates cells with properties of stem cells. Cell. 2008;133:704–15.
97. Matei D, Fang F, Shen C, Schilder J, Arnold A, Zeng Y, Berry WA, Huang T, Nephew KP. Epigenetic resensitization to platinum in ovarian cancer. Cancer Res. 2012;72:2197–205.
98. McCabe MT, Ott HM, Ganji G, Korenchuk S, Thompson C, Van Aller GS, Liu Y, Graves AP, Della Pietra A 3rd, Diaz E, et al. EZH2 inhibition as a therapeutic strategy for lymphoma with EZH2-activating mutations. Nature. 2012;492:108–12.
99. Melo SA, Kalluri R. Molecular pathways: microRNAs as cancer therapeutics. Clin Cancer Res. 2012;18:4234–9.
100. Miller TE, Ghoshal K, Ramaswamy B, Roy S, Datta J, Shapiro CL, Jacob S, Majumder S. MicroRNA-221/222 confers tamoxifen resistance in breast cancer by targeting p27Kip1. J Biol Chem. 2008;283:29897–903.
101. Mitchell PS, Parkin RK, Kroh EM, Fritz BR, Wyman SK, Pogosova-Agadjanyan EL, Peterson A, Noteboom J, O'Briant KC, Allen A, et al. Circulating microRNAs as stable blood-based markers for cancer detection. Proc Natl Acad Sci USA. 2008;105:10513–8.
102. Nagel R, le Sage C, Diosdado B, van der Waal M, Oude Vrielink JA, Bolijn A, Meijer GA, Agami R. Regulation of the adenomatous polyposis coli gene by the miR-135 family in colorectal cancer. Cancer Res. 2008;68:5795–802.
103. Nakshatri H, Bhat-Nakshatri P, Martin DA, Goulet RJ Jr, Sledge GW Jr. Constitutive activation of NF-kappaB during progression of breast cancer to hormone-independent growth. Mol Cell Biol. 1997;17:3629–39.
104. Nguyen LV, Vanner R, Dirks P, Eaves CJ. Cancer stem cells: an evolving concept. Nat Rev Cancer. 2012;12:133–43.
105. Osborne CK, Bardou V, Hopp TA, Chamness GC, Hilsenbeck SG, Fuqua SA, Wong J, Allred DC, Clark GM, Schiff R. Role of the estrogen receptor coactivator AIB1 (SRC-3) and HER-2/neu in tamoxifen resistance in breast cancer. J Natl Cancer Inst. 2003;95:353–61.
106. Ou Z, Wada T, Gramignoli R, Li S, Strom SC, Huang M, Xie W. MicroRNA hsa-miR-613 targets the human LXRalpha gene and mediates a feedback loop of LXRalpha autoregulation. Mol Endocrinol. 2011;25:584–96.
107. Ovcharenko D, Kelnar K, Johnson C, Leng N, Brown D. Genome-scale microRNA and small interfering RNA screens identify small RNA modulators of TRAIL-induced apoptosis pathway. Cancer Res. 2007;67:10782–8.
108. Paik S, Shak S, Tang G, Kim C, Baker J, Cronin M, Baehner FL, Walker MG, Watson D, Park T, et al. A multigene assay to predict recurrence of tamoxifen-treated, node-negative breast cancer. N Engl J Med. 2004;351:2817–26.
109. Pan YZ, Morris ME, Yu AM. MicroRNA-328 negatively regulates the expression of breast cancer resistance protein (BCRP/ABCG2) in human cancer cells. Mol Pharmacol. 2009;75:1374–9.
110. Patel JB, Appaiah HN, Burnett RM, Bhat-Nakshatri P, Wang G, Mehta R, Badve S, Thomson MJ, Hammond S, Steeg P, et al. Control of EVI-1 oncogene expression in metastatic breast cancer cells through microRNA miR-22. Oncogene. 2011;30:1290–301.
111. Pece S, Tosoni D, Confalonieri S, Mazzarol G, Vecchi M, Ronzoni S, Bernard L, Viale G, Pelicci PG, Di Fiore PP. Biological and molecular heterogeneity of breast cancers correlates with their cancer stem cell content. Cell. 2010;140:62–73.
112. Pichler M, Winter E, Stotz M, Eberhard K, Samonigg H, Lax S, Hoefler G. Down-regulation of KRAS-interacting miRNA-143 predicts poor prognosis but not response to EGFR-targeted agents in colorectal cancer. Br J Cancer. 2012;106:1826–32.
113. Pogribny IP, Filkowski JN, Tryndyak VP, Golubov A, Shpyleva SI, Kovalchuk O. Alterations of microRNAs and their targets are associated with acquired resistance of MCF-7 breast cancer cells to cisplatin. Int J Cancer. 2010;127:1785–94.
114. Polytarchou C, Iliopoulos D, Struhl K. An integrated transcriptional regulatory circuit that reinforces the breast cancer stem cell state. Proc Natl Acad Sci USA. 2012;109:14470–5.

115. Rai K, Takigawa N, Ito S, Kashihara H, Ichihara E, Yasuda T, Shimizu K, Tanimoto M, Kiura K. Liposomal delivery of microRNA-7-expressing plasmid overcomes epidermal growth factor receptor tyrosine kinase inhibitor-resistance in lung cancer cells. Mol Cancer Ther. 2011;10:1720–7.
116. Rao X, Di Leva G, Li M, Fang F, Devlin C, Hartman-Frey C, Burow ME, Ivan M, Croce CM, Nephew KP. MicroRNA-221/222 confers breast cancer fulvestrant resistance by regulating multiple signaling pathways. Oncogene. 2011;30:1082–97.
117. Rao E, Jiang C, Ji M, Huang X, Iqbal J, Lenz G, Wright G, Staudt LM, Zhao Y, McKeithan TW, et al. The miRNA-17 approximately 92 cluster mediates chemoresistance and enhances tumor growth in mantle cell lymphoma via PI3K/AKT pathway activation. Leukemia. 2012;26:1064–72.
118. Reardon JT, Vaisman A, Chaney SG, Sancar A. Efficient nucleotide excision repair of cisplatin, oxaliplatin, and Bis-aceto-ammine-dichloro-cyclohexylamine-platinum(IV) (JM216) platinum intrastrand DNA diadducts. Cancer Res. 1999;59:3968–71.
119. Rokavec M, Wu W, Luo JL. IL6-mediated suppression of miR-200c directs constitutive activation of inflammatory signaling circuit driving transformation and tumorigenesis. Mol Cell. 2012;45:777–89.
120. Ruby JG, Jan CH, Bartel DP. Intronic microRNA precursors that bypass Drosha processing. Nature. 2007;448:83–6.
121. Ruzzo A, Graziano F, Vincenzi B, Canestrari E, Perrone G, Galluccio N, Catalano V, Loupakis F, Rabitti C, Santini D, et al. High let-7a microRNA levels in KRAS-mutated colorectal carcinomas may rescue anti-EGFR therapy effects in patients with chemotherapy-refractory metastatic disease. Oncologist. 2012;17:823–9.
122. Sachdeva M, Wu H, Ru P, Hwang L, Trieu V, Mo YY. MicroRNA-101-mediated Akt activation and estrogen-independent growth. Oncogene. 2011;30:822–31.
123. Saito K, Kobayashi K, Mizuno Y, Fukuchi Y, Furihata T, Chiba K. Peroxisome proliferator-activated receptor alpha (PPARalpha) agonists induce constitutive androstane receptor (CAR) and cytochrome P450 2B in rat primary hepatocytes. Drug Metab Pharmacokinet. 2010;25:108–11.
124. San Jose-Eneriz E, Roman-Gomez J, Jimenez-Velasco A, Garate L, Martin V, Cordeu L, Vilas-Zornoza A, Rodriguez-Otero P, Calasanz MJ, Prosper F, et al. MicroRNA expression profiling in Imatinib-resistant chronic myeloid leukemia patients without clinically significant ABL1-mutations. Mol Cancer. 2009;8:69.
125. Schwarzenbach H, Hoon DS, Pantel K. Cell-free nucleic acids as biomarkers in cancer patients. Nat Rev Cancer. 2011;11:426–37.
126. Sergina NV, Rausch M, Wang D, Blair J, Hann B, Shokat KM, Moasser MM. Escape from HER-family tyrosine kinase inhibitor therapy by the kinase-inactive HER3. Nature. 2007;445:437–41.
127. Seton-Rogers S. Therapeutics: siRNAs jump the hurdle. Nat Rev Cancer. 2012;12:376–7.
128. Shen J, DiCioccio R, Odunsi K, Lele SB, Zhao H. Novel genetic variants in miR-191 gene and familial ovarian cancer. BMC Cancer. 2010;10:47.
129. Sheridan C, Kishimoto H, Fuchs RK, Mehrotra S, Bhat-Nakshatri P, Turner CH, Goulet R Jr, Badve S, Nakshatri H. CD44+/CD24− breast cancer cells exhibit enhanced invasive properties: an early step necessary for metastasis. Breast Cancer Res: BCR. 2006;8:R59.
130. Shi W, Gerster K, Alajez NM, Tsang J, Waldron L, Pintilie M, Hui AB, Sykes J, P'ng C, Miller N, et al. MicroRNA-301 mediates proliferation and invasion in human breast cancer. Cancer Res. 2011;71:2926–37.
131. Shimizu S, Takehara T, Hikita H, Kodama T, Miyagi T, Hosui A, Tatsumi T, Ishida H, Noda T, Nagano H, et al. The let-7 family of microRNAs inhibits Bcl-xL expression and potentiates sorafenib-induced apoptosis in human hepatocellular carcinoma. J Hepatol. 2010;52:698–704.
132. Siddik ZH. Cisplatin: mode of cytotoxic action and molecular basis of resistance. Oncogene. 2003;22:7265–79.

133. Song L, Liu L, Wu Z, Li Y, Ying Z, Lin C, Wu J, Hu B, Cheng SY, Li M, et al. TGF-beta induces miR-182 to sustain NF-kappaB activation in glioma subsets. J Clin Invest. 2012;122:3563–78.
134. Su H, Yang JR, Xu T, Huang J, Xu L, Yuan Y, Zhuang SM. MicroRNA-101, down-regulated in hepatocellular carcinoma, promotes apoptosis and suppresses tumorigenicity. Cancer Res. 2009;69:1135–42.
135. Sugimura K, Miyata H, Tanaka K, Hamano R, Takahashi T, Kurokawa Y, Yamasaki M, Nakajima K, Takiguchi S, Mori M, et al. Let-7 expression is a significant determinant of response to chemotherapy through the regulation of IL-6/STAT3 pathway in esophageal squamous cell carcinoma. Clin Cancer Res. 2012;18:5144–53.
136. Tanabe J, Tamasawa N, Yamashita M, Matsuki K, Murakami H, Matsui J, Sugimoto K, Yasujima M, Suda T. Effects of combined PPARgamma and PPARalpha agonist therapy on reverse cholesterol transport in the Zucker diabetic fatty rat. Diabetes Obes Metab. 2008;10:772–9.
137. Thiery JP, Acloque H, Huang RY, Nieto MA. Epithelial-mesenchymal transitions in development and disease. Cell. 2009;139:871–90.
138. Thomson S, Petti F, Sujka-Kwok I, Mercado P, Bean J, Monaghan M, Seymour SL, Argast GM, Epstein DM, Haley JD. A systems view of epithelial-mesenchymal transition signaling states. Clin Exp Metastasis. 2011;28:137–55.
139. Tsai NP, Lin YL, Wei LN. MicroRNA mir-346 targets the 5′-untranslated region of receptor-interacting protein 140 (RIP140) mRNA and up-regulates its protein expression. Biochem J. 2009;424:411–8.
140. Tsang WP, Kwok TT. Let-7a microRNA suppresses therapeutics-induced cancer cell death by targeting caspase-3. Apoptosis. 2008;13:1215–22.
141. Turk D, Hall MD, Chu BF, Ludwig JA, Fales HM, Gottesman MM, Szakacs G. Identification of compounds selectively killing multidrug-resistant cancer cells. Cancer Res. 2009;69:8293–301.
142. Uziel T, Karginov FV, Xie S, Parker JS, Wang YD, Gajjar A, He L, Ellison D, Gilbertson RJ, Hannon G, et al. The miR-17~92 cluster collaborates with the Sonic Hedgehog pathway in medulloblastoma. Proc Natl Acad Sci USA. 2009;106:2812–17.
143. Valeri N, Gasparini P, Braconi C, Paone A, Lovat F, Fabbri M, Sumani KM, Alder H, Amadori D, Patel T, et al. MicroRNA-21 induces resistance to 5-fluorouracil by down-regulating human DNA MutS homolog 2 (hMSH2). Proc Natl Acad Sci USA. 2010;107:21098–103.
144. Varambally S, Cao Q, Mani RS, Shankar S, Wang X, Ateeq B, Laxman B, Cao X, Jing X, Ramnarayanan K, et al. Genomic loss of microRNA-101 leads to overexpression of histone methyltransferase EZH2 in cancer. Science. 2008;322:1695–9.
145. Visone R, Veronese A, Rassenti LZ, Balatti V, Pearl DK, Acunzo M, Volinia S, Taccioli C, Kipps TJ, Croce CM. miR-181b is a biomarker of disease progression in chronic lymphocytic leukemia. Blood. 2011;118:3072–9.
146. Voutsadakis IA. Molecular predictors of gemcitabine response in pancreatic cancer. World J Gastrointest Oncol. 2011;3:153–64.
147. Wasik KA, Rebbeck CA. Keystone symposia 40th season: microRNAs and noncoding RNAs in cancer. Cancer Res. 2011;71:6102–5.
148. Williams KC, Renthal NE, Condon JC, Gerard RD, Mendelson CR. MicroRNA-200a serves a key role in the decline of progesterone receptor function leading to term and preterm labor. Proc Natl Acad Sci USA. 2012;109:7529–34.
149. Wilson TR, Fridlyand J, Yan Y, Penuel E, Burton L, Chan E, Peng J, Lin E, Wang Y, Sosman J, et al. Widespread potential for growth-factor-driven resistance to anticancer kinase inhibitors. Nature. 2012;487:505–9.
150. Xargay-Torrent S, Lopez-Guerra M, Saborit-Villarroya I, Rosich L, Campo E, Roue G, Colomer D. Vorinostat-induced apoptosis in mantle cell lymphoma is mediated by acetylation of proapoptotic BH3-only gene promoters. Clin Cancer Res. 2011;17:3956–68.

151. Xin FX, Li M, Balch C, Thomson M, Fan MY, Liu Y, Hammond SM, Kim S, Nephew KP. Computational analysis of microRNA profiles and their target genes suggests significant involvement in breast cancer antiestrogen resistance. Bioinformatics. 2009;25:430–4.
152. Xu N, Papagiannakopoulos T, Pan G, Thomson JA, Kosik KS. MicroRNA-145 regulates OCT4, SOX2, and KLF4 and represses pluripotency in human embryonic stem cells. Cell. 2009;137:647–58.
153. Xu S, Linher-Melville K, Yang BB, Wu D, Li J. Micro-RNA378 (miR-378) regulates ovarian estradiol production by targeting aromatase. Endocrinology. 2011;152:3941–51.
154. Yang M, Chen J, Su F, Yu B, Lin L, Liu Y, Huang JD, Song E. Microvesicles secreted by macrophages shuttle invasion-potentiating microRNAs into breast cancer cells. Mol Cancer. 2011;10:117.
155. Yu Y, Yang L, Zhao M, Zhu S, Kang R, Vernon P, Tang D, Cao L. Targeting microRNA-30a-mediated autophagy enhances imatinib activity against human chronic myeloid leukemia cells. Leukemia. 2012;26:1752–60.
156. Zhang J, Xiao Z, Lai D, Sun J, He C, Chu Z, Ye H, Chen S, Wang J. miR-21, miR-17 and miR-19a induced by phosphatase of regenerating liver-3 promote the proliferation and metastasis of colon cancer. Br J Cancer. 2012;107:352–9.
157. Zhang J, Sun Q, Zhang Z, Ge S, Han ZG, Chen WT. Loss of microRNA-143/145 disturbs cellular growth and apoptosis of human epithelial cancers by impairing the MDM2-p53 feedback loop. Oncogene. 2013;32:61–9.
158. Zhong M, Ma X, Sun C, Chen L. MicroRNAs reduce tumor growth and contribute to enhance cytotoxicity induced by gefitinib in non-small cell lung cancer. Chem Biol Interact. 2010;184:431–8.
159. Zhou AD, Diao LT, Xu H, Xiao ZD, Li JH, Zhou H, Qu LH. beta-Catenin/LEF1 transactivates the microRNA-371-373 cluster that modulates the Wnt/beta-catenin-signaling pathway. Oncogene. 2012;31:2968–78.
160. Zhuang G, Wu X, Jiang Z, Kasman I, Yao J, Guan Y, Oeh J, Modrusan Z, Bais C, Sampath D, et al. Tumour-secreted miR-9 promotes endothelial cell migration and angiogenesis by activating the JAK-STAT pathway. Embo J. 2012;31:3513–23.
161. Zunino F, Cassinelli G, Polizzi D, Perego P. Molecular mechanisms of resistance to taxanes and therapeutic implications. Drug Resist Updates. 1999;2:351–7.

Chapter 17
Microvesicular Transfer of MicroRNA in Tumor Microenvironment

Krishna C. Vallabhaneni, Patrice Penfornis, J. Griffin Orr, Krishna Chauhan and Radhika Pochampally

Abstract In recent years, the knowledge about the control of tumor microenvironment has increased and emerged as an important player in tumorigenesis. The role of normal stromal cells in the tumor cells initiation and development has brought our vision to the forefront of cell-to-cell communication. In this chapter, we focus on the novel mechanism of communication between stromal and tumor cells, which is based on the exchange of microvesicles. We describe several, ever-growing pieces of evidence that microvesicles transfer messages through their lipid, protein, or nucleic acid contents. microRNA exchange is emerging as a key player in this communication. A better understanding of this sophisticated method of communication between normal and cancerous cells may lead to developing novel, innovative approaches for cancer diagnostics and personalized therapy.

Keywords Tumor microenvironment · Mesenchymal stem cells · Stromal cells · Exosomes · Microvesicles · Cellular crosstalk · miRNA · ceRNA · Breast cancer · Metastasis

The tumor microenvironment includes normal cells surrounding the tumor, which is composed of an extracellular matrix (ECM) and numerous stromal cell types, including endothelial and inflammatory immune cells, fibroblasts, adipocytes, and tumor-associated vasculature [1]. Tumor malignancy is highly dependent on interactions between tumor cells and the tumor microenvironment [2]. In the last decade, the knowledge about the control of tumor microenvironment has become as important as that of the cancer cells. The cellular and molecular components of the tumor microenvironment contribute to tumor growth and progression, including invasion, metastasis, and angiogenesis [3–5]. Studies on comprehensive gene expression and genomic profile study of epithelial, myoepithelial, and stromal cells have revealed diverse microenvironments between normal breast tissue and breast carcinomas [6, 7]. Stromal elements secrete chemokines, which act as paracrine factors that

R. Pochampally (✉) · K. Chauhan
Department of Biochemistry, University of Mississippi Medical Center, Jackson, MS, USA
e-mail: rpochampally@umc.edu

R. Pochampally · K. C. Vallabhaneni · P. Penfornis · J. G. Orr · K. Chauhan
Cancer Institute, University of Mississippi Medical Center, Jackson, MS, USA

S. R. Singh, P. Rameshwar (eds.), *MicroRNA in Development and in the Progression of Cancer*, DOI 10.1007/978-1-4899-8065-6_17,

could induce ECM remodeling and enhance tumor cell proliferation and invasion [8, 9]. An example of paracrine mechanisms using tumor myoepithelial cells and myofibroblasts is the overexpression of CXCL14 and CXCL12, respectively. These bind to currently unidentified receptors on epithelial cells and enhance proliferation, migration, and invasion [8].

Tumor stroma is also known to associate with therapeutic resistance and relapse—a main reason for breast cancer treatment failure [10, 11]. A major component of tumor stroma is cancer-associated fibroblasts (CAFs), which play a role in the chemotherapy resistance by secreting collagen type I that reduces the drug uptake in tumors [12, 13]. One mechanism that tumor cells use for relapse is tumor dormancy due to failed neovascularization [14, 15]. The balance in expression of vascular endothelial growth factor (VEGF) and thrombospondin 1 (Tsp-1) by tumor stroma dictates the dormant state of the tumor cells [16]. In this chapter, various aspects of stromal support in tumors will be addressed with a focus on microvesicular mechanism of communication between stromal and tumor cells.

1 Factors Affecting Stromal Support

According to the soil and seed hypothesis, the cancerous cell must be in the right stroma to establish a niche and form a tumor [17]. Therefore, it is obvious that the soil, which in most cases is the stromal layer, supports tumor initiation and establishment. Supporting this hypothesis, various reports showed that the cross talk between cells within the tumor and associated stromal cells plays a major role in tumor initiation, progression, and metastasis [18]. Tumor cells in the process of establishing a supportive microenvironment for their progression secrete various cytokines, which includes basic fibroblast growth factor (bFGF), VEGF, platelet-derived growth factor (PDGF), epidermal growth factor receptor (EGFR) ligands, interleukins, colony-stimulating factors, transforming growth factor (TGFβ), and CXCL12/SDF-1 [19–22]. These cytokines modulate the tumor-associated stromal cells in a paracrine manner that results in angiogenesis [23] and inflammatory responses [24].

Physiological conditions such as nutrient deprivation and hypoxia are other critical factors of the tumor microenvironment that modulate tumor cell behavior. The regulation of these extrinsic factors is studied extensively and is attributed to expression of specific factors. For example, various hypoxia-induced growth factors (HIFs) secreted by tumor and stromal cells in response to hypoxia dictate the tumorigenic potential of HIFs [25]. HIF1α is essential for the proliferation, survival, and angiogenesis of both cancer cells and stromal cells [26]. Activation of the HIF1α pathway along with nuclear factor-kappa B (NFκB) in CAFs is shown to be involved in altering the metabolism of breast tumors [27]. HIF-1α and HIF-2α have antagonistic effects on nitric oxide production [28]. Several studies have shown that CAFs originate from mesenchymal stem cells (MSCs) [29–31]. Similarly, autophagy induced by nutrient deprivation—a phenomenon observed in rapidly growing tumor stroma—is shown to have both tumor-enhancing and tumor-inhibitory effects [32].

Autophagy in stromal cells was found to be tumor supportive when MSCs were used as stromal cells for breast cancers. In that study, MSCs were shown to survive by autophagy; the induced paracrine signaling was shown to be epigenetically regulated by serum deprivation [33–36].

2 Cellular Cross Talk in Tumor

An important factor in the support of the tumor microenvironment is the cell–cell communication between stromal cells and transformed cancer cells. The role of gap junctions in transport of cellular communicators [37] and juxtacrine regulation based on direct communication is well documented [38, 39]. Recently, Tsuyada et al. [40] demonstrated the cancer–stroma signaling circuit, in which breast cancer cells stimulate the expression of chemokine CCL2 in normal fibroblasts that become CAFs, which in turn stimulates the stemness of breast cancer cells constituting the cancer–stroma–cancer signaling circuit [8, 41].

One mechanism of communication between tumor and stromal cells is based on the exchange of microvesicles. Microvesicles are described to be exchanged between breast cancer cells and stromal cells [42, 43]. They are recognized as being involved in regulating a variety of extracellular signals and paracrine signaling [44, 45], including breast cancer invasiveness [46, 47]. A dynamic interaction between stromal cells, cancer cells, and the tumor microenvironment facilitates tumor progression (Fig. 17.1).

3 Microvesicles and Cancer

Cells are able to communicate with each other through the exchange of mobile, secreted cell membrane fragments known as microvesicles (MVs) in the extracellular space [48, 49]. MVs were first regarded as cellular debris [49–53] but starting in the late 1970s, studies began to show the presence and the secretion of MVs in both normal and cancer cells [54–57]. A correlation between elevated blood MV levels in cancer patients [58–60] and other studies has implicated MVs as potential diagnosis markers for cancer [61]. This has shifted recent research to focus on whether MVs play a supportive role in cancer pathology, including effects associated with cancer initiation, progression, angiogenesis, and metastasis [62].

The word "microvesicle (MV)" is actually a generic term that refers to a series of membrane-bound organelles, which are commonly distinguished by their size range. More specific nomenclature for MVs include exosomes (40–100 nm diameter), ectosomes (50–1,000 nm) [63], and apoptotic bodies (50–5,000 nm) [64]. There are problems establishing a standard terminology in this field of research [62] that have led to uncommon words such as "microparticles," [64] "endosomes," [62] and even organ-specific classifications such as "prostasomes" [65] (Table 17.1) used in the literature. Any discrepancies between the characteristics of specific

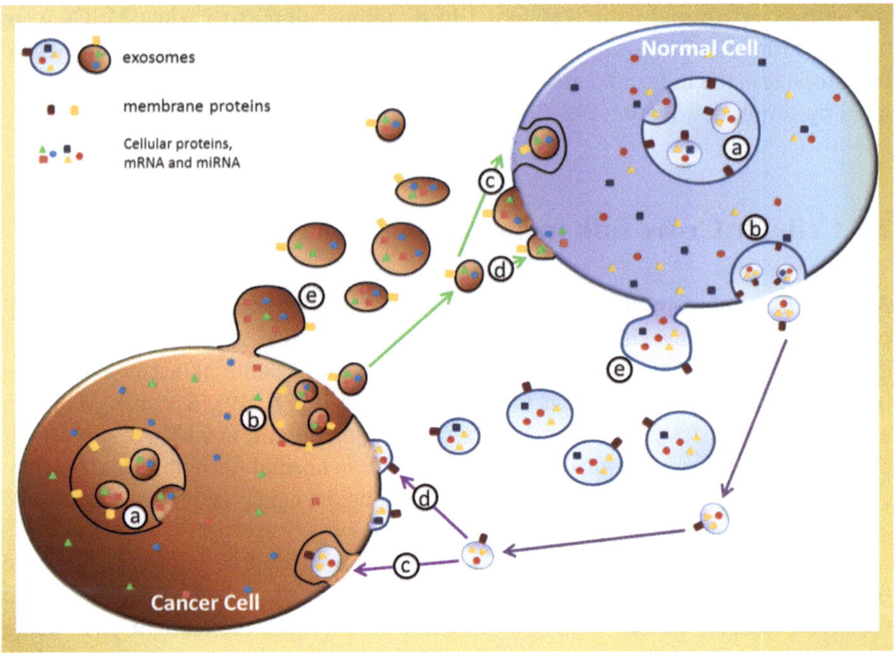

Fig. 17.1 Schematic to show cellular crosstalk in tumors: *a* Exosomes are generated through invagination of the cellular membrane, resulting in the formation of microvesicles (MVs) that contain proteins, mRNA, and miRNA from cytoplasm. *b* Exosomes are released through exocytosis when MVs fuse with the plasma membrane. *c* and *d* Exosomes released by one cell can enter the other in one of the two ways—by being taken up by cellular endocytic pathway *c* or by fusing to the plasma membrane and releasing the contents directly into the cytoplasm *d*. *e* MVs or exosomes are also secreted by cells by budding directly from the plasma membrane

Table 17.1 History and variety of microvesical names

Name	Size range	Source	Year	References
Prostasomes	N/A	Seminal plasma	1978	[174, 175]
Exosomes	40–100 nm	Cancer cells	1981	[55]
Epididymosomes	N/A	Epididymal fluid	1985	[176]
Argosomes	ND	*Drosophila* imaginal disc epithelium	2001	[177]
Archeosomes	ND	*Methanobrevibacter smithii*	2004	[178]
Dexosomes	60–90 nm	Tumor cell lines	2005	[179]
Prominosomes	50–80 nm	Neural stem cells	2005	[180]
Oncosomes	100–500 nm	Astrocytoma cells	2008	[181]

ND not defined, *N/A* not available

types of MVs are largely subject to debate, mainly due to the way these organelles are isolated (e.g., ultracentrifugation, use of a sucrose gradient, by biological markers) [66], the precise context of study, or vesicle-specific properties [62]. For the purpose of this chapter, MV will be used for all organelles in this general category between 40 and 1,000 nm in diameter unless explicitly noted.

MVs are evolutionarily conserved, which suggests that they carry out important biological functions [64]. Cells are known to secrete MVs due to factors such as environmental stress, cellular activation, or apoptosis [49, 62]. The composition of MVs varies between cell types and environmental conditions, and a formal classification based on vesicles components is still being actively debated [67, 68]. As an example, exosomes are usually characterized by membrane markers (CD63, CD81, CD9, TSG101, and Alix) that seem to be limited to this type of MVs [68].

One area of MV research is directed specifically at the understanding of how vesicles are formed and released, which may be a universal, conserved cellular response [69]. Unlike apoptotic bodies which are formed from randomized plasma membrane blebbing, exosomes are generally derived from the inward budding of late endosomes that leads to the formation of intraluminal vesicles that eventually fuse to the plasma membrane for extracellular release [70]. At least three mechanisms for MV generation have been proposed: (a) decay of dying cells into apoptotic bodies, (b) cellular plasma membrane blebbing ectosomes, and (c) emission of plasma membrane into exosomes [49, 62, 69, 71, 72]. The result is outward budding and fission of vesicles from the tumor cell surface (Fig. 17.2). Some observations have also described a direct formation and release of exosomes from cytoplasmic membrane budding of immune cells [73, 74].

Reports of MV isolation, size, density, and morphology should be interpreted with caution. Due to their small size and heterogeneity, conventional methods of classification for this type of biomolecule have proven to be difficult [67, 68]. MVs are hard to detect with basic light microscopy and flow cytometry, because they are generally less than 200 nm. Several methods have been in use for isolation and purification of MVs, ranging from centrifugation techniques to antibody precipitation [75, 76]. The method most commonly used is a differential ultracentrifugation including a sucrose density gradient [64, 68]. However, techniques such as these have been shown to change the size and morphology of MVs. For instance, while MVs are frequently described as cup shaped in the literature [77–80], Thery et al. demonstrated that this morphology was actually an artifact caused by the fixation process for transmission electron microscopy. In another study by Connor et al., repeated freeze–thaw cycles of plasma rich in platelets caused a considerable increase of annexin V+MVs [81]. The MV count in a sample can vary with storage time, temperature, dilution buffer used, and agitation. This shows that environmental factors affect the release of MVs and that cells often release MVs due to stress [64].

4 Microvesicle Cargo

Recent evidence shows that MVs can act as a unique vehicle for the release of soluble and insoluble molecules [49], including lipids, proteins, and nucleic acids [61, 82]. MV uptake into a target cell may allow the exchange of these molecules that are otherwise taken up by MV-producing cells. Such a mechanism would affect the target cell phenotype [49]. MVs are enriched in lipids like ceramides, cholesterol, and sphyngomyelin, which promote vesicle release and play important roles in

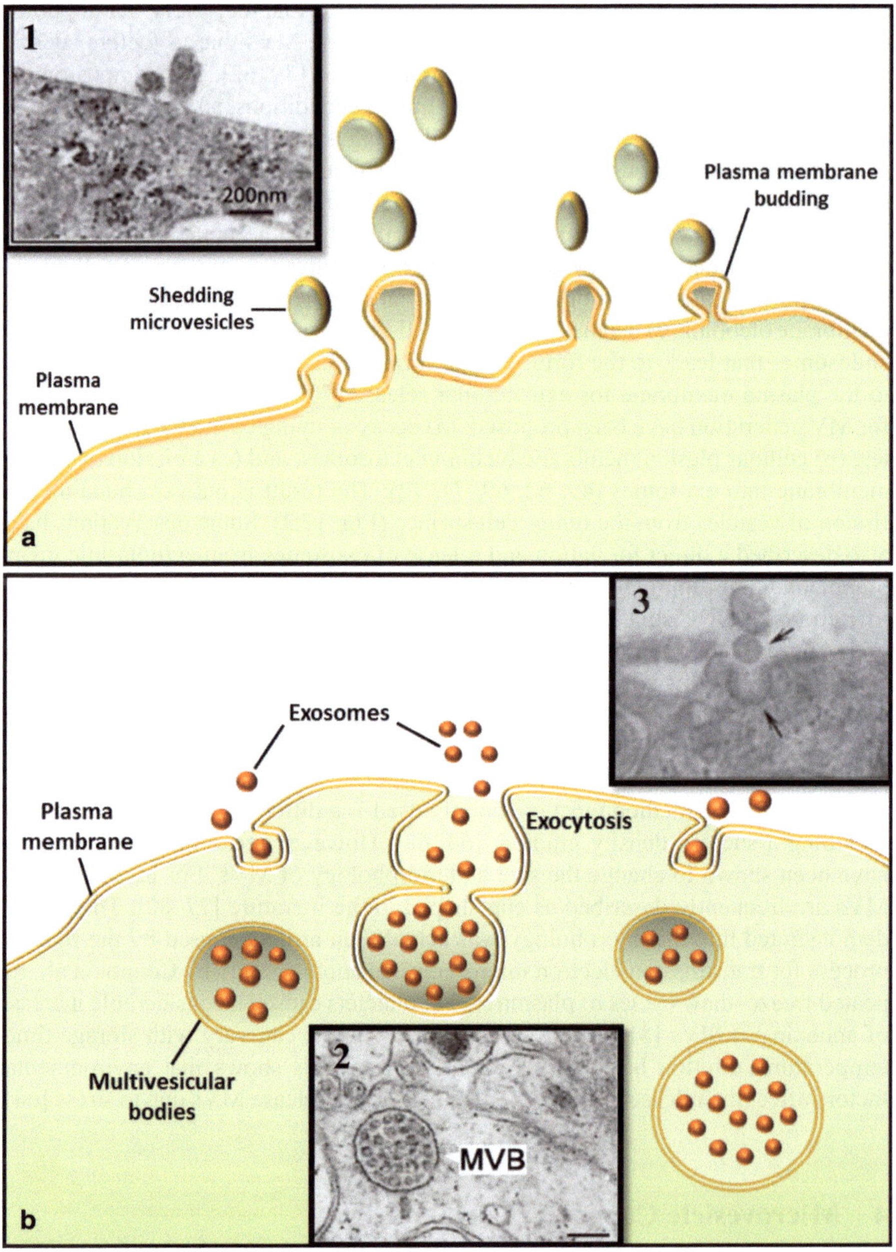

Fig. 17.2 Schematic representation of formation and release of microvesicles and exosomes: **a** Formation of tumor microvesicles (MVs). As a result of cell stimulation MVs are shed by budding of plasma membrane of the cancer cell. (*Inset 1*) Electron microscopy showing shedding MVs [173]. **b** Accumulation and release of exosomes. Exosomes are accumulated in the MV bodies (MVBs) and are released by exocytosis when the MVBs fuse with the plasma membrane as a result of cell stimulation. (*Inset 2*) TEM showing MVB containing many exosomes [174]. (*Inset 3*) TEM showing exosomal release [175]

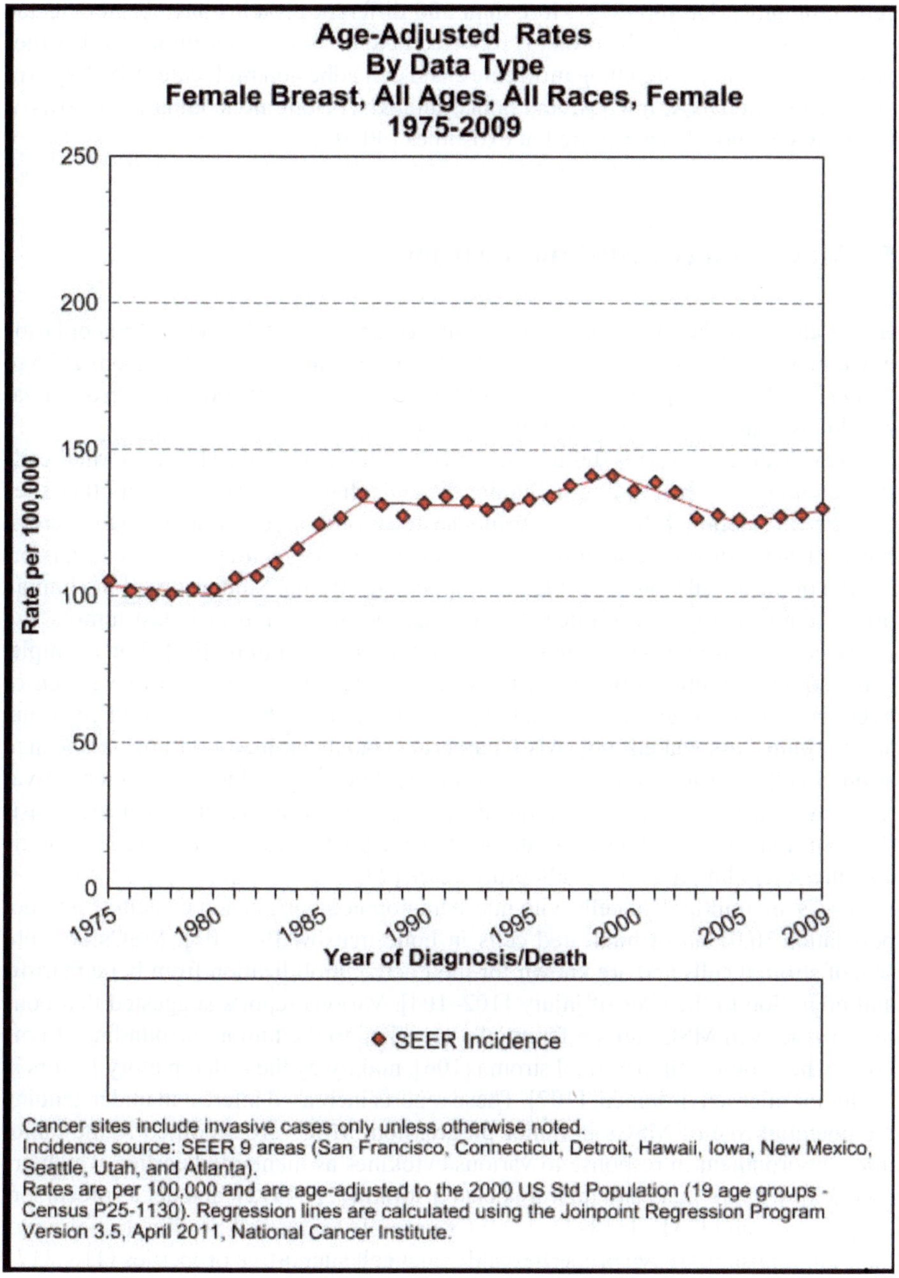

Fig. 17.3 Breast cancer rate of incidence per Surveillance Epidemiology and End Results (SEER) database between 1975 and 2009 in females from all ages and all races

cell communication [83–87]. More than 300 different proteins have been detected within exosomes [88]. The proteins reported belong, but are not limited to families of surface receptors, signaling molecules, and cell adhesion molecules [88, 89]. Nucleic acids like DNA, mRNA, and noncoding RNA (long noncoding RNA, tRNA, and miRNA) have been reported in exosomes [90–92].

5 MSCs and Breast Cancer Stroma

Breast cancer is the most commonly identified and one of the deadliest neoplasms in women worldwide. The rate of incidence has increased by 1.5-fold from 1975 to 2009 (Fig. 17.3). Approximately 230,000 diagnoses and 40,000 deaths from invasive breast cancer are expected in the USA in 2012 [93].

Breast cancer stroma is defined as the cells which are derived from normal cells surrounding the breast tumor, including the cells that are recruited from other sites through circulation [18, 94]. In response to stroma-modulating growth factors, homeostasis is disrupted in normal stromal cells. This eventually causes a transformation of these cells into breast tumor-supporting stroma. Studies involving human breast tumor xenografts in mice showed that the interaction of breast tumors and its associated stroma determines the phenotype of the tumor [95]. For example, gene expression studies on breast tumor stroma revealed increased expression of hypoxia- and angiogenesis-associated genes as compared to the stromal cells that are not tumor associated [96]. Also, paracrine signals emitted by tumor-associated stromal cells recruit distinct populations of leukocytes producing receptor activator of nuclear factor kappa-B ligand (RANKL)-promoting breast cancer metastasis [97]. Stromal cells modulate the tumor-invasive properties by secreting proteolytic enzymes including matrix metalloproteinases [98].

MSCs are multipotent cells with non-hematopoietic origin and constitute a minor population (0.01 %) of nucleated cells in bone marrow [99–101]. MSCs are subsets of stromal cells and are known for their active mobilization from bone marrow and migration to the sites of injury [102–104]. Various reports suggested that bone marrow-derived MSCs are preferentially recruited to the tumor-surrounding stroma [105] when compared to normal stroma [106], mainly by the inflammatory factors in the tumor microenvironment [107]. These reports increased interest in understanding the potential role of MSCs in tumor progression. MSCs are recruited to the tumor microenvironment in response to various cytokines as mentioned earlier, which are secreted by tumor cells and their associated stroma [108–112], and act as precursors for pericytes and CAFs [113–115]. MSCs promote tumor cell proliferation through their immunosuppressive properties and direct cell supportive properties [116, 117]. Earlier studies from our group suggest that under nutrient-deprived conditions, the MSCs associated with tumor stroma undergo autophagy, thereby secreting anti-apoptotic factors and facilitating breast tumor support [35] (Fig. 17.4). These studies suggest that targeting tumor-associated stromal cells along with tumor cells provides a more effective treatment strategy for breast cancer [10, 118].

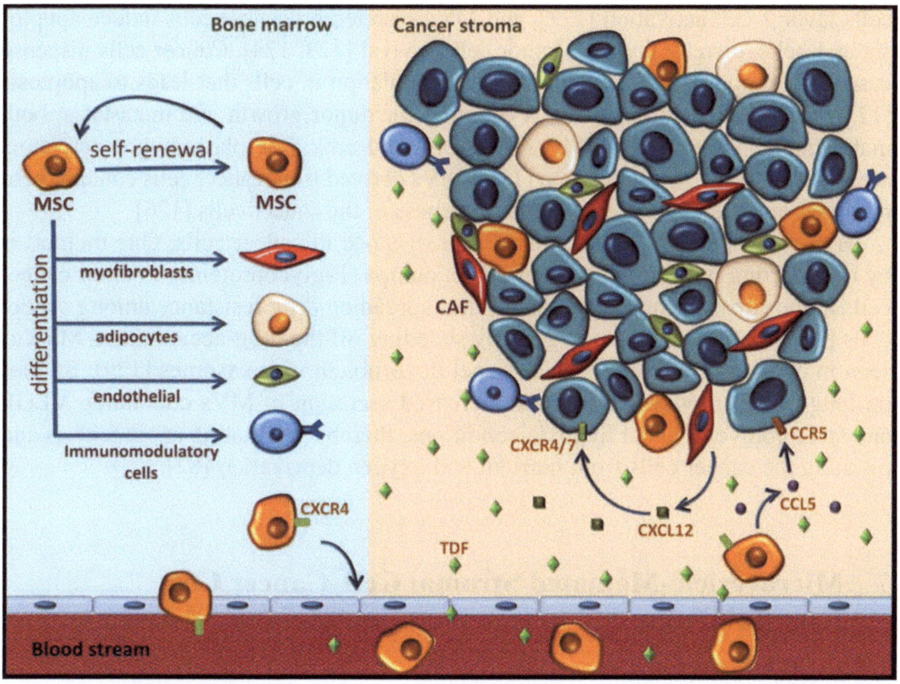

Fig. 17.4 Schematic depicting interactions between the bone marrow niche (*left*) and the cancer stroma. MSCs in bone marrow self-renew and also differentiate into other cell types that exhibit tumor stromal supportive properties—such as myofibroblasts, adipocytes, endothelial cells, and immunomodulatory cells. Cancer cells secrete tumor-derived soluble factors, which recruit MSCs to the cancer stroma. Recruited MSCs secreted various factors, for example CCL5 which interacts with its receptor CCR5 expressed on cancer cells. MSCs also differentiate to CAFs in cancer stroma, and these CAFs secrete CXCL12 which interacts with cancer cells through CXCR4/7 receptors

6 Microvesicular Transfer

MV secretion by most of the normal cell types is a regular physiological phenomenon and a mode of intercellular communication for cell growth and activation [64]. The evidence that MVs were involved in cancer was first documented in patients with Hodgkin's disease in the late 1970s [54]. Since then, various studies have revealed the active involvement of MVs in different stages of cancer progression [119]. In human breast cancer cell lines, there is a positive correlation between the amount of MVs released and the in vitro invasiveness of the cells [58]. Similar results were observed in in-vivo studies on ovarian cancer fluids [120]. MV secretion can provide either favorable or unfavorable features to cells, depending on the cell types that are released [121]. Cancer cells can use MVs to evade protective mechanisms of the organism by inducing immune tolerance, expression of pro-apoptotic signals, ECM remodeling, drug resistance, and in other various ways. MVs derived from antigen-presenting cells favor T cell activation [122], and MVs secreted by cancer cells induce apoptosis in T cells, thereby favoring tumor cell survival [123, 124]. Cancer cells dispense caspase 3 through MVs, preventing its accumulation in cells that leads to apoptosis [125]. Degradation of the ECM is essential for tumor growth and metastasis; both matrix metalloproteinase (MMP-2, MMP-9) and urokinase plasminogen activators are required for ECM degradation [126]. MVs derived from cancer cells contain these proteases and thereby increase the invasiveness of the cancer cells [126].

MVs are capable of developing drug resistance in cancer cells. One method is by transporting multidrug resistant efflux pumps (P-glycoproteins) to other cancer cells in the surrounding environment, thus spreading drug resistance among cancer cells [127, 128]. Another strategy is by shedding off the drug-accumulated MVs as seen in ovarian cancer cells which expel doxorubicin via exosomes [129]. Studies on lung cancer models showed the increased secretion of MVs containing VEGF and sphingomyelin under hypoxia conditions, thereby facilitating angiogenesis and rescuing the cancer cells from nutrient and oxygen deprivation [82].

7 Microvesicle-Mediated Stromal Cell–Cancer Cell Cros Talk

Cancer cells actively interact with stromal cells through MVs. One study on invasive prostate cancer cell lines showed that cancer cells not only activate fibroblasts in tumor stroma by secreting MVs, but also promote MV release from these activated fibroblasts to advance their own migration and invasion [130]. MVs contribute to the transformation of normal cells into cancer cells, as studies on breast carcinoma and glioma cells showed that MVs transfer tissue transglutaminase (tTG) from cancer cells to both normal fibroblasts and epithelial cells, as a result of which these cells transform to cancer cells [42].

Similar to cancer cells, normal cells also secrete MVs. Their function depends on the phenotype of the parent cells [131]. For example, MVs secreted by MSCs in kidney injury mice models have shown to have a potential renoprotective effect [132, 133]. In myocardial ischemia model and in a brain injury study, MVs secreted by MSCs provided cardioprotection and protection against stroke, respectively [134, 135].

8 microRNA

miRNA are noncoding small RNA (19–22 nucleotides in length) that play a wide spectrum of roles on both pre-transcriptional and posttranscriptional gene expression. miRNA are thought to regulate at least the third of the human genome; indeed, if a unique miRNA can target different mRNAs, it can also be targeted by different miRNA species. miRNA act principally by a specific binding to the 3′ untranslated region (UTR) coding mRNA [136, 137]. miRNA are themselves regulated by different factors such as epigenetics [138] or competitive endogenous RNA (ceRNA) [139].

Circulating miRNAs have been detected in various body fluids including serum, plasma, amniotic fluid, saliva, sweat, urine, and milk [140]. Data compilation shows that circulating miRNA originate mainly from MVs secreted by all kinds of cells or from MV-free miRNAs found in human plasma and serum. These miRNAs directly bind to Argonaute 2 (ago2) proteins and form a very stable complex. Because of the high stability of the miRNA–ago2 complex, it is difficult to trace the original source of these circulating miRNAs, since they can result from apoptotic or necrotic cells (e.g., short-life platelets) as well as from pathologies (e.g., cancer) [141–143].

miRNA levels are highly regulated in normal cells to maintain cell homeostasis. However, they are deregulated in cancerous cells and participate in the proliferation, dedifferentation, migration, and metastasis of these cells [144, 145]. Some miRNAs are more recurrently associated to cancer patterns. For example, miR-34a, miR-155, and miR-199a are found upregulated or downregulated in breast cancer [146–149].

Single miRNAs have been identified to regulate the balance between normal and cancer cells. For example, the transfer of secreted miR-143 from normal prostate cells induces the growth inhibition of prostate cancer cells, where miR1–143 is downregulated [150]. In another study, the MVs secreted by different cancer cell lines contain specific miRNAs (e.g., miR-9) that promote endothelial cell migration [151].

The presence of miRNA in cancer cell MVs seems to be driven selectively. For example, it has been shown that secreted breast cancer cell lines secrete a variety of MVs containing more abundant and more diverse miRNA species compared to those secreted by normal epithelial cells [152, 153].

9 microRNA-Mediated Cross Talk

The presence of functional mRNAs and miRNAs within mouse and human exosomes was first reported by Valadi et al. in 2007 [154]. Exosomal fractions were isolated from the supernatant of a mouse mast cell line, primary mouse bone marrow-derived mast cells, and a human mast cell line and then submitted for microarray analysis. More than 1,000 different miRNAs and 100 unique miRNAs were detected in these exosomes. Interestingly, most abundant miRNA in the exosomes were generally different from the most abundant miRNA found in the donor cells, suggesting that some miRNAs may be uniquely packed into exosomes. This study also demonstrated that mouse proteins could be synthesized within human mast cells when exosomes derived from mouse mast cells were shuttled into the cultured human mast cells in vitro [154].

Since then, the horizontal transfer of mRNA and miRNA has been reported in numerous studies between normal cells [85, 155–157], from virus-infected to normal cells [158], between embryonic stem cells [159, 160], from MSCs to cancer cells [161], between cancer cells [162, 163], from cancer to normal cells [164], and from normal to cancer cells [47, 162]. That exosomes can shuttle RNAs between cells suggest that RNAs may play a role in the regulation of gene expression in recipient cells. The ratio of RNA fragments found within exosomes varies depending on the cell type from which the exosomes originated [163].

Gibblings et al. demonstrated evidence that the loading of miRNAs into exosomes may not in fact be random, but instead controlled by specific proteins involved in the miRNA network. By showing the presence of ago2 protein and a noticeable enrichment of GW182 in purified exosomes-like vesicles, these findings suggested the selective sorting of GW182 into exosomes [165]. Ceramide is a major component of exosomes and plays an important role in their external secretion [83, 84]. It is synthesized from sphingomyelin by the enzyme sphingomyelinase 2 (nSMase2) [163]. nSMase2 also has a role in the external secretion of miRNAs within exosomes [85].

Additional studies demonstrating the uptake of secreted miRNA followed by induced function in the recipient cells have been reported. Katakowski et al. [166] transfected rat gliosarcoma cells to express miRNA that lacks homology in rat. The coculture of these cells with cells that expressed a gene reporter, which contained a complementary sequence of the miRNA, resulted in a reduction of the gene reporter expression. Carbenoxolone addition to cocultures prevented this effect, so the authors suggested that gap junction communication regulates intercellular transfer of miRNA. Other studies suggest the same miRNA transfer mechanism via gap junctions between cardiomyocytes in culture [167] and between bone marrow stromal cells and a breast cancer cell line [43].

Recent studies suggest the transfer of miRNAs between embryonic stem cells through MVs [159, 160]. Miltenbrunn et al. showed that exosomes of immune cells transfer functional miRNAs involving unidirectional and antigen-dependent driven mechanism. Interestingly, inhibition of exosome production by targeting neutral nSMase2 impairs transfer of miRNAs to recipient cells [85, 168]. miRNAs are well described to bind to the RNA-induced silencing complex (RISC) [169], but some miRNA have been reported to directly bind to other protein, act as a decoy, and prevent it from blocking translation of mRNAs [170]. Fabbri et al. showed that tumor-secreted exosomes from lung cancer lines contains miRNAs that can bind directly to Toll-like receptors (TLR), triggering a TLR-mediated prometastatic inflammatory response that may lead to tumor growth and metastasis [171]. Thus, the role of transferred miRNAs secreted by a donor cells not only can be limited on post-transcriptional effects in the recipient cells, but also can act as a paracrine signal.

10 Microvesicles and Metastasis

Metastasis is the leading cause of cancer death, yet it has been an enigma for researchers. It is considered a mechanistically inefficient process because of its dependence on very regulated and controlled systemic fueling. The harbinger to metastasis is the pre-metastatic niche formation. This niche is presumed to play a role in dormancy, relapse, and development of metastasis. An emerging potential of MVs is formulating the pre-metastatic niche. Ghasemi et al. have termed these MVs "metastasomes" and hypothesized that they may aid foundation of the secondary lesions via a "malignant trait" spreading system that regulates the interactions between tumor tissue-specific RNA being transferred via metastasomes and the cell-type/tissue-specific RNA within the target organ, thus serving as tumor–organ matchmakers [172]. Recent studies have shown that these MVs are actually "customized" to the cancers. In one study comparing MVs from breast cancer cells and normal cells, the selectively exported miRNAs, whose release is increased in malignant cells, are packaged in structures that are different from those that carry neutrally released miRNAs [152, 153].

In closing, the recent discoveries on the study of tumor-derived MVs reveal new insights into the cellular basis of tumor stromal support. There is potential to translate this information into developing novel innovative approaches for cancer diagnostics and personalized therapy. The complexity and variety of the MV cargo implicate them in a multipronged approach towards tumor support, and hijacking their functions to engineer tumor-inhibitory MVs seems plausible. Most of the current knowledge is on the molecular profiling of the circulating MVs as biomarkers for cancers, which induces multiple platforms for personalized diagnostics.

Acknowledgments The authors disclose no potential conflicts of interest

References

1. Joyce JA, Pollard JW. Microenvironmental regulation of metastasis. Nat Rev Cancer. 2009 Apr;9(4):239–52 (PubMed PMID: 19279573. Pubmed Central PMCID: 3251309).
2. Artacho-Cordon A, Artacho-Cordon F, Rios-Arrabal S, Calvente I, Nunez MI. Tumor microenvironment and breast cancer progression: a complex scenario. Cancer Biol Ther. 2012 Jan 1;13(1):14–24 (PubMed PMID: 22336584).
3. Polyak K, Haviv I, Campbell IG. Co-evolution of tumor cells and their microenvironment. Trends Genet. 2009 Jan;25(1):30–8 (PubMed PMID: 19054589).
4. Theocharis AD, Skandalis SS, Tzanakakis GN, Karamanos NK. Proteoglycans in health and disease: novel roles for proteoglycans in malignancy and their pharmacological targeting. FEBS J. 2010 Oct;277(19):3904–23 (PubMed PMID: 20840587).
5. Edwards IJ. Proteoglycans in prostate cancer. Nat Rev Urol. 2012 Apr;9(4):196–206 (PubMed PMID: 22349653).
6. Suryapranata H, Serruys PW, de Feyter PJ, van den Brand M, Beatt K, van Domburg R, et al. Coronary angioplasty immediately after thrombolysis in 115 consecutive patients with acute myocardial infarction. Am Heart J. 1988 Mar;115(3):519–29 (PubMed PMID: 2964182).
7. Louie E, Nik S, Chen JS, Schmidt M, Song B, Pacson C, et al. Identification of a stem-like cell population by exposing metastatic breast cancer cell lines to repetitive cycles of hypoxia and reoxygenation. Breast Cancer Res. 2010;12(6):R94 (PubMed PMID: 21067584. Pubmed Central PMCID: 3046435).
8. Allinen M, Beroukhim R, Cai L, Brennan C, Lahti-Domenici J, Huang H, et al. Molecular characterization of the tumor microenvironment in breast cancer. Cancer Cell. 2004 Jul;6(1):17–32 (PubMed PMID: 15261139).
9. Kalluri R, Zeisberg M. Fibroblasts in cancer. Nat Rev Cancer. 2006 May;6(5):392–401 (PubMed PMID: 16572188).
10. Mao Y, Keller ET, Garfield DH, Shen K, Wang J. Stromal cells in tumor microenvironment and breast cancer. Cancer Metastasis Rev. 2013 Jun;32(1–2):303–15 (PubMed PMID: 23114846).
11. Cuiffo BG, Karnoub AE. Mesenchymal stem cells in tumor development: emerging roles and concepts. Cell Adh Migr. 2012 May-Jun;6(3):220–30 (PubMed PMID: 22863739. Pubmed Central PMCID: 3427236).

12. Loeffler M, Kruger JA, Niethammer AG, Reisfeld RA. Targeting tumor-associated fibroblasts improves cancer chemotherapy by increasing intratumoral drug uptake. J Clin Invest. 2006 Jul;116(7):1955–62 (PubMed PMID: 16794736. Pubmed Central PMCID: 1481657).
13. Cirri P, Chiarugi P. Cancer-associated-fibroblasts and tumour cells: a diabolic liaison driving cancer progression. Cancer Metastasis Rev. 2012 Jun;31(1–2):195–208 (PubMed PMID: 22101652).
14. Holmgren L, O'Reilly MS, Folkman J. Dormancy of micrometastases: balanced proliferation and apoptosis in the presence of angiogenesis suppression. Nat Med. 1995 Feb;1(2):149–53 (PubMed PMID: 7585012).
15. Naumov GN, MacDonald IC, Weinmeister PM, Kerkvliet N, Nadkarni KV, Wilson SM, et al. Persistence of solitary mammary carcinoma cells in a secondary site: a possible contributor to dormancy. Cancer Res. 2002 Apr 1;62(7):2162–8 (PubMed PMID: 11929839).
16. Kang SY, Watnick RS. Regulation of tumor dormancy as a function of tumor-mediated paracrine regulation of stromal Tsp-1 and VEGF expression. APMIS. 2008 Jul-Aug;116(7–8):638–47 (PubMed PMID: 18834408).
17. Paget S. The distribution of secondary growths in cancer of the breast. 1889. Cancer Metastasis Rev. 1989 Aug;8(2):98–101 (PubMed PMID: 2673568).
18. Liotta LA, Kohn EC. The microenvironment of the tumour-host interface. Nature. 2001 May 17;411(6835):375–9 (PubMed PMID: 11357145).
19. Bronzert DA, Pantazis P, Antoniades HN, Kasid A, Davidson N, Dickson RB, et al. Synthesis and secretion of platelet-derived growth factor by human breast cancer cell lines. Proc Natl Acad Sci U S A. 1987 Aug;84(16):5763–7 (PubMed PMID: 3039506. Pubmed Central PMCID: 298943).
20. Shao ZM, Nguyen M, Barsky SH. Human breast carcinoma desmoplasia is PDGF initiated. Oncogene. 2000 Sep 7;19(38):4337–45 (PubMed PMID: 10980609).
21. Strutz F, Zeisberg M, Hemmerlein B, Sattler B, Hummel K, Becker V, et al. Basic fibroblast growth factor expression is increased in human renal fibrogenesis and may mediate autocrine fibroblast proliferation. Kidney Int. 2000 Apr;57(4):1521–38 (PubMed PMID: 10760088).
22. Kojima Y, Acar A, Eaton EN, Mellody KT, Scheel C, Ben-Porath I, et al. Autocrine TGF-beta and stromal cell-derived factor-1 (SDF-1) signaling drives the evolution of tumor-promoting mammary stromal myofibroblasts. Proc Natl Acad Sci U S A. 2010 Nov 16;107(46):20009–14 (PubMed PMID: 21041659. Pubmed Central PMCID: 2993333).
23. Bergers G, Benjamin LE. Tumorigenesis and the angiogenic switch. Nat Rev Cancer. 2003 Jun;3(6):401–10 (PubMed PMID: 12778130).
24. Coussens LM, Werb Z. Inflammation and cancer. Nature. 2002 Dec 19–26;420(6917):860–7 (PubMed PMID: 12490959. Pubmed Central PMCID: 2803035).
25. Krock BL, Skuli N, Simon MC. Hypoxia-induced angiogenesis: good and evil. Genes Cancer. 2011 Dec;2(12):1117–33 (PubMed PMID: 22866203. Pubmed Central PMCID: 3411127).
26. Li Z, Bao S, Wu Q, Wang H, Eyler C, Sathornsumetee S, et al. Hypoxia-inducible factors regulate tumorigenic capacity of glioma stem cells. Cancer Cell. 2009 Jun 2;15(6):501–13 (PubMed PMID: 19477429. Pubmed Central PMCID: 2693960).
27. Martinez-Outschoorn UE, Trimmer C, Lin Z, Whitaker-Menezes D, Chiavarina B, Zhou J, et al. Autophagy in cancer associated fibroblasts promotes tumor cell survival: Role of hypoxia, HIF1 induction and NFkappaB activation in the tumor stromal microenvironment. Cell Cycle. 2010 Sep 1;9(17):3515–33 (PubMed PMID: 20855962. Pubmed Central PMCID: 3047617).
28. Keith B, Johnson RS, Simon MC. HIF1alpha and HIF2alpha: sibling rivalry in hypoxic tumour growth and progression. Nat Rev Cancer. 2012 Jan;12(1):9–22 (PubMed PMID: 22169972. Pubmed Central PMCID: 3401912).
29. Hung SC, Pochampally RR, Hsu SC, Sanchez C, Chen SC, Spees J, et al. Short-term exposure of multipotent stromal cells to low oxygen increases their expression of CX3CR1 and CXCR4 and their engraftment in vivo. PLoS One. 2007;2(5):e416 (PubMed PMID: 17476338. Pubmed Central PMCID: 1855077).

30. Quante M, Tu SP, Tomita H, Gonda T, Wang SS, Takashi S, et al. Bone marrow-derived myofibroblasts contribute to the mesenchymal stem cell niche and promote tumor growth. Cancer Cell. 2011 Feb 15;19(2):257–72 (PubMed PMID: 21316604. Pubmed Central PMCID: 3060401).
31. Shibata W, Ariyama H, Westphalen CB, Worthley DL, Muthupalani S, Asfaha S, et al. Stromal cell-derived factor-1 overexpression induces gastric dysplasia through expansion of stromal myofibroblasts and epithelial progenitors. Gut. 2013 Feb;62(2):192–200 (PubMed PMID: 22362916).
32. Jin S, White E. Role of autophagy in cancer: management of metabolic stress. Autophagy. 2007 Jan-Feb;3(1):28–31 (PubMed PMID: 16969128. Pubmed Central PMCID: 2770734).
33. Pochampally RR, Smith JR, Ylostalo J, Prockop DJ. Serum deprivation of human marrow stromal cells (hMSCs) selects for a subpopulation of early progenitor cells with enhanced expression of OCT-4 and other embryonic genes. Blood. 2004 Mar 1;103(5):1647–52 (PubMed PMID: 14630823).
34. Sanchez C, Oskowitz A, Pochampally RR. Epigenetic reprogramming of IGF1 and leptin genes by serum deprivation in multipotential mesenchymal stromal cells. Stem Cells. 2009 Feb;27(2):375–82 (PubMed PMID: 19038795).
35. Sanchez CG, Penfornis P, Oskowitz AZ, Boonjindasup AG, Cai DZ, Dhule SS, et al. Activation of autophagy in mesenchymal stem cells provides tumor stromal support. Carcinogenesis. 2011 Jul;32(7):964–72 (PubMed PMID: 21317300. Pubmed Central PMCID: 3128555).
36. Oskowitz A, McFerrin H, Gutschow M, Carter ML, Pochampally R. Serum-deprived human multipotent mesenchymal stromal cells (MSCs) are highly angiogenic. Stem Cell Res. 2011 May;6(3):215–25 (PubMed PMID: 21421339).
37. Kandouz M, Batist G. Gap junctions and connexins as therapeutic targets in cancer. Expert Opin Ther Targets. 2010 Jul;14(7):681–92 (PubMed PMID: 20446866).
38. Mroue RM, El-Sabban ME, Talhouk RS. Connexins and the gap in context. Integr Biol (Camb). 2011 Apr;3(4):255–66 (PubMed PMID: 21437329).
39. Gilleron J, Carette D, Chevallier D, Segretain D, Pointis G. Molecular connexin partner remodeling orchestrates connexin traffic: from physiology to pathophysiology. Crit Rev Biochem Mol Biol. 2012 Sep;47(5):407–23 (PubMed PMID: 22551357).
40. Tsuyada A, Chow A, Wu J, Somlo G, Chu P, Loera S, et al. CCL2 mediates cross-talk between cancer cells and stromal fibroblasts that regulates breast cancer stem cells. Cancer Res. 2012 Jun 1;72(11):2768–79 (PubMed PMID: 22472119. Pubmed Central PMCID: 3367125).
41. Orimo A, Gupta PB, Sgroi DC, Arenzana-Seisdedos F, Delaunay T, Naeem R, et al. Stromal fibroblasts present in invasive human breast carcinomas promote tumor growth and angiogenesis through elevated SDF-1/CXCL12 secretion. Cell. 2005 May 6;121(3):335–48 (PubMed PMID: 15882617).
42. Antonyak MA, Li B, Boroughs LK, Johnson JL, Druso JE, Bryant KL, et al. Cancer cell-derived microvesicles induce transformation by transferring tissue transglutaminase and fibronectin to recipient cells. Proc Natl Acad Sci U S A. 2011 Mar 22;108(12):4852–7 (PubMed PMID: 21368175. Pubmed Central PMCID: 3064359).
43. Lim PK, Bliss SA, Patel SA, Taborga M, Dave MA, Gregory LA, et al. Gap junction-mediated import of microRNA from bone marrow stromal cells can elicit cell cycle quiescence in breast cancer cells. Cancer Res. 2011 Mar 1;71(5):1550–60 (PubMed PMID: 21343399).
44. Pap E. The role of microvesicles in malignancies. Adv Exp Med Biol. 2011;714:183–99 (PubMed PMID: 21506015).
45. Luga V, Zhang L, Viloria-Petit AM, Ogunjimi AA, Inanlou MR, Chiu E, et al. Exosomes mediate stromal mobilization of autocrine Wnt-PCP signaling in breast cancer cell migration. Cell. 2012 Dec 21;151(7):1542–56 (PubMed PMID: 23260141).
46. Janowska-Wieczorek A, Marquez-Curtis LA, Wysoczynski M, Ratajczak MZ. Enhancing effect of platelet-derived microvesicles on the invasive potential of breast cancer cells. Transfusion. 2006 Jul;46(7):1199–209 (PubMed PMID: 16836568).
47. Yang M, Chen J, Su F, Yu B, Su F, Lin L, et al. Microvesicles secreted by macrophages shuttle invasion-potentiating microRNAs into breast cancer cells. Mol Cancer. 2011;10:117 (PubMed PMID: 21939504. Pubmed Central PMCID: 3190352).

48. Howlett AR, Bissell MJ. The influence of tissue microenvironment (stroma and extracellular matrix) on the development and function of mammary epithelium. Epithelial Cell Biol. 1993 Apr;2(2):79–89 (PubMed PMID: 8353596).
49. Lee TH, D'Asti E, Magnus N, Al-Nedawi K, Meehan B, Rak J. Microvesicles as mediators of intercellular communication in cancer–the emerging science of cellular 'debris'. Semin Immunopathol. 2011 Sep;33(5):455–67 (PubMed PMID: 21318413).
50. Wolf P. The nature and significance of platelet products in human plasma. Br J Haematol. 1967 May;13(3):269–88 (PubMed PMID: 6025241).
51. Simons M, Raposo G. Exosomes–vesicular carriers for intercellular communication. Curr Opin Cell Biol. 2009 Aug;21(4):575–81 (PubMed PMID: 19442504).
52. Cocucci E, Racchetti G, Meldolesi J. Shedding microvesicles: artefacts no more. Trends Cell Biol. 2009 Feb;19(2):43–51 (PubMed PMID: 19144520).
53. Bang C, Thum T. Exosomes: new players in cell-cell communication. Int J Biochem Cell Biol. 2012 Nov;44(11):2060–4 (PubMed PMID: 22903023).
54. Friend C, Marovitz W, Henie G, Henie W, Tsuei D, Hirschhorn K, et al. Observations on cell lines derived from a patient with Hodgkin's disease. Cancer Res. 1978 Aug;38(8):2581–91 (PubMed PMID: 78764).
55. Trams EG, Lauter CJ, Salem N Jr, Heine U. Exfoliation of membrane ecto-enzymes in the form of micro-vesicles. Biochim Biophys Acta. 1981 Jul 6;645(1):63–70 (PubMed PMID: 6266476).
56. Harding C, Heuser J, Stahl P. Receptor-mediated endocytosis of transferrin and recycling of the transferrin receptor in rat reticulocytes. J Cell Biol. 1983 Aug;97(2):329–39 (PubMed PMID: 6309857. Pubmed Central PMCID: 2112509).
57. Pan BT, Teng K, Wu C, Adam M, Johnstone RM. Electron microscopic evidence for externalization of the transferrin receptor in vesicular form in sheep reticulocytes. J Cell Biol. 1985 Sep;101(3):942–8 (PubMed PMID: 2993317. Pubmed Central PMCID: 2113705).
58. Ginestra A, La Placa MD, Saladino F, Cassara D, Nagase H, Vittorelli ML. The amount and proteolytic content of vesicles shed by human cancer cell lines correlates with their in vitro invasiveness. Anticancer Res. 1998 Sep-Oct;18(5A):3433–7 (PubMed PMID: 9858920).
59. Kim HK, Song KS, Park YS, Kang YH, Lee YJ, Lee KR, et al. Elevated levels of circulating platelet microparticles, VEGF, IL-6 and RANTES in patients with gastric cancer: possible role of a metastasis predictor. Eur J Cancer. 2003 Jan;39(2):184–91 (PubMed PMID: 12509950).
60. Zwicker JI, Liebman HA, Neuberg D, Lacroix R, Bauer KA, Furie BC, et al. Tumor-derived tissue factor-bearing microparticles are associated with venous thromboembolic events in malignancy. Clin Cancer Res. 2009 Nov 15;15(22):6830–40 (PubMed PMID: 19861441. Pubmed Central PMCID: 2783253).
61. D'Souza-Schorey C, Clancy JW. Tumor-derived microvesicles: shedding light on novel microenvironment modulators and prospective cancer biomarkers. Genes Dev. 2012 Jun 15;26(12):1287–99 (PubMed PMID: 22713869. Pubmed Central PMCID: 3387656).
62. Rak J. Microparticles in cancer. Semin Thromb Hemost. 2010 Nov;36(8):888–906 (PubMed PMID: 21049390).
63. Cocucci E, Meldolesi J. Ectosomes. Curr Biol. 2011 Dec 6;21(23):R940–1 (PubMed PMID: 22153157).
64. Gyorgy B, Szabo TG, Pasztoi M, Pal Z, Misjak P, Aradi B, et al. Membrane vesicles, current state-of-the-art: emerging role of extracellular vesicles. Cell Mol Life Sci. 2011 Aug;68(16):2667–88 (PubMed PMID: 21560073. Pubmed Central PMCID: 3142546).
65. Ronquist G. Prostasomes are mediators of intercellular communication: from basic research to clinical implications. J Intern Med. 2012 Apr;271(4):400–13 (PubMed PMID: 22112042).
66. Tauro BJ, Greening DW, Mathias RA, Ji H, Mathivanan S, Scott AM, et al. Comparison of ultracentrifugation, density gradient separation, and immunoaffinity capture methods for isolating human colon cancer cell line LIM1863-derived exosomes. Methods. 2012 Feb;56(2):293–304 (PubMed PMID: 22285593).
67. Mathivanan S, Ji H, Simpson RJ. Exosomes: extracellular organelles important in intercellular communication. J Proteomics. 2010 Sep 10;73(10):1907–20 (PubMed PMID: 20601276).
68. van der Pol E, Boing AN, Harrison P, Sturk A, Nieuwland R. Classification, functions, and clinical relevance of extracellular vesicles. Pharmacol Rev. 2012 Jul;64(3):676–705 (PubMed PMID: 22722893).

69. Thery C, Ostrowski M, Segura E. Membrane vesicles as conveyors of immune responses. Nat Rev Immunol. 2009 Aug;9(8):581–93 (PubMed PMID: 19498381).
70. Ludwig AK, Giebel B. Exosomes: small vesicles participating in intercellular communication. Int J Biochem Cell Biol. 2012 Jan;44(1):11–5 (PubMed PMID: 22024155).
71. Al-Nedawi K, Meehan B, Rak J. Microvesicles: messengers and mediators of tumor progression. Cell Cycle. 2009 Jul 1;8(13):2014–8 (PubMed PMID: 19535896).
72. Simpson RJ, Lim JW, Moritz RL, Mathivanan S. Exosomes: proteomic insights and diagnostic potential. Expert Rev Proteomics. 2009 Jun;6(3):267–83 (PubMed PMID: 19489699).
73. Booth AM, Fang Y, Fallon JK, Yang JM, Hildreth JE, Gould SJ. Exosomes and HIV Gag bud from endosome-like domains of the T cell plasma membrane. J Cell Biol. 2006 Mar 13;172(6):923–35 (PubMed PMID: 16533950. Pubmed Central PMCID: 2063735).
74. Lenassi M, Cagney G, Liao M, Vaupotic T, Bartholomeeusen K, Cheng Y, et al. HIV Nef is secreted in exosomes and triggers apoptosis in bystander CD4 + T cells. Traffic. 2010 Jan;11(1):110–22 (PubMed PMID: 19912576. Pubmed Central PMCID: 2796297).
75. Gyorgy B, Modos K, Pallinger E, Paloczi K, Pasztoi M, Misjak P, et al. Detection and isolation of cell-derived microparticles are compromised by protein complexes resulting from shared biophysical parameters. Blood. 2011 Jan 27;117(4):e39–48 (PubMed PMID: 21041717).
76. Mrvar-Brecko A, Sustar V, Jansa V, Stukelj R, Jansa R, Mujagic E, et al. Isolated microvesicles from peripheral blood and body fluids as observed by scanning electron microscope. Blood Cells Mol Dis. 2010 Apr 15;44(4):307–12 (PubMed PMID: 20199878).
77. Thery C, Amigorena S, Raposo G, Clayton A. Isolation and characterization of exosomes from cell culture supernatants and biological fluids. Curr Protoc Cell Biol. 2006 Apr;Chap. 3:Unit 3 22 (PubMed PMID: 18228490).
78. Kesimer M, Scull M, Brighton B, DeMaria G, Burns K, O'Neal W, et al. Characterization of exosome-like vesicles released from human tracheobronchial ciliated epithelium: a possible role in innate defense. FASEB J. 2009 Jun;23(6):1858–68 (PubMed PMID: 19190083. Pubmed Central PMCID: 2698655).
79. Hosseini-Beheshti E, Pham S, Adomat H, Li N, Tomlinson Guns ES. Exosomes as biomarker enriched microvesicles: characterization of exosomal proteins derived from a panel of prostate cell lines with distinct AR phenotypes. Mol Cell Proteomics. 2012 Oct;11(10):863–85 (PubMed PMID: 22723089. Pubmed Central PMCID: 3494141).
80. Lau CS, Wong DT. Breast cancer exosome-like microvesicles and salivary gland cells interplay alters salivary gland cell-derived exosome-like microvesicles in vitro. PLoS One. 2012;7(3):e33037 (PubMed PMID: 22448232. Pubmed Central PMCID: 3308964).
81. Connor DE, Exner T, Ma DD, Joseph JE. The majority of circulating platelet-derived microparticles fail to bind annexin V, lack phospholipid-dependent procoagulant activity and demonstrate greater expression of glycoprotein Ib. Thromb Haemost. 2010 May;103(5):1044–52 (PubMed PMID: 20390225).
82. Muralidharan-Chari V, Clancy JW, Sedgwick A, D'Souza-Schorey C. Microvesicles: mediators of extracellular communication during cancer progression. J Cell Sci. 2010 May 15;123(Pt 10):1603–11 (PubMed PMID: 20445011. Pubmed Central PMCID: 2864708).
83. Marsh M, van Meer G. Cell biology. No ESCRTs for exosomes. Science. 2008 Feb 29;319(5867):1191–2 (PubMed PMID: 18309064).
84. Trajkovic K, Hsu C, Chiantia S, Rajendran L, Wenzel D, Wieland F, et al. Ceramide triggers budding of exosome vesicles into multivesicular endosomes. Science. 2008 Feb 29;319(5867):1244–7 (PubMed PMID: 18309083).
85. Kosaka N, Iguchi H, Yoshioka Y, Takeshita F, Matsuki Y, Ochiya T. Secretory mechanisms and intercellular transfer of microRNAs in living cells. J Biol Chem. 2010 Jun 4;285(23):17442–52 (PubMed PMID: 20353945. Pubmed Central PMCID: 2878508).
86. Wang G, Dinkins M, He Q, Zhu G, Poirier C, Campbell A, et al. Astrocytes secrete exosomes enriched with proapoptotic ceramide and prostate apoptosis response 4 (PAR-4): potential mechanism of apoptosis induction in Alzheimer disease (AD). J Biol Chem. 2012 Jun 15;287(25):21384–95 (PubMed PMID: 22532571. Pubmed Central PMCID: 3375560).
87. Beloribi S, Ristorcelli E, Breuzard G, Silvy F, Bertrand-Michel J, Beraud E, et al. Exosomal lipids impact notch signaling and induce death of human pancreatic tumoral SOJ-6 cells. PLoS One. 2012;7(10):e47480 (PubMed PMID: 23094054. Pubmed Central PMCID: 3477155).

88. Mathivanan S, Fahner CJ, Reid GE, Simpson RJ. ExoCarta 2012: database of exosomal proteins, RNA and lipids. Nucleic Acids Res. 2012 Jan;40(Database issue):D1241–4 (PubMed PMID: 21989406. Pubmed Central PMCID: 3245025).
89. Kim HS, Choi DY, Yun SJ, Choi SM, Kang JW, Jung JW, et al. Proteomic analysis of microvesicles derived from human mesenchymal stem cells. J Proteome Res. 2012 Feb 3;11(2):839–49 (PubMed PMID: 22148876).
90. Khatua AK, Taylor HE, Hildreth JE, Popik W. Inhibition of LINE-1 and Alu retrotransposition by exosomes encapsidating APOBEC3G and APOBEC3F. Virology. 2010 Apr 25;400(1):68–75 (PubMed PMID: 20153011. Pubmed Central PMCID: 2851184).
91. Peters DL, Pretorius PJ. Origin, translocation and destination of extracellular occurring DNA–a new paradigm in genetic behaviour. Clin Chim Acta. 2011 May 12;412(11–12):806–11 (PubMed PMID: 21277292).
92. Vickers KC, Remaley AT. Lipid-based carriers of microRNAs and intercellular communication. Curr Opin Lipidol. 2012 Apr;23(2):91–7 (PubMed PMID: 22418571).
93. Siegel R, Naishadham D, Jemal A. Cancer statistics, 2012. CA Cancer J Clin. 2012 Jan-Feb;62(1):10–29 (PubMed PMID: 22237781).
94. Rosenberg SA. Progress in human tumour immunology and immunotherapy. Nature. 2001 May 17;411(6835):380–4 (PubMed PMID: 11357146).
95. Kuperwasser C, Chavarria T, Wu M, Magrane G, Gray JW, Carey L, et al. Reconstruction of functionally normal and malignant human breast tissues in mice. Proc Natl Acad Sci U S A. 2004 Apr 6;101(14):4966–71 (PubMed PMID: 15051869. Pubmed Central PMCID: 387357).
96. Finak G, Bertos N, Pepin F, Sadekova S, Souleimanova M, Zhao H, et al. Stromal gene expression predicts clinical outcome in breast cancer. Nat Med. 2008 May;14(5):518–27 (PubMed PMID: 18438415).
97. Tan W, Zhang W, Strasner A, Grivennikov S, Cheng JQ, Hoffman RM, et al. Tumour-infiltrating regulatory T cells stimulate mammary cancer metastasis through RANKL-RANK signalling. Nature. 2011 Feb 24;470(7335):548–53 (PubMed PMID: 21326202. Pubmed Central PMCID: 3166217).
98. De Wever O, Mareel M. Role of tissue stroma in cancer cell invasion. J Pathol. 2003 Jul;200(4):429–47 (PubMed PMID: 12845611).
99. Pittenger MF, Mackay AM, Beck SC, Jaiswal RK, Douglas R, Mosca JD, et al. Multilineage potential of adult human mesenchymal stem cells. Science. 1999 Apr 2;284(5411):143–7 (PubMed PMID: 10102814).
100. Smith JR, Pochampally R, Perry A, Hsu SC, Prockop DJ. Isolation of a highly clonogenic and multipotential subfraction of adult stem cells from bone marrow stroma. Stem Cells. 2004;22(5):823–31 (PubMed PMID: 15342946).
101. Wolfe M, Pochampally R, Swaney W, Reger RL. Isolation and culture of bone marrow-derived human multipotent stromal cells (hMSCs). Methods Mol Biol. 2008;449:3–25 (PubMed PMID: 18370080).
102. !!! INVALID CITATION!!!
103. Vallabhaneni KC, Tkachuk S, Kiyan Y, Shushakova N, Haller H, Dumler I, et al. Urokinase receptor mediates mobilization, migration, and differentiation of mesenchymal stem cells. Cardiovasc Res. 2011 Apr 1;90(1):113–21 (PubMed PMID: 21088115).
104. Shoji M, Oskowitz A, Malone CD, Prockop DJ, Pochampally R. Human mesenchymal stromal cells (MSCs) reduce neointimal hyperplasia in a mouse model of flow-restriction by transient suppression of anti-inflammatory cytokines. J Atheroscler Thromb. 2011;18(6):464–74 (PubMed PMID: 21307612).
105. Hall B, Andreeff M, Marini F. The participation of mesenchymal stem cells in tumor stroma formation and their application as targeted-gene delivery vehicles. Handb Exp Pharmacol. 2007;180:263–83 (PubMed PMID: 17554513).
106. Studeny M, Marini FC, Champlin RE, Zompetta C, Fidler IJ, Andreeff M. Bone marrow-derived mesenchymal stem cells as vehicles for interferon-beta delivery into tumors. Cancer Res. 2002 Jul 1;62(13):3603–8 (PubMed PMID: 12097260).

107. Reddy BY, Lim PK, Silverio K, Patel SA, Won BW, Rameshwar P. The microenvironmental effect in the progression, metastasis, and dormancy of breast cancer: a model system within bone marrow. Int J Breast Cancer. 2012;2012:721659 (PubMed PMID: 22482060. Pubmed Central PMCID: 3296216).
108. Schichor C, Birnbaum T, Etminan N, Schnell O, Grau S, Miebach S, et al. Vascular endothelial growth factor A contributes to glioma-induced migration of human marrow stromal cells (hMSC). Exp Neurol. 2006 Jun;199(2):301–10 (PubMed PMID: 16574102).
109. Birnbaum T, Roider J, Schankin CJ, Padovan CS, Schichor C, Goldbrunner R, et al. Malignant gliomas actively recruit bone marrow stromal cells by secreting angiogenic cytokines. J Neurooncol. 2007 Jul;83(3):241–7 (PubMed PMID: 17570034).
110. Dwyer RM, Potter-Beirne SM, Harrington KA, Lowery AJ, Hennessy E, Murphy JM, et al. Monocyte chemotactic protein-1 secreted by primary breast tumors stimulates migration of mesenchymal stem cells. Clin Cancer Res. 2007 Sep 1;13(17):5020–7 (PubMed PMID: 17785552).
111. Spaeth E, Klopp A, Dembinski J, Andreeff M, Marini F. Inflammation and tumor microenvironments: defining the migratory itinerary of mesenchymal stem cells. Gene Ther. 2008 May;15(10):730–8 (PubMed PMID: 18401438).
112. Feng B, Chen L. Review of mesenchymal stem cells and tumors: executioner or coconspirator? Cancer Biother Radiopharm. 2009 Dec;24(6):717–21 (PubMed PMID: 20025552).
113. Kidd S, Spaeth E, Klopp A, Andreeff M, Hall B, Marini FC. The (in) auspicious role of mesenchymal stromal cells in cancer: be it friend or foe. Cytotherapy. 2008;10(7):657–67 (PubMed PMID: 18985472).
114. Spaeth EL, Dembinski JL, Sasser AK, Watson K, Klopp A, Hall B, et al. Mesenchymal stem cell transition to tumor-associated fibroblasts contributes to fibrovascular network expansion and tumor progression. PLoS One. 2009;4(4):e4992 (PubMed PMID: 19352430. Pubmed Central PMCID: 2661372).
115. Hogan NM, Dwyer RM, Joyce MR, Kerin MJ. Mesenchymal stem cells in the colorectal tumor microenvironment: recent progress and implications. Int J Cancer. 2012 Jul 1;131(1):1–7 (PubMed PMID: 22290082).
116. Djouad F, Plence P, Bony C, Tropel P, Apparailly F, Sany J, et al. Immunosuppressive effect of mesenchymal stem cells favors tumor growth in allogeneic animals. Blood. 2003 Nov 15;102(10):3837–44 (PubMed PMID: 12881305).
117. Zhu W, Xu W, Jiang R, Qian H, Chen M, Hu J, et al. Mesenchymal stem cells derived from bone marrow favor tumor cell growth in vivo. Exp Mol Pathol. 2006 Jun;80(3):267–74 (PubMed PMID: 16214129).
118. Early Breast Cancer Trialists' Collaborative Group (EBCTCG), Davies C, Godwin J, Gray R, Clarke M, Cutter D, et al. Relevance of breast cancer hormone receptors and other factors to the efficacy of adjuvant tamoxifen: patient-level meta-analysis of randomised trials. Lancet. 2011 Aug 27;378(9793):771–84 (PubMed PMID: 21802721. Pubmed Central PMCID: 3163848).
119. van Doormaal FF, Kleinjan A, Di Nisio M, Buller HR, Nieuwland R. Cell-derived microvesicles and cancer. Neth J Med. 2009 Jul-Aug;67(7):266–73 (PubMed PMID: 19687520).
120. Ginestra A, Miceli D, Dolo V, Romano FM, Vittorelli ML. Membrane vesicles in ovarian cancer fluids: a new potential marker. Anticancer Res. 1999 Jul-Aug;19(4C):3439–45 (PubMed PMID: 10629632).
121. Hakulinen J, Junnikkala S, Sorsa T, Meri S. Complement inhibitor membrane cofactor protein (MCP; CD46) is constitutively shed from cancer cell membranes in vesicles and converted by a metalloproteinase to a functionally active soluble form. Eur J Immunol. 2004 Sep;34(9):2620–9 (PubMed PMID: 15307194).
122. van Niel G, Porto-Carreiro I, Simoes S, Raposo G. Exosomes: a common pathway for a specialized function. J Biochem. 2006 Jul;140(1):13–21 (PubMed PMID: 16877764).
123. Andreola G, Rivoltini L, Castelli C, Huber V, Perego P, Deho P, et al. Induction of lymphocyte apoptosis by tumor cell secretion of FasL-bearing microvesicles. J Exp Med. 2002 May 20;195(10):1303–16 (PubMed PMID: 12021310. Pubmed Central PMCID: 2193755).

124. Huber V, Fais S, Iero M, Lugini L, Canese P, Squarcina P, et al. Human colorectal cancer cells induce T-cell death through release of proapoptotic microvesicles: role in immune escape. Gastroenterology. 2005 Jun;128(7):1796–804 (PubMed PMID: 15940614).
125. Abid Hussein MN, Boing AN, Sturk A, Hau CM, Nieuwland R. Inhibition of microparticle release triggers endothelial cell apoptosis and detachment. Thromb Haemost. 2007 Nov;98(5):1096–107 (PubMed PMID: 18000616).
126. Hotary KB, Allen ED, Brooks PC, Datta NS, Long MW, Weiss SJ. Membrane type I matrix metalloproteinase usurps tumor growth control imposed by the three-dimensional extracellular matrix. Cell. 2003 Jul 11;114(1):33–45 (PubMed PMID: 12859896).
127. Levchenko A, Mehta BM, Niu X, Kang G, Villafania L, Way D, et al. Intercellular transfer of P-glycoprotein mediates acquired multidrug resistance in tumor cells. Proc Natl Acad Sci U S A. 2005 Feb 8;102(6):1933–8 (PubMed PMID: 15671173. Pubmed Central PMCID: 545583).
128. Ambudkar SV, Sauna ZE, Gottesman MM, Szakacs G. A novel way to spread drug resistance in tumor cells: functional intercellular transfer of P-glycoprotein (ABCB1). Trends Pharmacol Sci. 2005 Aug;26(8):385–7 (PubMed PMID: 15978680. Pubmed Central PMCID: 1350964).
129. Shedden K, Xie XT, Chandaroy P, Chang YT, Rosania GR. Expulsion of small molecules in vesicles shed by cancer cells: association with gene expression and chemosensitivity profiles. Cancer Res. 2003 Aug 1;63(15):4331–7 (PubMed PMID: 12907600).
130. Castellana D, Zobairi F, Martinez MC, Panaro MA, Mitolo V, Freyssinet JM, et al. Membrane microvesicles as actors in the establishment of a favorable prostatic tumoral niche: a role for activated fibroblasts and CX3CL1-CX3CR1 axis. Cancer Res. 2009 Feb 1;69(3):785–93 (PubMed PMID: 19155311).
131. Katsuda T, Kosaka N, Takeshita F, Ochiya T. The therapeutic potential of mesenchymal stem cell-derived extracellular vesicles. Proteomics. 2013;13(10–11):1637–53 (PubMed PMID: 23335344).
132. Bruno S, Grange C, Deregibus MC, Calogero RA, Saviozzi S, Collino F, et al. Mesenchymal stem cell-derived microvesicles protect against acute tubular injury. J Am Soc Nephrol. 2009 May;20(5):1053–67 (PubMed PMID: 19389847. Pubmed Central PMCID: 2676194).
133. He J, Wang Y, Sun S, Yu M, Wang C, Pei X, et al. Bone marrow stem cells-derived microvesicles protect against renal injury in the mouse remnant kidney model. Nephrology (Carlton). 2012 Jul;17(5):493–500 (PubMed PMID: 22369283).
134. Timmers L, Lim SK, Arslan F, Armstrong JS, Hoefer IE, Doevendans PA, et al. Reduction of myocardial infarct size by human mesenchymal stem cell conditioned medium. Stem Cell Res. 2007 Nov;1(2):129–37 (PubMed PMID: 19383393).
135. Dreyer JL. New insights into the roles of microRNAs in drug addiction and neuroplasticity. Genome Med. 2010;2(12):92 (PubMed PMID: 21205279. Pubmed Central PMCID: 3025434).
136. He L, Hannon GJ. MicroRNAs: small RNAs with a big role in gene regulation. Nat Rev Genet. 2004 Jul;5(7):522–31 (PubMed PMID: 15211354).
137. Bushati N, Cohen SM. microRNA functions. Annu Rev Cell Dev Biol. 2007;23:175–205 (PubMed PMID: 17506695).
138. Fabbri M, Calore F, Paone A, Galli R, Calin GA. Epigenetic regulation of miRNAs in cancer. Adv Exp Med Biol. 2013;754:137–48 (PubMed PMID: 22956499).
139. Salmena L, Poliseno L, Tay Y, Kats L, Pandolfi PP. A ceRNA hypothesis: the Rosetta Stone of a hidden RNA language? Cell. 2011 Aug 5;146(3):353–8 (PubMed PMID: 21802130. Pubmed Central PMCID: 3235919).
140. Blondal T, Jensby Nielsen S, Baker A, Andreasen D, Mouritzen P, Wrang Teilum M, et al. Assessing sample and miRNA profile quality in serum and plasma or other biofluids. Methods. 2013;59(1):S1–6 (PubMed PMID: 23036329).
141. Arroyo JD, Chevillet JR, Kroh EM, Ruf IK, Pritchard CC, Gibson DF, et al. Argonaute2 complexes carry a population of circulating microRNAs independent of vesicles in human plasma. Proc Natl Acad Sci U S A. 2011 Mar 22;108(12):5003–8 (PubMed PMID: 21383194. Pubmed Central PMCID: 3064324).
142. Turchinovich A, Weiz L, Langheinz A, Burwinkel B. Characterization of extracellular circulating microRNA. Nucleic Acids Res. 2011 Sep 1;39(16):7223–33 (PubMed PMID: 21609964. Pubmed Central PMCID: 3167594).

143. Russo F, Di Bella S, Nigita G, Macca V, Lagana A, Giugno R, et al. miRandola: extracellular circulating microRNAs database. PLoS One. 2012;7(10):e47786 (PubMed PMID: 23094086. Pubmed Central PMCID: 3477145).
144. Lages E, Ipas H, Guttin A, Nesr H, Berger F, Issartel JP. MicroRNAs: molecular features and role in cancer. Front Biosci. 2012;17:2508–40 (PubMed PMID: 22652795).
145. Iorio MV, Croce CM. Causes and consequences of microRNA dysregulation. Cancer J. 2012 May-Jun;18(3):215–22 (PubMed PMID: 22647357. Pubmed Central PMCID: 3528102).
146. Mudduluru G, Ceppi P, Kumarswamy R, Scagliotti GV, Papotti M, Allgayer H. Regulation of Axl receptor tyrosine kinase expression by miR-34a and miR-199a/b in solid cancer. Oncogene. 2011 Jun 23;30(25):2888–99 (PubMed PMID: 21317930).
147. Harquail J, Benzina S, Robichaud GA. MicroRNAs and breast cancer malignancy: an overview of miRNA-regulated cancer processes leading to metastasis. Cancer Biomark. 2012 Jan 1;11(6):269–80 (PubMed PMID: 23248185).
148. Yang S, Li Y, Gao J, Zhang T, Li S, Luo A, et al. MicroRNA-34 suppresses breast cancer invasion and metastasis by directly targeting Fra-1. Oncogene. 2013;32(36):4294–303 (PubMed PMID: 23001043).
149. Neilsen PM, Noll JE, Mattiske S, Bracken CP, Gregory PA, Schulz RB, et al. Mutant p53 drives invasion in breast tumors through up-regulation of miR-155. Oncogene. 2013;32(24):2992–3000 (PubMed PMID: 22797073).
150. Kosaka N, Iguchi H, Yoshioka Y, Hagiwara K, Takeshita F, Ochiya T. Competitive interactions of cancer cells and normal cells via secretory microRNAs. J Biol Chem. 2012 Jan 6;287(2):1397–405 (PubMed PMID: 22123823. Pubmed Central PMCID: 3256909).
151. Zhuang G, Wu X, Jiang Z, Kasman I, Yao J, Guan Y, et al. Tumour-secreted miR-9 promotes endothelial cell migration and angiogenesis by activating the JAK-STAT pathway. EMBO J. 2012 Aug 29;31(17):3513–23 (PubMed PMID: 22773185. Pubmed Central PMCID: 3433782).
152. Pigati L, Yaddanapudi SC, Iyengar R, Kim DJ, Hearn SA, Danforth D, et al. Selective release of microRNA species from normal and malignant mammary epithelial cells. PLoS One. 2010;5(10):e13515 (PubMed PMID: 20976003. Pubmed Central PMCID: 2958125).
153. Palma J, Yaddanapudi SC, Pigati L, Havens MA, Jeong S, Weiner GA, et al. MicroRNAs are exported from malignant cells in customized particles. Nucleic Acids Res. 2012 Oct;40(18):9125–38 (PubMed PMID: 22772984. Pubmed Central PMCID: 3467054).
154. Valadi H, Ekstrom K, Bossios A, Sjostrand M, Lee JJ, Lotvall JO. Exosome-mediated transfer of mRNAs and microRNAs is a novel mechanism of genetic exchange between cells. Nat Cell Biol. 2007 Jun;9(6):654–9 (PubMed PMID: 17486113).
155. Ratajczak J, Miekus K, Kucia M, Zhang J, Reca R, Dvorak P, et al. Embryonic stem cell-derived microvesicles reprogram hematopoietic progenitors: evidence for horizontal transfer of mRNA and protein delivery. Leukemia. 2006 May;20(5):847–56 (PubMed PMID: 16453000).
156. Zernecke A, Bidzhekov K, Noels H, Shagdarsuren E, Gan L, Denecke B, et al. Delivery of microRNA-126 by apoptotic bodies induces CXCL12-dependent vascular protection. Sci Signal. 2009 Dec 8;2(100):ra81 (PubMed PMID: 19996457).
157. Weber C, Schober A, Zernecke A. MicroRNAs in arterial remodelling, inflammation and atherosclerosis. Curr Drug Targets. 2010 Aug;11(8):950–6 (PubMed PMID: 20415650).
158. Zomer A, Vendrig T, Hopmans ES, van Eijndhoven M, Middeldorp JM, Pegtel DM. Exosomes: Fit to deliver small RNA. Commun Integr Biol. 2010 Sep;3(5):447–50 (PubMed PMID: 21057637. Pubmed Central PMCID: 2974077).
159. Yuan A, Farber EL, Rapoport AL, Tejada D, Deniskin R, Akhmedov NB, et al. Transfer of microRNAs by embryonic stem cell microvesicles. PLoS One. 2009;4(3):e4722 (PubMed PMID: 19266099. Pubmed Central PMCID: 2648987).
160. Koh W, Sheng CT, Tan B, Lee QY, Kuznetsov V, Kiang LS, et al. Analysis of deep sequencing microRNA expression profile from human embryonic stem cells derived mesenchymal stem cells reveals possible role of let-7 microRNA family in downstream targeting of hepatic nuclear factor 4 alpha. BMC Genomics. 2010;11 Suppl 1:S6 (PubMed PMID: 20158877. Pubmed Central PMCID: 2822534).

161. Gregory LA, Ricart RA, Patel SA, Lim PK, Rameshwar P. microRNAs, gap junctional intercellular communication and mesenchymal stem cells in breast cancer metastasis. Curr Cancer Ther Rev. 2011 Aug;7(3):176–83 (PubMed PMID: 21886602. Pubmed Central PMCID: 3163384).
162. Skog J, Wurdinger T, van Rijn S, Meijer DH, Gainche L, Sena-Esteves M, et al. Glioblastoma microvesicles transport RNA and proteins that promote tumour growth and provide diagnostic biomarkers. Nat Cell Biol. 2008 Dec;10(12):1470–6 (PubMed PMID: 19011622. Pubmed Central PMCID: 3423894).
163. Chiba M, Kimura M, Asari S. Exosomes secreted from human colorectal cancer cell lines contain mRNAs, microRNAs and natural antisense RNAs, that can transfer into the human hepatoma HepG2 and lung cancer A549 cell lines. Oncol Rep. 2012 Nov;28(5):1551–8 (PubMed PMID: 22895844).
164. Zhang Y, Liu D, Chen X, Li J, Li L, Bian Z, et al. Secreted monocytic miR-150 enhances targeted endothelial cell migration. Mol Cell. 2010 Jul 9;39(1):133–44 (PubMed PMID: 20603081).
165. Gibbings DJ, Ciaudo C, Erhardt M, Voinnet O. Multivesicular bodies associate with components of miRNA effector complexes and modulate miRNA activity. Nat Cell Biol. 2009 Sep;11(9):1143–9 (PubMed PMID: 19684575).
166. Katakowski M, Buller B, Wang X, Rogers T, Chopp M. Functional microRNA is transferred between glioma cells. Cancer Res. 2010 Nov 1;70(21):8259–63 (PubMed PMID: 20841486. Pubmed Central PMCID: 2970756).
167. Kizana E, Cingolani E, Marban E. Non-cell-autonomous effects of vector-expressed regulatory RNAs in mammalian heart cells. Gene Ther. 2009 Sep;16(9):1163–8 (PubMed PMID: 19516277).
168. Mittelbrunn M, Gutierrez-Vazquez C, Villarroya-Beltri C, Gonzalez S, Sanchez-Cabo F, Gonzalez MA, et al. Unidirectional transfer of microRNA-loaded exosomes from T cells to antigen-presenting cells. Nat Commun. 2011;2:282. (PubMed PMID: 21505438. Pubmed Central PMCID: 3104548).
169. Fabian MR, Sonenberg N. The mechanics of miRNA-mediated gene silencing: a look under the hood of miRISC. Nat Struct Mol Biol. 2012 Jun;19(6):586–93 (PubMed PMID: 22664986).
170. Eiring AM, Harb JG, Neviani P, Garton C, Oaks JJ, Spizzo R, et al. miR-328 functions as an RNA decoy to modulate hnRNP E2 regulation of mRNA translation in leukemic blasts. Cell. 2010 Mar 5;140(5):652–65 (PubMed PMID: 20211135. Pubmed Central PMCID: 2924756).
171. Fabbri M, Paone A, Calore F, Galli R, Gaudio E, Santhanam R, et al. MicroRNAs bind to Toll-like receptors to induce prometastatic inflammatory response. Proc Natl Acad Sci U S A. 2012 Jul 31;109(31):E2110–6 (PubMed PMID: 22753494. Pubmed Central PMCID: 3412003).
172. Ghasemi R, Grassadonia A, Tinari N, Piccolo E, Natoli C, Tomao F, et al. Tumor-derived microvesicles: the metastasomes. Med Hypotheses. 2013 Jan;80(1):75–82 (PubMed PMID: 23177570).
173. Ohno SI, Ishikawa A, Kuroda M. Roles of exosomes and microvesicles in disease pathogenesis. Adv Drug Deliv Rev. 2013;65(3):398–401 (PubMed PMID: 22981801).
174. Von Bartheld CS, Altick AL. Multivesicular bodies in neurons: distribution, protein content, and trafficking functions. Prog Neurobiol. 2011 Mar;93(3):313–40 (PubMed PMID: 21216273. Pubmed Central PMCID: 3055956).
175. Camussi G, Deregibus MC, Bruno S, Cantaluppi V, Biancone L. Exosomes/microvesicles as a mechanism of cell-to-cell communication. Kidney Int. 2010 Nov;78(9):838–48 (PubMed PMID: 20703216).

Chapter 18
MicroRNA Cancer Therapeutics and the Challenge of Drug Delivery

Steven J. Greco, Jessian L. Munoz and Pranela Rameshwar

Abstract MicroRNAs (miRNAs) have progressed to a separate field in cancer biology. As cancer cells undergo investigations to build a hierarchy, the dissection of miRNA in the developmental process occurs in parallel. There is no doubt that RNA interference (RNAi) will be in the clinic. However, the efficiency of the process remains a challenge. This chapter discusses the potential for RNAi in therapy and summarizes some of the different methods to package RNAi for delivery to the regions of cancer. The ability of mesenchymal stem cells to migrate to the site of tumors has been explored as a method to deliver RNAi. However, these stem cells can influence tumor growth and the immune response to tumors. This chapter briefly discusses the potential of using mesenchymal stem cells in RNAi delivery. This topic would require a separate chapter on stem cells and RNAi delivery. Confounds discussed in this chapter are applicable to any method developed with stem cells to target cancer cells with RNAi.

Keywords MicroRNA · Stem cells · Cancer · RNAi · Drug delivery

1 Introduction

The identification of microRNAs (miRNAs), small single-stranded nucleic acid molecules, with diverse roles in cellular biology, can be considered one of the most influential breakthroughs in the field of genetics. First identified in *Caenorhabditis elegans* as regulators of larval development [11], these noncoding molecules exert important regulatory effects, specifically posttranscriptional modification [50]. Ultimately, miRNAs bind to endogenous messenger RNAs (mRNAs), leading to either mRNA silencing or degradation [50]. miRNAs are functionally relevant in development, apoptosis, cell proliferation, cancer progression, and a variety of other biological processes [18].

P. Rameshwar (✉) · S. J. Greco · J. L. Munoz
Department of Medicine, Hematology/Oncology,
New Jersey Medical School, Rutgers, Room E-579, MSB,
185 South Orange Ave, Newark, NJ 07103, USA
e-mail: rameshwa@njms.rutgers.edu

S. R. Singh, P. Rameshwar (eds.), *MicroRNA in Development and in the Progression of Cancer*, DOI 10.1007/978-1-4899-8065-6_18,

Endogenous miRNA synthesis occurs through a highly orchestrated set of events involving polymerases, ribonucleases, and nuclear shuttles [32]. RNA polymerase II-mediated transcription leads to the production of a hairpin structure known as primary miRNA (pri-miRNA), which is then processed by the nuclease RNase III Drosha into a 70-nucleotide precursor miRNA (pre-miRNA) [32]. The pre-miRNA is exported from the nucleus via Exportin-5 and enters the cytoplasm, where it encounters an RNase III known as Dicer. The product of this reaction is a mature 19–22-nucleotide miRNA. miRNA hairpins can complex with a variety of Argonaute proteins, forming the miRNA-induced silencing complex (miRISC). The core components of the mammalian miRISC consists of one of four Argonaute proteins (AGO1–4) and Argonaute-bound GW182 [10]. AGOs are essential for miRNA-mediated gene silencing, since depletion of AGOs from cells of multiple organisms impairs miRNA-mediated silencing [1, 37]. Indeed, AGO proteins repress protein synthesis when artificially bound to reporter mRNA 3′ untranslated regions (UTRs) lacking miRNA target sites [33]. These findings suggest that miRNA–mRNA interaction is dispensable for the activity of miRNA-loaded AGO proteins, but determines endogenous mRNA specificity.

Unlike short interfering RNAs (siRNAs), miRNAs bind to the 3′ UTR of mRNAs through imperfect Watson–Crick base pairing. Binding to the 5′ UTR and protein-encoding sequence can also occur, although this is less likely than binding to the 3′ UTR [35]. The mechanisms by which miRNAs lead to gene silencing are under investigation, but thus far, include suppression of translational initiation or elongation, protein degradation, and premature chain termination [18]. Importantly, the seed sequence of the miRNA, which is a 2–7-nucleotide sequence located at the 5′ end, is necessary for binding of miRNAs to 3′ UTRs [18]. This results in silencing of endogenous mRNA transcripts [32]. In addition to silencing, miRNAs can lead to degradation of transcripts [18]. miRNAs are vital to life in that Dicer knockouts have been shown to result in embryonic lethality [50].

The miRNAome, which refers to the collection of all miRNAs, is currently being sequenced, using next-generation technologies [35]. Deep sequencing can be used to characterize various miRNA expression profiles [4]. At the time of writing this chapter, there were greater than 1,000 known human miRNAs sequences deposited in databases [11]. The miRNAome for various cancers are under investigation with the goal of determining miRNA expression profiles and polymorphisms that may be linked to disease. miRNAs map to approximately 100 different clusters in the genome, and miRNA located in the same cluster are generally coexpressed [11], for example, onco-miRs, which are involved in the progression of cancer, and expressed from the miR-21 family/cluster [22].

An important concept regarding miRNAs is that they have pleiotropic effects. One miRNA can regulate multiple endogenous messages, and multiple miRNAs can exert a synergistic or antagonist effect on one gene [35]. For example, miR-130a and miR-206 have been shown to bind to the 3′ UTR of *Tac1,* which encodes substance P [16]. These miRNAs act in synergy to suppress *Tac1* gene expression [16]. Thus, miRNA regulatory networks are complex, and algorithms have been developed in order to predict putative interactions [11]. In addition, miRNAs can be interdependent in that miRNAs located in the same cluster are subject to similar regulation [11].

Once the molecular target for a miRNA has been ascribed and its cellular function identified, the potential for utilization of miRNAs or other siRNAs in therapy exists. In theory, if aberrant expression of a protein were linked to a disease state, then in vivo delivery of a small RNA that can bind the overexpressed protein's mRNA transcript would target the transcript for degradation and reduce amounts of the protein. This type of RNA interference (RNAi) therapeutic is an active area of research in the pharmaceutical industry and holds a great deal of promise. However, before this type of therapy can be tested in humans, safety and efficacy studies must be conducted in animal models of the disease.

One of the greatest challenges in utilizing RNAi in therapy is delivery of the molecule to the target tissue. In general, there are two types of delivery of RNA molecules in vivo: systemic and local [15]. Systemic delivery involves administration of relatively large doses of the RNA molecule into the whole organism to achieve the desired downregulation, whereas local delivery into a specific site requires much less RNA and has a lower chance of an off-target or a toxic effect since it is not delivered to the entire organism. A large number of delivery methodologies have been tested for both systemic and local delivery. Some of these methods include cholesterol conjugates, liposomes, polymer-based nanoparticles, antibodies, peptides, and viral particles. Currently, many pharmaceutical companies are either organizing or already performing RNAi clinical trials for a multitude of indications, including macular degeneration and inhibition of tumor growth/reduction of tumor size in pancreatic adenocarcinoma [9]. RNAi therapeutics are still in their infancy; however, successful clinical application of this approach will likely lead to an expansion of the field and a growing number of clinical trials for other indications.

Technology involving miRNAs has diverse applications, including detection, expression profiling and functional analysis. miRNA technology can be employed as a powerful tool that allows researchers to understand human disease at the genetic and molecular levels. They have the potential to be used as biomarkers for the prediction of cancer risk and for prognosis. miRNA profiling may become a key component of many basic science and clinical applications [4]. For example, it has been shown that tumor microenvironments can be distinguished based on expression patterns of miRNAs [4]. In the case of breast cancer, miRNA patterns are diverse among different disease subtypes, such as luminal, basal-like, and HER2-positive cancers [39]. In the case of cardiovascular disease, dysregulation of miRNA expression has been shown in damaged heart tissue. Similarly, in hepatocellular carcinoma, miRNA networks have been defined, increasing our understanding of the molecular basis of this disease [24].

In addition to the potential diagnostic use of miRNAs, they may also have therapeutic applications, such as the use of anti-miRs to abrogate the undesirable effects of aberrant miRNA expression. For example, miR-21 has been strongly associated with chemotherapy resistance in a variety of cancers, such as pancreatic cancer, hepatocellular carcinoma, and chronic myelogenous leukemia [19, 31, 44]. Scientists have suggested that miR-21 may serve as a target in cancer therapy based on its powerful role in oncogenesis [2]. Anti-miR nucleotide therapy may allow for the resensitization of cancers to chemotherapy [31]. By extrapolation, miRNAs that are associated with protection against diseases may be employed with pre-miR therapy.

One of the major challenges to therapeutic modalities involving miRNAs is the mechanism of in vivo delivery. It has been proposed that single-stranded circular DNA molecules can serve as the vehicle through which small nucleotides can be delivered [38]. Special formulations may need to be developed for intravenous or oral (*p.o.*) delivery. Besides delivery considerations, immunological phenomena must also be considered. If miRNAs are recognized as foreign antigens, they may be rejected by the innate or adaptive immune system. A final consideration is the stability of these molecules upon entry into target cell types, considering endogenous nucleases that may be present. These obstacles must be overcome prior to clinical implementation of small nucleic acid therapy. These prospects merit further research into miRNA mechanisms and therapeutic applications with the hope of alleviating disease.

2 miRNA and Cancer

The body of literature linking miRNA to cancer is growing at an exponential rate. Here, we focused on glioblastoma multiforme (GBM). Several miRNAs were shown to have oncogenic functions, referred to as onco-miRs. The role of miRNA includes diagnostic and perhaps prognostic value since the profile of miRNAs have been reported in cerebrospinal fluid [43]. GBM is the most common adult intracranial malignancy and the most aggressive with an ~11-month medial survival rate. The treatment options for patients with GBM are limited to surgery, radiation, and limitation with temozolomide [8]. This combination of treatment is the result of years of clinical trials with different types of drugs [12]. In 2009, the US Food and Drug Administration approved bevacizumab, also known as Avastin [5]. However, Avastin was only approved for cases that continue to progress after the standard care, which included surgery, radiotherapy, and temozolomide [5]. GBM comprises of oligodendroglioma and oligoastrocytoma [41]. Patients with both subtypes of GBM respond to temozolomide, but with little effect during recurrence [41].

Intercellular communication among tumor cells as well as the involvement of adhesion molecules have been linked as mediators of metastasis [17]. Gap junctional intercellular communication between cancer cells, including GBM, has been associated in the pathophysiology of cancer development [40]. Small second messenger molecules and miRNAs can pass through gap junction to regulate intercellular communications [25].

Specific miRNAs have been shown to regulate the invasion, migration, and proliferation of GBM, leading to poor prognosis [27, 29]. Interestingly, signal transducer and activator of transcription 3 (STAT3), through miRNA21, maintained telomerase reverse transcriptase (TERT), which suggested an antiaging mechanism by miRNA21 in GBM [46]. miRNAs are important in a number of cancer-related signaling pathways, specifically epidermal growth factor (EGF) and its receptor (EGFR), TP53, and phosphatase and tensin homolog (PTEN).

A number of parallels exist between normal human developmental pathways and those leading to oncogenic transformation/maintenance. The EGF/EGFR pathway is one common example of these overlaps. EGF is a 6-kDa protein with 53 amino acids, stabilized by three disulfide bonds. EGF binds with very high affinity to EGFR and induced cellular proliferations, differentiation, and survival. The over-expression of EGFR has been reported in ~60% of primary GBMs. These changes in *EGFR* expression have been correlated with the worst clinical prognosis for patients. EGFR serves as an attractive target for pharmacological inhibition of GBM signaling. Recent clinical trials have shown some improvement in patients treated with tyrosine-kinase inhibitors. EGFR has intrinsic protein tyrosine kinase activity, which leads to canonical receptor tyrosine kinase pathways. Prior to EGF ligand binding, EGFR is a nonactive monomeric transmembrane protein, although some evidence has suggested the existence of ligand-independent dimerization. After EGF binding to EGFR, the monomeric structure undergoes homodimerization, which results in the activation of the intrinsic protein tyrosine kinase activity. This results in the autotransphosphorylation of at least five residues (Y992, Y1045, Y1068, Y1148, and Y1173). Downstream proteins interact with the phosphorylated EGFR through SH2-phosphotyrosine-binding domains. EGFR binds to phospholipase C (PLC) and activates this lipase through phosphorylation. PLC activation results in the formation of inositol trisphosphate (IP3) and diacylglycerol (DAG). PLC downstream signaling leads to the release of stored calcium and the activation of the Akt/PKB pathway, a known antiapoptotic mechanism. Phosphorylated EGFR also binds to the Grb2/SOS complex, which can activate the p21 RAS pathway resulting in mitogen-activated protein kinase (MAPK)/extracellular signal-related kinase (ERK) activation. Pathway "cross talk" and activation of MAPK/AKT/PLC as well as other EGFR-induced pathways leads to profound effects of EGF secretion on recipient cells.

The *TP53* gene is one of the most frequently mutated genes in human malignancy. The *TP53* gene product p53 serves a dual role within cells. It can inhibit cellular proliferation and can activate apoptosis to regulate proper cellular function. The functions of p53 are tightly regulated by a feedback loop mediated through the p53 target *Mdm2*. Mdm2 functions as an E3 ubiquitin ligase that recognizes the amino terminal of p53. Following ubiquitination, p53 is degraded via the proteasome. miRNAs provide an additional mechanism for p53 regulation. This section will focus on not only p53-targeting miRNA but also p53-induced/inhibited miRNA.

To screen for p53-regulated miRNAs, GBM cells (U87) were profiled with and without the addition of the Mdm2/p53 complex inhibitor, Nutilin-3a. Using this approach, miR-25 was the most strongly repressed (>2-fold) by p53 activation (cells + Nutilin-3a). Yet, miR-32, which shares the same seeding sequence as miR-25, was also repressed by p53. Through in situ hybridization, it was confirmed that p53-expressing GBM cells lacked miR-25/-32. miRNA-25 is part of a larger cluster of miRNAs located in the 13th intronic region of the *MCM7* gene. Transcriptional regulation of the miRNA cluster has been shown to be E2F1- and MYC- dependent. Not surprisingly, p53-expressing GBM cells showed reduced expression of E2F1 and MYC, as well as miR-25/-32. But, the miR-32 host gene *C9orf5* did not show reduced expression. This led to the discovery of a novel miR-32 promoter, which is regulated by MYC. Interestingly enough, exogenous addition of miR-25/-32 led to prolonged p53 protein expression through stabilization; later, it was uncovered that miR-25/-32 functionally targets Mdm2. In vivo data, using intracranial xenograft models, show that miR-25/-32 expression increased overall animal survival through the accumulation of cytoplasmic p53. Taken together, the aforementioned discussion showed a feedback loop in which miR-25/-32 are repressed by p53 activation, resulting in increased Mdm2, which allows for p53 degradation and impaired function.

miRNA-21 is one of the most commonly upregulated miRNAs in oncogenesis and the maintenance of malignancies. Recent data showed miR-21 expression to be specific for tumors of glial origin, and this resulted in a decrease in cellular apoptosis. Knockdown of miR-21 in these tumors resulted in the activation of caspase 3/9-dependent cellular death. miR-21 can repress p53-mediated apoptosis and detained the cell cycle, thus inducing chemoresistance to a number of antineoplastic treatments.

Research studies supported a role for miR-21 in the regulation of a number of proteins. Chen et al. [4] showed that programmed cell death 4 (PDCD4) is a direct and functional target of miR-21 through binding to the 3′ UTR of PDCD4. GBM is a malignancy with extensive migration and infiltration of surrounding tissues. Gabriely et al. also reported on a role for miR-21 in the migratory properties of GBM by regulating a number of matrix metalloprotease inhibitors including TIMP3 and RECK. *PTEN* is a well-characterized and commonly mutated human tumor suppressor gene located on chromosome 10, which has been confirmed to be regulated by miR-21 in GBM cells.

3 miRNA Therapeutics and Confounds

As discussed earlier, there is no question that aberrant expression of miRNAs can result in different diseases, including malignancy. miRNAs are found in the plasma and are overwhelmingly proposed as biomarkers [6]. The goal is to identify if the miRNAs found in plasma are relevant to the disease and if they can be targeted. The next major challenge is to identify the source of the miRNA and to directly target them in the tissues. While there are discussions on the mechanisms by which miRNAs are involved in cancer, there is a need to establish methods to deliver miRNAs or antagomiRs for treatment. Indeed, biopharmaceutical companies have been actively involved in the development of new technologies to treat with RNAi [23]. The delivery of RNAi can occur by siRNA, short hairpin RNA (shRNA), bifunctional shRNA, and miRNA. These categories of RNAi can suppress translation or degrade mRNA, resulting in the suppression of gene expression through post-transcriptional intervention.

Recently, reports indicated that shRNA can be processed by methods that do not involve Dicer [26]. Studies with shRNAs with various stem length and loop sizes indicated that these parameters were important for the method to process the shRNA to RNAi [26]. The alternate method was Dicer-independent. Another point noted by the investigators is that the short stem length prevented processing by Dicer, and this resulted in the alternate method of processing, although similar incorporation occurred into RISC. The major difference was shown by the processing through AGO2-mediated slicing. This resulted in a single RNA strand as compared to the canonical processing to achieve siRNA duplexes. Going forward, this new information and more likely other mechanisms in the near future are important for the effective design of RNAi molecules for therapies. In the best cases where the ideal RNAi is established, the challenge is to identify methods for targeted delivery. There will also be a need to find a balance to prevent prolonged lingering of the RNAi without rapid excretion. Also, the challenge is to identify a delivery system that is stable, nontoxic, does not activate the immune system, and simultaneously shows improved efficacy. There are ongoing studies and challenges to make modifications of current molecules to achieve the ideal goals [21].

Despite the obstacles involved in targeted delivery of RNAi, the research continues to be active, with the goal of developing and identifying methods to treat different disorders. There are clinical trials for macular degeneration, cancer, viral infection, and other disorders [23]. The use of RNAi for viral infection is not new since this method has been largely used by arthropods to prevent infections from RNA viruses [36]. There were several trials with RNAi but relatively few remained in the clinic. We will discuss some of the methods in RNAi and what we believe could be major obstacles. We will also provide solutions in this section.

A growing body of literature identified aptamers for targeted delivery of RNAi [51]. Although these are synthetic single-stranded nucleic acids, the molecules can be arranged into various shapes for specific binding. The analogy of the specificity for the aptamer has been linked to antibodies [3]. The aptamers were adapted from those found endogenously from misfolding of proteins and later proposed as biomarkers for diseases such as cancer. Although the application of aptamers as biomarkers has not proved to be fruitful for cancer, the research on the technology could benefit RNAi treatment. Despite the potential application for aptamers, their use in RNAi delivery requires intensive investigation. This will require a strong collaboration among engineers, chemists, and biologists.

Intuitively, the liver should be one of the first organs to target RNA since this organ is the site of clearance. Translational scientists have taken advantage of the liver and have developed chemical molecules to prevent the replication of the hepatitis B virus. This was achieved by RNAi to prevent the translation of viral proteins through chemical delivery or adenovirus [47, 49]. As the therapy for RNAi develops, the method will replace the current treatments by relatively toxic methods [34]. The studies are promising since there is no evidence of liver chemistry or aberrant cytokine production. The methods to deliver RNAi would benefit miRNA since modified techniques might be used to deliver pre-miRs or antagonize their functions, referred to as antagomiR.

The development of methods to directly target cells through receptors is an active area of investigation. Conjugates to deliver RNAi through the folate receptor for cancer treatment appeared to be specific [7]. There is no evidence that the folate receptors are different in normal and malignant cells. This brings up the specificity for the treatment of cancer without untoward effects on the healthy cells. This concern is not eliminated with targeted treatments such as the folate receptor. As with other drugs that can target cancer and normal cells, preclinical studies will be able to find a dose to deliver the RNAi to cells with minimal toxicity.

Hyaluronic acid, which binds to CD44, is highly expressed in cancer cells and was, therefore, studied as a targeted method to deliver RNAi. A nano-based system with hyaluronic acid was tested with cancer cell lines [13]. The proof of principle studies showed a promising outcome for this nanoparticle. A major concern to target the hyaluronic acid receptor is bone marrow toxicity because hyaluronic acid is a part of the bone marrow microenvironment [14, 42]. If the particles go to different cavities within the bone marrow, the particles would bind to the same hematopoietic cells that interact with the extracellular matrix of bone marrow. The interacting cells could be hematopoietic stem cells and progenitors that interact with hyaluronic acid. This could lead to major toxicity since the RNAi could directly enter hematopoietic stem cells.

In addition to the receptor-targeted delivery of RNAi to cancer cells, there is evidence of passive entry of RNAi into cells. There are studies to identify methods to deliver RNAi for fractures which can be caused by osteoporosis. Investigators have developed a material to prevent bone resorption caused by osteoclasts [45]. The studies targeted RANK, which is involved in the development of osteoclasts [42]. Resorbable polymer, poly(lactic-co-glycolic acid) (PLGA) microparticles were developed for the purpose of delivering RNAi to the bone to reduce the development of osteoclasts. The studies investigated passive entry of the microparticles into osteoclast precursors and differentiated osteoclasts to deliver RANK siRNA. The studies showed specific internalization into osteoclasts. These studies on osteoclasts are discussed in this chapter because this could be a strategy to prevent osteoporosis that occurs with cancers such as those of the prostate and the breast [28].

Here, we also discuss an option to deliver RNAi by extrapolating on what is known about mesenchymal stem cells (MSCs) and their attraction to the site of cancer. MSCs are attracted to tumor sites [48]. This property of MSCs has been explored to deliver anticancer drugs [48]. This could be applied by loading MSCs with miRNA-containing exosomes since the MSCs have been shown to release exosomes when in contact with glioblastoma [20]. Another method could be the exploration of intercellular communication between MSCs and cancer through gap junction for the exchange of miRNA exchange [25]. This method might benefit the elimination of the cancer cells that is currently difficult to target, that is, the cancer stem cell subset [30].

References

1. Behm-Ansmant I, Rehwinkel J, et al. mRNA degradation by miRNAs and GW182 requires both CCR4:NOT deadenylase and DCP1:DCP2 decapping complexes. Genes Dev. 2006;20(14):1885–98.
2. Bonci D. MicroRNA-21 as therapeutic target in cancer and cardiovascular disease. Recent Pat Cardiovasc Drug Discov. 2010;5(3):156–61.

3. Chang YM, Donovan MJ, et al. Using aptamers for cancer biomarker discovery. J Nucleic Acids. 2013;2013:817350.
4. Chen SJ, Chen GH, et al. Characterization of Epstein-Barr virus miRNAome in nasopharyngeal carcinoma by deep sequencing. PLoS One. 2010;5(9):e12745.
5. Cohen MH, Shen YL, et al. FDA drug approval summary: bevacizumab (Avastin) as treatment of recurrent glioblastoma multiforme. Oncologist. 2009;14(11):1131–8.
6. Dickinson BA, Semus HM, et al. Plasma microRNAs serve as biomarkers of therapeutic efficacy and disease progression in hypertension-induced heart failure. Eur J Heart Fail. 2013;15(6):650–9.
7. Dohmen C, Frohlich T, et al. Defined folate-PEG-siRNA conjugates for receptor-specific gene silencing. Mol Ther Nucleic Acids. 2012;1:e7.
8. Dunn GP, Rinne ML, et al. Emerging insights into the molecular and cellular basis of glioblastoma. Genes Dev. 2012;26(8):756–84.
9. Duxbury MS, Ito H, et al. RNA interference targeting the M2 subunit of ribonucleotide reductase enhances pancreatic adenocarcinoma chemosensitivity to gemcitabine. Oncogene. 2004;23(8):1539–48.
10. Fabian MR, Sonenberg N. The mechanics of miRNA-mediated gene silencing: a look under the hood of miRISC. Nat Struct Mol Biol. 2012;19(6):586–93.
11. Farazi TA, Spitzer JI, et al. miRNAs in human cancer. J Pathol. 2011;223(2):102–15.
12. Galanis E, Wu W, et al. Phase 2 trial design in neuro-oncology revisited: a report from the RANO group. Lancet Oncol. 2012;13(5):e196–204.
13. Ganesh S, Iyer AK, et al. Hyaluronic acid based self-assembling nanosystems for CD44 target mediated siRNA delivery to solid tumors. Biomaterials. 2013;34(13):3489–502.
14. Goncharova V, Serobyan N, et al. Hyaluronan expressed by the hematopoietic microenvironment is required for bone marrow hematopoiesis. J Biol Chem. 2012;287(30):25419–33.
15. Gondi CS, Rao JS. Concepts in in vivo siRNA delivery for cancer therapy. J Cell Physiol. 2009;220(2):285–91.
16. Greco SJ, Rameshwar P. MicroRNAs regulate synthesis of the neurotransmitter substance P in human mesenchymal stem cell-derived neuronal cells. Proc Natl Acad Sci U S A. 2007;104(39):15484–9.
17. Hotz-Wagenblatt A, Shalloway D. Gap junctional communication and neoplastic transformation. Crit Rev Oncog. 1993;4(5):541–58.
18. Huntzinger E, Izaurralde E. Gene silencing by microRNAs: contributions of translational repression and mRNA decay. Nat Rev Genet. 2001;12(2):99–110.
19. Hwang JH, Voortman J, et al. Identification of microRNA-21 as a biomarker for chemoresistance and clinical outcome following adjuvant therapy in resectable pancreatic cancer. PLoS One. 2010;5(5):e10630.
20. Katakowski M, Buller B, et al. Exosomes from marrow stromal cells expressing miR-146b inhibit glioma growth. Cancer Lett. 2013;335(1):201–4.
21. Kenski DM, Butora G, et al. siRNA-optimized modifications for enhanced in vivo activity. Mol Ther Nucleic Acids. 2012;1:e5.
22. Krichevsky AM, Gabriely G. miR-21: a small multi-faceted RNA. J Cell Mol Med. 2009;13(1):39–53.
23. Kubowicz P, Zelaszczyk D, et al. RNAi in clinical studies. Curr Med Chem. 2013;20(14)1801–16.
24. Law PT, Wong N. Emerging roles of microRNA in the intracellular signaling networks of hepatocellular carcinoma. J Gastroenterol Hepatol. 2011;26(3):437–49.
25. Lim PK, Bliss SA, et al. Gap junction-mediated import of microRNA from bone marrow stromal cells can elicit cell cycle quiescence in breast cancer cells. Cancer Res. 2011;71(5):1550–60.
26. Liu YP, Schopman NC, et al. Dicer-independent processing of short hairpin RNAs. Nucleic Acids Res. 2013;41(6):3723–33.
27. Loftus JC, Ross JT, et al. miRNA expression profiling in migrating glioblastoma cells: regulation of cell migration and invasion by miR-23b via targeting of Pyk2. PloS One. 2012;7(6):e39818.

28. Lustberg MB, Reinbolt RE, et al. Bone health in adult cancer survivorship. J Clin Oncol. 2012;30(30):3665–74.
29. Ma R, Yan W, et al. Upregulation of miR-196b confers a poor prognosis in glioblastoma patients via inducing a proliferative phenotype. PloS One. 2012;7(6):e38096.
30. Patel SA, Ramkissoon SH, et al. Delineation of breast cancer cell hierarchy identifies the subset responsible for dormancy. Sci Rep. 2012;2:906.
31. Perin EC, Silva GV, et al. A randomized study of transendocardial injection of autologous bone marrow mononuclear cells and cell function analysis in ischemic heart failure (FOCUS-HF). Am Heart J. 2011;161(6):1078–87e1073.
32. Persson H, Kvist A, et al. Identification of new microRNAs in paired normal and tumor breast tissue suggests a dual role for the ERBB2/Her2 gene. Cancer Res. 71(1):78–86.
33. Pillai RS, Artus CG, et al. Tethering of human Ago proteins to mRNA mimics the miRNA-mediated repression of protein synthesis. RNA. 2004;10(10):1518–25.
34. Poordad F, Lawitz E, et al. Exploratory study of oral combination antiviral therapy for hepatitis C. N Engl J Med. 2013;368(1):45–53.
35. Ramsingh G, Koboldt DC, et al. Complete characterization of the microRNAome in a patient with acute myeloid leukemia. Blood. 2010;116(24):5316–26.
36. Sabin LR, Cherry S. Small creatures use small RNAs to direct antiviral defenses. Eur J Immunol. 2013;43(1):27–33.
37. Schmitter D, Filkowski J, et al. Effects of Dicer and Argonaute down-regulation on mRNA levels in human HEK293 cells. Nucleic Acids Res. 2006;34(17):4801–15.
38. Seidl CI, Ryan K. Circular single-stranded synthetic DNA delivery vectors for microRNA. PLoS One 2011;6(2):e16925.
39. Sotiriou C, Pusztai L. Gene-expression signatures in breast cancer. N Engl J Med. 2009;360(8):790–800.
40. Strale PO, Clarhaut J, et al. Down-regulation of Connexin43 expression reveals the involvement of caveolin-1 containing lipid rafts in human U251 glioblastoma cell invasion. Mol Carcinog. 2012;51(11):845–60.
41. Stupp R, Hegi ME, et al. Changing paradigms–an update on the multidisciplinary management of malignant glioma. Oncologist. 2006;11(2):165–80.
42. Tanaka S. Regulation of bone destruction in rheumatoid arthritis through RANKL-RANK pathways. World J Orthop. 2013;4(1):1–6.
43. Teplyuk NM, Mollenhauer B, et al. MicroRNAs in cerebrospinal fluid identify glioblastoma and metastatic brain cancers and reflect disease activity. Neuro Oncol. 2012;14(6):689–700.
44. Tomimaru Y, Eguchi H, et al. MicroRNA-21 induces resistance to the anti-tumour effect of interferon-alpha/5-fluorouracil in hepatocellular carcinoma cells. Br J Cancer. 2010;103(10):1617–26.
45. Wang Y, Tran KK, et al. Selective local delivery of RANK siRNA to bone phagocytes using bone augmentation biomaterials. Biomaterials. 2012;33(33):8540–7.
46. Wang YY, Sun G, et al. MiR-21 modulates hTERT through a STAT3-dependent manner on glioblastoma cell growth. CNS Neurosci Ther. 2012;18(9):722–8.
47. Wooddell CI, Rozema DB, et al. Hepatocyte-targeted RNAi therapeutics for the treatment of chronic hepatitis B virus infection. Mol Ther. 2013;21(5):973–85.
48. Yang X, Hou J, et al. One cell, multiple roles: contribution of mesenchymal stem cells to tumor development in tumor microenvironment. Cell Biosci. 2013;3(1):5.
49. Yang X, Marcucci K, et al. Preclinical evaluation of an anti-HCV miRNA cluster for treatment of HCV infection. Mol Ther. 2013;21(3):588–601.
50. Zhang Q, Kandic I, et al. Dysregulation of angiogenesis-related microRNAs in endothelial progenitor cells from patients with coronary artery disease. Biochem Biophys Res Commun. 2011;405(1):42–6.
51. Zhou J, Bobbin ML, et al. Current progress of RNA aptamer-based therapeutics. Front Genet. 2012;3:234.

Chapter 19
MicroRNA and Drug Delivery

Joseph S. Fernandez-Moure, Jeffrey Van Eps, Bradley K. Weiner, Mauro Ferrari and Ennio Tasciotti

Abstract The human genome was once thought to be a redundant sequence containing few functional regions coding for proteins. This teaching is being rewritten as we continue to understand the vast complexity of the noncoding regions of the genetic code. These regions we now understand are transcribed into small single-stranded segments or microRNAs (miRNAs) that participate in the regulation of gene expression. miRNAs interact across many pathways and thus have the potential as targets for oncologic therapies. Their efficacy is limited because methods to traverse the many biologic barriers are yet to be developed. In order to achieve effective therapeutic levels at the site of interest, the tumor, the miRNA must be shuttled to the site and simultaneously be protected from the body's defensive mechanisms. To this end, scientists have developed many vehicles for delivery at both the micro- and nanoscale using both synthetic and biologically derived vectors. Viral vectors continue to be the most commonly used vehicles, but are plagued by complications related to the vector itself. These inadequacies led researchers to explore synthetic materials such aspolylactic co-glycolic-acid (PLGA), silicon, gold, and liposomes to overcome the biobarriers of our body. While these vehicles have shown promise, problems such as high clearance rates, poor tumor accumulation and targeting, and adverse reactions have limited their translation into the clinic. In order to overcome these problems, a multistage theory was developed. By decoupling the tasks required of the carrier system, the multistage delivery system is able to simultaneously protect the payload, target the site of interest, and deliver the payload in therapeutic concentrations. This presents a paradigm shift in the concept of drug delivery and may provide the solution to the limited translational gene therapy in oncology.

Keywords miRNA · Drug delivery · Nanotechnology · Multistage delivery · Viral vector · Nanoparticles · Silicon · Liposomes

---

J. S. Fernandez-Moure (✉). J. Van Eps · B. K. Weiner
Surgical Advanced Technologies Lab, Houston Methodist Hospital and Houston Methodist Research Institute, 6560 Fannin St. Suite 2060, Houston, TX, 77030, USA
e-mail: jsfernandez-moure@tmhs.org

J. Van Eps · B. K. Weiner · M. Ferrari · E. Tasciotti
Department of Nanomedicine, Houston Methodist Hospital and Houston Methodist Research Institute, Houston, TX, USA

S. R. Singh, P. Rameshwar (eds.), *MicroRNA in Development and in the Progression of Cancer,* DOI 10.1007/978-1-4899-8065-6_19,

1 Introduction

In the past, it was believed that the only functional product of any given gene was a protein, and that the noncoding sequences in the genome were nothing but remnants of evolutionary redundancy. This principle of the central dogma taught for years was revolutionized with the discovery of *lin-4,* a gene of the *Caenorhabditis elegans* larvae found in development that represses LIN-14 protein expression and thus developmental timing [1, 2]. The *lin-4* gene does not encode a protein, but instead a pair of small RNA molecules with sequence complementarity to *lin-14,* and in fact represses its expression through RNA–RNA interaction [1].

The functionality of therapeutic RNAs is based on the catalytic process of naturally occurring 15–22-nt single-stranded RNA that couples with the cytoplasmic multiprotein complex RNA-inducing silencing complex (RISC) to pair with messenger RNAs (mRNAs) carrying complementary sequences to repress gene expression [3, 4]. The discovery of these small RNAs, later known as microRNA (miRNA), by Victor Ambrose represented the first example of gene expression regulation by endogenously produced RNA molecules.

Since the discovery of *lin-4,* miRNAs have been identified in organisms ranging from plants to humans. It has been shown that up to 3 % of human genes encode for miRNAs and that up to 30 % of human protein-coding genes are regulated by miRNAs [5, 6]. The interconnectivity of the regulation of protein expression and miRNA regulation has yet to be fully elucidated. What is clear is the opportunity for miRNAs to be used as both diagnostic and therapeutic entities. The frequent aberrant expression and functional implication of miRNAs in human disease have elevated these molecules from the ranks of redundant cellular components to preferred drug targets [7–9]. With approximately 1,400 human miRNAs known to date, a new pool of therapeutic targets with a new mechanism has emerged [6]. Additionally, acting not only as targets, but also as agents themselves, miRNAs possess the capability to suppress a broad range of oncogenes and oncogenic pathways deregulated in cancer [8]. As our view of cancer evolves as a heterogeneous disease process unable to be successfully treated by attacking a single gene or gene product, so does the therapeutic potential of miRNA as multifaceted multitarget agents capable of disabling systems rather than select cellular components [7, 10–12].

Because cancer genesis and progression involve the loss of endogenous regulatory controls of essential cellular processes, the treatment of malignancies is extremely challenging. Current chemotherapeutic regimens continue to rely on heavy dosing to overcome their low cancer specificity and poor penetration [13–15]. Immunotargeting strategies have aided tumor targeting, but they still fall short of being substantially efficacious across a wide range of tumor types and are often limited in their scope because of the specificity of their targets [16, 17]. Because they interact with multiple pathways and convergence points of tumor suppression, miRNAs are ideal candidates for delivery.

Several advancements in cancer treatment have been developed over the last century to control and eradicate the disease [18]. Currently, the leading treatments for cancer are surgery, chemotherapy, and radiotherapy [11, 19, 20]. Further innovations have been made over the years to overcome weaknesses and side effects that may arise with these treatments [21]. While many novel approaches to the noninvasive diagnosis and treatment of cancer have emerged, effective penetration of the tumor and specific cancer cell targeting remains an obstacle, resulting in multiple administrations of current therapeutics, and for prolonged periods. This could lead to renal, cardiac, hepatic, and neurologic toxicities.

To overcome these disadvantages and enable a greater efficacy of treatment, novel drug delivery systems have emerged as prospective vehicles to carry and deliver the therapeutic payload. Vectors derived from natural and synthetic materials have been used to deliver oligonucleotide fragments with inhibitory and restorative functions to cancer cells in hopes of achieving greater therapeutic efficacy than the currently commercially available drugs, which are fraught with side effects and shortcomings.

2 MicroRNA as Pharmaceutical Agents

2.1 MicroRNA Characteristics, Comparisons, and Biogenesis

The discovery and characterization of several classes of small (~20–30 nt), noncoding RNAs during the last decade changed the field of molecular biology. This family of RNA is capable of regulating gene expression and can be classified into three main categories: miRNAs, short-interfering RNAs (siRNAs), and piwi-interacting RNAs (piRNAs). The piRNAs are single strands found mainly in animals; they act primarily in the germline and will not be discussed [22, 23]. In contrast, siRNAs and miRNAs show similarities with regard to structure, biogenesis, and repression of translation. However, it is the diminutive differences between siRNAs and miRNAs that point to disparities in function and therapeutic potential.

Both siRNAs and miRNAs are ~21–23 nt long, have double-stranded precursors (double-stranded RNA, dsRNA) of a guide strand (miRNA strand) and passenger strand (miRNA* strand) [22, 24]. These precursors undergo extensive splicing and refinement via their respective RISC before interacting with target mRNA [22, 25]. Both classes' biogenesis is utterly dependent upon two protein families composed of specific domains: dicer enzymes (PAZ, 0–2 dsRNA-binding domains (dsRBDs), and tandem RNase III domains) to separate them from their precursors, and Argonaut (Ago) proteins (PAZ, PIWI, N-terminal (N), and middle (Mid) domains) that are central to the RISC and allow the complex unwinding, base pairing, and cleavage described later (Fig. 19.1) [22, 26–28]. Additionally, both classes have highly reprogrammable silencing effects due to predictable Watson–Crick base pairing recognition of gene targets, and both can dynamically redirect such silencing depending on changing genomic demands (miRNA) or threats (siRNA) [22, 29–31]. Both siRNAs and miRNAs have nuclear and cytoplasmic phases of refinement, but they preferentially exist/function in the cytoplasm, lose their aforementioned

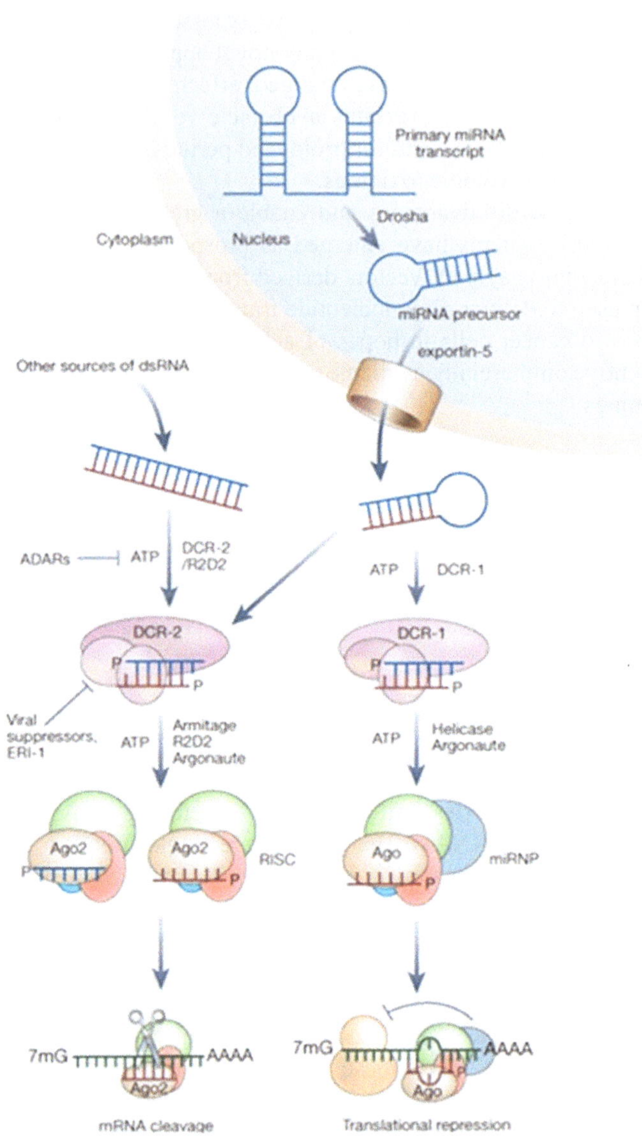

Fig. 19.1 Schematic of miRNA processing in the cell cytoplasm. As the primary mRNA transcript is shuttled from the nucleus to the cytoplasm, it associates with the DCR complex, where the double strand is extensively spliced. Following splicing and refinement, the precursors then associate with the RISC complex before interacting with their target mRNA. (Reproduced from [28] with permission from Nature Publishing Group)

passenger strand during unwinding in the RISC, and can act by direct nucleolytic degradation, histone methylation, or heterochromatin formation [22, 25, 26, 28, 30, 32–36]. These two classes do, however, have notable differences.

Differentiation of miRNAs/siRNAs has classically and simply been based on their origin and precursor characteristics. The sources of siRNAs are extensive, but the majority are exogenous linear, base-paired dsRNA like viruses, transgenes, transposons, etc. that are either directly introduced into the cytoplasm or actively taken into the cell as an endosome from the environment and are thought to function as guardians of genomic integrity [22, 24, 37, 38]. This was classically viewed in stark contrast to miRNAs—perceived as intentional, endogenous products of a host genome in response to an inherent genomic need, and far more conserved across related organisms [3]. However, recent evidence shows that endogenous siRNAs can be derived from hairpin RNAs, centromeres/telomeres, naturally occurring sense–antisense pairs like convergent mRNA transcripts or pseudogene antisense/mRNA sense transcript duplexes [22, 37, 39, 40]. While siRNAs are cleaved from long, complementary dsRNAs, miRNAs are refined from incomplete dsRNA primary miRNAs (pri-miRNA) in what is called a "stem loop" with important flanking segments [3, 22, 23]. Additional differences have been discovered in the structure, processing, and mechanisms of repression by siRNAs and miRNAs.

A central difference between most siRNAs and miRNAs is the exactness of their 3′ and 5′ ends, as miRNA duplexes have very precise ends, with siRNAs showing much greater variability [3, 22]. It is this feature that is believed to afford miRNA its enhanced target specificity despite often imperfect central base pairing. Conversely, siRNA typically depends on perfect substrate complementarity for silencing, but may still function in the face of mismatches in an miRNA-like manner, despite suppressed endonucleolytic cleavage activity [22, 25, 26, 37, 40]. Often, siRNAs undergo amplification, with secondary siRNA generation via RNA-dependent RNA polymerase (RdRP) enzymes, causing a sustained response [22, 28]. This has been studied most extensively in other eukaryotes. miRNAs are unique in the fact that they may arise from either a dedicated transcription unit or one that makes several products, they may be cotranscribed with a protein within an intron, and they exhibit combinatorial regulation—target specificity across a variable number of mRNAs and the sharing of a common mRNA target by several different miRNAs [22, 33, 41, 42]. The broad possible effect on multiple cellular pathways allowed by miRNA is quite different than the target specificity required for RNA interference (RNAi) by siRNA. Finally, miRNA–RISC (miRISC) complexes require additional proteins like GW182 for silencing activity and are believed to repress genes by additional mechanisms to the canonical cleavage and nuclease degradation preferred by siRNA, such as deadenylation/decapping, inhibition of mRNA circularization, and premature ribosomal dissociation [25, 31, 36, 43, 44].

2.2 Mechanisms of Silencing

Although the steps and machinery required for miRNA biogenesis and refinement have been well characterized, more robust theoretical debate has surrounded its functional mechanisms of genomic silencing. We know that a pri-miRNA first travels to the nucleus for a critically precise cleavage of its stem loop from surrounding RNA by an RNase III/dsRBD complex (Droscha/DGCR8 in mammals) before travelling to the cytoplasm, where the Dicer enzyme separates the terminal loop from its pre-miRNA stem, making a duplex of mature miRNA that is ~22 bp in length [3, 22, 45, 46]. This miRNA is quickly unwound when assembled with Ago and associated proteins in the miRISC, which guide it to mRNA targets with variable complementarity, so they may be silenced and/or degraded [31, 47]. Protein expression is repressed by miRNA through degradation and decay, inhibition of translation initiation, inhibition of translation elongation, and translation abruption via ribosomal dissociation (Fig. 19.2) [28, 31, 48].

Target mRNAs with exact complementarity to the guide miRNA strand can undergo Ago-catalyzed cleavage/degradation, while those containing centrally located mismatches have their translation repressed by another mechanism. These mismatched targets may be sequestered for decay within cytoplasmic P bodies—containing Argonaut proteins, GW182, deadenylase complexes, decapping machinery, and an RNA helicase [30, 31, 49, 50]. Recent evidence suggests that such a destabilization of mRNA target transcripts accounts for a greater percentage of the diminished protein production observed [51, 52]. Studies have shown that miRNA can halt translation before it begins initiation by binding the essential cap-binding protein eIF4E to central phenylalanine residues on Argonaut, or competitive inhibition at the methylated cap structure [31, 53]. Finally, miRNA is believed to inhibit early translation by the binding of eIF4E, subunit recruitment, causing their early dissociation, as well as interfering with translation elongation [22, 31, 48]. Until recently, the timing and contribution of these mechanisms were unknown. Ribosomal and initiation effects are now confirmed to occur early before decay, while translational repression is the primary mechanism observed on new targets and confirmed to occur prior to deadenylation or decapping causes of degradation, which are prominent mechanisms at steady state [48, 50, 54–56].

2.3 Dysregulation of miRNA in Cancer and Its Potential Clinical Implications

Given their ability to affect protein production at the genomic level, it is not surprising that miRNAs play significant roles in nearly every cellular process from development through apoptotic death. Indeed, several in vitro and in vivo studies have proved this to be true, demonstrating key regulatory roles for miRNA in: embryological development, stem cell proliferation and differentiation, organogenesis, metabolism, apoptosis, angiogenesis, muscular, neuronal and immune cell development and function, and more [3, 47, 57–60]. Such intertwining of miRNA and normal cellular function thus makes their dysregulation a prime suspect in the genesis of both benign and malignant diseases. Experimental evidence has confirmed this suspicion and generated useful databases, revealing links between miRNA dysfunction and many benign diseases, from cardiovascular disease to myasthenia gravis, and from Tourette syndrome to scleroderma

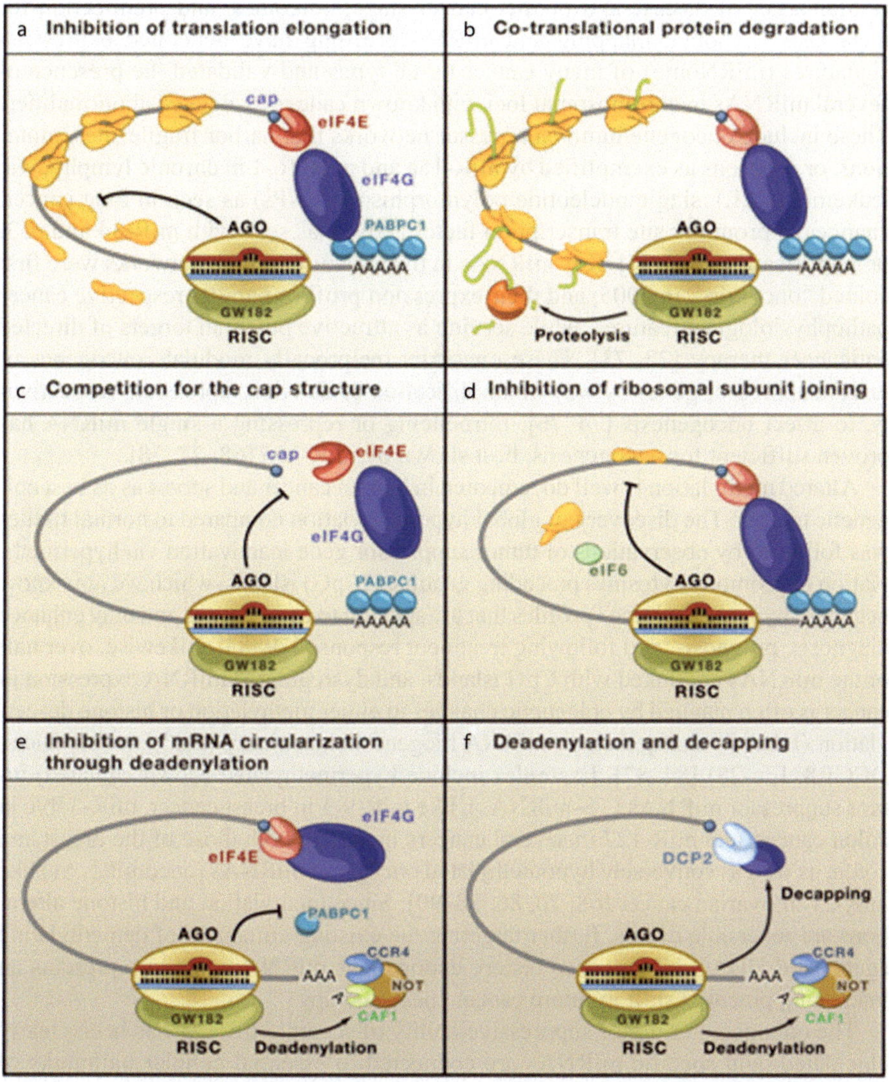

Fig. 19.2 Mechanisms of miRNA silencing. Silencing by miRNA can occur through several mechanisms at different points throughout the protein translation phase. Posttranslational inhibition of protein elongation or posttranslational protein degradation may be affected (**a, b**). Interference with the formation of the ribosomal translational unit will inhibit translation (**c, d**). Pretranslation modifications, such as deadenylation or decapping, will prevent proper translational processing and result in early degradation (**e, f**). (Reproduced from [31] with permission from Elsevier)

[61–66]. Likewise, altered miRNA expression in human cancer is the rule rather than the exception.

Numerous associations exist between various cancer subtypes and specific miRNA aberrances, and we now know the potential of miRNA to serve as an early biomarker of disease and predict cancer stage, outcomes, and predilection for metastasis. Genomic mapping and miRNA profiling have generated expression signatures (miRNome) of many cancer tissue types and validated the presence of several miRNAs in chromosomal loci with known cancer-associated abnormalities. These include oncogene/tumor suppressor networks that harbor fragile sites, mutations, or deletions as exemplified by miR-15a and miR-16-1 in chronic lymphocytic leukemia (CLL), single nucleotide polymorphisms (SNPs) as seen in lung cancer, changes in promoter site transcription factor activity as seen with miR-34 and p53, or amplifications [67–71]. Key miRNAs in the structure of such networks were first coined "oncomirs" in 2005, and their expression profiles parallel respective cancer pathophysiological features, while serving as attractive potential targets of directed anticancer therapy [72, 73]. These oncomirs reciprocally modulate oncogenes or tumor suppressor genes by way of amplification or deletion/repression, respectively, to affect oncogenesis [74–76]. Introducing or repressing a single miRNA has proven sufficient for oncogenesis, best shown by miR-155 [68, 77, 78].

Altered methylation is well documented in human cancer and serves as its best epigenetic marker. The discovery of global hypomethylation compared to normal tissues was followed by observations of tumor suppressor gene inactivation via hypermethylation of promoter cytosines preceding guanines (CpG) islands, which we now know occur as "hypermethylome" profiles that are specific to cancer types and may enhance diagnosis, prognosis, and following treatment response [79–83]. Likewise, over half of the miRNAs are linked with CpG islands, and dysregulated miRNA expression in cancer is often plagued by epigenetic changes in either methylation or histone deacetylation (HDAC) and defects in miRNA biogenesis machinery (e.g., Drosha, Dicer, DCGR8, Lin-28) [84–87]. Examples include hypermethylated (downregulated) tumor suppressor miRNAs (TS-miRNAs), like miR-9-1 in breast cancer, miR-34b/c in colon cancer, and miR-127 in several cancers including both those of the breast and colon, as well as conversely hypomethylated oncogenic miRNAs (oncomiRNAs) like miR-21 in ovarian cancer [68, 70, 86, 88–90]. Since methylation and histone alterations are reversible events, further targeting the existing armament of demethylating agents and HDAC inhibitors to restore endogenous miRNA function represents an interesting potential shift in future cancer chemotherapy.

The oncogenic or tumor suppressive ability of several miRNAs has been clearly elucidated, and specific miRNAs are connected to essential cellular hallmarks of cancer (Fig. 19.3), thus establishing the need for these small entities as central substrates in our knowledge pursuit and therapeutic goals within the cancer paradigm going forward. The original TS-miRNAs described were members of the let-7 family, found to be modulated by p53 and downregulated in many cancer types with compromised interactions with several putative targets (MYC, RAS, HMGA2, FOS) [84, 87, 91]. The archetypal oncomiRNA is represented by miRNA-17-92 cluster, nicknamed "oncomiR-1." This oncomiR is linked to several cancer types with putative targets, such as MYC, Bim, tsp-1, and phosphatase and tensin homolog (PTEN) [68, 71, 72, 85]. Several other oncomirs have subsequently been identified studied to provide "signatures" for human cancers (Tab. 19.1) [8]. Breast cancer is one of the best characterized for miRNA signatures [68, 73, 91–93]. A similar signa-

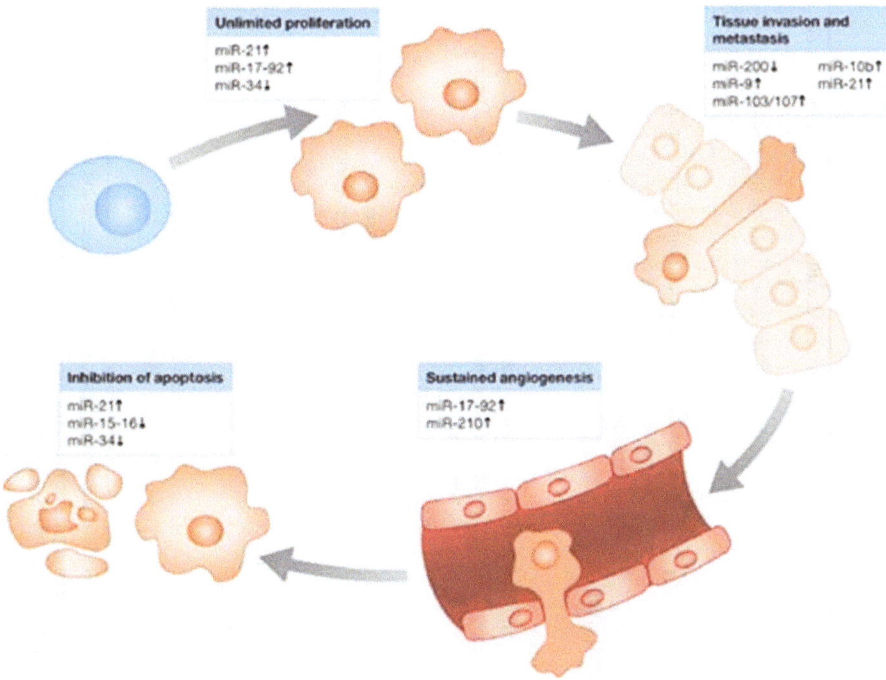

Fig. 19.3 MicroRNAs involved in targeting the hallmarks of cancer. (Reproduced from [68] with permission from John Wiley and Sons Publishing Group)

ture has also been developed for lung cancer [68, 73, 91, 92, 94, 95]. These targets show great promise for therapy and biomarkers for diagnosis, prognosis, treatment response, and predicting invasion/metastasis.

The stability of miRNA in vitro, degree of conservation in vivo, and verified presence in circulating serum make it a good theoretical agent for diagnosis [91]. miRNA profiling has proven to be more accurate than mRNA in reliably classifying tissues of origin in both primary and metastatic tumors due to their similarly conserved miRNA expression, showing great potential for diagnosing the origin of metastasis of unknown primary origin [68, 96]. Many miRNAs have demonstrated links with cancer aggressiveness and outcomes as well. The first link was shown in 2005 between prognosis and progression of CLL and a unique 13-gene miRNA signature [97]. High miR-193a and low miR-191 levels are associated with shorter survival in melanomas [68, 98]. Additionally, it was later discovered that downregulation of miR-17-5p, -381, and let-7 and upregulation of miR-155 are predictive of poor prognosis/survival in lung cancer [99–101]. Certain miRNAs either promote (miR-10b and -21) or reduce (miR-200 family, -205) invasion and metastasis in breast cancer, and silencing miR-10b in mice successfully inhibits metastasis [68, 94, 102–104]. Both miR-126 and miR-335 suppress the "epithelial–mesenchymal transition" (EMT) necessary for metastasis in breast and lung cancer [92].

Table 19.1 Known patterns of microRNA up or downregulation associated with common cancers, their putative gene targets, and downstream effect(s). (Reprinted with permissions from [73])

miRNAs	Cancer cell type												Putative targets	Notes
	Br	*Co*	*Ce*	*Gl*	*Hp*	*Ov*	*Pn*	*LL*	*Ln*	*Pr*	*Te*	*Thy*		
let-7	↓	↓			↓	↓							c-MYC, RAS, HMGA2	Level inversely correlated with survival in NSCLCs; modulated by p53
miR-10b	↑												HOXD10	Regulates metastatic/invasive function in breast cancer; modulated by p53
miR-15a, miR-16	↓							↓	↓	↓			Bcl-2	
miR-17-92		↑						↑	↑				Tsp-1, CTGF	miR-18, miR19 bind to antiangiogenic proteins in myc-overexpressing tumors, modulated by p53
miR-20a								↓					E2F1, TGFβR2	Modulated by p53, has antiapoptotic role
miR-21	↑	↑	↑	↑	↑	↑		↑					PTEN, TPM1, PDCD4	
miR-26a												↓	PLAG1	
miR-30c, Mir-d	↓											↓		
miR-34a, -34b, -34bc	↑iR	↓					↓		↓				E2F3, NOTCH1, DLL1	p53-Dependent induction by DNA damage and oncogenic stress; expression leads to apoptosis and cellular senescence
miR-106a		↑					↑	↑		↑			RB-1	Regulated by c-myc
miR-122a					↓								Cyclin G1	Hepato-specific expression
miR-125b	↓		↓			↓			↓	↓		↓	ErbB2/3, EIF4EBP1	Targets TNFα production in macrophages
miR-127	↓	↓	↓				↓	↓	↓				Bcl-6	Epigenetically downregulated in the majority of human cancer lines
miR-128				↓										
miR-143, miR-145	↓	↓	↓		↓	↓			↓	↓			ERK5, Raf1, G-protein 7	
miR-155	↑						↑	↑	↑				AT1R, TP53INP 1	
miR-200a, -200b, -200c						↑			↓				TCF8 for miR-200c	
miR-221/222				↑	↑	↑				↓↔↑		↑	P27Kip1, C-KIT	Directly correlated with NSCLC survival, miR-221 induced by MYCN in neuroblastoma.
miR-372, miR-373											↑		LATS2	

AT1R angiotensin II type I receptor, *Bcl-2* B cell lymphoma 2, *Br* breast, *Ce* cervical, *Co* colon, *CTGF,* connective tissue growth factor, *E2F1* E2F transcription factor 1, *EIF4EBP1* eukaryotic elongation initiation factor 4E binding protein, *ERK5* extracellular signal-regulated kinase-5, *G-protein* guanine nucleotide-binding protein, *Gl*, glioblastoma, *Hp* hepatocellular, *c-kit* tyrosine kinase, *LATS2* large tumor suppressor homolog 2, *LL* Leukemia and lymphoma, *Ln* lung, *miRNA* microRNA, *c-myc* myelocytomatosis viral oncogene, *NSCLC* nonsmall cell lung cancer, *Ov* ovarian, *PDCD4* programmed cell death 4 tumor suppressor protein, *PLAG1* pleomorphic adenoma gene 1, *Pn*, pancreatic, *Pr* prostate, *PTEN* phosphatase and tensin homolog, *Raf1* seine/threonine protein kinase, *Ras* rat sarcoma viral oncogene homolog, *RB-1* retinoblastoma 1, *TCF8* (ZEB1, deltaEF1) transcription factor 8, *Te* testes, *TGFβ2R* transforming growth factor beta-2 receptor, *Thy* thyroid carcinomas, *TP53INP 1* tumor protein p53-induced nuclear protein 1, *TPM1* tropomyosin, *tsp-1* thrombospondin-1.

Even hormone receptor (ER) status and predictive response (PR) to hormone-based therapies in breast cancer can be predicted by miRNA. Various patterns of decreased miR-145/-125b and increased miR-21/-34/-155/-375 together can mirror tumor ER/PR expression, stage, proliferation index, or vascular invasion [73, 105–107]. Overexpression of miR-221/-222 is responsible for resistance to antiestrogenic therapies, while improved responsiveness to tyrosine kinase inhibitors via HER3 targeting is seen with ectopic expression of TSmiR-205 [108, 109]. These far-reaching and "awe-inspiring" data provide a wide base of validity and enumerable targets for pursuing novel miRNA-based cancer therapies or diagnostic modalities. However, as is often the case, the question looms just how far behind our technical ability trails our current knowledge base.

3 Mimicry vs. Antagonism

Theoretically, miRNA therapy represents a superior approach to multi-pathway disease such as cancer, since multiple genes can be affected by targeting a single miRNA. Given the nature of miRNA as a natural inhibitor of gene/protein production and function, it follows that there are two possible approaches for miRNA-based therapies: mimicry (replacement) or antagonism. At the phenotypic level of cellular machinery, these are experienced as loss or gain of function effects, respectively; and pertaining specifically to cancer, they represent repression of oncogene or restoration of tumor suppressor activity. Technically speaking, antagonists act upstream from the RISC outside of P bodies and mechanistically resemble siRNA or small-molecule inhibition, while mimicry is likened to traditional gene therapy and acts via miRNA replacement downstream of the RISC [7, 110]. The effect(s) of either mimics or antagonists can be followed by functional studies of their target genes and individual or genome-wide miRNA expression profiling [91].

miRNA mimics are either single- or double-stranded RNA molecules, which form a complex with RISC to suppress the target genes on the intended target of native miRNA. Double-stranded mimics are preferred due to 100–1000-fold greater potency as compared to single-stranded versions, as the passenger (miRNA*) strand may interact with either the mature miRNA or its precursor form [77, 111]. Although likened to gene therapy, miRNA mimicry has potential advantages due to its smaller size, ability for systemic administration, and the need for cytosolic entry for functional changes [77]. In addition, miRNA profiling of cancers indicated changes in miRNAs, suggesting their roles in oncogenesis. miRNAs could be tumor suppressors since their inhibition resulted in enhanced tumorigenesis. Since the miR mimics are identical to endogenous miRNA, the mimics should target the same mRNAs, making off-target cytotoxic cellular effects less likely. One mimic has been shown to modulate multiple oncogenic targets and pathways [9, 77, 112].

Evidence from animal studies has demonstrated the potential validity of miRNA replacement therapy. The miRNA let-7 suppresses RAS and is decreased in breast and non-small cell lung cancer (NSCLC), and replacement therapy with mimics in mouse models reduces existing tumor growth and stops proliferation of cancer cells [77, 84, 113–116]. Similar mimicry of miR-34a, a transcriptional target of p53 that is lost or repressed in most cancer types, showed inhibition of lung tumor growth in existing tumors with low miR-34a levels and protection from tumorigenesis in cells with normal levels, implying a protective role [7, 77, 117, 118]. Numerous other examples exist, including miR-16 and prostate cancer, miR-205 and breast cancer, and miR-26a and hepatocellular cancer [93, 119, 120].

Antagonists inhibit oncomiRNA that has acquired a gain of function. They are collectively known as anti-miR oligonucleotides (AMOs) and may take the form of anti-miRs, antagomiRs, or locked nucleic acids (LNAs) and are also useful in functional studies confirming miRNA target sites [121–123]. Similar to RNAi using siRNA, efficient antagonism depends on sufficient binding specificity for inhibition/degradation. The most common method uses antisense single-stranded oligonucleotides containing complementary, sugar/phosphate chemically modified backbone analogs (e.g., 2′-O-methyl, 2′NH_2, phosphorothioates) with enhanced RNase protection and affinity for endogenous miRNA that sequester it from its RISC processing and promote degradation [7, 52, 121, 123]. Their length and composition can be fine-tuned for improved circulation time, cellular internalization, etc., and endogenous miRNA targets can act as biomarkers of efficacy and optimization [7]. “AntagomiRs” are analogs with a 3′ cholesterol moiety that were first coined in 2005 with evidence of their inhibition in mice [124]. Important antagomiR characterization has been done using antagomiR-16 and its target miR-122 in vivo in mice, demonstrating: dose and time dependency, a minimum length of 19 nt required for silencing, extensive target degradation that is “independent” of exonucleases, and interaction with miRNA in cytosol “outside” of P bodies [110]. AntagomiRs also display high sequence specificity with impaired silencing possible by only a single position-dependent mismatch [110]. AMOs are capable of downregulating the abundant miR-122 in the liver, which has two target sites within the hepatitis C viral (HCV) genome, and the first human clinical trial has been conducted using LNA anti-miR [68]. However, effective delivery is a main impediment to therapeutic mimicry/antagonism, with great potential for improvement.

4 Negotiating Biological Barriers in miRNA Delivery

The promise of in vitro efficacy of miRNA for cancers is dampened by the limitation to effectively deliver the miRNA to the site of interest [125–128]. The body’s natural protective barriers such as serum nucleases, vascular endothelium, interstitial and oncotic pressure, and the cell wall have made the targeting and treatment of tumor cells particularly challenging (Fig. 19.4) [128, 129]. The great obstacle in engineering therapeutics, though, is often not the synthesis and characterization of the agent, but rather overcoming the many biobarriers of the body without damaging the integrity of the payload, in this case miRNA. By understanding the critical barriers to overcome, one could exploit these mechanisms to

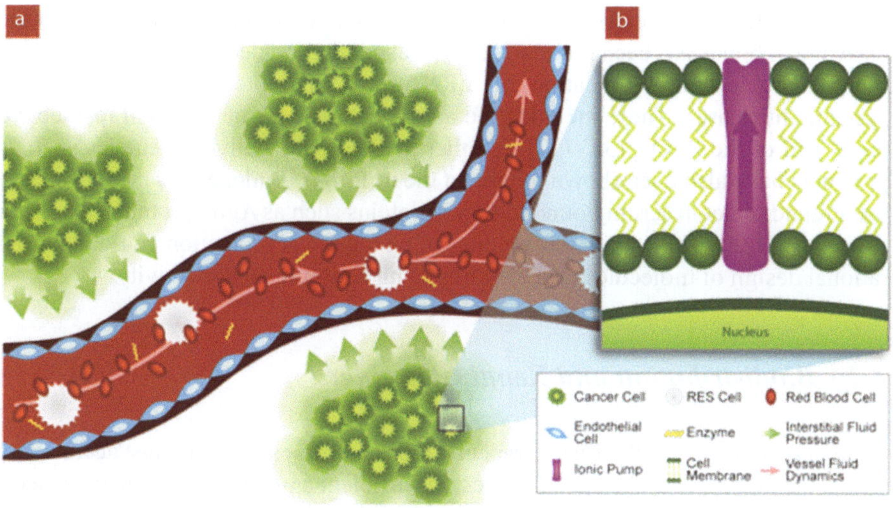

Fig. 19.4 The highly efficient biological barriers of the body present themselves as obstacles to the delivery of injected chemicals, biomolecules, nanoparticles, and any other therapeutic foreign agents from reaching their site of action. The barriers include the reticulo-–endothelial system, endothelial/epithelial membranes, complex networks of blood vessels, abnormal flow of blood, and interstitial pressure gradients (**a**). The tumor cells also pose additional barriers: cellular/nuclear membranes and ionic/molecular pumps that can expel drugs from the cancerous cells creating drug resistance (**b**). (Reproduced from [129] with permission from Nature Publishing Group)

develop solutions that preferentially direct the activity of therapeutic agents and sequentially negotiate those barriers for effective targeting and treatment. Because of these hurdles, the successful application of miRNAs as a cancer therapy has been limited [119, 130, 131].

The site of action of miRNA therapeutics occurs in the cytosol. The barriers to successful delivery of miRNA to the point of action in the cytosol or the nucleus are many and depend on the targeted organ and the route of administration. Locoregional administration of miRNA has fewer barriers compared to systemic delivery. For example, intranasal delivery of viral particles with *let-7* miRNA in a *Kras*$^{G1D/+}$ autochronous NSCLC mouse reduced tumor burden [115]. The efficacy of this delivery route was comparable to systemic delivery of *let-7* in a neutral lipid emulsion in the same NSCLC model [132]. Many times this is not the case and one requires an understanding of the complex barriers to be able to achieve effective systemic delivery of miRNA.

4.1 Challenges: Serum Nucleases

The intravenous injection of miRNA poses significant challenges, beginning at the site of intravascular injection where the miRNA can be degraded. The first of many barriers encountered by delivered miRNA therapeutics are serum nucleases [133].

The major class of these nucleases includes 3′ exonuclease, although internucleotide cleavage of bonds has been reported [134]. Naked oligonucleotides can be degraded by serum nucleases. Modifications of miRNAs were attempted to protect them from systemic nucleases. These included the addition of adenosine or uracil residues to the 3′ end of the RNA [135–137]. In other methods, the delivery vehicles such as polymers have also been shown to protect the miRNAs. Endogenous miRNAs have been found intact in serum bound to RISC proteins such as Ago2 protein and within exosomes [138]. Thus, an understanding of endogenous protection could lead to a rational design of molecules capable of evading serum nuclease activity.

4.2 *Biodistribution and Renal Clearance*

In order to achieve an effective therapeutic index, a proposed agent must accumulate within the system or at the site of interest in sufficient amounts so as to have a therapeutic effect [139]. In the design of oligonucleotides administered as individual molecules or encapsulated within polymers, consideration must be given to the clearance mechanisms of the liver and kidney. The liver's role in clearance is more prominent when the oligonucleotides are associated with nanoparticles [140]. Clearance by the kidney, on the other hand, has been demonstrated in nanoparticle and non-nanoparticle-based miRNA therapeutics [141]. Studies that characterized the clearance of antisense oligonucleotides were extensively reported during the 1990s [142–144]. Many of these studies were focused on protein-bound oligonucleotides or those with a phosphorothioate backbone. While protein bound, the main route of elimination was found to be in the urine, and the major organ of accumulation was found to be the kidney. Important features of oligonucleotide pharmacokinetics emerged from these studies. These include the following: (1) The biphasic plasma half-lives of oligonucleotides are in the range of minutes for the distributional phase and in the range of minutes to hours in the elimination phase. This indicates that without targeting or protection from nonspecific organ accumulation, miRNAs may be eliminated before any desired effect. (2) Oligonucleotides are accumulated in the liver and kidney but not the central nervous system. (3) In general, the information on the pharmacokinetic studies of phosphorothioate oligonucleotides in lower animal models such as rodents can be applicable to humans [145]. Animal models reported that the biodistribution of siRNA duplexes was similar to single-stranded antisense molecules, with highest uptake in kidney followed by liver [146, 147]. Interestingly, although found circulating in the blood bound to protein or within exosomic vesicles, these circulating miRNAs are not cleared by hemodialysis, suggesting these mechanisms as potential vehicles for effective protection of antisense oligonucleotides in the blood [148]. Clearly, uptake by the reticuloendothelial system (RES) plays a critical role in the delivery of miRNAs and other oligonucleotides. Further understanding the mechanisms that govern uptake into these RES-rich organs will allow us to design and deliver oligonucleotide carriers to achieve greater efficacy.

4.3 Clearance by the RES

In addition to circulating nucleases and renal clearance, a major barrier to effective in vivo delivery of antisense drugs is clearance by the RES [133]. Phagocytic cells, including circulating monocytes and tissue macrophages, whose physiological functions are to clear the body of foreign pathogens, remove cellular debris that is generated during tissue remodeling, and clear cells that have undergone apoptosis comprise the RES [149]. These cells, named Kupffer cells in the liver, or known simply as splenic macrophages, can detect and eliminate antisense oligonucleotides, such as miRNAs or siRNAs. Further, these same cells are capable of eliminating the nanoparticle carriers into which antisense drugs may be loaded [150]. This clearance may occur through various opsonization mechanisms including immunoglobulins, complement components, and other serum proteins.

Following administration, nanoparticles are detected by the immune system and tagged by proteins identifying them as foreign bodies and sequestering them for elimination. Several factors such as carrier surface charge, size, and surface characteristics may affect RES uptake [151, 152]. These characteristics have been used to target drug carriers to RES organs [153, 154]. Sites outside this system, though, continue to be elusive in their targeting. Immunoglobulin opsonization leads to Fc receptor activation and internalization characterized by cytoskeletal rearrangement, pseudopodia extension, and engulfment of the opsonized particle [155]. Complementopsonized particles are internalized by complement receptors such as CR1, CR3, and CR4, members of the integrin family of heterotrimeric membrane proteins. While CR1 is involved in particle binding to the macrophage membrane, CR3 and CR4 mediate internalization. Lastly, lipopolysaccharide (LPS) and modified lipoprotein opsonization not only targets particles, but also targets naked oligonucleotides. The scavenger receptors that recognize these ligands found in the liver and spleen are suggested to be responsible for the uptake of monomolecular oligonucleotides [156, 157].

4.4 Endothelial Barrier

The endothelial cells lining the vascular lumen present both a barrier and an opportunity for oligonucleotide-based therapeutics [133]. Endothelial cells line the vasculature, adhering tightly to the underlying extracellular matrix largely via integrins and forming junctions with each other via several types of cell–cell adhesion molecules, including vascular endothelial (VE)-cadherin, junctional adhesion molecule (JAM), occludins, claudins, and platelet endothelial cell adhesion molecule (PE-CAM) [158]. The integrity of the endothelial junctions is influenced by complex signal transduction processes that respond to a variety of mediators. For example, microvascular transport of macromolecules increases during inflammation [159]. This process, often driven by diffusion, is due to shifts in the concentration gradients between the inside and outside of the vessel wall. Transvascular fluid transport is driven by the difference in hydrostatic and osmotic pressures between the blood vessel and interstitial space. When delivering particulates, the upper size limit of diffusion is limited by the pore size of the capillary. In the case of inflammation, these pore sizes can increase, altering the normal gradients and allowing greater diffusion [160–162].

Angiogenesis is a known and widely accepted mechanism of tumor growth and survival. A mixture of heterogeneous precursor cells and preexisting normal vasculature makes up the new sprouting vessels [163]. The tumor blood vessels generated from aberrant mechanisms of tumor survival are structurally and architecturally different from the normal vasculature [164]. The combination of constitutively activated genes and heterogeneous cell populations develops abnormal vasculature marked by precocious capillary sprouting, convoluted and excessive vessel branching, distorted and enlarged vessels, erratic blood flow, microhemorrhaging, leakiness, and abnormal levels of endothelial cell proliferation and apoptosis [165]. Structurally, the dense vascular networks are not organized into venules, arterioles, and capillaries as seen in normal anatomy, but rather are a chaotic organization of loops and arcs that encircle clusters of cancer cells. Architecturally, the vessels themselves have an atypical basement membrane, a high percentage of hyperproliferative endothelial-like cells, and a decreased number of pericytes [166]. Directed by the cancer cells, the resulting vessels have large intercellular openings between the endothelial cells that are not present in normal continuous vasculature [167–169]. These fenestrations are known to range in size from 100 nm to 10 μm and are exploited in many nanomaterial-based therapies [170, 171]. Due to the leakiness of these vessels, the major pathway of drug transport across tumor microvasculature is by extravasation via diffusion and/or convection through the discontinuous endothelial junctions, whereas transcytosis plays a relatively minor role. This leakiness promotes not only drug carrier extravasation, but also protein transport, leading to high interstitial compartment pressures and reduced transvascular transport [169, 172].

4.5 Extracellular Matrix Compartment and Subcellular Distribution

The interstitial compartment of solid tumors is mainly composed of extracellular matrix proteins, interstitial fluid, and a random organization of poorly formed vessels. The distinguishing feature of the interstitium is the high pressure resulting from the lack of sufficient lymphatic drainage. The movement of small molecules in the interstitial space is governed by diffusion, while the movement of large molecules, such as drug carriers, is governed by convection [173]. While diffusion relies on diffusion coeffecients and concentration gradient, the convection depends on hydraulic conductivity and pressure difference. While the enhanced permeability and retention (EPR) effect allows the particles to accumulate to the tumor loci, it is the high interstitium pressure that allows extravasated particles to be pushed into the margin of the tumor [174, 175]. This characteristic presents an additional biobarrier to the penetration of therapeutic agents into the tumor core as the transport is governed by the physical properties (charge, pH, lipophilicity, size, structure) of the subcompartment and of the drug [173, 176]. This phenomenon of EPR within tumor stroma elucidated by Maeda et al. supports the use of silicon microparticles as drug carriers because of the passive accumulation in solid tumor after intravenous administration [139, 169, 177, 178].

4.6 Inefficient Endocytosis and Endosomal Release

The cellular subcompartment represents the final barrier to therapeutic delivery. Efficient endocytosis and endosomal release are required to bring the payload to the site of action in the cytosol. Mechanisms like phospholipid bilayer, lysozomal and endosomal degradation, and efflux glycoprotein channels are particularly crucial to intracellular delivery of agents as small interfering RNAs and miRNAs, where cytoplasmic delivery is critical and endogenous degradation pathways prevent that delivery. Targets such as the previously mentioned TP53 pathway, Ras/Raf pathway, and other various growth factor signal transduction pathways require this intracellular delivery.

The plasma membrane is a significant barrier for miRNA uptake [179]. Despite their small size, the charge and hydrophilicity of miRNA molecules prevent them from readily crossing the plasma membrane [180]. While complexing to nanoparticles or other delivery mechanism may target them to the cell itself, the entry of the system into the cell is critical for function. The integration of the payload into nanoparticles that are uptaking by cells may present an efficient of innovative mechanism for delivery.

The final barrier to effective targeting of miRNA is the release form the endosome [180]. Whether delivery by cationic lipids, nanoparticles, or cell-type-specific delivery reagents, the intracellular trafficking of miRNA begins in the early endosomal vesicles [180]. These early vesicles fuse with late endosomes with increasingly acidic environments until, ultimately, they fuse with the lysosome [181]. The lysosomal compartment is the most acidic environment and will lead to the degradation of any nucleic acids that may find themselves in that environment. The fate of the internalized molecules inside the fused vesicle depends on the specific type of bound receptors and includes the following: recycling back to the cell surface, degradation in lysosomes, or release into the cytosol to interact with other intracellular components [182–185]. The endosomal entrapment contributes to the low transfection efficacy of non-viral carriers and represents a final but significant hurdle to overcome in delivery. Various strategies such as liposomal delivery, hydrogen sponges, and cell-penetrating peptides have been developed to facilitate the release of oligonucleotides from the endosomal compartment in order to avoid lysosomal degradation [186–188].

Table 19.2 In vivo barriers to successful miRNA delivery

Degradation by serum and tissue proteases
Renal clearance through filtration of particles < 50 kDa
Failure to negotiate endothelium in organs other than liver and spleen
Phagocytosis by macrophages
Limited movement through polysaccharide rich extracellular matrix
Inefficient endocytosis by target cells
Unsuccessful endosomal release

5 Vehicles of Delivery

5.1 Non-Nanoparticle Therapeutic Delivery

Successfully delivering miRNA mimics or antagonists to the tissue of interest while preserving their structural integrity remains a principal challenge [189, 190]. Several technologies have proven effective in delivering miRNAs to tissues: vector based (adenoviral and lentiviral (let-7)), atelocollagen, liposomes, porous silica, gold, and polylactic co-glycolic-acid (PLGA). Development of clinically relevant miRNA formulations frequently involves a thorough evaluation of existing technologies to identify those that are amenable to miRNA and its chemistry. Although synthetic modifications provide some degree of protection from nucleases and promotion of necessary cellular internalization as stated above, naked RNAs in biological fluids rapidly undergo nuclease destruction. This fact limits many therapeutic RNAi platforms to local administration only, allowing a restricted number of amenable tissues and, often, only a fraction of diseased cells get adequate drug exposure. Systemic administration has many theoretical advantages, but besides nucleases, a horde of other in vivo barriers, from macrophages to renal clearance, impedes successful miRNA delivery (Tab. 19.2). Critical criteria in the evaluation process are: (1) sufficient delivery to induce a therapeutic effect and (2) a significant safety margin at therapeutic levels. Since the size and base chemical structure of miRNA are virtually identical to siRNA, many of the same therapeutic delivery platforms pioneered by siRNA-based RNAi show great promise for miRNA therapy. Methods of in vivo delivery have witnessed varied levels of success and side effects and are basically classified as either viral or non-viral, including but not limited to conjugated RNAs, viral vector-based systems, and liposomal or polymer-based systems (Fig. 19.5).

The degree of nuclease degradation limits naked dsRNA use to local injection in the eye, skin, or tumor, and even this local application has largely given way to newer conjugated or packaged iterations [191]. The cholesterol conjugation used with some AMOs is a simple mechanism to improve in vivo performance. Cholesterol enhances cell surface membrane receptor binding via associations with high-density (BI receptor) and low-density (LDL receptor) lipoproteins [191]. In addition to examples previously cited, such formulations are capable of inhibiting miRNA in liver, lung, kidney, heart, intestine, fat, skin, bone marrow, muscle, ovaries, and adrenal glands, albeit at high doses [110, 123, 124,

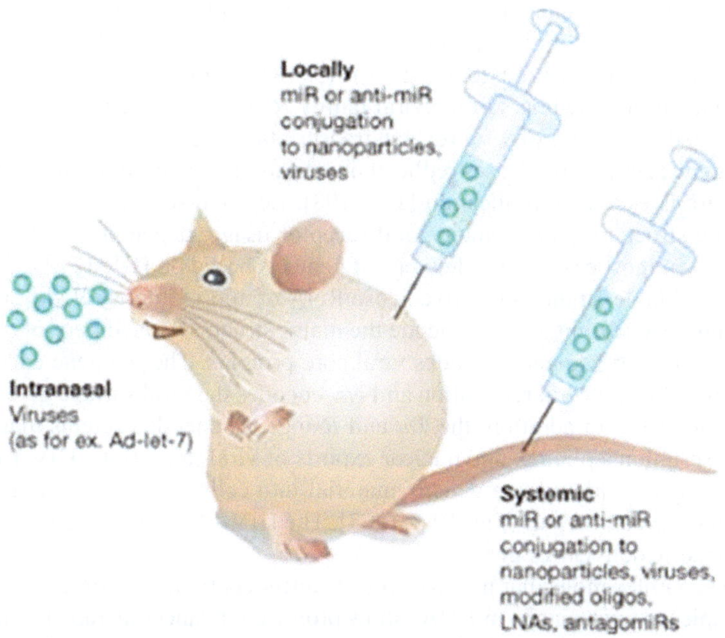

Fig. 19.5 Current approaches to in vivo miRNA targeting. Because of the body's multiple levels of biobarriers, researchers have investigated multiple animal models for delivery. miRNA targeting has evolved to include local injection, systemic, and intranasal delivery. (Reproduced from [68] with permission from John Wiley and Sons Publishing Group)

191]. Such need for higher doses may be secondary to the propensity for these molecules to be sequestered from the cytoplasm by phagolysosomes [191]. AMOs also have difficulty reaching targets beyond certain biological barriers, illustrated by antagomiR-16's ability to specifically inhibit miR-16 in brain cortex when directly applied, but inability to reach these tissues via intravascular injection [124, 192].

5.2 Vector-Based Delivery: Lentiviruses and Adenoviruses

Vector-based therapies exhibit significant anti-cancer activity both locally and systemically. There are essentially two vectors for the delivery of genetic material: viral based and non-viral based. Because many of the constructs used to deliver miRNA have been viral based, we will focus on those vectors. Because viruses have evolved to develop machinery to enter the cell, deliver, and integrate genetic material, they are excellent candidates for a vehicle to deliver miRNA [193–195]. Several viruses have been explored to deliver genetic material: retroviruses or lentiviruses like human immunodeficiency virus 1 (HIV-1), adeno- and adeno-associated viral vectors, herpes simplex, and poxviruses to name a few. Viruses with miRNA activity can be used as mimics or antagonists, and miRNA transcripts can augment classic gene therapy. Currently, the most commonly used vector for transport of miRNA or other oligonucleotides are adeno-associated virus (AAV) vectors and lentiviral vectors.

Lentiviruses are members of the viral family *Retroviridae* (retroviruses) that are characterized by the use of viral reverse transcriptase and integrase to stably insert into the gene [196]. Lentiviruses replicate in nondividing cells and can lyse the host cell if there is excessive replication [197, 198]. Lentiviruses can show differences in genome structure, pathogenicity, and receptor usage, depending on the species [199–201]. In most cases, viral delivery of genes is based on HIV-1.

HIV is a single-stranded positive-sense RNA of approximately 9 kb [202]. The three main open reading frames encode the major structural machinery of the virus: *gag, pol, env*. The *gag* gene encodes viral core proteins. The *pol* gene encodes the enzyme required for viral replication and *env* encodes the viral surface glycoprotein gp160 [196, 197]. In addition, the *Tat* and *Rev* genes encode proteins that control viral transcription, splicing, and nuclear exports of viral transcripts. HIV-1 is effective with regard to delivering genetic material into cells as compared to its nurine leukemia virus counterpart (Fig. 19.6) [203]. The HIV-1 has since become a vehicle for therapeutic delivery of genes.

As the complex molecular mechanisms of cancer continue to evolve, so do targets and therapies to treat cancer. miRNAs show promise for cancer therapy because they can target multi-pathways with the possibility of reduced toxicity [42, 204]. As discussed above, the mimic or antagomiR can easily be incorporated into virus for cellular delivery [124, 205–207]. Experimentally, lentiviral vector overexpressing miRNAs was shown to demonstrate loss of function of specific genes in a myelomoncytic cell line [208, 209]. Transgenes engineered to express miRNA-binding transcripts are called “miRNA sponges” or “decoys,” to compete with endogenous miRNAs [68, 191, 210]. The sponges can be inserted into chromosomes and then induced to be expressed (“miRZips”) [191]. The expression of miRNA in rats caused organ failure and death, indicating that toxicity could be confound in miRNA therapy [191, 211].

In addition to the potential of using lentiviral vector-based miRNA for cancer, there is potential for other pathologies, such as abdominal aortic aneurysms and hypercholesterolemia [205, 212, 213]. Lentiviral delivery of miR-33 in mice was shown to lower HDL by repression adenosine triphosphate-binding cassette (ABC) transporter, ABCA1 [213]. Systemic injection of a locked nucleic acid-modified antagomiR targeting miR-21 diminished the pro-proliferative impact of downregulated PTEN, leading to a marked increase in the size of AAA. Similar results were seen in mice with AAA augmented by nicotine and in human aortic tissue samples from patients undergoing surgical repair of AAA (with more pronounced effects observed in smokers) [205, 214, 215].

AAV vectors belong to the Parvoviridae family and are part of the *Dependovirus* genus. Twelve human serotypes (AAV serotype 1 (AAV-1) to AAV-12 have been reported in addition to the more than 100 serotypes from nonhuman primates [215–217]. The nonenveloped virus that packages a single-stranded DNA genome can only occur in the presence of another virus, such as adenovirus or herpesvirus [218]. In the absence of a helper virus, the AAV and specifically the serotype 2 can set up latency

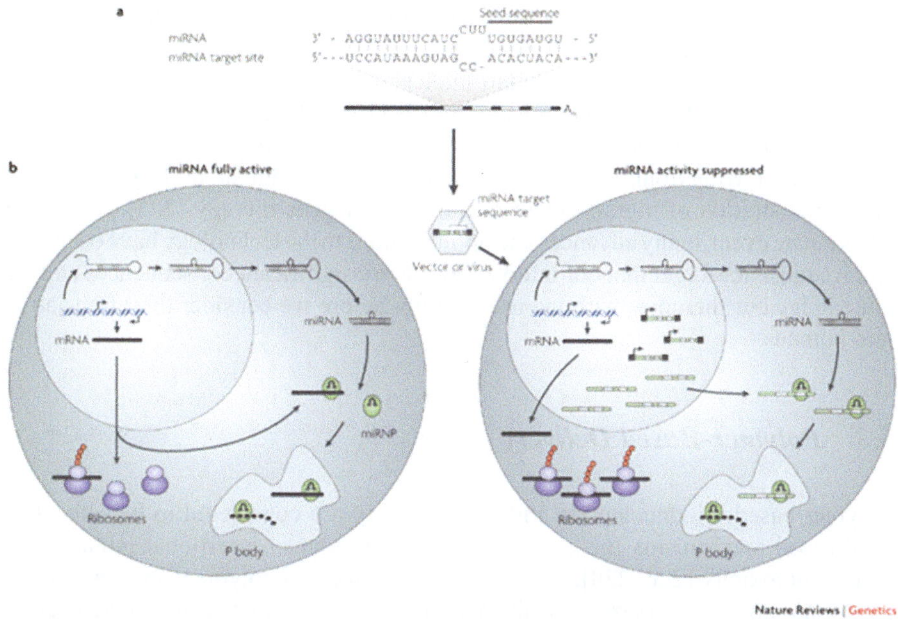

Fig. 19.6 Viral vector antagonism. Schematic of a vector or virus to inhibit a miRNA. (**a**) The vector encodes multiple copies of perfectly complementary or bulged target sites (as shown) for a miRNA or miRNA family. The target sites are expressed at high levels by a strong RNA polymerase II or III promoter or by introducing a large number of vectors into a cell. (**b**) The target containing transcripts bind to the cognate miRNA, and because of excess target concentration saturate the miRNA. This limits the availability of the miRNA, and inhibits the regulation of its natural target mRNAs. The processing body (*P body*) is a region rich in enzymes involved in miRNA turnover. niRNP, niRNA ribonucleoprotein complex. Reproduced from [29] with permission from Nature Publishing Group.

by integrating into the chromosome 19q13.4. This characteristic has made it the only mammalian DNA virus known to be capable of site-specific integration. Due to its lack of pathogenicity, its persistence, and availability of many serotypes, the translational potential of AAVs has soared for a variety of therapeutic applications [218].

Tumors are characterized by the global miRNA patterns [8, 89, 219, 220]. This presents a unique opportunity for therapeutic intervention that may span several tumor types at once, as many of the mechanisms that induce the tumorigenesis have commonalities in the miRNA patterns [221]. For example, therapeutic miRNA delivery of miR-26 through AAV in a murine model of liver tumorigenesis found a decrease in the proliferation of cancer, induced apoptosis with no evidence of toxicity [119]. Adenovirus-expressing miR-133 was shown to reduce cardiomegaly and intranasal administration of adenovirus-expressing let-7a reduced tumor formation in a K-ras mouse lung cancer model [130, 222].

Retroviruses and lentiviruses incorporate their DNA into the host genome without specificity, making insertional mutagenesis or oncogene activation serious theoretical off-target effects of gene therapy, in addition to their immunogenicity

[68]. The method of using virus for gene delivery can be adopted for miRNA by incorporating the transcripts or target sites into the 3′ untranslated region (UTR) of vectors. This has been demonstrated in studies using oncolytic viruses and miR-122 in the liver, and miR-124 in neurons/glial cells [191, 223, 224].

The path for the successful use of viruses as vectors for gene therapy has been fraught with troubles, such as the death of a recipient in 1999 following the administration of high doses of adenovirus for gene therapy [225]. Since that unfortunate event, many advances and optimizations to the technology have occurred. But we must not forget that our interventions carry significant consequences and we must judge our therapies with intense scrutiny before we consider their translation into humans.

5.3 Polymer-Based Delivery

Polymer-based oligonucleotide and gene delivery are considered to be safer than viral-based mechanisms for gene silencing with reduced insertional mutagenesis and phenotoxicity [226–228]. The advent of polymers/copolymers and lipid-based platforms expanded horizons in the world of biotechnology and bioengineering due to their favorable plasticity. Applied to RNAi, they overcome inherent (naked) miRNA physicochemical deficiencies and present advantages of biocompatibility/degradability, easy tailoring of particle size, functional targeting ability, protection from nucleases, and release from endosomes.

The most widely studied/used polymer for RNAi is polyethyleneimine (PEI), but many others exist (e.g., PLGA, chitosan, gelatin) [229]. PEI is a polycation polymer capable of complexing with DNA and protecting nucleic acids from nuclease degradation. PEI polymers can be synthesized in the nanoscale allowing for cellular endocytosis and also allowing for pH-buffering capability [230]. The number of positive charges in the polymer and negative phosphate groups within the RNA determines the size and structure of the resulting complexes (termed "polyplexes" or "polymersomes"). The polyplexes act as a "proton sponge" to protect the RNA from degradation and to stimulate necessary endosomal rupture within the cytosol [229]. Decreased cancer cell proliferation and increased apoptosis have been achieved in vivo using atellocollagen/miR-34a and PEI/miR-145/-33a formulations, respectively [231, 232]. In other studies, combinations of the cationic effects of PEI with the biocompatibility and physicochemical properties of polyurethanes (PUs) formed a PU–PEI polymer complex to deliver miR145 to treat lung adenocarcinoma (LAC) [233]. MiR145 was also identified as a candidate for therapy for LAC-associated cancer stem cells (CSCs). Experimental studies with nude mice treated LAC with PU–PEI-miR145 in combination with radiotherapy and cisplatin and this resulted in maximal inhibition of tumor growth [234].

Other biodegradable polymers have been investigated to deliver miRNAs. PLGA seemed to be one of the most successful biodegradable polymers currently used in

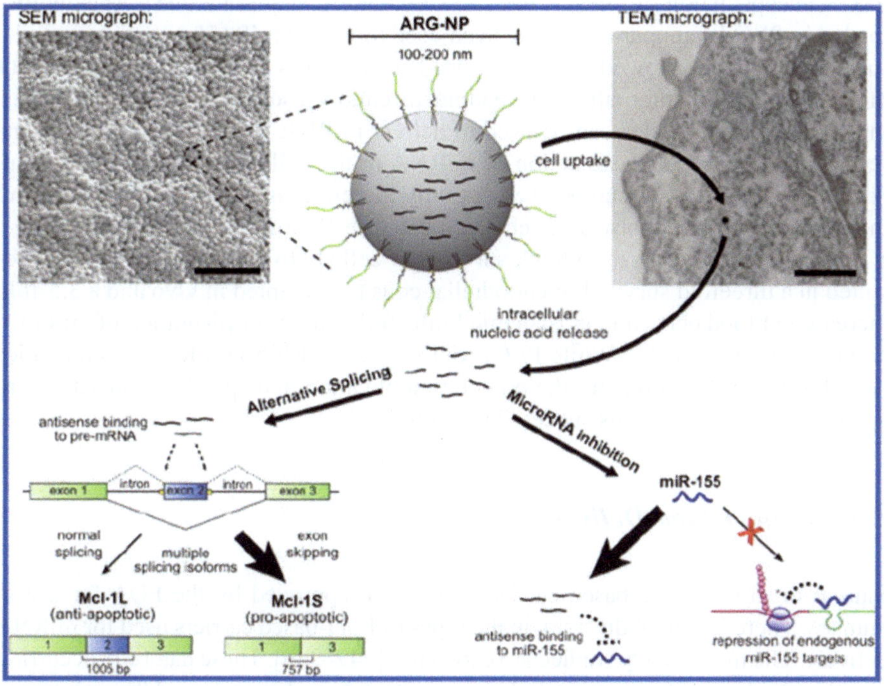

Fig. 19.7 Workflow schematic of alternative splicing and miRNA inhibition induced by non-arginine-coated nanoparticles (ARG-NPs) that deliver charge-neutral oligonucleotide analogs. ARG-NPs comprise a spherical PLGA core coated with PEGylated ARG. Dehydrated ARG-NPs were visualized by scanning electron microscopy (SEM), intracellular uptake of osmium tetroxide-loaded ARG-NPs was visualized by transmission electron microscopy (TEM); *scale bars* for SEM and TEM micrographs represent 1 μm. (Adapted with permission from [316]. Copyright 2013 American Chemical Society)

translational research (Fig. 19.7). PLGA is approved by the Federal Drug Administration (FDA) and the European Medicine Agency (EMA) for drug delivery [235]. One advantage of PLGA is the control attainable by altering the co-polymeric ratios to achieve long-term release [236]. PLGA, which is 50% lactic acid and 50% glycolic acid, is released when PLGA undergoes biodegradation. These monomers are quickly metabolized through the Krebs cycle resulting in minimal toxicity [237]. While extremely biocompatible, the hydrophobic residues on the surface as well as the nanoparticle size make PLGA particle targets for the RES. Cells in the RES organs sequester PLGA particles by opsonization, leading to payload loss. Surface modifications with other polymers such as polyethylene glycol (PEG) have led to increased circulating half-life by several orders of magnitude [238]. PLGA drug delivery has been used with antibiotics to treat infections, growth factors for tissue engineering applications, and polymeric stent coverage in cardiovascular disease [239–243].

PLGA has also served as an effective delivery system for miRNA in cancer therapeutics. Because of its favorable biocompatibility and ability to be targeted,

PLGA encapsulation of miRNA has shown great efficacy [244, 245]. AntagomiRs loaded in PLGA nanoparticles have been used to inhibit miR155 in a mouse model for lymphoma and showed slowing of pre-B cell tumors in vivo [245]. Because of the mechanism of internalization leading to endolysosomal compartment targeting, PLGA nanoparticles can be used to deliver miRNA to the regulatory proteins responsible for the ultimate action of miRNA in the cell [237, 246]. Gomes et al. have shown that PLGA can be used not only to deliver miR132 to endothelial and mononuclear cells, but also as a vehicle to carry perfluoro-1,5-crown ether (PFCE), a labeling compound used to track survival of cells [246]. Delivery of miR132 resulted in a threefold survival of endothelial cells transplanted in vivo and a 3.5-fold increase in blood perfusion in ischemic limbs following transplantation of miR132-treated endothelial cells. While PLGA delivery of miRNA and RNAi therapeutics is still early in development, the potential application and speed to translation are apparent given its previous success in the clinic.

5.4 Lipid-Based Delivery

Similar to PLGA, lipid-based carriers have been approved by the FDA for use in humans. There is a great diversity in the types of lipid-based carriers used for miRNA delivery. The most widely studied are exosomes [247–249]. These naturally occurring vesicles are derived from intracellular compartments and are released through fusion of the multi-vesicular endosomes with the plasma membrane [249]. These lipid-based carriers are capable of both transport and protection of the payload from the harsh extracellular environment [249]. Our understanding of exosomes has given us insight into the nature of intercellular miRNA communications and has led researchers to investigate other lipid-based vehicles for miRNA delivery, namely liposomes. Liposomes represent a major sector of nanomedicines and consist of spherical lipid vesicles 30–200 nm composed of lipid bilayers encapsulating hydrophilic chemotherapeutics, such as doxorubicin or danorubucin [250]. Doxil® and Myocet®DaunoXome® are three anthracycline-carrying anticancer liposome formulations approved by the FDA and have been used in clinics for more than a decade [251, 252]. The lipid bilayer of liposomal formulations permits avoidance of renal clearance and improved intracellular delivery. Examples abound as RNAi delivery vehicles, including PEGylated liposomes, lipoplexes, lipoids, and lipidoids [229, 253]. Liposomal carriers alter the biodistribution of the drug, increasing the circulation time and decreasing the cardiotoxicity commonly associated with chemotherapeutics [254, 255]. Micelles are spheres, roughly 20 nm in diameter, consisting of a single lipid or amphiphilic polymer layer, with a hydrophilic exterior and hydrophobic lumen (Fig. 19.8) [256]. Lipid- and polymer-based nanoparticles for systemic delivery of siRNAs have been developed and tested. While lipid-based delivery of miRNA is efficient, it tends to induce an inflammatory response. On the other hand, biodegradable polymers induce less inflammation, but deliver less efficiently and have shorter effects.

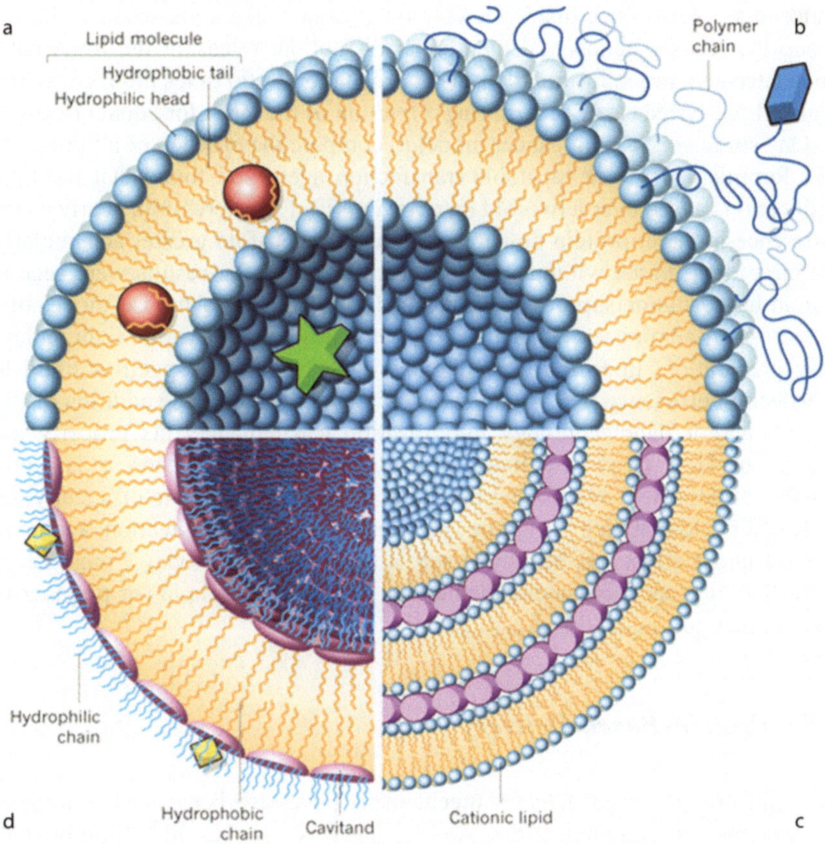

Fig. 19.8 The evolution of liposomes. Simple liposomal vesicles are constructed of a lipid bilayer shell. **a** Liposomes can trap molecules (*red sphere*) up to a few nanometers in diameter within the hydrophobic region, while the hydrophilic region can trap molecules up to several hundred nanometers in size (*green star*). In order to negotiate biobarriers, "stealth" liposomes were constructed for drug delivery applications. (b) Polymers attached to the lipid bilayer can be modified with targeting moieties (*blue rectangle*) that enable specific cellular targeting. (c) Cationic liposomes, able to carry DNA (*purple circles*), are constructed in an onion-like fashion with the DNA sandwiches between the cationic membranes. (d) In the most recent evolution, the bilayer is assembled from cavitands where hydrophobic and hydrophilic chains can be attached. Within the cavitands angstrom-sized molecules (*yellow diamond*) can be trapped on the surface. This enables molecules of different sizes or configurations, such as miRNA, to be carried and protected. (Reproduced from [256] with permission from Nature Publishing Group)

Lipoplexes contain cationic lipids in their bilayer that interact with the negatively charged anionic nucleic acid molecules. The hydrophilicity of the nucleic acids is offset by the cationic lipids and a net positive charge results, which allows the liposomes to bind to anionic cell surface molecules [191]. These particles can be further modified to facilitate fusion with extra/intracellular membranes and foster endosomal release once inside the cell via attributes like a pH-sensitive lipid head group that causes endosomal membrane disruption, releasing the RNA payload. Such later-generation formulations have proven to be 100 times more effective than typical lipid-based carriers, requiring a fraction of the RNA for equal effect [191].

One study used a lung cancer model to compare the efficacy of a lipoplex/miR-133 formulation with a reputable transfection agent and found that the lipoplex application caused over a twofold increase in miR-133 expression, nearly a twofold difference in target protein downregulation, and a 50-fold greater accumulation in the target lung tissue without toxic effects [257]. Similar results were seen using a lipid-based tumor suppressor miR-34a formulation in a mouse model of lung cancer when administered locally or systemically [118]. Despite their safety and easy preparation, liposomal formulations have less tumor specificity and lower transfection efficiency than viral vector-based therapies, but the addition of surface ligands and/or nanotechnology hybridization can enhance this property greatly [258]. The Holy Grail in delivery of miRNA-based cancer therapy is a product capable of delicately negotiating the many biological barriers to efficiently deliver an RNAi payload with great tissue specificity, complete transfection, and no negative off-target effects on healthy tissue. A combination of today's most advanced biotechnology, such as nanomedicine, and biocompatible lipid- or polymer-based carriers may just make this dream a reality.

5.5 Protein-Based Delivery

Nonlipid protein-based delivery mechanisms have also been used as drug delivery systems for oligonucleotides. Atelocollagen is a processed natural biomaterial produced from bovine type I collagen [259]. Because it is derived from naturally occurring collagen, it retains the characteristics of high biocompatibility and high biodegradability [260]. The mixture of atelocollagen and oligonucleotides results in complexes, 200 nm–10 μm [260]. Atelocollagen protects the nucleic material from serum nucleases, which is a major biological barrier known to degrade nucleic acids. [261]. Atelocollagen forms bonds with the nucleic acids in the miRNA or siRNA, where hydrogen bonds are formed between the phosphate groups of the oligonucleotides and the CH groups of the collagen (Fig. 19.9) [262]. By controlling the size as potentially adding targeting moieties to the atelocollagen itself, this system presents itself as an appealing and biologically inspired alternative to synthetic polymer nanoparticle delivery systems.

Atelocollagen has found utility as a drug delivery system in both oncologic and tissue engineering applications [263–269]. As an oncologic drug delivery system, atelocollagen has proven effective in delivering miRNAs to target tissues. In one model of lung metastasis of osteosarcoma, Osaki et al. seeded luciferase-tagged human osteosarcoma subclone cell line in the knee of athymic mice and allowed lung metastasis to develop. Once metastasis was confirmed, the investigators complexed miR-143 with atelocollagen for systemic delivery every 3 days for 19

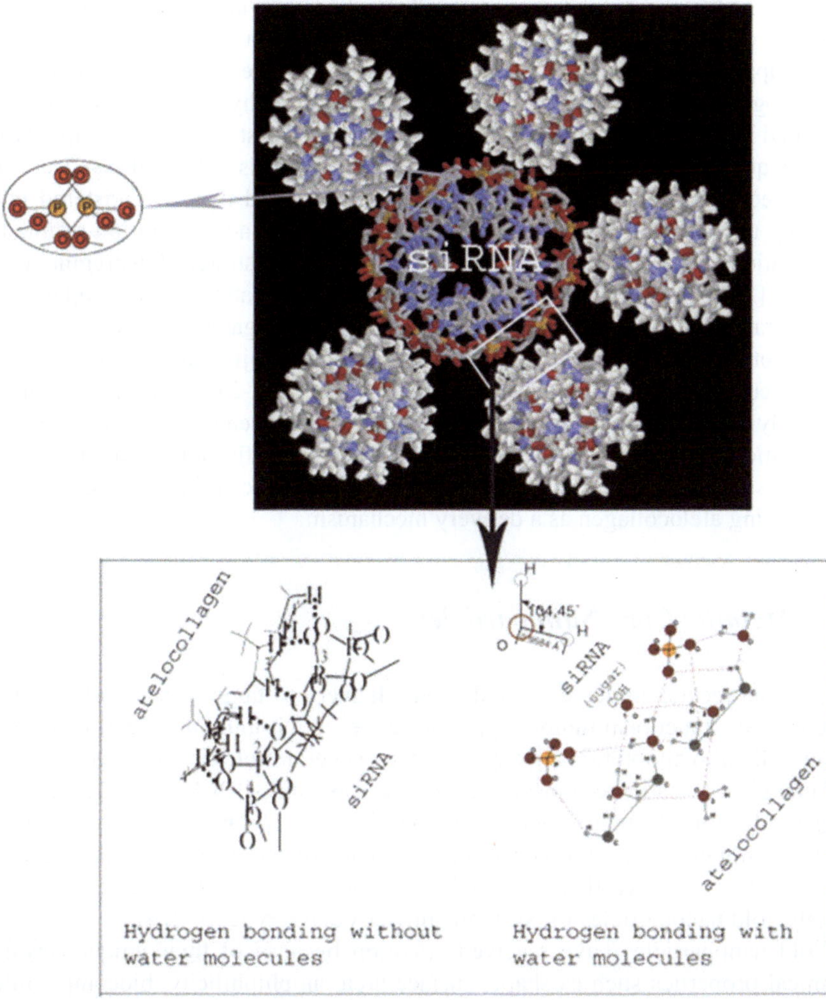

Fig. 19.9 The molecular model of atelocollagen–siRNA complex. Hydrogen bonds arise between CH and phosphate groups. 1,3 phosphate groups are situated on the first chain of ds-siRNA and the other 2,4 phosphate groups (which are below the 1,3 phosphate groups) are situated on another chain. H-bonds are created between 1,3 ↔ $1^1,3^1$ and 2,4 ↔ $2^1,4^1$ groups $1^1,3^1,2^1,4^1$ are CH groups of Gly and Pro amino acids situated on the triple helix of collagen and $2^1,4^1$ CH groups are situated below of $1^1,3^1$ CH groups ("hydrogen bonding without water molecules"). On the "hydrogen bonding with water molecules" are shown water-mediated intermolecular contacts between atelocollagen and siRNA which are restricted to the backbone phosphates and sugar 2′-OH groups (are shown hydrogen bonding between one layer of phosphate groups and CH groups of collagen). (Reproduced from [262] with permission from Elsevier)

days. At 1 week, the only lesions detected were those at the primary osteosarcoma site in the knee. Following the course of treatment, only two often mice injected with miR-143 atelocollagen showed lung metastasis [268]. Similarly, in a model of metastatic prostate cancer, miR-16 delivery via atelocollagen systemic delivery resulted in significant reduction of bone metastatic growth [264]. The tumor-suppressive effects of atelocollagen miRNA have also been demonstrated in a xenograft tumor model of human colon cancer in athymic mice. miR-34a was identified as a potential miRNA capable of growth arrest and growth suppression through quantitative polymerase chain reaction analysis. The miRNA was then complexed with atelocollagen for systemic delivery and then administered intravenously to the xenograft mice. miR-34a atelocollagen administration resulted in upregulation of p53 and 36% of cancer tissues demonstrated downregulation of miR-34a [265]. As a delivery mechanism in oncologic treatments, atelocollagen has demonstrated efficacy, but its utility has also been investigated in tissue engineering applications. In a rat model of ligamentous healing, Shoji et al. demonstrated that intra-articular injection of atelocollagen complexed miR-210 prompted healing of a partially torn anterior cruciate ligament through enhancement of angiogenesis via upregulation of vascular endothelial growth factor and fibroblast growth factor-2 [269]. In sum, the delivery of miRNA and other RNAi nucleotides has been demonstrated using atelocollagen as a delivery mechanism.

5.6 Metallic Core Nanoparticles

Cancer therapeutic strategies include gene delivery to target cancer cells in order to replace dysfunctional tumor suppressor genes, elicit immune rejection, or drive tumor cells into apoptotic pathways. To date, several biological delivery vehicles, including DNA, cationic liposomes, viral vectors, and siRNA nanoparticles, have been used with advantages and limitations [270–272]. Other vehicles have thus been investigated to overcome the inefficiencies of the previously described vehicles. To this end, investigators have begun to look into metal-based nanoparticles, namely gold nanoparticles for systemic miRNA delivery [273, 274].

Gold nanoparticles have received attention because of their unique physicochemical properties such as shape, surface area, amphiphilicity, biocompatibility, and safe carrier capabilities essential for effective nucleotide delivery [275–278]. However, the nucleotide cargo, often siRNA, had to be modified to facilitate delivery [279–281]. These modifications would likely alter the activity of miRNA and would therefore not be possible in their delivery. Recognizing the need to efficiently deliver unmodified miRNA investigators has used gold nanoparticle–miRNA complexing to form stable delivery vehicles. Crew et al. demonstrated that through conjugation miRNA could be immobilized to the surface of gold nanoparticles and efficiently transfected into cells for gene knockdown [282]. Similarly, Ghosh et al. developed a system where unmodified miRNA would be bound to the gold nanoparticles, then covered in a PEG layer to allow for protection from serum nucleases (Fig. 19.10). In vitro functional characterization demonstrated that this system was capable of efficiently delivering miRNA into cells resulting in significant target gene downregulation and decreased cell proliferation [276]. While this delivery system appears promising, true in vivo translation has yet to be realized.

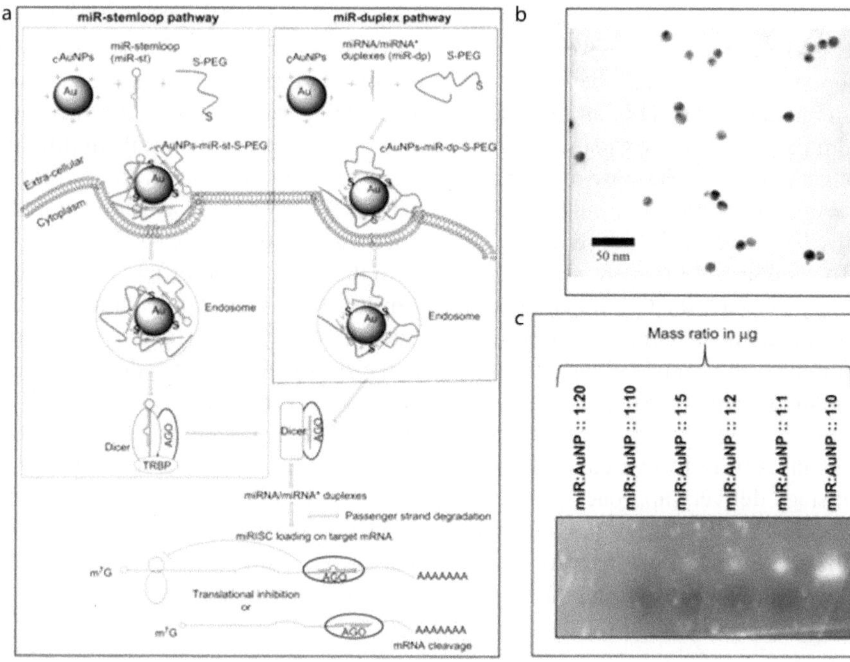

Fig. 19.10 Synthesis, delivery, and characterization of gold nanoparticles (AuNPs) and miR-AuNP-S-PEG polyelectrolyte complexes. **a** Schematic diagram of the miR-AuNP synthesis and uptake through endocytosis pathway. **b** TEM image of dialyzed cationic AuNPs that were prepared by chemical reduction of chloroauric acid ($HAuCl_4$) using sodium borohydride ($NaBH_4$) in the presence of cysteamine hydrochloride ($HS–CH_2CH_2–NH_3^+Cl^-$) followed by dialysis against RNAse-free water in a Spectra/Por dialysis membrane with a molecular weight cutoff of 10 kDa. **c** Gel retardation assay: TBE-Urea gel electrophoresis of miR-AuNP polyelectrolyte complexes. In each case, 1 µg of miRNA was loaded to observe polyelectrolyte complex formation with AuNPs. (Reproduced from [276] with permission from Elsevier)

6 Future Perspectives on Drug Delivery: Multistage Theory

The targeting of genetic materials, since their inception, has found difficulty in efficient targeting. Naked oligonucleotides are rapidly cleared by nucleases and encapsulation in lipid-based carriers, while effective, also suffer from rapid clearance in addition to their immunogenicity [283, 284]. Opsonin recognition to lipid nanoparticles and complement-mediated clearance can trigger toll-like receptor activation and transgene silencing. While viral vehicles for delivery have the potential for mass reproduction of carried genes and have shown some efficacy, they too are troubled with limitations that have prevented their full translation into the clinic [285, 286]. In some cases, the inflammatory response generated from the vector itself may enhance its clearance and thus decrease the payload delivery and therapeutic efficacy [287]. Polymeric nanoparticles similarly are troubled by clearance from the system [288]. While the mechanisms of clearance vary, the end result is the same in decreased payload delivery and increased immunogenic response. Because of the ability to effectively target polymer nanoparticles to the cell, they are capable of entry but release from the endosome results in endosomal content leakage and intracellular damage [289]. Further, their accumulation in RES organs such as the lung, liver, and spleen makes them difficult to target elsewhere. Because of their recognition by the host as foreign, antibodies are made and further dosing results in more decreased delivery as well as stimulating the body to reject any cells transduced by the foreign viral particles [290–293].

To summarize, there are many limitations to the success and efficacy of current delivery systems for RNAi and specifically miRNA delivery. These shortcomings include inadequate targeting, poor tumor accumulation, systemic toxicity, and most importantly, inadequate clinical translation and efficacy. A paradigm shift is needed in the fundamental principles of how to target and deliver miRNA and other RNAi oligonucleotides effectively to the site of interest. Since no single agent truly possesses the ability to overcome the plethora of obstacles to therapeutic delivery, a multistage delivery approach proposes the use a nested multicomponent construct, engineered to sequentially avoid biological barriers [178, 294]. According to the multistage dogma, biobarrier avoidance, recognition, and cytotoxicity interdependence are reduced, yet acting in efficacious operational synchrony [295, 296].

7 Mesoporous Silicon and Therapeutic Applications

The rationale of the multistage theory is to decouple the multiplicity of tasks in targeting to preferentially negotiate bio-barriers, sequester the vehicle at the site of interest, and then deliver the payload. By enabling the slow, controlled, and persistent release of a drug, one could potentially reduce toxicities and heighten the therapeutic efficacy of the molecule. Through this mechanism, drugs with great therapeutic efficacy, but previously deemed too toxic for systemic therapy, can be considered for clinical use. Additionally, other barriers such as the blood-—brain -barriers and the RES, previously thought to be unavoidable, may be negotiated. These barriers have presented the most challenging obstacles to current oligonucleotide- based therapies and present opportunities for improvement through multistage delivery.

Despite the vast assortment of nanomaterials, the majority of the nanocarriers rely on molecules functionalized on their surface for biological recognition or protection from the RES. Polyethylene glycol-coated nanoparticles have been used to hide the payload from elimination, but in doing so lose their targeting abilities. Conversely, targeted particles, while preferentially finding their site of action, are quickly cleared by the body's defensive mechanisms. Unfortunately, the overall activity of biological barriers ultimately prevents these carriers from localizing at a specific site in adequate therapeutic concentrations [297, 298]. As nanocarriers evolved from generation to generation, so did their ability to in overcome biobarriers. First-generation nanocarriers, relying on passively homing to diseased sites through the enhanced permeability and retention, resulted in extravasation through

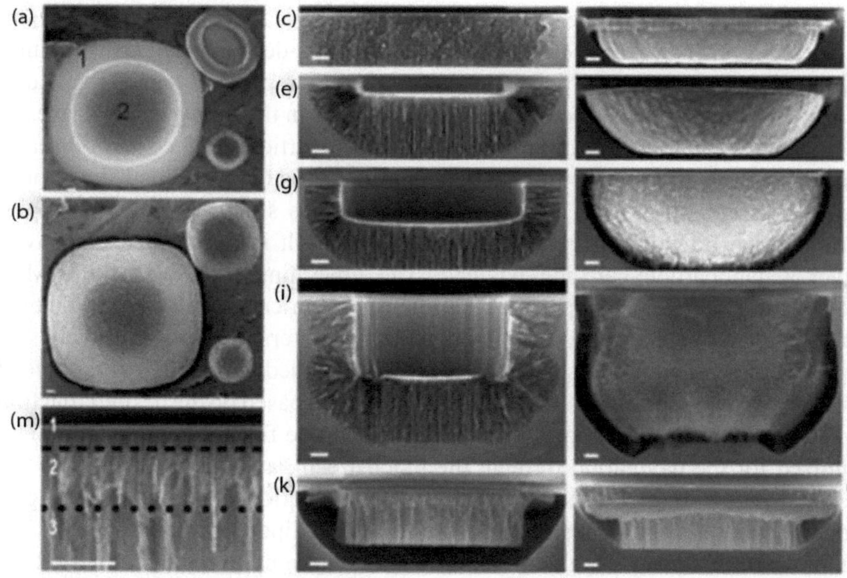

Fig. 19.11 SEM micrographs of PSPs. **a** Digital composition of three distinct SEM micrographs showing the nucleation side of a 3.2 μm, 1.6 μm, and 0.97 μm PSP:the external corona (*1*) and the nucleation site (*2*). **b** Digital composition of three distinct SEM micrographs showing the release side of a 3.2 μm, 1.6 μm, and 0.97 μm PSP. Section along the diameter and the lateral view of: a flat disk PSP obtained by wet etch of the masking layer (**c**, **d**, respectively); a discoidal PSP obtained by trench formation by CF4 RIE (**e**, **f**, respectively); a hemispherical PSP obtained by trench formation by SF6 RIE (**g**, **h**, respectively); a tubular PSP obtained by trench formation by combination of HBr and SF6 RIE (**i**, **j**, respectively); a XLP1 PSP (**k**, **l**, respectively). (m) Close-up view of the multilayer structure of an XLP1 particle, from *top* to *bottom*: *1* SP layer, *2* transitional layer, *3* XLP layer. (**c–m**) Nucleation side is at the *top* of the figure and release side at the *bottom*. All *scale bars* are 200 nm. (Reproduced from [314] with permission from John Wiley and Sons Publishing Group)

the tumor's diseased vasculature. Second-generation nanocarriers possessed greater functionality through the conjugation of targeting moieties and thus could selectively concentrate at the site of interest. Liposomes and other targeted nanomaterials such as gold are examples of delivery systems [299–301]. Third-generation nanocarriers are characterized by their ability to carry both therapeutic and diagnostic multifunctional components with logic-embedded functions. This system focuses on negotiating the body's biobarriers, accumulating at the target site, and delivering the protected payload to the target site. To accomplish these combined operations, third-generation nanocarriers are comprised of nanoparticles loaded into a nanostructured carrier [302–304]. Third-generation systems are comprised of multiple components assembled with embedded instructions to act in a synergistic, preprogrammable, and sequential manner. The multistage delivery system (MDS) is an emblematic system for third-generation carriers and may be a potential solution to the shortcomings of current miRNA delivery mechanisms.

Multistage Delivery System In order to negotiate biological barriers and fully utilize the advantages of a third-generation nanocarrier, our laboratory envisioned and engineered the MDS (Fig. 19.11). The MDS aimed to decouple the tasks required of the carrier system into vehicle, nanocarrier, and therapeutic agent. By shielding the nanocarrier and payload from the serum nucleases in the intravascular space, the MDS vehicles can transport millions of loaded nanoparticles to the site of diseased tissue. The rational design, based on mathematical modeling of blood flow within diseased vessels, allows the micron-sized mesoporous silicon particles to exploit aberrant blood flow of tumor vasculature and land itself in the tissue to selectively deliver a nanoparticle-carrying payload. The governing principle of the MDS involves the loading of nanoparticles into first-stage micron-sized particle that can be finely tuned to achieve precise targeting. Tuning refers to changing the physical properties of the particle to specialize it for the intended application and payload. Pore size, porosity, shape, and size can all be calibrated based on the desired function, target, and ultimate destination. With this approach, the first-stage silicon carrier is optimized for the navigation through the bloodstream, avoidance of the RES and nucleases, recognition of a diseased site, as well as retention and protection of a payload. The second-stage nanoparticle, which can be a liposome, gold nanoparticle, or a viral vector, is then capable of extravasation through the tumor's fenestrations and into the target microenvironment. Silicon, the precursor element to mesoporous silicon and one of Earth's richest elements, was strategically chosen to fabricate this first-stage carrier because of its biodegradability and biocompatibility [305–307]. Porous silicon particles have shown considerable advantageous properties, such as increased drug solubility, bioavailability, biocompatibility, and sustained localized release. Through the direct release of the therapeutic agent at the site, one could potentially increase the amount of drug actually penetrating the tumor.

To effectively target therapeutic agents, it is necessary to develop a delivery system that is simultaneously able to negotiate biological barriers, while protecting the payload, and then only at the site of interest effectively release its payload [178, 308–310]. Designed to exploit the body's natural environment, the discoid shape was chosen for the MDS to allow for greater blood margination and improved cell surface adhesion to the leaky vasculature of a tumor [308, 311]. The pores' tunable size enables control of surface area, density, the degradation rate, and thus of the dose of drug released [312, 313]. The ability to customize the pore size also permits the loading of nanoparticles of different sizes within the porous matrix [314]. Through the employment of these unique features, therapeutic and imaging agents can be loaded into the MDS and transported in an effective manner. As this technology evolves and the capabilities of first-stage nanoparticle loading improve, so does the potential for therapeutic and diagnostic applications.

8 Conclusion

Since the discovery of *lin-4,* miRNA has been shown to be a novel and multifunctional target in many disease processes. Although the interconnectivity of gene expression, protein synthesis, and miRNA interaction has yet to be elucidated, the utility of miRNA targeting or delivery has been demonstrated [315]. Many obstacles, though, confront the true translation of these ubiquitously expressed small, noncoding RNAs to the clinic. Effective negotiation of the body's biological barriers of serum nucleases, renal and hepatic clearance, vascular endothelium, cellular membranes, and endosomal release is paramount to the success of these molecules and effective therapeutic agents. To that end, investigators have developed and tested many delivery systems ranging from synthetic to naturally occurring polymers, as well as viral vectors for targeted delivery of the oligonucleotide payload. These delivery systems have shown some success, but are limited by their poor clinical translation. Although many advances in polymer chemistry and interdisciplinary approaches have been utilized, a new paradigm for drug delivery could potentially revolutionize the effective delivery of miRNA. Through multistage delivery, miRNA could be shielded from the degradation and sequestration, while not sacrificing targeting or organ specificity. Taken together, there is great potential for miRNA to have a great impact in the clinic, and with a proper understanding of the obstacles to overcome and tools in our armament to negotiate these barriers, it will find true translation into the clinic.

References

1. Lee RC, RL Feinbaum, Ambros V. The C. elegans heterochronic gene lin-4 encodes small RNAs with antisense complementarity to lin-14. Cell. 1993;75(5):843–54.
2. Wightman B, Ha I, Ruvkun G. Posttranscriptional regulation of the heterochronic gene lin-14 by lin-4 mediates temporal pattern formation in C. elegans. Cell. 1993;75(5):855–62.
3. Bartel DP. MicroRNAs: genomics, biogenesis, mechanism, and function. Cell. 2004; 116(2):281–97.
4. Baek D, et al. The impact of microRNAs on protein output. Nature. 2008;455(7209):64–71.
5. Lewis BP, Burge CB, Bartel DP. Conserved seed pairing, often flanked by adenosines, indicates that thousands of human genes are microRNA targets. Cell. 2005;120(1):15–20.
6. Lewis BP, et al. Prediction of mammalian microRNA targets. Cell. 2003;115(7):787–98.
7. Bader AG et al. Developing therapeutic microRNAs for cancer. Gene Ther. 2011;18(12):1121–6.
8. Calin GA, Croce CM. MicroRNA signatures in human cancers. Nat Rev Cancer. 2006;6(11):857–66.
9. Esquela-Kerscher A, Slack FJ. Oncomirs – microRNAs with a role in cancer. Nature reviews. Cancer. 2006;6(4):259–69.
10. Hanahan D, Weinberg RA. The hallmarks of cancer. Cell. 2000;100(1):57–70.
11. Hanahan D, Weinberg RA. Hallmarks of cancer: the next generation. Cell. 2011;144(5):646–74.
12. Check Hayden E. Cancer complexity slows quest for cure. Nature. 2008;455(7210):148.
13. Hogberg T. Chemotherapy: current drugs still have potential in advanced ovarian cancer. Nat Rev Clin Oncol. 2010;7(4):191–3.
14. Nahleh Z. Neoadjuvant chemotherapy for "triple negative" breast cancer: a review of current practice and future outlook. Med Oncol. 2010;27(2):531–9.
15. Ohtsu A. Current status and future prospects of chemotherapy for metastatic gastric cancer: a review. Gastric cancer: official journal of the International Gastric Cancer Association and the Japanese Gastric Cancer Association. 2005;8(2):95–102.
16. Imai K, Takaoka A. Comparing antibody and small-molecule therapies for cancer. Nat Rev Cancer. 2006;6(9):714–27.
17. Aggarwal S, et al. A dimeric peptide that binds selectively to prostate-specific membrane antigen and inhibits its enzymatic activity. Cancer Res. 2006;66(18):9171–7.

18. Le Pechoux C. Role of postoperative radiotherapy in resected non-small cell lung cancer: a reassessment based on new data. Oncologist. 2011;16(5):672–81.
19. Buchholz TA. Radiation therapy for early-stage breast cancer after breast-conserving surgery. N Engl J Med. 2009;360(1):63–70.
20. Recht A, et al. The sequencing of chemotherapy and radiation therapy after conservative surgery for early-stage breast cancer. N Engl J Med. 1996;334(21):1356–61.
21. McCahill L, Ferrell B. Palliative surgery for cancer pain. West J Med. 2002;176(2):107–10.
22. Carthew RW, Sontheimer EJ. Origins and mechanisms of miRNAs and siRNAs. Cell. 2009;136(4):642–55.
23. Couvillion MT, et al. Sequence, biogenesis, and function of diverse small RNA classes bound to the Piwi family proteins of Tetrahymena thermophila. Genes Dev. 2009;23(17):2016–32.
24. Sontheimer EJ, Carthew RW. Silence from within: endogenous siRNAs and miRNAs. Cell. 2005;122(1):9–12.
25. Hammond SM, et al. An RNA-directed nuclease mediates post-transcriptional gene silencing in Drosophila cells. Nature. 2000;404(6775):293–6.
26. Meister G, et al. Human Argonaute2 mediates RNA cleavage targeted by miRNAs and siRNAs. Mol Cell. 2004;15(2):185–97.
27. Meister G, et al. Identification of novel argonaute-associated proteins. Curr Biol. 2005;15(23):2149–55.
28. Meister G, Tuschl T. Mechanisms of gene silencing by double-stranded RNA. Nature. 2004;431(7006):343–9.
29. Brown BD, Naldini L. Exploiting and antagonizing microRNA regulation for therapeutic and experimental applications. Nature reviews. Genetics. 2009;10(8):578–85.
30. Cai Y, et al. A brief review on the mechanisms of miRNA regulation. Genomics, Proteomics Bioinformatics. 2009;7(4):147–54.
31. Eulalio A, Huntzinger E, Izaurralde E. Getting to the root of miRNA-mediated gene silencing. Cell. 2008;132(1):9–14.
32. Guang S, et al. An Argonaute transports siRNAs from the cytoplasm to the nucleus. Science. 2008;321(5888):537–41.
33. Lu J, et al. The birth and death of microRNA genes in Drosophila. Nat Genet. 2008;40(3):351–5.
34. Moazed D. Small RNAs in transcriptional gene silencing and genome defence. Nature. 2009;457(7228):413–20.
35. Ohrt T, et al. Fluorescence correlation spectroscopy and fluorescence cross-correlation spectroscopy reveal the cytoplasmic origination of loaded nuclear RISC in vivo in human cells. Nucleic Acids Res. 2008;36(20):6439–49.
36. Wu L, Fan J, Belasco JG. MicroRNAs direct rapid deadenylation of mRNA. Proc Natl Acad Sci USA. 2006;103(11):4034–9.
37. Golden DE, Gerbasi VR, Sontheimer EJ. An inside job for siRNAs. Mol Cell. 2008;31(3):309–12.
38. Mello CC, Conte D Jr. Revealing the world of RNA interference. Nature. 2004;431(7006):338–42.
39. Dunoyer P, et al. An endogenous, systemic RNAi pathway in plants. EMBO J. 2010;29(10):1699–712.
40. Fagegaltier D, et al. The endogenous siRNA pathway is involved in heterochromatin formation in Drosophila. Proc Natl Acad Sci USA. 2009;106(50):21258–63.
41. Krek A, et al. Combinatorial microRNA target predictions. Nat Genet. 2005;37(5):495–500.
42. Rajewsky N. microRNA target predictions in animals. Nat Genet. 2006;38(Suppl):S8–13.
43. Ding XC, Grosshans H. Repression of C. elegans microRNA targets at the initiation level of translation requires GW182 proteins. EMBO J. 2009;28(3):213–22.
44. Wakiyama M, et al. Let-7 microRNA-mediated mRNA deadenylation and translational repression in a mammalian cell-free system. Genes Dev. 2007;21(15):1857–62.
45. Han J, et al. Molecular basis for the recognition of primary microRNAs by the Drosha-DGCR8 complex. Cell. 2006;125(5):887–901.

46. Kim VN. MicroRNA biogenesis: coordinated cropping and dicing. Nat Rev Mol Cell Biol. 2005;6(5):376–85.
47. Bushati N, Cohen SM. microRNA functions. Annu Rev Cell Dev Biol. 2007;23:175–205.
48. Fabian MR, Sonenberg N. The mechanics of miRNA-mediated gene silencing: a look under the hood of miRISC. Nat Struct Mol Biol. 2012;19(6):586–93.
49. Behm-Ansmant I, et al. mRNA degradation by miRNAs and GW182 requires both CCR4:NOT deadenylase and DCP1:DCP2 decapping complexes. Genes Dev. 2006;20(14):1885–98.
50. Mishima Y, et al. Translational inhibition by deadenylation-independent mechanisms is central to microRNA-mediated silencing in zebrafish. Proc Natl Acad Sci USA. 2012;109(4):1104–9.
51. Guo H, et al. Mammalian microRNAs predominantly act to decrease target mRNA levels. Nature. 2010;466(7308):835–40.
52. Huntzinger E, Izaurralde E. Gene silencing by microRNAs: contributions of translational repression and mRNA decay. Nat Rev Genet. 2011;12(2):99–110.
53. Kiriakidou M, et al. An mRNA m7G cap binding-like motif within human Ago2 represses translation. Cell. 2007;129(6):1141–51.
54. Bazzini AA, Lee MT, Giraldez AJ. Ribosome profiling shows that miR-430 reduces translation before causing mRNA decay in zebrafish. Science. 2012;336(6078):233–7.
55. Bethune J, Artus-Revel CG, Filipowicz W. Kinetic analysis reveals successive steps leading to miRNA-mediated silencing in mammalian cells. EMBO Rep. 2012;13(8):716–23.
56. Djuranovic S, Nahvi A, Green R. miRNA-mediated gene silencing by translational repression followed by mRNA deadenylation and decay. Science. 2012. 336(6078):237–40.
57. Baltimore D, et al. MicroRNAs: new regulators of immune cell development and function. Nat Immunol. 2008;9(8):839–45.
58. Huang Y, et al. Biological functions of microRNAs: a review. J Physiology Biochem. 2011;67(1):129–39.
59. Kloosterman WP, et al. Targeted inhibition of miRNA maturation with morpholinos reveals a role for miR-375 in pancreatic islet development. PLoS Biol. 2007;5(8):e203.
60. Neilson JR, et al. Dynamic regulation of miRNA expression in ordered stages of cellular development. Genes Dev. 2007;21(5):578–589.
61. Cheng Y, Zhang C. MicroRNA-21 in cardiovascular disease. J Cardiovasc Transl Res. 2010;3(3):251–5.
62. Cheng Z, et al. MiR-320a is downregulated in patients with myasthenia gravis and modulates inflammatory cytokines production by targeting mitogen-activated protein kinase 1. J Clin Immunol. 2013;33(3):567–76.
63. Esau CC, Monia BP. Therapeutic potential for microRNAs. Adv Drug Deliv Rev. 2007;59-(2–3):101–14.
64. Jazbutyte V, Thum T. MicroRNA-21: from cancer to cardiovascular disease. Curr Drug Targets. 2010;11(8):926–35.
65. Lu M, et al. An analysis of human microRNA and disease associations. PLoS One. 2008;3(10):e3420.
66. Small EM, Frost RJ, Olson EN. MicroRNAs add a new dimension to cardiovascular disease. Circulation. 2010;121(8):1022–32.
67. Calin GA, et al. Human microRNA genes are frequently located at fragile sites and genomic regions involved in cancers. Proc Natl Acad Sci USA. 2004;101(9):2999–3004.
68. Iorio MV, Croce CM. MicroRNA dysregulation in cancer: diagnostics, monitoring and therapeutics. A comprehensive review. EMBO Mol Med. 2012;4(3):143–59.
69. Zhang L, et al. microRNAs exhibit high frequency genomic alterations in human cancer. Proc Natl Acad Sci USA. 2006;103(24):9136–41.
70. Brueckner B, et al. The human let-7a-3 locus contains an epigenetically regulated microRNA gene with oncogenic function. Cancer Res. 2007;67(4):1419–23.
71. Ota A, et al. Identification and characterization of a novel gene, C13orf25, as a target for 13q31-q32 amplification in malignant lymphoma. Cancer Res. 2004;64(9):3087–95.
72. He L, et al. A microRNA polycistron as a potential human oncogene. Nature. 2005;435 (7043):828–33.

73. Tong AW, Nemunaitis J. Modulation of miRNA activity in human cancer: a new paradigm for cancer gene therapy? Cancer Gene Ther. 2008;15(6):341–55.
74. Hammond SM. MicroRNAs as oncogenes. Curr Opin Genet Dev. 2006;16(1):4–9.
75. Zhang B, et al. microRNAs as oncogenes and tumor suppressors. Dev Biol. 2007;302(1):1–12.
76. Croce CM. Oncogenes and cancer. N Engl J Med. 2008. 358(5):502–11.
77. Bader AG, Brown D, Winkler M. The promise of microRNA replacement therapy. Cancer Res. 2010;70(18):7027–30.
78. Costinean S, et al. Pre-B cell proliferation and lymphoblastic leukemia/high-grade lymphoma in E(mu)-miR155 transgenic mice. Proc Natl Acad Sci USA. 2006;103(18):7024–9.
79. Esteller M. Epigenetics in cancer. N Engl J Med. 2008;358(11):1148–59.
80. Feinberg AP, Vogelstein B. Hypomethylation distinguishes genes of some human cancers from their normal counterparts. Nature. 1983;301(5895):89–92.
81. Gonzalez-Zulueta M, et al. Methylation of the 5' CpG island of the p16/CDKN2 tumor suppressor gene in normal and transformed human tissues correlates with gene silencing. Cancer Res. 1995;55(20):4531–5.
82. Herman JG, et al. Inactivation of the CDKN2/p16/MTS1 gene is frequently associated with aberrant DNA methylation in all common human cancers. Cancer Res. 1995;55(20):4525–30.
83. Sakai T, et al. Allele-specific hypermethylation of the retinoblastoma tumor-suppressor gene. Am J Hum Gen. 1991;48(5):880–8.
84. Johnson SM, et al. RAS is regulated by the let-7 microRNA family. Cell. 2005;120(5):635–47.
85. Garzon R, et al. MicroRNA signatures associated with cytogenetics and prognosis in acute myeloid leukemia. Blood. 2008;111(6):3183–9.
86. Toyota M, et al. Epigenetic silencing of microRNA-34b/c and B-cell translocation gene 4 is associated with CpG island methylation in colorectal cancer. Cancer Res. 2008;68(11):4123–32.
87. Wang T, et al. A micro-RNA signature associated with race, tumor size, and target gene activity in human uterine leiomyomas. Genes Chromosomes Cancer. 2007;46(4):336–47.
88. Iorio MV, et al. MicroRNA gene expression deregulation in human breast cancer. Cancer Res. 2005;65(16):7065–70.
89. Iorio MV, et al. MicroRNA signatures in human ovarian cancer. Cancer Res. 2007;67(18):8699–707.
90. Lehmann U, et al. Epigenetic inactivation of microRNA gene hsa-mir-9–1 in human breast cancer. J Pathol. 2008;214(1):17–24.
91. Blenkiron C, Miska EA. miRNAs in cancer: approaches, aetiology, diagnostics and therapy. Human Mol Genet. 2007;16(Spec No 1):R106–13.
92. Tavazoie SF, et al. Endogenous human microRNAs that suppress breast cancer metastasis. Nature. 2008;451(7175):147–52.
93. Tang J, Ahmad A, Sarkar FH. The Role of MicroRNAs in Breast Cancer Migration, Invasion and Metastasis. Int J Mol Sci. 2012;13(10):13414–37.
94. Wu H, Zhu S, Mo YY. Suppression of cell growth and invasion by miR-205 in breast cancer. Cell Res. 2009;19(4):439–48.
95. Mishra PJ, Merlino G. MicroRNA reexpression as differentiation therapy in cancer. J Clin Invest. 2009;119(8):2119–23.
96. Rosenfeld N, et al. MicroRNAs accurately identify cancer tissue origin. Nat Biotechnol. 2008;26(4):462–9.
97. Calin GA, et al. A MicroRNA signature associated with prognosis and progression in chronic lymphocytic leukemia. N Engl J Med. 2005;353(17):1793–801.
98. Caramuta S, et al. MicroRNA expression profiles associated with mutational status and survival in malignant melanoma. J Invest Dermatol. 2010;130(8):2062–70.
99. Chen Q, et al. Prognostic significance of serum miR-17–5p in lung cancer. Med Oncol. 2013;30(1):353.
100. Rothschild SI, et al. MicroRNA-381 represses ID1 and is deregulated in lung adenocarcinoma. J Thorac Oncol: official publication of the International Association for the Study of Lung Cancer. 2012;7(7):1069–77.

101. Yanaihara N, et al. Unique microRNA molecular profiles in lung cancer diagnosis and prognosis. Cancer Cell. 2006;9(3):189–98.
102. Gregory PA, et al. MicroRNAs as regulators of epithelial-mesenchymal transition. Cell Cycle. 2008;7(20):3112–8.
103. Ma L, et al. Therapeutic silencing of miR-10b inhibits metastasis in a mouse mammary tumor model. Nat Biotechnol. 2010;28(4):341–7.
104. Zhu S, et al. MicroRNA-21 targets tumor suppressor genes in invasion and metastasis. Cell Res. 2008;18(3):350–9.
105. Ma L, Teruya-Feldstein J, Weinberg RA. Tumour invasion and metastasis initiated by microRNA-10b in breast cancer. Nature. 2007;449(7163):682–8.
106. Sahab ZJ, et al. Putative biomarkers and targets of estrogen receptor negative human breast cancer. Int J Mol Sci. 2011;12(7):4504–4521.
107. Simonini PdSR, et al. Epigenetically deregulated microRNA-375 is involved in a positive feedback loop with estrogen receptor α in breast cancer cells. Cancer Res. 2010;70(22):9175–9184.
108. Miller TE, et al. MicroRNA-221/222 confers tamoxifen resistance in breast cancer by targeting p27Kip1. J Biol Chem. 2008;283(44):29897–903.
109. Zhao JJ, et al. MicroRNA-221/222 negatively regulates estrogen receptor alpha and is associated with tamoxifen resistance in breast cancer. J Biol Chem. 2008;283(45):31079–86.
110. Krutzfeldt J, et al. Specificity, duplex degradation and subcellular localization of antagomirs. Nucleic Acids Res. 2007;35(9):2885–92.
111. Schwarz DS, et al. Asymmetry in the assembly of the RNAi enzyme complex. Cell. 2003;115(2):199–208.
112. Kumar MS, et al. Impaired microRNA processing enhances cellular transformation and tumorigenesis. Nat Genet. 2007;39(5):673–7.
113. Johnson CD, et al. The let-7 microRNA represses cell proliferation pathways in human cells. Cancer Res. 2007;67(16):7713–22.
114. Kumar MS, et al. Suppression of non-small cell lung tumor development by the let-7 microRNA family. Proc Natl Acad Sci USA. 2008;105(10):3903–8.
115. Trang P, et al. Regression of murine lung tumors by the let-7 microRNA. Oncogene. 2010;29(11):1580–7.
116. Yu F, et al. let-7 regulates self renewal and tumorigenicity of breast cancer cells. Cell. 2007;131(6):1109–23.
117. He L, et al. microRNAs join the p53 network–another piece in the tumour-suppression puzzle. Nature reviews. Cancer. 2007;7(11):819–22.
118. Wiggins JF, et al. Development of a lung cancer therapeutic based on the tumor suppressor microRNA-34. Cancer Res. 2010;70(14):5923–30.
119. Kota J, et al. Therapeutic microRNA delivery suppresses tumorigenesis in a murine liver cancer model. Cell. 2009;137(6):1005–17.
120. Takeshita F, et al. Systemic delivery of synthetic microRNA-16 inhibits the growth of metastatic prostate tumors via downregulation of multiple cell-cycle genes. Mol Ther. 2010;18(1):181–7.
121. Hutvagner G, et al. Sequence-specific inhibition of small RNA function. PLoS Biol. 2004;2(4):E98.
122. Meister G, et al. Sequence-specific inhibition of microRNA- and siRNA-induced RNA silencing. RNA. 2004;10(3):544–50.
123. Orom UA, Kauppinen S, Lund AH. LNA-modified oligonucleotides mediate specific inhibition of microRNA function. Gene. 2006;372:137–41.
124. Krutzfeldt J, et al. Silencing of microRNAs in vivo with 'antagomirs'. Nature. 2005;438 (7068):685–9.
125. Shen H, Sun T, Ferrari M. Nanovector delivery of siRNA for cancer therapy. Cancer Gene Ther. 2012;19(6):367–373.
126. Ganta S, et al. A review of multifunctional nanoemulsion systems to overcome oral and CNS drug delivery barriers. Molecular Membrane Biology. 2010;27(7):260–273.

127. Tanaka T, et al. Nanotechnology for breast cancer therapy. Biomed Microdevices. 2009;11(1):49–63.
128. Wakai Y, et al. Effective cancer targeting using an anti-tumor tissue vascular endothelium-specific monoclonal antibody (TES-23). Jpn J Cancer Res. 2000;91(12):1319–1325.
129. Sanhai WR, et al. Seven challenges for nanomedicine. Nat Nanotechnol. 2008;3(5):242–4.
130. Esquela-Kerscher A, et al. The let-7 microRNA reduces tumor growth in mouse models of lung cancer. Cell Cycle. 2008;7(6):759–64.
131. Liu C, et al. The microRNA miR-34a inhibits prostate cancer stem cells and metastasis by directly repressing CD44. Nat Med. 2011;17(2):211–5.
132. Trang P, et al. Systemic delivery of tumor suppressor microRNA mimics using a neutral lipid emulsion inhibits lung tumors in mice. Mol Ther: the journal of the American Society of Gene Therapy. 2011;19(6):1116–22.
133. Juliano R, et al. Biological barriers to therapy with antisense and siRNA oligonucleotides. Mol Pharm. 2009;6(3):686–95.
134. Yang XC, et al. Studies of the 5' exonuclease and endonuclease activities of CPSF-73 in histone pre-mRNA processing. Mol Cell Biol. 2009;29(1):31–42.
135. Krol J, Loedige I, Filipowicz W. The widespread regulation of microRNA biogenesis, function and decay. Nat Rev Genet. 2010;11(9):597–610.
136. Katoh T, et al. Selective stabilization of mammalian microRNAs by 3' adenylation mediated by the cytoplasmic poly(A) polymerase GLD-2. Genes Dev. 2009;23(4):433–8.
137. Ramachandran V, Chen X. Degradation of microRNAs by a family of exoribonucleases in Arabidopsis. Science. 2008;321(5895):1490–2.
138. Turchinovich A, et al. Characterization of extracellular circulating microRNA. Nucleic Acids Res. 2011;39(16):7223–33.
139. Matsumura Y, Maeda H. A new concept for macromolecular therapeutics in cancer chemotherapy: mechanism of tumoritropic accumulation of proteins and the antitumor agent smancs. Cancer Res. 1986;46(12 Pt 1):6387–92.
140. Shinohara N., Danno N., Ichinose T., Sasaki T., Fukui H., Honda K., and Gamo M. Tissue distribution and clearance of intravenously administered titanium dioxide (TiO2) nanoparticles. Nanotoxicology. 2013;8(2):132–141.
141. Sa LT, et al. Biodistribution of nanoparticles: initial considerations. J Pharm Biomed Anal. 2012. 70:602–4.
142. Kang YS, Boado RJ, Pardridge WM. Pharmacokinetics and organ clearance of a 3'-biotinylated, internally [32P]-labeled phosphodiester oligodeoxynucleotide coupled to a neutral avidin/monoclonal antibody conjugate. Drug Metab Dispos. 1995;23(1):55–9.
143. Rifai A, et al. Clearance kinetics, biodistribution, and organ saturability of phosphorothioate oligodeoxynucleotides in mice. Am J Pathol. 1996;149(2):717–25.
144. Gutierrez-Puente Y, et al. Safety, pharmacokinetics, and tissue distribution of liposomal P-ethoxy antisense oligonucleotides targeted to Bcl-2. J Pharm Exp Ther. 1999;291(2):865–9.
145. Juliano RL, et al. Antisense pharmacodynamics: critical issues in the transport and delivery of antisense oligonucleotides. Pharm Res. 1999;16(4):494–502.
146. Merkel OM, et al. In vivo SPECT and real-time gamma camera imaging of biodistribution and pharmacokinetics of siRNA delivery using an optimized radiolabeling and purification procedure. Bioconjug Chem. 2009;20(1):174–82.
147. van de Water FM, et al. Intravenously administered short interfering RNA accumulates in the kidney and selectively suppresses gene function in renal proximal tubules. Drug Metab. Dispos. 2006;34(8):1393–7.
148. Martino, F., et al., Circulating microRNAs are not eliminated by hemodialysis. PloS one. 2012;7(6):e38269.
149. Mosser DM, Edwards JP. Exploring the full spectrum of macrophage activation. Nature Rev Immunol. 2008;8(12):958–69.
150. Alexis F, et al. Factors affecting the clearance and biodistribution of polymeric nanoparticles. Mol Pharm. 2008;5(4):505–15.

151. Yadav KS, et al. Effect of size on the biodistribution and blood clearance of etoposide-loaded PLGA nanoparticles. PDA J Pharm Sci Technol. 2011;65(2):131–9.
152. Xiao K, et al. The effect of surface charge on in vivo biodistribution of PEG-oligocholic acid based micellar nanoparticles. Biomaterials. 2011;32(13):3435–46.
153. Talegaonkar S, Vyas SP. Inverse targeting of diclofenac sodium to reticuloendothelial system-rich organs by sphere-in-oil-in-water (s/o/w) multiple emulsion containing poloxamer 403. J Drug Target. 2005;13(3):173–8.
154. Li P, et al. Targeting study of gelatin adsorbed clodronate in reticuloendothelial system and its potential application in immune thrombocytopenic purpura of rat model. J Control Release: official journal of the Controlled Release Society. 2006;114(2):202–8.
155. Underhill DM, Ozinsky A. Phagocytosis of microbes: complexity in action. Annu Rev Immunol. 2002;20:825–52.
156. Nolting A, et al. Hepatic distribution and clearance of antisense oligonucleotides in the isolated perfused rat liver. Pharm Res. 1997;14(4):516–21.
157. Thomas LL, Price CM. Cellular uptake, distribution, and degradation of phosphorothioate oligonucleotides in Euplotes crassus. Ann N Y Acad Sci. 1992;660:328–30.
158. Dejana E, Simionescu M, Wolburg H. Endothelial cell biology and pathology. Cell Tissue Res. 2009;335(1):1–3.
159. Szekanecz Z, Koch AE. Vascular endothelium and immune responses: implications for inflammation and angiogenesis. Rheum Dis Clin North Am. 2004;30(1):97–114.
160. Chorianopoulos E, et al. The role of endothelial cell biology in endocarditis. Cell Tissue Res. 2009;335(1):153–63.
161. Pate M, et al. Endothelial cell biology: role in the inflammatory response. Adv Clin Chem. 2010;52:109–30.
162. Cid MC. Endothelial cell biology, perivascular inflammation, and vasculitis. Cleve Clin J Med. 2002;69(Suppl 2):SII45–9.
163. Furuya M, Yonemitsu Y. Cancer neovascularization and proinflammatory microenvironments. Current Cancer Drug Targets. 2008;8(4):253–65.
164. van der Wijk T, et al. Osmotic swelling-induced activation of the extracellular-signal-regulated protein kinases Erk-1 and Erk-2 in intestine 407 cells involves the Ras/Raf-signalling pathway. Biochem J. 1998;331(Pt 3):863–9.
165. Baeriswyl V, Christofori G. The angiogenic switch in carcinogenesis. Semin Cancer Biol. 2009;19(5):329–37.
166. Hendrix MJ, et al. Vasculogenic mimicry and tumour-cell plasticity: lessons from melanoma. Nature Rev Cancer. 2003;3(6):411–21.
167. Hashizume H, et al. Openings between defective endothelial cells explain tumor vessel leakiness. Am J Clin Pathol. 2000;156(4):1363–80.
168. Dvorak AM, Feng D. The vesiculo-vacuolar organelle (VVO). A new endothelial cell permeability organelle. J Histochemistry Cytochem: official journal of the Histochemistry Society. 2001;49(4):419–32.
169. Dreher MR, et al. Tumor vascular permeability, accumulation, and penetration of macromolecular drug carriers. J Natl Cancer Inst. 2006;98(5):335–44.
170. Folkman J. Fundamental concepts of the angiogenic process. Curr Mol Med. 2003;3(7):643–51.
171. Caron WP, et al. Interpatient pharmacokinetic and pharmacodynamic variability of carrier-mediated anticancer agents. Clin Pharmacol Ther. 2012;91(5):802–12.
172. Dvorak HF, et al. Identification and characterization of the blood vessels of solid tumors that are leaky to circulating macromolecules. Am J Pathol. 1988;133(1):95–109.
173. Jain RK. Transport of molecules in the tumor interstitium: a review. Cancer Res. 1987;47(12):3039–51.
174. Prabhakar U, et al. Challenges and key considerations of the enhanced permeability and retention effect for nanomedicine drug delivery in oncology. Cancer Res. 2013;73(8):2412–7.
175. Azzopardi EA, Ferguson EL, Thomas DW. The enhanced permeability retention effect: a new paradigm for drug targeting in infection. J Antimicrob Chemother. 2013;68(2):257–74.

176. Netti PA, et al. Role of extracellular matrix assembly in interstitial transport in solid tumors. Cancer Res. 2000;60(9):2497–503.
177. Liu W, et al. Tumor accumulation, degradation and pharmacokinetics of elastin-like polypeptides in nude mice. J Control Release: official journal of the Controlled Release Society. 2006;116(2):170–8.
178. Tasciotti E, et al. Mesoporous silicon particles as a multistage delivery system for imaging and therapeutic applications. Nat Nanotechnol. 2008;3(3):151–7.
179. Khatri N, et al. In vivo delivery aspects of miRNA, shRNA and siRNA. Crit Rev Ther Drug Carrier Syst. 2012;29(6):487–527.
180. Dominska M, Dykxhoorn DM. Breaking down the barriers: siRNA delivery and endosome escape. J Cell Sci. 2010;123(Pt 8):1183–9.
181. Park B, et al. Proteolytic cleavage in an endolysosomal compartment is required for activation of Toll-like receptor 9. Nat Immunol. 2008;9(12):1407–14.
182. Thompson DB, et al. Cellular uptake mechanisms and endosomal trafficking of supercharged proteins. Chem Biol. 2012;19(7):831–43.
183. Hierro A, et al. Structure of the ESCRT-II endosomal trafficking complex. Nature. 2004;431(7005):221–5.
184. Kelly EE, et al. The role of endosomal-recycling in long-term potentiation. Cell Mol Life Sci. 2011;68(2):185–94.
185. Puthenveedu MA, et al. Sequence-dependent sorting of recycling proteins by actin-stabilized endosomal microdomains. Cell. 2010;143(5):761–73.
186. Endoh T, Ohtsuki T. Cellular siRNA delivery using cell-penetrating peptides modified for endosomal escape. Adv Drug Deliv Rev. 2009;61(9):704–9.
187. Endoh T, Ohtsuki T. Cellular siRNA delivery using TatU1A and photo-induced RNA interference. Methods Mol Biol. 2010;623:271–81.
188. Tseng YC, Huang L. Self-assembled lipid nanomedicines for siRNA tumor targeting. J Biomed Nanotechnol. 2009;5(4):351–63.
189. Akhtar S, Benter IF. Nonviral delivery of synthetic siRNAs in vivo. J Clin Invest. 2007;117(12):3623–32.
190. Castanotto D, Rossi JJ. The promises and pitfalls of RNA-interference-based therapeutics. Nature. 2009;457(7228):426–33.
191. Broderick JA, Zamore PD. MicroRNA therapeutics. Gene Ther. 2011;18(12):1104–10.
192. Krutzfeldt J, Poy MN, Stoffel M. Strategies to determine the biological function of microRNAs. Nat Genet. 2006;38 Suppl:S14–9.
193. Varble A, ten Oever BR. Implications of RNA virus-produced miRNAs. RNA Biol. 2011;8(2):190–4.
194. Miyazaki Y, et al. Viral delivery of miR-196a ameliorates the SBMA phenotype via the silencing of CELF2. Nat Med. 2012;18(7):1136–41.
195. Chistiakov DA, Sobenin IA, Orekhov AN. Strategies to deliver microRNAs as potential therapeutics in the treatment of cardiovascular pathology. Drug Deliv. 2012;19(8):392–405.
196. Sakuma T, Barry MA, Ikeda Y. Lentiviral vectors: basic to translational. Biochem J. 2012;443(3):603–18.
197. Bukrinsky MI, et al. A nuclear localization signal within HIV-1 matrix protein that governs infection of non-dividing cells. Nature. 1993;365(6447):666–9.
198. Freed EO, Martin MA. HIV-1 infection of non-dividing cells. Nature. 1994;369(6476):107–8.
199. Klimatcheva E, Rosenblatt JD, Planelles V. Lentiviral vectors and gene therapy. Front Biosci: a journal and virtual library. 1999;4:D481–96.
200. Modlich U, Baum C. Preventing and exploiting the oncogenic potential of integrating gene vectors. J Clin Invest. 2009;119(4):755–8.
201. Zufferey R, et al. Multiply attenuated lentiviral vector achieves efficient gene delivery in vivo. Nat Biotechnol. 1997;15(9):871–5.
202. Lehmann-Che J, Saib A. Early stages of HIV replication: how to hijack cellular functions for a successful infection. AIDS Rev. 2004;6(4):199–207.
203. Naldini L, et al. In vivo gene delivery and stable transduction of nondividing cells by a lentiviral vector. Science. 1996;272(5259):263–7.

204. Kong YW, et al. microRNAs in cancer management. Lancet Oncol. 2012;13(6):e249–58.
205. Maegdefessel L, et al. MicroRNA-21 blocks abdominal aortic aneurysm development and nicotine-augmented expansion. Sci Transl Med. 2012;4(122):122ra22.
206. Craig VJ, et al. Epigenetic silencing of microRNA-203 dysregulates ABL1 expression and drives Helicobacter-associated gastric lymphomagenesis. Cancer Res. 2011;71(10):3616–24.
207. Liu QS, et al. Lentiviral-mediated miRNA against liver-intestine cadherin suppresses tumor growth and invasiveness of human gastric cancer. Cancer Sci. 2010;101(8):1807–12.
208. Gentner B, et al. Stable knockdown of microRNA in vivo by lentiviral vectors. Nat Methods. 2009;6(1):63–6.
209. Brown BD, et al. Endogenous microRNA can be broadly exploited to regulate transgene expression according to tissue, lineage and differentiation state. Nat Biotechnol. 2007; 25(12):1457–67.
210. Ebert MS, Neilson JR, Sharp PA. MicroRNA sponges: competitive inhibitors of small RNAs in mammalian cells. Nat methods. 2007;4(9):721–6.
211. Grimm D, et al. Argonaute proteins are key determinants of RNAi efficacy, toxicity, and persistence in the adult mouse liver. J Clin Invest. 2010;120(9):3106–19.
212. Cavazzana-Calvo M, et al. Transfusion independence and HMGA2 activation after gene therapy of human beta-thalassaemia. Nature. 2010;467(7313):318–22.
213. Rayner KJ, et al. MiR-33 contributes to the regulation of cholesterol homeostasis. Science. 2010;328(5985):1570–3.
214. Stevenson SC, et al. Human adenovirus serotypes 3 and 5 bind to two different cellular receptors via the ol head domain. J Virol. 1995;69(5):2850–7.
215. Shashkova EV, May SM, Barry MA. Characterization of human adenovirus serotypes 5, 6, 11, and 35 as anticancer agents. Virology. 2009;394(2):311–20.
216. Fu J, Li L, Bouvier M. Adenovirus E3–19K proteins of different serotypes and subgroups have similar, yet distinct, immunomodulatory functions toward major histocompatibility class I molecules. J Biol Chem. 2011;286(20):17631–9.
217. Grimm D, et al. Fatality in mice due to oversaturation of cellular microRNA/short hairpin RNA pathways. Nature. 2006;441(7092):537–41.
218. Daya S, Berns KI. Gene therapy using adeno-associated virus vectors. Clin Microbiol Rev. 2008;21(4):583–93.
219. Han Y, et al. MicroRNA expression signatures of bladder cancer revealed by deep sequencing. PloS one. 2011;6(3):e18286.
220. Grady WM, Tewari M. The next thing in prognostic molecular markers: microRNA signatures of cancer. Gut. 2010;59(6):706–8.
221. Calin GA, Croce CM. MicroRNAs and chromosomal abnormalities in cancer cells. Oncogene. 2006;25(46):6202–10.
222. Care A, et al. MicroRNA-133 controls cardiac hypertrophy. Nat Med. 2007;13(5):613–8.
223. Colin A, et al. Engineered lentiviral vector targeting astrocytes in vivo. Glia. 2009;57(6): 667–79.
224. Leja J, et al. Double-detargeted oncolytic adenovirus shows replication arrest in liver cells and retains neuroendocrine cell killing ability. PLoS One. 2010;5(1):e8916.
225. Thomas CE, Ehrhardt A, Kay MA. Progress and problems with the use of viral vectors for gene therapy. Nat Rev Genet. 2003;4(5):346–58.
226. Mingozzi F, High KA. Therapeutic in vivo gene transfer for genetic disease using AAV: progress and challenges. Nat Rev Genet. 2011;12(5):341–55.
227. Kang EM, et al. Retrovirus gene therapy for X-linked chronic granulomatous disease can achieve stable long-term correction of oxidase activity in peripheral blood neutrophils. Blood. 2010;115(4):783–91.
228. Hacein-Bey-Abina S, et al. Insertional oncogenesis in 4 patients after retrovirus-mediated gene therapy of SCID-X1. J Clin Invest. 2008;118(9):3132–42.
229. Wang J, et al. Delivery of siRNA therapeutics: barriers and carriers. The AAPS J. 2010; 12(4):492–503.

230. Kline KK, Tucker SA. Spectroscopic characterization of core-based hyperbranched poly(ethyleneimine) and dendritic poly(propyleneimine) as selective uptake devices. J Phys Chem A. 2010;114(27):7338–44.
231. Ibrahim AF, et al. MicroRNA replacement therapy for miR-145 and miR-33a is efficacious in a model of colon carcinoma. Cancer Res. 2011;71(15):5214–24.
232. Tazawa H, et al. Tumor-suppressive miR-34a induces senescence-like growth arrest through modulation of the E2F pathway in human colon cancer cells. Proc Natl Acad Sci USA. 2007;104(39):15472–7.
233. Chiou GY, et al. Cationic polyurethanes-short branch PEI-mediated delivery of Mir145 inhibited epithelial-mesenchymal transdifferentiation and cancer stem-like properties and in lung adenocarcinoma. J Control Release: official journal of the Controlled Release Society. 2012;159(2):240–50.
234. Chiou SH, et al. Coexpression of Oct4 and Nanog enhances malignancy in lung adenocarcinoma by inducing cancer stem cell-like properties and epithelial-mesenchymal transdifferentiation. Cancer Res. 2010;70(24):10433–44.
235. Danhier F, et al. PLGA-based nanoparticles: an overview of biomedical applications. J Control Release: official journal of the Controlled Release Society. 2012;161(2):505–22.
236. Prokop A, Davidson JM. Nanovehicular intracellular delivery systems. J Pharm Sci. 2008;97(9):3518–90.
237. Kumari A, Yadav SK, Yadav SC. Biodegradable polymeric nanoparticles based drug delivery systems. Colloids Surf B Biointerfaces. 2010;75(1):1–18.
238. Owens DE 3rd, Peppas NA. Opsonization, biodistribution, and pharmacokinetics of polymeric nanoparticles. Int J Pharm. 2006;307(1):93–102.
239. Klugherz BD, et al. Gene delivery from a DNA controlled-release stent in porcine coronary arteries. Nat Biotechnol. 2000;18(11):1181–4.
240. Gupta H, et al. Sparfloxacin-loaded PLGA nanoparticles for sustained ocular drug delivery. Nanomedicine. 2010;6(2):324–33.
241. Pillai RR, et al. Nafcillin-loaded PLGA nanoparticles for treatment of osteomyelitis. Biomed Mater. 2008;3(3):034114.
242. Golub JS, et al. Sustained VEGF delivery via PLGA nanoparticles promotes vascular growth. Am J Physiol Heart Circ Physiol. 2010;298(6):H1959–65.
243. Yilgor P, Hasirci N, Hasirci V. Sequential BMP-2/BMP-7 delivery from polyester nanocapsules. J Biomed Mater Res A. 2010;93(2):528–36.
244. Liang G. F., Zhu Y. L., Sun B., Hu F. H., Tian T., Li S. C., and Xiao Z. D. PLGA-based gene delivering nanoparticle enhance suppression effect of miRNA in HePG2 cells. Nanoscale research letters. 2011;6(1):1–9.
245. Babar IA, et al. Nanoparticle-based therapy in an in vivo microRNA-155 (miR-155)-dependent mouse model of lymphoma. Proc Natl Acad Sci USA. 2012;109(26):E1695–704.
246. Gomes RS, et al. Efficient pro-survival/angiogenic miRNA delivery by an MRI-detectable nanomaterial. ACS Nano. 2013;7(4):3362–72.
247. Kosaka N, et al. Secretory mechanisms and intercellular transfer of microRNAs in living cells. J Biol Chem. 2010;285(23):17442–52.
248. Ohshima K, et al. Let-7 microRNA family is selectively secreted into the extracellular environment via exosomes in a metastatic gastric cancer cell line. PloS one. 2010;5(10): e13247.
249. Théry C. Exosomes: secreted vesicles and intercellular communications. F1000 Biol Rep. 2011;3(15):130.
250. Torchilin VP. Lipid-core micelles for targeted drug delivery. Curr Drug Deliv. 2005; 2(4):319–27.
251. Markman M. Pegylated liposomal doxorubicin in the treatment of cancers of the breast and ovary. Expert Opin Pharmacother. 2006;7(11):1469–74.
252. Strother R, Matei D. Pegylated liposomal doxorubicin in ovarian cancer. Ther Clin Risk Manag. 2009;5(3):639–50.
253. Akinc A, et al. A combinatorial library of lipid-like materials for delivery of RNAi therapeutics. Nat Biotechnol. 2008;26(5):561–9.

254. Muggia FM. Clinical efficacy and prospects for use of pegylated liposomal doxorubicin in the treatment of ovarian and breast cancers. Drugs. 1997;54(Suppl 4):22–9.
255. Muggia FM, et al. Phase II study of liposomal doxorubicin in refractory ovarian cancer: antitumor activity and toxicity modification by liposomal encapsulation. J Clin Oncol: official journal of the American Society of Clinical Oncology. 1997;15(3):987–93.
256. Safinya CR, Ewert KK. Materials chemistry: Liposomes derived from molecular vases. Nature. 2012;489(7416):372–4.
257. Wu Y, et al. MicroRNA delivery by cationic lipoplexes for lung cancer therapy. Mol Pharm. 2011;8(4):1381–9.
258. DeSano JT, Xu L. MicroRNA regulation of cancer stem cells and therapeutic implications. AAPS J. 2009;11(4):682–92.
259. Hanai K, et al. Atelocollagen-mediated systemic DDS for nucleic acid medicines. Ann N Y Acad Sci. 2006;1082:9–17.
260. Miyata T, Taira T, Noishiki Y. Collagen engineering for biomaterial use. Clin Mater. 1992; 9(3–4):139–48.
261. Minakuchi Y, et al. Atelocollagen-mediated synthetic small interfering RNA delivery for effective gene silencing in vitro and in vivo. Nucleic Acids Res. 2004;32(13):e109.
262. Svintradze DV, Mrevlishvili GM. Fiber molecular model of atelocollagen-small interfering RNA (siRNA) complex. Int J Biol Macromol. 2005;37(5):283–6.
263. Nakasa T, et al. Acceleration of muscle regeneration by local injection of muscle-specific microRNAs in rat skeletal muscle injury model. J Cell Mol Med. 2010;14(10):2495–505.
264. Takeshita F, et al. Systemic delivery of synthetic microRNA-16 inhibits the growth of metastatic prostate tumors via downregulation of multiple cell-cycle genes. Mol Ther: the journal of the American Society of Gene Therapy. 2010;18(1):181–7.
265. Tazawa H, et al. Tumor-suppressive miR-34a induces senescence-like growth arrest through modulation of the E2F pathway in human colon cancer cells. Proc Natl Acad Sci USA. 2007;104(39):15472–7.
266. Nagata Y, et al. Induction of apoptosis in the synovium of mice with autoantibody-mediated arthritis by the intraarticular injection of double-stranded MicroRNA-15a. Arthritis Rheum. 2009;60(9):2677–83.
267. Takei Y, et al. The metastasis-associated microRNA miR-516a-3p is a novel therapeutic target for inhibiting peritoneal dissemination of human scirrhous gastric cancer. Cancer Res. 2011;71(4):1442–53.
268. Osaki M, et al. MicroRNA-143 regulates human osteosarcoma metastasis by regulating matrix metalloprotease-13 expression. Mol Ther: the journal of the American Society of Gene Therapy. 2011;19(6):1123–30.
269. Shoji T, et al. The effect of intra-articular injection of microRNA-210 on ligament healing in a rat model. Am J Sports Med. 2012;40(11):2470–8.
270. Kaasgaard T, Andresen TL. Liposomal cancer therapy: exploiting tumor characteristics. Expert Opin Drug Deliv. 2010;7(2):225–43.
271. Kato Y, Sawata SY, Inoue A. A lentiviral vector encoding two fluorescent proteins enables imaging of adenoviral infection via adenovirus-encoded miRNAs in single living cells. J Biochem. 2010;147(1):63–71.
272. Wang Y, et al. Nanoparticle-based delivery system for application of siRNA in vivo. Curr Drug Metab. 2010;11(2):182–96.
273. Bergen JM, et al. Gold nanoparticles as a versatile platform for optimizing physicochemical parameters for targeted drug delivery. Macromol Biosci. 2006;6(7):506–16.
274. Han G, Ghosh P, Rotello VM. Multi-functional gold nanoparticles for drug delivery. Adv Exp Med Biol. 2007;620:48–56.
275. Papasani MR, Wang G, Hill RA. Gold nanoparticles: the importance of physiological principles to devise strategies for targeted drug delivery. Nanomedicine. 2012;8(6):804–14.
276. Ghosh R, et al. A gold nanoparticle platform for the delivery of functional microRNAs into cancer cells. Biomaterials. 2013;34(3):807–16.

277. Shan Y, et al. Gene delivery using dendrimer-entrapped gold nanoparticles as nonviral vectors. Biomaterials. 2012;33(10):3025–35.
278. Shukla R, et al. Biocompatibility of gold nanoparticles and their endocytotic fate inside the cellular compartment: a microscopic overview. Langmuir: the ACS journal of surfaces and colloids. 2005;21(23):10644–54.
279. Rosi NL, et al. Oligonucleotide-modified gold nanoparticles for intracellular gene regulation. Science. 2006;312(5776):1027–30.
280. Demers LM, et al. A fluorescence-based method for determining the surface coverage and hybridization efficiency of thiol-capped oligonucleotides bound to gold thin films and nanoparticles. Anal Chem. 2000;72(22):5535–41.
281. Giljohann DA, et al. Oligonucleotide loading determines cellular uptake of DNA-modified gold nanoparticles. Nano Lett. 2007;7(12):3818–21.
282. Crew E, et al. MicroRNA conjugated gold nanoparticles and cell transfection. Anal Chem. 2012;84(1):26–9.
283. Bravo V, et al. Instability of miRNA and cDNAs derivatives in RNA preparations. Biochem Biophys Res Commun. 2007;353(4):1052–5.
284. van Etten EW, et al. Administration of liposomal agents and blood clearance capacity of the mononuclear phagocyte system. Antimicrob Agents Chemother. 1998;42(7):1677–81.
285. Baum C, et al. Mutagenesis and oncogenesis by chromosomal insertion of gene transfer vectors. Hum Gene Ther. 2006;17(3):253–63.
286. Bouard D, Alazard-Dany D, Cosset FL. Viral vectors: from virology to transgene expression. Br J Pharmacol. 2009;157(2):153–65.
287. Brown BD, et al. In vivo administration of lentiviral vectors triggers a type I interferon response that restricts hepatocyte gene transfer and promotes vector clearance. Blood. 2007;109(7):2797–805.
288. Lesina E, Dames P, Rudolph C. The effect of CpG motifs on gene expression and clearance kinetics of aerosol administered polyethylenimine (PEI)-plasmid DNA complexes in the lung. J Control Release: official journal of the Controlled Release Society. 2010;143(2):243–50.
289. Mizrak A, et al. Genetically engineered microvesicles carrying suicide mRNA/protein inhibit schwannoma tumor growth. Mol Ther: the journal of the American Society of Gene Therapy. 2013;21(1):101–8.
290. Zintchenko A, et al. Drug nanocarriers labeled with near-infrared-emitting quantum dots (quantoplexes): imaging fast dynamics of distribution in living animals. Mol Ther: the journal of the American Society of Gene Therapy. 2009;17(11):1849–56.
291. Merdan T, et al. PEGylation of poly(ethylene imine) affects stability of complexes with plasmid DNA under in vivo conditions in a dose-dependent manner after intravenous injection into mice. Bioconjug Chem. 2005;16(4):785–92.
292. Kang E, Yun CO. Current advances in adenovirus nanocomplexes: more specificity and less immunogenicity. BMB Rep. 2010;43(12):781–8.
293. Arnberg N. Adenovirus receptors: implications for tropism, treatment and targeting. Rev Med Virol. 2009;19(3):165–78.
294. Godin B, et al. Multistage nanovectors: from concept to novel imaging contrast agents and therapeutics. Acc Chem Res. 2011;44(10):979–89.
295. Serda RE, et al. Multi-stage delivery nano-particle systems for therapeutic applications. Biochim Biophys Acta. 2011;1810(3):317–29.
296. Godin B, et al. multistage nanovectors: from concept to novel imaging contrast agents and therapeutics. Acc Chem Res. 2011;44(10):979–989.
297. Ferrari M. Cancer nanotechnology: opportunities and challenges. Nature reviews. Cancer. 2005;5(3):161–71.
298. Sakamoto J, et al. Antibiological barrier nanovector technology for cancer applications. Expert Opin Drug Deliv. 2007;4(4):359–69.
299. Geers B., De Wever O., Demeester J., Bracke M., De Smedt S. C., and Lentacker I. Targeted Liposome-Loaded Microbubbles for Cell-Specific Ultrasound-Triggered Drug Delivery. Small. 2013;9(23):4027–4035.

300. Vader P, et al. Targeted delivery of small interfering RNA to angiogenic endothelial cells with liposome-polycation-DNA particles. J Control Release: official journal of the Controlled Release Society. 2012;160(2):211–6.
301. Geng J, et al. Conjugated polymer and gold nanoparticle co-loaded PLGA nanocomposites with eccentric internal nanostructure for dual-modal targeted cellular imaging. Small. 2012;8(15):2421–9.
302. Ferrari M. Vectoring siRNA therapeutics into the clinic. Nat Rev Clin Oncol. 2010;7(9):485–6.
303. Sengupta S, et al. Temporal targeting of tumour cells and neovasculature with a nanoscale delivery system. Nature. 2005;436(7050):568–72.
304. Chen JD, Wang Y, Chen X. In situ fabrication of nano-hydroxyapatite in a macroporous chitosan scaffold for tissue engineering. J Biomater Sci Polym Ed. 2009;20(11):1555–65.
305. Martin FJ, et al. Acute toxicity of intravenously administered microfabricated silicon dioxide drug delivery particles in mice: preliminary findings. Drugs R D. 2005;6(2):71–81.
306. Low SP, et al. Evaluation of mammalian cell adhesion on surface-modified porous silicon. Biomaterials. 2006. 27(26):4538–46.
307. Goh AS, et al. A novel approach to brachytherapy in hepatocellular carcinoma using a phosphorous32 (32P) brachytherapy delivery device–a first-in-man study. Int J Radiat Oncol Biol Phys. 2007;67(3):786–92.
308. Decuzzi P, Ferrari M. Design maps for nanoparticles targeting the diseased microvasculature. Biomaterials. 2008;29(3):377–84.
309. Decuzzi P, et al. A theoretical model for the margination of particles within blood vessels. Ann Biomed Eng. 2005;33(2):179–90.
310. Decuzzi P, et al. Intravascular delivery of particulate systems: does geometry really matter? Pharm Res. 2009;26(1):235–43.
311. Godin B, et al. An integrated approach for the rational design of nanovectors for biomedical imaging and therapy. Adv Genet. 2010;69:31–64.
312. Anderson SHC, et al. Dissolution of different forms of partially porous silicon wafers under simulated physiological conditions. physica status solidi (a). 2003;197(2):331–335.
313. Canham LT, et al. Derivatized mesoporous silicon with dramatically improved stability in simulated human blood plasma. Adv Mater. 1999;11(18):1505–1507.
314. Chiappini C, et al. Tailored porous silicon microparticles: fabrication and properties. Chemphyschem. 2010;11(5):1029–35.
315. Kasinski AL, Slack FJ. Epigenetics and genetics. MicroRNAs en route to the clinic: progress in validating and targeting microRNAs for cancer therapy. Nat Rev Cancer. 2011;11(12): 849–64.
316. Cheng CJ, Saltzman WM. Polymer nanoparticle-mediated delivery of microRNA inhibition and alternative splicing. Mol Pharm. 2012 May 7;9(5):1481–8.

Index

S. R. Singh, P. Rameshwar (eds.), *MicroRNA in Development and in the Progression of Cancer,* DOI 10.1007/978-1-4899-8065-6,